FIFTH EDITION

INTRODUCTION TO SPECTROSCOPY

Donald L. Pavia

Gary M. Lampman

George S. Kriz

James R. Vyvyan

Department of Chemistry
Western Washington University
Bellingham, Washington

CENGAGE
Learning

Australia • Brazil • Mexico • Singapore • United Kingdom • United States

TO ALL OF OUR "O-SPEC" STUDENTS

**Introduction to Spectroscopy,
Fifth Edition**
**Donald L. Pavia, Gary M. Lampman,
George S. Kriz, and James R. Vyvyan**

Product Director: Mary Finch

Product Manager: Maureen Rosener

Content Developer: Alyssa White

Content Coordinator: Karolina Kiwak

Product Development Manager:
   Rebecca Berardy-Schwartz

Media Developer: Lisa Weber

Marketing Director: Jason Sakos

Marketing Development Manager:
   Julie Shuster

Content Project Manager: Teresa L. Trego

Art Director: Maria Epes

Manufacturing Planner: Judy Inouye

Rights Acquisitions Specialist:
   Thomas McDonough

Production Service: MPS Limited

Photo Researcher: PreMedia Global

Text Researcher: PreMedia Global

Copy Editor: MPS Limited

Illustrator: MPS Limited

Cover Designer: Bartay Studio

Cover Image: Expansion of the C-4 methine
   proton of ethyl 2-methyl-4-pentenoate
   acquired at 500 MHz. Data acquired and
   processed by James Vyvyan.

Compositor: MPS Limited

For product information and technology assistance, contact us at
Cengage Learning Customer & Sales Support, 1-800-354-9706

For permission to use material from this text or product,
submit all requests online at **www.cengage.com/permissions**
Further permissions questions can be e-mailed to
permissionrequest@cengage.com

Library of Congress Control Number: 2013950396

ISBN-13: 978-1-285-46012-3

ISBN-10: 1-285-46012-X

Cengage Learning
200 First Stamford Place, 4th Floor
Stamford, CT 06902
USA

Cengage Learning is a leading provider of customized learning solutions with office locations around the globe, including Singapore, the United Kingdom, Australia, Mexico, Brazil, and Japan. Locate your local office at **www.cengage.com/global**.

Cengage Learning products are represented in Canada by Nelson Education, Ltd.

To learn more about Cengage Learning Solutions, visit **www.cengage.com**.

Purchase any of our products at your local college store or at our preferred online store **www.cengagebrain.com**.

Printed in the United States of America
   4 5 6   18 17 16

PREFACE

This is the fifth edition of a textbook in spectroscopy intended for students of organic chemistry. Our textbook can serve as a supplement for the typical organic chemistry lecture textbook or as a stand-alone textbook for an advanced undergraduate or first-year graduate course in spectroscopic methods. This book is also a useful tool for students engaged in research. Our aim is not only to teach students to interpret spectra, but also to present basic theoretical concepts. As with the previous editions, we have tried to focus on the important aspects of each spectroscopic technique without dwelling excessively on theory or complex mathematical analyses.

This book is a continuing evolution of materials that we use in our own courses, both as a supplement to our organic chemistry lecture course series and also as the principal textbook in our upper division and graduate courses in spectroscopic methods and advanced NMR techniques. Explanations and examples that we have found to be effective in our courses have been incorporated into this edition.

NEW TO THIS EDITION

This fifth edition of *Introduction to Spectroscopy* contains some important changes. The material on mass spectrometry has been moved closer to the front of the text and divided into two more easily digested chapters. Material on some newer sampling and ionization methods is included, as are additional methods of structural analysis using fragmentation patterns. All of the chapters dealing with nuclear magnetic resonance have been gathered together into sequential chapters. Expanded discussions of diastereotopic systems and heteronuclear coupling are included, as is a revised discussion of solvent effects in NMR.

Additional practice problems have been added to each of the chapters. We have included some additional solved problems, too, so that students can better develop strategies and skills for solving spectroscopy problems. The problems that are marked with an asterisk (*) have solutions included in the Answers to Selected Problems following Chapter 11.

We wish to alert persons who adopt this book that answers to all of the problems are available online from the publisher. Authorization to gain access to the website may be obtained through the local Cengage textbook representative.

ADVICE FOR STUDENTS

Success in working out the solutions to spectroscopy problems comes more easily and is more enjoyable by following some simple suggestions:

1. Carefully study the solved examples that may be found at the end of each chapter. Do not attempt to work on additional problems until you are comfortable with the approach that is being demonstrated with the solved examples.

2. There is great value to be gained in working collaboratively to solve spectroscopy problems. Try standing around a blackboard to exchange ideas. You will find it to be fun, and you will learn more!

3. Don't be afraid to struggle. It is too easy to look up the answer to a difficult problem, and you won't learn much. You need to train your brain to think like a scientist, and there is no substitute for hard work.

4. Work problems concurrently as you study each chapter. That will solidify the concepts in your mind.

Although this book concentrates on organic chemistry examples, be aware that the study of spectroscopy crosses over into many areas, including biochemistry, inorganic chemistry, physical chemistry, materials chemistry, and analytical chemistry. Spectroscopy is an indispensible tool to support all forms of laboratory research.

ACKNOWLEDGMENTS

The authors are very grateful to Mr. Charles Wandler, without whose expert help this project could not have been accomplished. We also acknowledge numerous contributions made by our students, who use the textbook and who provide us careful and thoughtful feedback.

Finally, once again we must thank our wives, Neva-Jean Pavia, Marian Lampman, and Cathy Vyvyan, for their support and patience. They endure a great deal in order to support us as we write, and they deserve to be part of the celebration when the textbook is completed! We honor the memory of Carolyn Kriz; we miss her and the love and encouragement that she provided.

Donald L. Pavia
Gary M. Lampman
George S. Kriz
James R. Vyvyan

INTRO TO SPECTROSCOPY FIFTH EDITION SUMMARY OF CHANGES

The order of the chapters was rearranged to better reflect the requests and practices of our users. Mass Spectroscopy was moved to an earlier position, causing the renumbering.

Fourth edition chapter number/title	Fifth edition chapter number/title	Notes
1 Molecular Formulas and What Can Be Learned from Them	1 Molecular Formulas and What Can Be Learned from Them	Section 1.6, A Quick Look Ahead to Simple Uses of Mass Spectra, was deleted. (Mass Spectra were moved earlier into Chapters 3 and 4.) A new Section 1.6 is now titled: "The Nitrogen Rule." References were revised/updated.
2 Infrared Spectroscopy	2 Infrared Spectroscopy	Section 2.6, the solid samples subsection was updated to include ATR techniques. Several figures were revised/updated. Section 2.21, Alkyl and Aryl Halides, was revised. Section 2.23, How to Solve Infrared Spectral Problems, is a new section. The sections that followed were renumbered. Problems were revised. References were revised/updated.
3 Nuclear Magnetic Resonance Spectroscopy Part One: Basic Concepts.	5 Nuclear Magnetic Resonance Spectroscopy Part One: Basic Concepts.	New Section 5.20 References were revised/updated. New online resources were referenced and/or updated.
4 Nuclear Magnetic Resonance Spectroscopy Part Two: Carbon-13 etc.	6 Nuclear Magnetic Resonance Spectroscopy Part Two: Carbon-13 etc.	Section 6.4 introduces a new decoupling notation. New Section 6.12. Sections following 6.12 are renumbered. Several new problems were added. Some spectra replaced/improved. References were revised/updated. New online resources referenced and/or updated.
5 Nuclear Magnetic Resonance Spectroscopy Part Three: Spin-Spin Coupling	7 Nuclear Magnetic Resonance Spectroscopy Part Three: Spin-Spin Coupling	New discussion of splitting in diastereotopic systems. New discussion of heteronuclear splitting between $^1H-^{19}F$ and $S-^{31}P$ Addition of solved example problems. New and revised end-of-chapter problems using coupling constant information and chemical shift calculations. References were revised/updated.
6 Nuclear Magnetic Resonance Spectroscopy Part Four: Other Topics in One-Dimensional NMR	8 Nuclear Magnetic Resonance Spectroscopy Part Four: Other Topics in One-Dimensional NMR	New discussion and examples of solvent effects. Addition of solved example problems. New and revised end-of-chapter problems. References were revised/updated.

(Continued)

Fourth edition chapter number/title	Fifth edition chapter number/title	Notes
7 Ultraviolet Spectroscopy	10 Ultraviolet Spectroscopy	Few changes.
8 Mass Spectrometry (first half) Chapter was split.	3 Mass Spectrometry Part One: Basic Theory, Instrumentation, and Sampling Techniques	To highlight the continued development and importance of mass spectrometry (MS) methods, we have moved this material to the early part of the text and split it into two chapters, one on theory and instrumentation (Chapter 3) and the other on detailed structural analysis using characteristic fragmentation patterns of common functional groups (Chapter 4). Expanded and refined discussion of sampling and ionization methods, including atmospheric pressure chemical ionization techniques. Examples of applications for different MS techniques and instrumentation, including pros and cons of different methods.
8 Mass Spectrometry (second half)	4 Mass Spectrometry Part Two: Fragmentation and Structural Analysis	Refined discussion of fragmentations in EI-MS for common functional groups. New examples of use of MS in structure determination. Additional solved example problems. New and revised end-of-chapter problems.
9 Combined Structure Problems	11 Combined Structure Problems	Several new problems were introduced. Two-dimensional spectra were replaced with new, improved ones. References were revised/updated. Online resources were updated.
10 Nuclear Magnetic Resonance Spectroscopy Part Five: Advanced NMR Techniques.	9 Nuclear Magnetic Resonance Spectroscopy Part Five: Advanced NMR Techniques	Sections 9.4 and 9.7 were extensively revised. Many of the two-dimensional spectra were replaced with new, improved ones.
Appendices	Appendices	Old Appendix 11 was removed. Values in some of the tables were updated or revised.

CONTENTS

CHAPTER 3

MASS SPECTROMETRY

PART ONE: BASIC THEORY, INSTRUMENTATION, AND SAMPLING TECHNIQUES   107

CHAPTER 4

MASS SPECTROMETRY

PART TWO: FRAGMENTATION AND STRUCTURAL ANALYSIS   139

CHAPTER 5
NUCLEAR MAGNETIC RESONANCE SPECTROSCOPY

PART ONE: BASIC CONCEPTS   215

CHAPTER 6

NUCLEAR MAGNETIC RESONANCE SPECTROSCOPY

PART TWO: CARBON-13 SPECTRA, INCLUDING HETERONUCLEAR COUPLING WITH OTHER NUCLEI   290

CHAPTER 7
NUCLEAR MAGNETIC RESONANCE SPECTROSCOPY

CHAPTER 10

ULTRAVIOLET SPECTROSCOPY   577

CHAPTER 11
COMBINED STRUCTURE PROBLEMS   614

ANSWERS TO SELECTED PROBLEMS   ANS-1

APPENDICES   A-1

INDEX   I-1

CHAPTER 1

MOLECULAR FORMULAS AND WHAT CAN BE LEARNED FROM THEM

Before attempting to deduce the structure of an unknown organic compound from an examination of its spectra, we can simplify the problem somewhat by examining the molecular formula of the substance. The purpose of this chapter is to describe how the molecular formula of a compound is determined and how structural information may be obtained from that formula. The chapter reviews both the modern and classical *quantitative methods* of determining the molecular formula. While use of the mass spectrometer (Chapter 3) can supplant many of these quantitative analytical methods, they are still in use. Many journals still require that a satisfactory quantitative elemental analysis (Section 1.1) be obtained prior to the publication of research results.

1.1 ELEMENTAL ANALYSIS AND CALCULATIONS

The classical procedure for determining the molecular formula of a substance involves three steps:

1. A **qualitative elemental analysis** to find out what types of atoms are present: C, H, N, O, S, Cl, and so on.

2. A **quantitative elemental analysis** (or **microanalysis**) to find out the relative numbers (percentages) of each distinct type of atom in the molecule.

3. A **molecular mass** (or **molecular weight**) **determination**.

The first two steps establish an **empirical formula** for the compound. When the results of the third procedure are known, a **molecular formula** is found.

Virtually all organic compounds contain carbon and hydrogen. In most cases, it is not necessary to determine whether these elements are present in a sample: their presence is assumed. However, if it should be necessary to demonstrate that either carbon or hydrogen is present in a compound, that substance may be burned in the presence of excess oxygen. If the combustion produces carbon dioxide, carbon must be present; if combustion produces water, hydrogen atoms must be present. Today, the carbon dioxide and water can be detected by gas chromatographic methods. Sulfur atoms are converted to sulfur dioxide; nitrogen atoms are often chemically reduced to nitrogen gas following their combustion to nitrogen oxides. Oxygen can be detected by the ignition of the compound in an atmosphere of hydrogen gas; the product is water. Currently, all such analyses are performed by gas chromatography, a method that can also determine the relative amounts of each of these gases. If the amount of the original sample is known, it can be entered, and the computer can calculate the **percentage composition** of the sample.

Unless you work in a large company or in one of the larger universities, it is quite rare to find a research laboratory in which elemental analyses are performed on site. It requires too much time to set up the apparatus and keep it operating within the limits of suitable accuracy and precision. Usually, samples are sent to a commercial **microanalytical laboratory** that is prepared to do this work routinely and to vouch for the accuracy of the results.

1

Before the advent of modern instrumentation, the combustion of the precisely weighed sample was carried out in a cylindrical glass tube, contained within a furnace. A stream of oxygen was passed through the heated tube on its way to two other sequential, unheated tubes that contained chemical substances that would absorb first water ($MgClO_4$) and then carbon dioxide (NaOH/silica). These preweighed absorption tubes were detachable and were removed and reweighed to determine the amounts of water and carbon dioxide formed. The percentages of carbon and hydrogen in the original sample were calculated by simple stoichiometry. Table 1.1 shows a sample calculation.

Notice in this calculation that the amount of oxygen was determined by difference, a common practice. In a sample containing only C, H, and O, one needs to determine the percentages of only C and H; oxygen is assumed to be the unaccounted-for portion. You may also apply this practice in situations involving elements other than oxygen; if all but one of the elements is determined, the last one can be determined by difference. Today, most calculations are carried out automatically by the computerized instrumentation. Nevertheless, it is often useful for a chemist to understand the fundamental principles of the calculations.

Table 1.2 shows how to determine the **empirical formula** of a compound from the percentage compositions determined in an analysis. Remember that the empirical formula expresses the simplest whole-number ratios of the elements and may need to be multiplied by an integer to obtain the true **molecular formula.** To determine the value of the multiplier, a molecular mass is required. Determination of the molecular mass is discussed in the next section.

For a totally unknown compound (unknown chemical source or history) you will have to use this type of calculation to obtain the suspected empirical formula. However, if you have prepared the compound from a known precursor by a well-known reaction, you will have an idea of the structure of the compound. In this case, you will have calculated the expected percentage composition of your

TABLE 1.1
CALCULATION OF PERCENTAGE COMPOSITION FROM COMBUSTION DATA

$C_xH_yO_z$ + excess $O_2 \longrightarrow$ $x\,CO_2$ + $y/2\,H_2O$
9.83 mg   23.26 mg   9.52 mg

millimoles $CO_2 = \dfrac{23.26 \text{ mg } CO_2}{44.01 \text{ mg/mmole}} = 0.5285$ mmoles CO_2

mmoles CO_2 = mmoles C in original sample
(0.5285 mmoles C)(12.01 mg/mmole C) = 6.35 mg C in original sample

millimoles $H_2O = \dfrac{9.52 \text{ mg } H_2O}{18.02 \text{ mg/mmole}} = 0.528$ mmoles H_2O

$(0.528 \text{ mmoles } H_2O)\left(\dfrac{2 \text{ mmoles H}}{1 \text{ mmole } H_2O}\right) = 1.056$ mmoles H in original sample

(1.056 mmoles H)(1.008 mg/mmole H) = 1.06 mg H in original sample

$\% \text{ C} = \dfrac{6.35 \text{ mg C}}{9.83 \text{ mg sample}} \times 100 = 64.6\%$

$\% \text{ H} = \dfrac{1.06 \text{ mg H}}{9.83 \text{ mg sample}} \times 100 = 10.8\%$

$\% \text{ O} = 100 - (64.6 + 10.8) = 24.6\%$

TABLE 1.2
CALCULATION OF EMPIRICAL FORMULA

Using a 100-g sample:

64.6% of C $= 64.6$ g

10.8% of H $= 10.8$ g

24.6% of O $= \dfrac{24.6 \text{ g}}{100.0 \text{ g}}$

moles C $= \dfrac{64.6 \text{ g}}{12.01 \text{ g/mole}} = 5.38$ moles C

moles H $= \dfrac{10.8 \text{ g}}{1.008 \text{ g/mole}} = 10.7$ moles H

moles O $= \dfrac{24.6 \text{ g}}{16.0 \text{ g/mole}} = 1.54$ moles O

giving the result

$C_{5.38}H_{10.7}O_{1.54}$

Converting to the simplest ratio:

$C_{\frac{5.38}{1.54}} H_{\frac{10.7}{1.54}} O_{\frac{1.54}{1.54}} = C_{3.49}H_{6.95}O_{1.00}$

which approximates

$C_{3.50}H_{7.00}O_{1.00}$

or

$C_7H_{14}O_2$

sample in advance (from its postulated structure) and will use the analysis to verify your hypothesis. When you perform these calculations, be sure to use the full molecular weights as given in the periodic chart and do not round off until you have completed the calculation. The final result should be good to two decimal places: four significant figures if the percentage is between 10 and 100; three figures if it is between 0 and 10. If the analytical results do not agree with the calculation, the sample may be impure, or you may have to calculate a new empirical formula to discover the identity of the unexpected structure. To be accepted for publication, most journals require the percentages found to be *less than 0.4% off from the calculated value*. Most microanalytical laboratories can easily obtain accuracy well below this limit provided the sample is pure.

In Figure 1.1, a typical situation for the use of an analysis in research is shown. Professor Amyl Carbon, or one of his students, prepared a compound believed to be the epoxynitrile with the structure shown at the bottom of the first form. A sample of this liquid compound (25 μL) was placed in a small vial correctly labeled with the name of the submitter and an identifying code (usually one that corresponds to an entry in the research notebook). Only a small amount of the sample is required, usually a few milligrams of a solid or a few microliters of a liquid. A Request for Analysis form must be filled out and submitted along with the sample. The sample form on the left side of the figure shows the type of information that must be submitted. In this case, the professor calculated the expected results for C, H, and N and the expected formula and molecular weight. Note that the compound also contains oxygen, but that there was no request for an oxygen analysis. Two other samples were also submitted at the same time. After a short time, typically within a week, the

Microanalytical Company, Inc.

REQUEST FOR ANALYSIS FORM

Date: _October 30, 2006_

Report To: _Professor Amyl Carbon_
Department of Chemistry
Western Washington University
Bellingham, WA 98225

Sample No: _PAC599A_   P.O. No : _PO 2349_

Report By: AirMail Phone   (Email)
(circle one)   _pac@www.edu_

Elements to Analyze: _C, H, N_

Other Elements Present : _O_

[X] Single Analysis   [ ] Duplicate Analysis

[ ] Duplicate only if results are not in range

M.P. _____   B.P. _69_ °C @ _2.3_ mmHg

Sensitive to : _____ Weigh under N? Y(N)

Dry the Sample? Y(N) Details: _____

[ ] Hygroscopic [ ] Volatile [ ] Explosive

THEORY OR RANGE

%C _67.17_   Amount Provided _25 μL_

%H _8.86_   Stucture:

%N _11.19_

%O _____

%Other _____ Comments: _C₇H₁₁NO_

Mol. Wt. _125.17_   _____

Microanalytical Company, Inc.

November 25, 2006

Professor Amyl Carbon
Department of Chemistry
Western Washington University
Bellingham, WA

RESULTS OF ANALYSIS

Sample ID	Carbon (%)	Hydrogen (%)	Nitrogen (%)
PAC599A	67.39	9.22	11.25
PAC589B	64.98	9.86	8.03
PAC603	73.77	8.20	—

Dr. B. Grant Poohbah,
_Ph.D._   _____

Director of Analytical Services
Microanalytical Company, Inc

FIGURE 1.1 Sample microanalysis forms. Shown on the left is a typical submission form that is sent with the samples. (The three shown here in labeled vials were all sent at the same time.) Each sample needs its own form. In the background on the right is the formal letter that reported the results. Were the results obtained for sample PAC599A satisfactory?

results were reported to Professor Carbon as an email (see the request on the form). At a later date, a formal letter (shown in the background on the right-hand side) is sent to verify and authenticate the results. Compare the values in the report to those calculated by Professor Carbon. Are they within the accepted range? If not, the analysis will have to be repeated with a freshly purified sample, or a new possible structure will have to be considered.

Keep in mind that in an actual laboratory situation, when you are trying to determine the molecular formula of a totally new or previously unknown compound, you will have to allow for some variance in the quantitative elemental analysis. Other data can help you in this situation since infrared (Chapter Two) and nuclear magnetic resonance (NMR) (Chapters Five to Nine) data will also suggest a possible structure or at least some of its prominent features. Many times, these other data will be less sensitive to small amounts of impurities than the microanalysis.

☐ 1.2 DETERMINATION OF MOLECULAR MASS

The next step in determining the molecular formula of a substance is to determine the weight of one mole of that substance. This may be accomplished in a variety of ways. Without knowledge of the molecular mass of the unknown, there is no way of determining whether the empirical formula, which is determined directly from elemental analysis, is the true formula of the substance or whether the empirical formula must be multiplied by some integral factor to obtain the molecular formula. In the example cited in Section 1.1, without knowledge of the molecular mass of the unknown, it is impossible to tell whether the molecular formula is $C_7H_{14}O_2$ or $C_{14}H_{28}O_4$.

In a modern laboratory, the molecular mass is determined using mass spectrometry. The details of this method and the means of determining molecular mass can be found in Chapter 3, Section 3.6. This section reviews some classical methods of obtaining the same information.

An old method that is used occasionally is the **vapor density method.** In this method, a known volume of gas is weighed at a known temperature. After converting the volume of the gas to standard temperature and pressure, we can determine what fraction of a mole that volume represents. From that fraction, we can easily calculate the molecular mass of the substance.

Another method of determining the molecular mass of a substance is to measure the freezing-point depression of a solvent that is brought about when a known quantity of test substance is added. This is known as a **cryoscopic method.** Another method, which is used occasionally, is **vapor pressure osmometry,** in which the molecular weight of a substance is determined through an examination of the change in vapor pressure of a solvent when a test substance is dissolved in it.

If the unknown substance is a carboxylic acid, it may be titrated with a standardized solution of sodium hydroxide. By use of this procedure, a **neutralization equivalent** can be determined. The neutralization equivalent is identical to the equivalent weight of the acid. If the acid has only one carboxyl group, the neutralization equivalent and the molecular mass are identical. If the acid has more than one carboxyl group, the neutralization equivalent is equal to the molecular mass of the acid divided by the number of carboxyl groups. Many phenols, especially those substituted by electron-withdrawing groups, are sufficiently acidic to be titrated by this same method, as are sulfonic acids.

☐ 1.3 MOLECULAR FORMULAS

Once the molecular mass and the empirical formula are known, we may proceed directly to the **molecular formula.** Often, the empirical formula weight and the molecular mass are the same. In such cases, the empirical formula is also the molecular formula. However, in many cases, the empirical formula weight is less than the molecular mass, and it is necessary to determine how many times the empirical formula weight can be divided into the molecular mass. The factor determined in this manner is the one by which the empirical formula must be multiplied to obtain the molecular formula.

Ethane provides a simple example. After quantitative element analysis, the empirical formula for ethane is found to be **CH_3.** A molecular mass of 30 is determined. The empirical formula weight of ethane, 15, is half of the molecular mass, 30. Therefore, the molecular formula of ethane must be $2(CH_3)$ or C_2H_6.

For the sample unknown introduced earlier in this chapter, the empirical formula was found to be $C_7H_{14}O_2$. The formula weight is 130. If we assume that the molecular mass of this substance was determined to be 130, we may conclude that the empirical formula and the molecular formula are identical, and that the molecular formula must be $C_7H_{14}O_2$.

☐ 1.4 INDEX OF HYDROGEN DEFICIENCY

Frequently, a great deal can be learned about an unknown substance simply from knowledge of its molecular formula. This information is based on the following general molecular formulas:

alkane	C_nH_{2n+2}	} Difference of 2 hydrogens
cycloalkane or alkene	C_nH_{2n}	} Difference of 2 hydrogens
alkyne	C_nH_{2n-2}	

Notice that each time a ring or π bond is introduced into a molecule, the number of hydrogens in the molecular formula is reduced by *two*. For every *triple bond* (*two* π bonds), the molecular formula is reduced by four. This is illustrated in Figure 1.2.

When the molecular formula for a compound contains noncarbon or nonhydrogen elements, the ratio of carbon to hydrogen may change. Following are three simple rules that may be used to predict how this ratio will change:

1. To convert the formula of an open-chain, saturated hydrocarbon to a formula containing Group V elements (N, P, As, Sb, Bi), one additional hydrogen atom must be *added* to the molecular formula for each such Group V element present. In the following examples, each formula is correct for a two-carbon acyclic, saturated compound:

$$C_2H_6, \qquad C_2H_7N, \qquad C_2H_8N_2, \qquad C_2H_9N_3$$

2. To convert the formula of an open-chain, saturated hydrocarbon to a formula containing Group VI elements (O, S, Se, Te), *no change* in the number of hydrogens is required. In the following examples, each formula is correct for a two-carbon, acyclic, saturated compound:

$$C_2H_6, \qquad C_2H_6O, \qquad C_2H_6O_2, \qquad C_2H_6O_3$$

FIGURE 1.2 Formation of rings and double bonds. Formation of each ring or double bond causes the loss of 2H.

3. To convert the formula of an open-chain, saturated hydrocarbon to a formula containing Group VII elements (F, Cl, Br, I), one hydrogen must be *subtracted* from the molecular formula for each such Group VII element present. In the following examples, each formula is correct for a two-carbon, acyclic, saturated compound:

$$C_2H_6, \qquad C_2H_5F, \qquad C_2H_4F_2, \qquad C_2H_3F_3$$

Table 1.3 presents some examples that should demonstrate how these correction numbers were determined for each of the heteroatom groups.

The **index of hydrogen deficiency** (sometimes called the **unsaturation index**) is the number of π bonds and/or rings a molecule contains. It is determined from an examination of the molecular formula of an unknown substance and from a comparison of that formula with a formula for a corresponding acyclic, saturated compound. The difference in the number of hydrogens between these formulas, when divided by 2, gives the index of hydrogen deficiency.

The index of hydrogen deficiency can be very useful in structure determination problems. A great deal of information can be obtained about a molecule before a single spectrum is examined. For example, a compound with an index of **one** must have one double bond or one ring, but it cannot have both structural features. A quick examination of the infrared spectrum could confirm the presence of a double bond. If there were no double bond, the substance would have to be cyclic and saturated. A compound with an index of **two** could have a triple bond, or it could have two double bonds, two rings, or one of each. Knowing the index of hydrogen deficiency of a substance, the chemist can proceed directly to the appropriate regions of the spectra to confirm the presence or absence of π bonds or rings. Benzene contains one ring and three "double bonds" and thus has an index of hydrogen deficiency of **four.** Any substance with an index of *four* or more may contain a benzenoid ring; a substance with an index less than *four* cannot contain such a ring.

To determine the index of hydrogen deficiency for a compound, apply the following steps:

1. Determine the formula for the saturated, acyclic hydrocarbon containing the same number of carbon atoms as the unknown substance.

2. Correct this formula for the nonhydrocarbon elements present in the unknown. Add one hydrogen atom for each Group V element present and subtract one hydrogen atom for each Group VII element present.

3. Compare this formula with the molecular formula of the unknown. Determine the number of hydrogens by which the two formulas differ.

4. Divide the difference in the number of hydrogens by **two** to obtain the index of hydrogen deficiency. This equals the number of π bonds and/or rings in the structural formula of the unknown substance.

T A B L E   1 . 3
**CORRECTIONS TO THE NUMBER OF HYDROGEN ATOMS
WHEN GROUP V AND VII HETEROATOMS ARE INTRODUCED
(GROUP VI HETEROATOMS DO NOT REQUIRE A CORRECTION)**

Group	Example	Correction	Net Change
V	C—H → C—NH$_2$	+1	Add nitrogen, add 1 hydrogen
VI	C—H → C—OH	0	Add oxygen (no hydrogen)
VII	C—H → C—CI	−1	Add chlorine, lose 1 hydrogen

The following examples illustrate how the index of hydrogen deficiency is determined and how that information can be applied to the determination of a structure for an unknown substance.

■ EXAMPLE 1

The unknown substance introduced at the beginning of this chapter has the molecular formula $C_7H_{14}O_2$.

1. Using the general formula for a saturated, acyclic hydrocarbon (C_nH_{2n+2}, where $n = 7$), calculate the formula C_7H_{16}.

2. Correction for oxygens (no change in the number of hydrogens) gives the formula $C_7H_{16}O_2$.

3. The latter formula differs from that of the unknown by two hydrogens.

4. The index of hydrogen deficiency equals **one.** There must be one ring or one double bond in the unknown substance.

Having this information, the chemist can proceed immediately to the double-bond regions of the infrared spectrum. There, she finds evidence for a carbon–oxygen double bond (carbonyl group). At this point, the number of possible isomers that might include the unknown has been narrowed considerably. Further analysis of the spectral evidence leads to an identification of the unknown substance as **isopentyl acetate.**

$$CH_3-\overset{\overset{\displaystyle O}{\|}}{C}-O-CH_2-CH_2-\underset{\underset{\displaystyle CH_3}{|}}{CH}-CH_3$$

■ EXAMPLE 2

Nicotine has the molecular formula $C_{10}H_{14}N_2$.

1. The formula for a 10-carbon, saturated, acyclic hydrocarbon is $C_{10}H_{22}$.

2. Correction for the two nitrogens (add two hydrogens) gives the formula $C_{10}H_{24}N_2$.

3. The latter formula differs from that of nicotine by 10 hydrogens.

4. The index of hydrogen deficiency equals **five.** There must be some combination of five π bonds and/or rings in the molecule. Since the index is greater than *four,* a benzenoid ring could be included in the molecule.

Analysis of the spectrum quickly shows that a benzenoid ring is indeed present in nicotine. The spectral results indicate no other double bonds, suggesting that another ring, this one saturated, must be present in the molecule. More careful refinement of the spectral analysis leads to a structural formula for nicotine:

■ **EXAMPLE 3**

Chloral hydrate ("knockout drops") is found to have the molecular formula $C_2H_3Cl_3O_2$.

1. The formula for a two-carbon, saturated, acyclic hydrocarbon is C_2H_6.

2. Correction for oxygens (no additional hydrogens) gives the formula $C_2H_6O_2$.

3. Correction for chlorines (subtract three hydrogens) gives the formula $C_2H_3Cl_3O_2$.

4. This formula and the formula of chloral hydrate correspond exactly.

5. The index of hydrogen deficiency equals **zero.** Chloral hydrate cannot contain rings or double bonds.

Examination of the spectral results is limited to regions that correspond to singly bonded structural features. The correct structural formula for chloral hydrate follows. You can see that all of the bonds in the molecule are single bonds.

$$Cl_3C-\underset{\underset{OH}{|}}{\overset{\overset{OH}{|}}{C}}-H$$

1.5 THE RULE OF THIRTEEN

High-resolution mass spectrometry provides molecular mass information from which the user can determine the exact molecular formula directly. The discussion on exact mass determination in Chapter 3 explains this process in detail. When such molar mass information is not available, however, it is often useful to be able to generate all the possible molecular formulas for a given mass. By applying other types of spectroscopic information, it may then be possible to distinguish among these possible formulas. A useful method for generating possible molecular formulas for a given molecular mass is the **Rule of Thirteen.**[1]

As a first step in the Rule of Thirteen, we generate a **base formula,** which contains only carbon and hydrogen. The base formula is found by dividing the molecular mass M by 13 (the mass of one carbon plus one hydrogen). This calculation provides a numerator n and a remainder r.

$$\frac{M}{13} = n + \frac{r}{13}$$

The base formula thus becomes

$$C_nH_{n+r}$$

which is a combination of carbons and hydrogens that has the desired molecular mass M.

The **index of hydrogen deficiency** (unsaturation index) U that corresponds to the preceding formula is calculated easily by applying the relationship

$$U = \frac{(n - r + 2)}{2}$$

[1] Bright, J. W., and E. C. M. Chen, "Mass Spectral Interpretation Using the 'Rule of 13,'" *Journal of Chemical Education, 60* (1983): 557.

Of course, you can also calculate the index of hydrogen deficiency using the method shown in Section 1.4.

If we wish to derive a molecular formula that includes other atoms besides carbon and hydrogen, then we must subtract the mass of a combination of carbons and hydrogens that equals the masses of the other atoms being included in the formula. For example, if we wish to convert the base formula to a new formula containing one oxygen atom, then we subtract one carbon and four hydrogens at the same time that we add one oxygen atom. Both changes involve a molecular mass equivalent of 16 ($O = CH_4 = 16$). Table 1.4 includes a number of C/H equivalents for replacement of carbon and hydrogen in the base formula by the most common elements likely to occur in an organic compound.[2]

To comprehend how the Rule of Thirteen might be applied, consider an unknown substance with a molecular mass of 94 amu. Application of the formula provides

$$\frac{94}{13} = 7 + \frac{3}{13}$$

According to the formula, $n = 7$ and $r = 3$. The base formula must be

$$C_7H_{10}$$

The index of hydrogen deficiency is

$$U = \frac{(7 - 3 + 2)}{2} = 3$$

TABLE 1.4
CARBON/HYDROGEN EQUIVALENTS FOR SOME COMMON ELEMENTS

Add Element	Subtract Equivalent	Add ΔU	Add Element	Subtract Equivalent	Add ΔU
C	H_{12}	7	^{35}Cl	C_2H_{11}	3
H_{12}	C	-7	^{79}Br	C_6H_7	-3
O	CH_4	1	^{79}Br	C_5H_{19}	4
O_2	C_2H_8	2	F	CH_7	2
O_3	C_3H_{12}	3	Si	C_2H_4	1
N	CH_2	$\frac{1}{2}$	P	C_2H_7	2
N_2	C_2H_4	1	I	C_9H_{19}	0
S	C_2H_8	2	I	$C_{10}H_7$	7

[2] In Table 1.4, the equivalents for chlorine and bromine are determined assuming that the isotopes are ^{35}Cl and ^{79}Br, respectively. Always use this assumption when applying this method.

A substance that fits this formula must contain some combination of three rings or multiple bonds. A possible structure might be

C_7H_{10}

$U = 3$

If we were interested in a substance that had the same molecular mass but that contained one oxygen atom, the molecular formula would become C_6H_6O. This formula is determined according to the following scheme:

1. Base formula = C_7H_{10} $U = 3$

2. Add: $+ O$

3. Subtract: $- CH_4$

4. Change the value of U: $\Delta U = 1$

5. New formula = C_6H_6O

6. New index of hydrogen deficiency: $U = 4$

A possible substance that fits these data is

C_6H_6O

$U = 4$

There are additional possible molecular formulas that conform to a molecular mass of 94 amu:

$C_5H_2O_2$ $U = 5$ C_5H_2S $U = 5$

C_6H_8N $U = 3\frac{1}{2}$ CH_3Br $U = 0$

As the formula C_6H_8N shows, any formula that contains an even number of hydrogen atoms but an odd number of nitrogen atoms leads to a fractional value of U, an unlikely choice.

Any compound with a value of U less than zero (i.e., negative) is an impossible combination. Such a value is often an indicator that an oxygen or nitrogen atom must be present in the molecular formula.

When we calculate formulas using this method, if there are not enough hydrogens, we can subtract 1 carbon and add 12 hydrogens (and make the appropriate correction in U). This procedure works only if we obtain a positive value of U. Alternatively, we can obtain another potential molecular formula by adding 1 carbon and subtracting 12 hydrogens (and correcting U).

☐ 1.6 THE NITROGEN RULE

Another fact that can be used in determining the molecular formula is expressed as the **Nitrogen Rule.** This rule states that when the number of nitrogen atoms present in the molecule is odd, the molecular mass will be an odd number; when the number of nitrogen atoms present in the molecule is even (or zero), the molecular mass will be an even number. The Nitrogen Rule is explained further in Chapter 3, Section 3.6.

PROBLEMS

*1. Researchers used a combustion method to analyze a compound used as an antiknock additive in gasoline. A 9.394-mg sample of the compound yielded 31.154 mg of carbon dioxide and 7.977 mg of water in the combustion.

   (a) Calculate the percentage composition of the compound.

   (b) Determine its empirical formula.

*2. The combustion of an 8.23-mg sample of unknown substance gave 9.62 mg CO_2 and 3.94 mg H_2O. Another sample, weighing 5.32 mg, gave 13.49 mg AgCl in a halogen analysis. Determine the percentage composition and empirical formula for this organic compound.

*3. An important amino acid has the percentage composition C 32.00%, H 6.71%, and N 18.66%. Calculate the empirical formula of this substance.

*4. A compound known to be a pain reliever had the empirical formula $C_9H_8O_4$. When a mixture of 5.02 mg of the unknown and 50.37 mg of camphor was prepared, the melting point of a portion of this mixture was determined. The observed melting point of the mixture was 156 °C. What is the molecular mass of this substance?

*5. An unknown acid was titrated with 23.1 mL of 0.1 N sodium hydroxide. The weight of the acid was 120.8 mg. What is the equivalent weight of the acid?

*6. Determine the index of hydrogen deficiency for each of the following compounds:

   (a) C_8H_7NO

   (b) $C_3H_7NO_3$

   (c) $C_4H_4BrNO_2$

   (d) $C_5H_3ClN_4$

   (e) $C_{21}H_{22}N_2O_2$

*7. A substance has the molecular formula C_4H_9N. Is there any likelihood that this material contains a triple bond? Explain your reasoning.

*8. (a) A researcher analyzed an unknown solid, extracted from the bark of spruce trees, to determine its percentage composition. An 11.32-mg sample was burned in a combustion apparatus. The carbon dioxide (24.87 mg) and water (5.82 mg) were collected and weighed. From the results of this analysis, calculate the percentage composition of the unknown solid.

   (b) Determine the empirical formula of the unknown solid.

   (c) Through mass spectrometry, the molecular mass was found to be 420 g/mole. What is the molecular formula?

   (d) How many aromatic rings could this compound contain?

*9. Calculate the molecular formulas for possible compounds with molecular masses of 136; use the Rule of Thirteen. You may assume that the only other atoms present in each molecule are carbon and hydrogen.

(a) A compound with two oxygen atoms

(b) A compound with two nitrogen atoms

(c) A compound with two nitrogen atoms and one oxygen atom

(d) A compound with five carbon atoms and four oxygen atoms

*10. An alkaloid was isolated from a common household beverage. The unknown alkaloid proved to have a molecular mass of 194. Using the Rule of Thirteen, determine a molecular formula and an index of hydrogen deficiency for the unknown. Alkaloids are naturally occurring organic substances that contain **nitrogen.** (*Hint:* There are four nitrogen atoms and two oxygen atoms in the molecular formula. The unknown is **caffeine.** Look up the structure of this substance in *The Merck Index* and confirm its molecular formula.)

*11. The Drug Enforcement Agency (DEA) confiscated a hallucinogenic substance during a drug raid. When the DEA chemists subjected the unknown hallucinogen to chemical analysis, they found that the substance had a molecular mass of 314. Elemental analysis revealed the presence of carbon and hydrogen only. Using the Rule of Thirteen, determine a molecular formula and an index of hydrogen deficiency for this substance. (*Hint:* The molecular formula of the unknown also contains two oxygen atoms. The unknown is **tetrahydrocannabinol,** the active constituent of marijuana. Look up the structure of tetrahydrocannabinol in *The Merck Index* and confirm its molecular formula.)

12. A carbohydrate was isolated from a sample of cow's milk. The substance was found to have a molecular mass of 342. The unknown carbohydrate can be hydrolyzed to form two isomeric compounds, each with a molecular mass of 180. Using the Rule of Thirteen, determine a molecular formula and an index of hydrogen deficiency for the unknown and for the hydrolysis products. (*Hint:* Begin by solving the molecular formula for the 180-amu hydrolysis products. These products have one oxygen atom for every carbon atom in the molecular formula. The unknown is **lactose.** Look up its structure in *The Merck Index* and confirm its molecular formula.)

---

*Answers are provided in the chapter, *Answers to Selected Problems*

REFERENCES

O'Neil, M. J., et al., eds. *The Merck Index,* 15th ed., London, England: Royal Society of Chemistry, 2013.

Pavia, D. L., G. M. Lampman, G. S. Kriz, and R. G. Engel, *Introduction to Organic Laboratory Techniques: A Small Scale Approach,* 3rd ed., Belmont, CA: Brooks-Cole Cengage Learning, 2011.

Pavia, D. L., G. M. Lampman, G. S. Kriz, and R. G. Engel, *Introduction to Organic Laboratory Techniques: A Micro-scale Approach,* 5th ed., Belmont, CA: Brooks-Cole Cengage Learning, 2013.

Shriner, R. L., C. K. F. Hermann, T. C. Morrill, D. Y. Curtin, and R. C. Fuson, *The Systematic Identification of Organic Compounds,* 8th ed., New York: John Wiley, 2004.

INFRARED SPECTROSCOPY

Almost any compound having covalent bonds, whether organic or inorganic, absorbs various frequencies of electromagnetic radiation in the infrared region of the electromagnetic spectrum. This region lies at wavelengths longer than those associated with visible light, which range from approximately 400 to 800 nm (1 nm = 10^{-9} m), but lies at wavelengths shorter than those associated with microwaves, which are longer than 1 mm. For chemical purposes, we are interested in the **vibrational** portion of the infrared region. It includes radiation with wavelengths (λ) between 2.5 μm and 25 μm (1μm = 10^{-6} m). Although the more technically correct unit for wavelength in the infrared region of the spectrum is the micrometer (μm), you will often see the micron (μ) used on infrared spectra. Figure 2.1 illustrates the relationship of the infrared region to others included in the electromagnetic spectrum.

Figure 2.1 shows that the wavelength λ is inversely proportional to the frequency v and is governed by the relationship $v = c/\lambda$, where c = speed of light. Also observe that the energy is directly proportional to the frequency: $E = hv$, where h = Planck's constant. From the latter equation, you can see qualitatively that the highest energy radiation corresponds to the X-ray region of the spectrum, where the energy may be great enough to break bonds in molecules. At the other end of the electromagnetic spectrum, radio-frequencies have very low energies, only enough to cause nuclear or electronic spin transitions within molecules—that is, nuclear magnetic resonance (NMR) or electron spin resonance (ESR), respectively.

Table 2.1 summarizes the regions of the spectrum and the types of energy transitions observed there. Several of these regions, including the infrared, give vital information about the structures of organic molecules. Nuclear magnetic resonance, which occurs in the radiofrequency part of the spectrum, is discussed in Chapters 5 through 9, whereas ultraviolet and visible spectroscopy are described in Chapter 10.

Most chemists refer to the radiation in the vibrational infrared region of the electromagnetic spectrum in terms of a unit called a **wavenumber** ($\overline{v}$), rather than wavelength (μ or μm).

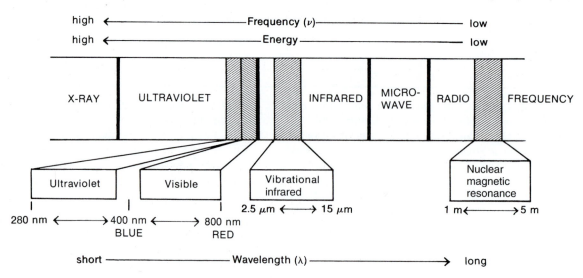

F I G U R E   2 . 1 A portion of the electromagnetic spectrum showing the relationship of the vibrational infrared to other types of radiation.

TABLE 2.1
TYPES OF ENERGY TRANSITIONS IN EACH REGION
OF THE ELECTROMAGNETIC SPECTRUM

Region of Spectrum	Energy Transitions
X-rays	Bond breaking
Ultraviolet/visible	Electronic
Infrared	Vibrational
Microwave	Rotational
Radiofrequencies	Nuclear spin (nuclear magnetic resonance)
	Electronic spin (electron spin resonance)

Wavenumbers are expressed as reciprocal centimeters (cm^{-1}) and are easily computed by taking the reciprocal of the wavelength expressed in centimeters. Convert a wavenumber $\overline{v}$ to a frequency v by multiplying it by the speed of light (expressed in centimeters per second).

$$\overline{v}\,(\text{cm}^{-1}) = \frac{1}{\lambda\,(\text{cm})} \qquad v\,(\text{Hz}) = \overline{v}c = \frac{c\,(\text{cm/sec})}{\lambda\,(\text{cm})}$$

The main reason chemists prefer to use wavenumbers as units is that they are directly proportional to energy (*a higher wavenumber corresponds to a higher energy*). Thus, in terms of wavenumbers, the vibrational infrared extends from about 4000 to 400 cm^{-1}. This range corresponds to wavelengths of 2.5 to 25 μm. *We will use wavenumber units exclusively in this textbook.* You may encounter wavelength values in older literature. Convert wavelengths (μ or μm) to wavenumbers (cm^{-1}) by using the following relationships:

$$\text{cm}^{-1} = \frac{1}{(\mu\text{m})} \times 10,000 \qquad \text{and} \qquad \mu\text{m} = \frac{1}{(\text{cm}^{-1})} \times 10,000$$

INTRODUCTION TO INFRARED SPECTROSCOPY

2.1 THE INFRARED ABSORPTION PROCESS

As with other types of energy absorption, molecules are excited to a higher energy state when they absorb infrared radiation. The absorption of infrared radiation is, like other absorption processes, a quantized process. A molecule absorbs only selected frequencies (energies) of infrared radiation. The absorption of infrared radiation corresponds to energy changes on the order of 8 to 40 kJ/mole. Radiation in this energy range corresponds to the range encompassing the stretching and bending vibrational frequencies of the bonds in most covalent molecules. In the absorption process, those frequencies of infrared radiation that match the natural vibrational frequencies of the molecule in question are absorbed, and the energy absorbed serves to increase the **amplitude** of the vibrational motions of the bonds in the molecule. Note, however, that not all bonds in a molecule are capable of absorbing infrared energy, even if the frequency of the radiation exactly matches that of the bond motion. Only those bonds that have a **dipole moment** that changes as a function of time are capable of absorbing infrared radiation. Symmetric bonds, such as those of H_2 or Cl_2, do not absorb infrared radiation. A bond must present an electrical dipole that is changing at the same frequency as the incoming radiation for energy to be transferred. The changing electrical dipole of the bond can then couple with the sinusoidally changing electromagnetic field of the incoming radiation. Thus, a symmetric bond that has identical or nearly identical groups on each end will not absorb in the infrared. For the purposes of an organic chemist, the bonds most likely to be affected by this restraint are those of symmetric or pseudosymmetric alkenes (C=C) and alkynes (C≡C).

$$CH_3\!\!\diagdown\!\!\underset{CH_3\diagup}{C}\!=\!\underset{\diagdown CH_3}{C}\!\!\diagup\!\!CH_3 \qquad\qquad CH_3\!-\!CH_2\!\!\diagdown\!\!\underset{CH_3\diagup}{C}\!=\!\underset{\diagdown CH_3}{C}\!\!\diagup\!\!CH_3$$

$$CH_3\!-\!C\!\equiv\!C\!-\!CH_3 \qquad\qquad\qquad CH_3\!-\!CH_2\!-\!C\!\equiv\!C\!-\!CH_3$$

Symmetric   Pseudosymmetric

2.2 USES OF THE INFRARED SPECTRUM

Since every type of bond has a different natural frequency of vibration, and because two of the same type of bond in two different compounds are in two slightly different environments, no two molecules of different structure have exactly the same infrared absorption pattern, or **infrared spectrum.** Although some of the frequencies absorbed in the two cases might be the same, in no case of two different molecules will their infrared spectra (the patterns of absorption) be identical. Thus, the infrared spectrum can be used for molecules much as a fingerprint can be used for humans. By comparing the infrared spectra of two substances thought to be identical, you can establish whether they are, in fact, identical. If their infrared spectra coincide peak for peak (absorption for absorption), in most cases the two substances will be identical.

A second and more important use of the infrared spectrum is to determine structural information about a molecule. The absorptions of each type of bond (N−H, C−H, O−H, C−X, C=O, C−O, C−C, C=C, C≡C, C≡N, and so on) are regularly found only in certain small portions of the vibrational infrared region. A small range of absorption can be defined for each type of bond. Outside this range, absorptions are normally due to some other type of bond. For instance, any absorption in the range 3000 ± 150 cm^{-1} is almost always due to the presence of a C−H bond in the molecule; an absorption in the range 1715 ± 100 cm^{-1} is normally due to the presence of a C=O bond (carbonyl group) in the molecule. The same type of range applies to each type of bond. Figure 2.2 illustrates schematically how these are spread out over the vibrational infrared. Try to fix this general scheme in your mind for future convenience.

FREQUENCY (cm^{-1})

4000		2500	2000	1800	1650	1550		650
O−H C−H			C≡C	VERY FEW BANDS	C=O	C=N		C−Cl C−O
N−H			C≡N			C=C	C−N	C−C
			X=C=Y (C,O,N,S)				N=O N=O	

2.5	4	5	5.5	6.1	6.5	15.4

WAVELENGTH (μ)

FIGURE 2.2 The approximate regions where various common types of bonds absorb (stretching vibrations only; bending, twisting, and other types of bond vibrations have been omitted for clarity).

2.3 THE MODES OF STRETCHING AND BENDING

The simplest types, or **modes,** of vibrational motion in a molecule that are **infrared active**—those, that give rise to absorptions—are the stretching and bending modes.

C—H

Stretching

Bending

However, other, more complex types of stretching and bending are also active. The following illustrations of the normal modes of vibration for a methylene group introduce several terms. In general, asymmetric stretching vibrations occur at higher frequencies than symmetric stretching vibrations; also, stretching vibrations occur at higher frequencies than bending vibrations. The terms **scissoring, rocking, wagging,** and **twisting** are commonly used in the literature to describe the origins of infrared bands.

In any group of three or more atoms, at least two of which are identical, there are *two* modes of stretching: symmetric and asymmetric. Examples of such groupings are $-CH_3$, $-CH_2-$, $-NO_2$, $-NH_2$, and anhydrides. The methyl group gives rise to a symmetric stretching vibration at about $2872 \ cm^{-1}$ and an asymmetric stretch at about $2962 \ cm^{-1}$. The anhydride functional group gives two absorptions in the C=O region because of the asymmetric and symmetric stretching modes. A similar phenomenon occurs in the amino group, where a primary amine (NH_2) usually has two absorptions in the N—H stretch region, while a secondary amine (R_2NH) has only one absorption peak. Amides exhibit similar bands. There are two strong N=O stretch peaks for a nitro group, with the symmetric stretch appearing at about $1350 \ cm^{-1}$ and the asymmetric stretch appearing at about $1550 \ cm^{-1}$.

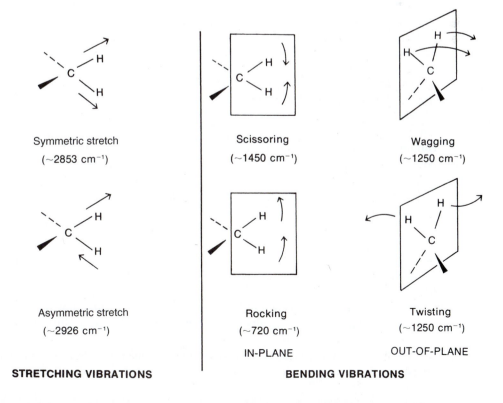

Symmetric stretch
(~2853 cm⁻¹)

Scissoring
(~1450 cm⁻¹)

Wagging
(~1250 cm⁻¹)

Asymmetric stretch
(~2926 cm⁻¹)

Rocking
(~720 cm⁻¹)

IN-PLANE

Twisting
(~1250 cm⁻¹)

OUT-OF-PLANE

STRETCHING VIBRATIONS

BENDING VIBRATIONS

	Symmetric Stretch	Asymmetric Stretch

Methyl

~ 2872 cm^{-1} ~ 2962 cm^{-1}

Anhydride

~ 1760 cm^{-1} ~ 1800 cm^{-1}

Amino

~ 3300 cm^{-1} ~ 3400 cm^{-1}

Nitro

~ 1350 cm^{-1} ~ 1550 cm^{-1}

The vibrations we have been discussing are called **fundamental absorptions.** They arise from excitation from the ground state to the lowest-energy excited state. Usually, the spectrum is complicated because of the presence of weak overtone, combination, and difference bands. **Overtones** result from excitation from the ground state to higher energy states, which correspond to integral multiples of the frequency of the fundamental (ν). For example, you might observe weak overtone bands at $2\bar{\nu}, 3\bar{\nu}, \ldots$ Any kind of physical vibration generates overtones. If you pluck a string on a cello, the string vibrates with a fundamental frequency. However, less-intense vibrations are also set up at several overtone frequencies. An absorption in the infrared at 500 cm^{-1} may well have an accompanying peak of lower intensity at 1000 cm^{-1}—an overtone.

When two vibrational frequencies ($\bar{\nu}_1$ and $\bar{\nu}_2$) in a molecule couple to give rise to a vibration of a new frequency within the molecule, and when such a vibration is infrared active, it is called a **combination band.** This band is the sum of the two interacting bands ($\bar{\nu}_{comb} = \bar{\nu}_1 + \bar{\nu}_2$). Not all possible combinations occur. The rules that govern which combinations are allowed are beyond the scope of our discussion here.

Difference bands are similar to combination bands. The observed frequency in this case results from the difference between the two interacting bands ($\nu_{diff} = \bar{\nu}_1 - \bar{\nu}_2$).

One can calculate overtone, combination, and difference bands by directly manipulating frequencies in wavenumbers via multiplication, addition, and subtraction, respectively. When a fundamental vibration couples with an overtone or combination band, the coupled vibration is called **Fermi resonance.** Again, only certain combinations are allowed. Fermi resonance is often observed in carbonyl compounds.

Although rotational frequencies of the whole molecule are not infrared active, they often couple with the stretching and bending vibrations in the molecule to give additional fine structure to these absorptions, thus further complicating the spectrum. One of the reasons a band is broad rather than sharp in the infrared spectrum is rotational coupling, which may lead to a considerable amount of unresolved fine structure.

2.4 BOND PROPERTIES AND ABSORPTION TRENDS

Let us now consider how bond strength and the masses of the bonded atoms affect the infrared absorption frequency. For the sake of simplicity, we will restrict the discussion to a simple heteronuclear diatomic molecule (two *different* atoms) and its stretching vibration.

A diatomic molecule can be considered as two vibrating masses connected by a spring. The bond distance continually changes, but an equilibrium or average bond distance can be defined. Whenever the spring is stretched or compressed beyond this equilibrium distance, the potential energy of the system increases.

As for any harmonic oscillator, when a bond vibrates, its energy of vibration is continually and periodically changing from kinetic to potential energy and back again. The total amount of energy is proportional to the frequency of the vibration,

$$E_{osc} \propto h\nu_{osc}$$

which for a harmonic oscillator is determined by the force constant K of the spring, or its stiffness, and the masses (m_1 and m_2) of the two bonded atoms. The natural frequency of vibration of a bond is given by the equation

$$\bar{\nu} = \frac{1}{2\pi c}\sqrt{\frac{K}{\mu}}$$

which is derived from Hooke's Law for vibrating springs. The **reduced mass** μ of the system is given by

$$\mu = \frac{m_1 m_2}{m_1 + m_2}$$

K is a constant that varies from one bond to another. As a first approximation, the force constants for triple bonds are three times those of single bonds, whereas the force constants for double bonds are twice those of single bonds.

Two things should be noticeable immediately. One is that stronger bonds have a larger force constant K and vibrate at higher frequencies than weaker bonds. The second is that bonds between atoms of higher masses (larger reduced mass, μ) vibrate at lower frequencies than bonds between lighter atoms.

In general, triple bonds are stronger than double or single bonds between the same two atoms and have higher frequencies of vibration (higher wavenumbers):

C≡C	C=C	C—C
2150 cm^{-1}	1650 cm^{-1}	1200 cm^{-1}

⟵ Increasing K

The C—H stretch occurs at about 3000 cm^{-1}. As the atom bonded to carbon increases in mass, the reduced mass (μ) increases, and the frequency of vibration decreases (wavenumbers get smaller):

C—H	C—C	C—O	C—Cl	C—Br	C—I
3000 cm^{-1}	1200 cm^{-1}	1100 cm^{-1}	750 cm^{-1}	600 cm^{-1}	500 cm^{-1}

⟶ Increasing μ

Bending motions occur at lower energy (lower frequency) than the typical stretching motions because of the lower value for the bending force constant K.

C—H stretching   C—H bending

~3000 cm^{-1} ~1340 cm^{-1}

Hybridization affects the force constant K, also. Bonds are stronger in the order $sp > sp^2 > sp^3$, and the observed frequencies of C—H vibration illustrate this nicely.

sp sp^2 sp^3

≡C—H   =C—H   —C—H

3300 cm^{-1} 3100 cm^{-1} 2900 cm^{-1}

Resonance also affects the strength and length of a bond and hence its force constant K. Thus, while a normal ketone has its C=O stretching vibration at 1715 cm^{-1}, a ketone that is conjugated with a C=C double bond absorbs at a lower frequency, near 1675 to 1680 cm^{-1}. That is because resonance lengthens the C=O bond distance and gives it more single-bond character:

Resonance has the effect of reducing the force constant K, and the absorption moves to a lower frequency.

The Hooke's Law expression given earlier may be transformed into a very useful equation as follows:

$$\bar{v} = \frac{1}{2\pi c}\sqrt{\frac{K}{\mu}}$$

$\bar{v}$ = frequency in cm^{-1}

c = velocity of light = 3×10^{10} cm/sec

K = force constant in dynes/cm

$$\mu = \frac{m_1 m_2}{m_1 + m_2}, \quad \text{masses of atoms in grams,}$$

or $\dfrac{M_1 M_2}{(M_1 + M_2)(6.02 \times 10^{23})}$, masses of atoms in amu

Removing Avogadro's number (6.02×10^{23}) from the denominator of the reduced mass expression (μ) by taking its square root, we obtain the expression

$$\bar{v} = \frac{7.76 \times 10^{11}}{2\pi c}\sqrt{\frac{K}{\mu}}$$

TABLE 2.2

CALCULATION OF STRETCHING FREQUENCIES FOR DIFFERENT TYPES OF BONDS

C=C bond:

$$\overline{v} = 4.12 \sqrt{\frac{K}{\mu}}$$

$K = 10 \times 10^5$ dynes/cm

$$\mu = \frac{M_C M_C}{M_C + M_C} = \frac{(12)(12)}{12 + 12} = 6$$

$$\overline{v} = 4.12 \sqrt{\frac{10 \times 10^5}{6}} = 1682 \text{ cm}^{-1} \text{ (calculated)}$$

$\overline{v} = 1650 \text{ cm}^{-1}$ (experimental)

C—H bond:

$$\overline{v} = 4.12 \sqrt{\frac{K}{\mu}}$$

$K = 5 \times 10^5$ dynes/cm

$$\mu = \frac{M_C M_H}{M_C + M_H} = \frac{(12)(1)}{12 + 1} = 0.923$$

$$\overline{v} = 4.12 \sqrt{\frac{5 \times 10^5}{0.923}} = 3032 \text{ cm}^{-1} \text{ (calculated)}$$

$\overline{v} = 3000 \text{ cm}^{-1}$ (experimental)

C—D bond:

$$\overline{v} = 4.12 \sqrt{\frac{K}{\mu}}$$

$K = 5 \times 10^5$ dynes/cm

$$\mu = \frac{M_C M_D}{M_C + M_D} = \frac{(12)(2)}{12 + 2} = 1.71$$

$$\overline{v} = 4.12 \sqrt{\frac{5 \times 10^5}{1.71}} = 2228 \text{ cm}^{-1} \text{ (calculated)}$$

$\overline{v} = 2206 \text{ cm}^{-1}$ (experimental)

A new expression is obtained by inserting the actual values of π and c:

$$\overline{v}(\text{cm}^{-1}) = 4.12\sqrt{\frac{K}{\mu}}$$

$$\mu = \frac{M_1 M_2}{M_1 + M_2}, \quad \text{where } M_1 \text{ and } M_2 \text{ are atomic weights}$$

$$K = \text{force constant in dynes/cm } (1 \text{ dyne} = 1.020 \times 10^{-3} \text{ g})$$

This equation may be used to calculate the approximate position of a band in the infrared spectrum by assuming that K for single, double, and triple bonds is 5, 10, and 15×10^5 dynes/cm, respectively. Table 2.2 gives a few examples. Notice that excellent agreement is obtained with the experimental values given in the table. However, experimental and calculated values may vary considerably owing to resonance, hybridization, and other effects that operate in organic molecules. Nevertheless, good *qualitative* values are obtained by such calculations.

2.5 THE INFRARED SPECTROMETER

The instrument that determines the absorption spectrum for a compound is called an **infrared spectrometer** or, more precisely, a **spectrophotometer.** Two types of infrared spectrometers are in common use in the organic laboratory: dispersive and Fourier transform (FT) instruments. Both of these types of instruments provide spectra of compounds in the common range of 4000 to 400 cm^{-1}. Although the two provide nearly identical spectra for a given compound, FT infrared spectrometers provide the infrared spectrum much more rapidly than the dispersive instruments.

A.   Dispersive Infrared Spectrometers

Figure 2.3a schematically illustrates the components of a simple dispersive infrared spectrometer. The instrument produces a beam of infrared radiation from a hot wire and, by means of mirrors, divides it into two parallel beams of equal-intensity radiation. The sample is placed in one beam, and the other beam is used as a reference. The beams then pass into the **monochromator,** which disperses each into a continuous spectrum of frequencies of infrared light. The monochromator consists of a rapidly rotating sector (beam chopper) that passes the two beams alternately to a diffraction grating (a prism in older instruments). The slowly rotating diffraction grating varies the frequency or wavelength of radiation reaching the thermocouple detector. The detector senses the ratio between the intensities of the reference and sample beams. In this way, the detector determines which frequencies have been absorbed by the sample and which frequencies are unaffected by the light passing through the sample. After the signal from the detector is amplified, the recorder draws the resulting spectrum of the sample on a chart. It is important to realize that the spectrum is recorded as the frequency of infrared radiation changes by rotation of the diffraction grating. Dispersive instruments are said to record a spectrum in the **frequency domain.**

F I G U R E 2 . 3 Schematic diagrams of (a) dispersive and (b) Fourier transform infrared spectrophotometers.

Note that it is customary to plot frequency (wavenumber, cm^{-1}) versus light transmitted, not light absorbed. This is recorded as **percent transmittance (%*T*)** because the detector records the ratio of the intensities of the two beams, and

$$\text{percent transmittance} = \frac{I_s}{I_r} \times 100$$

where I_s is the intensity of the sample beam, and I_r is the intensity of the reference beam. In many parts of the spectrum, the transmittance is nearly 100%, meaning that the sample is nearly transparent to radiation of that frequency (does not absorb it). Maximum absorption is thus represented by a *minimum* on the chart. Even so, the absorption is traditionally called a **peak.**

The chemist sometimes obtains the spectrum of a compound by dissolving it in a solvent (Section 2.6). The solution is then placed in the **sample beam,** while pure solvent is placed in the **reference beam** in an identical cell. The instrument automatically "subtracts" the spectrum of the solvent from that of the sample. The instrument also cancels out the effects of the infrared-active atmospheric gases, carbon dioxide and water vapor, from the spectrum of the sample (they are present in both beams). This convenience feature is the reason most dispersive infrared spectrometers are double-beam (sample + reference) instruments that measure intensity ratios; since the solvent absorbs in both beams, it is in both terms of the ratio I_s / I_r and cancels out. If a pure liquid is analyzed (no solvent), the compound is placed in the sample beam, and nothing is inserted into the reference beam. When the spectrum of the liquid is obtained, the effects of the atmospheric gases are automatically canceled since they are present in both beams.

B.   Fourier Transform Spectrometers

The most modern infrared spectrometers (spectrophotometers) operate on a different principle. The design of the optical pathway produces a pattern called an **interferogram.** The interferogram is a complex signal, but its wave-like pattern contains all the frequencies that make up the infrared spectrum. An interferogram is essentially a plot of intensity versus time (a **time-domain spectrum**). However, a chemist is more interested in a spectrum that is a plot of intensity versus frequency (a **frequency-domain spectrum**). A mathematical operation known as a **Fourier transform (FT)** can separate the individual absorption frequencies from the interferogram, producing a spectrum virtually identical to that obtained with a dispersive spectrometer. This type of instrument is known as a **Fourier transform infrared spectrometer, or FT-IR.**[1] The advantage of an FT-IR instrument is that it acquires the interferogram in less than a second. It is thus possible to collect dozens of interferograms of the same sample and accumulate them in the memory of a computer. When a Fourier transform is performed on the sum of the accumulated interferograms, a spectrum with a better signal-to-noise ratio can be plotted. An FT-IR instrument is therefore capable of greater speed and greater sensitivity than a dispersion instrument.

A schematic diagram of an FT-IR is shown in Figure 2.3b. The FT-IR uses an **interferometer** to process the energy sent to the sample. In the interferometer, the source energy passes through a **beam splitter,** a mirror placed at a 45° angle to the incoming radiation, which allows the incoming radiation to pass through but separates it into two perpendicular beams, one undeflected, the other oriented at a 90° angle. One beam, the one oriented at 90° in Figure 2.3b, goes to a stationary or "fixed" mirror and is returned to the beam splitter. The undeflected beam goes to a moving mirror and is also returned to the beam splitter. The motion of the mirror causes the pathlength that the second beam traverses to vary. When the two beams meet at the beam splitter, they recombine, but the pathlength differences (differing wavelength content) of the two beams cause both constructive and destructive interferences. The combined beam containing these interference patterns is called the interferogram. This interferogram contains all of the radiative energy coming from the source and has a wide range of wavelengths.

The interferogram generated by combining the two beams is oriented toward the sample by the beam splitter. As it passes through the sample, the sample *simultaneously* absorbs all of the wavelengths (frequencies) that are normally found in its infrared spectrum. The modified interferogram signal that reaches the detector contains information about the amount of energy that was absorbed at every wavelength (frequency). The computer compares the modified interferogram to a reference laser beam to have a standard of comparison. The final interferogram contains all of the information in one time-domain signal, a signal that cannot be read by a human. A mathematical process called a Fourier transform must be implemented by computer to extract the individual frequencies that were absorbed and to reconstruct and plot what we recognize as a typical infrared spectrum.

---

[1] The principles of interferometry and the operation of an FT-IR instrument are explained in two articles by W. D. Perkins: "Fourier Transform–Infrared Spectroscopy, Part 1: Instrumentation," *Journal of Chemical Education, 63* (January 1986): A5–A10, and "Fourier Transform–Infrared Spectroscopy, Part 2: Advantages of FT-IR," *Journal of Chemical Education, 64* (November 1987): A269–A271.

Computer-interfaced FT-IR instruments operate in a single-beam mode. To obtain a spectrum of a compound, the chemist first obtains an interferogram of the "background," which consists of the infrared-active atmospheric gases, carbon dioxide and water vapor (oxygen and nitrogen are not infrared active). The interferogram is subjected to a Fourier transform, which yields the spectrum of the background. Then, the chemist places the compound (sample) into the beam and obtains the spectrum resulting from the Fourier transform of the interferogram. This spectrum contains absorption bands for *both the compound and the background*. The computer software automatically subtracts the spectrum of the background from the sample spectrum, yielding the spectrum of the compound being analyzed. The subtracted spectrum is essentially identical to that obtained from a traditional double-beam dispersive instrument. See Section 2.22 for more detailed information about the background spectrum.

2.6 PREPARATION OF SAMPLES FOR INFRARED SPECTROSCOPY

To determine the infrared spectrum of a compound, one must place the compound in a sample holder, or cell. In infrared spectroscopy, this immediately poses a problem. Glass and plastics absorb strongly throughout the infrared region of the spectrum. Cells must be constructed of ionic substances—typically sodium chloride or potassium bromide. Potassium bromide plates are more expensive than sodium chloride plates but have the advantage of usefulness in the range of 4000 to 400 cm^{-1}. Sodium chloride plates are used widely because of their relatively low cost. The practical range for their use in spectroscopy extends from 4000 to 650 cm^{-1}. Sodium chloride begins to absorb at 650 cm^{-1}, and any bands with frequencies less than this value will not be observed. Since few important bands appear below 650 cm^{-1}, sodium chloride plates are in most common use for routine infrared spectroscopy.

Liquids. A drop of a liquid organic compound is placed between a pair of polished sodium chloride or potassium bromide plates, referred to as **salt plates.** When the plates are squeezed gently, a thin liquid film forms between them. A spectrum determined by this method is referred to as a **neat** spectrum since no solvent is used. Salt plates break easily and are water soluble. Organic compounds analyzed by this technique must be free of water. The pair of plates is inserted into a holder that fits into the spectrometer.

Solids. There are several methods for determining infrared spectra for solids. One method of choice has been to mix a finely ground sample with powdered potassium bromide and press the mixture under high pressure. Under pressure, the potassium bromide melts and seals the sample into a matrix. The resulting **KBr pellet** is inserted in the instrument. If a good pellet is prepared, the spectrum obtained will have no interfering bands since potassium bromide is transparent down to 400 cm^{-1}.

Another method, a **Nujol mull,** involves grinding the compound with mineral oil (Nujol) to create a suspension of the finely divided sample dispersed in mineral oil. The thick suspension is placed between two salt plates. The main disadvantage of this method is that the mineral oil obscures bands that may be present in the analyzed compound. Nujol hands appear at 2924, 1462, and 1377 cm^{-1} (Fig. 2.8).

The best method of determining a spectrum of a solid is to make use of an **attenuated total reflectance (ATR) accessory**. Modern FT-IR instruments now offer this accessory along with the typical transmittance module. The ATR method provides a powerful sampling technique that virtually eliminates sample preparation with both liquids and solids, thus leading to rapid analysis of samples. Although manufacturers offer multiple crystal options, the diamond ATR offers the best option for maximum durability in the organic chemistry laboratory. With the ATR accessory, one

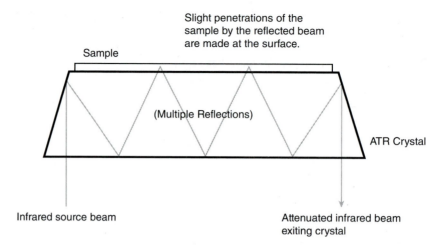

simply places a small amount of a liquid or solid directly on the diamond without any previous preparation. The diamond is cut and mounted at precise angles so that the incoming beam of infrared radiation bounces back and forth off the internal surfaces of the crystal. When the beam reflects off the surface where the sample has been applied, it penetrates the sample slightly, and the vibrational frequencies of the sample are somewhat absorbed, thereby "attenuating" the beam.

The ATR accessory has revolutionized the ease of infrared spectral analysis of solids. For example, it is often unnecessary to make use of KBr pellets and Nujol mulls. The spectrum obtained with an ATR FT-IR is nearly identical to that obtained with an FT-IR operating in the transmittance mode. One may observe some differences in the relative intensities of the peaks, but the peak position in wavenumbers is identical in both modes.[2] ATR FT-IR does not require a clear sample that allows light to pass through the sample, such as is common with transmittance instruments. There are some limitations with a diamond ATR instrument. Some materials such as coatings on metal and very dark samples do not analyze satisfactorily, but there are few other limitations.

2.7 WHAT TO LOOK FOR WHEN EXAMINING INFRARED SPECTRA

An infrared spectrometer determines the positions and relative sizes of all the absorptions, or peaks, in the infrared region and plots them on a piece of paper. This plot of absorption intensity versus wavenumber (or sometimes wavelength) is referred to as the **infrared spectrum** of the compound. Figure 2.4 shows a typical infrared spectrum, that of 3-methyl-2-butanone. The spectrum exhibits at least two strongly absorbing peaks at about 3000 and 1715 cm^{-1} for the C—H and C=O stretching frequencies, respectively.

The strong absorption at 1715 cm^{-1} that corresponds to the carbonyl group (C=O) is quite intense. In addition to the characteristic position of absorption, the *shape* and *intensity* of this peak are also unique to the C=O bond. This is true for almost every type of absorption peak; both shape and intensity characteristics can be described, and these characteristics often enable the chemist to

[2] Shuttlefield, J. D., and V. H. Grassian, "ATR-FTIR in the Undergraduate Chemistry Laboratory, Part 1: Fundamentals and Examples." *Journal of Chemical Education*, 85 (2008): 279–281.

distinguish the peak in potentially confusing situations. For instance, to some extent C=O and C=C bonds absorb in the same region of the infrared spectrum:

$$C=O \qquad 1850–1630 \text{ cm}^{-1}$$

$$C=C \qquad 1680–1620 \text{ cm}^{-1}$$

However, the C=O bond is a strong absorber, whereas the C=C bond generally absorbs only weakly (Fig. 2.5). Hence, trained observers would not interpret a strong peak at 1670 cm^{-1} to be a C=C double bond or a weak absorption at this frequency to be due to a carbonyl group.

The shape and fine structure of a peak often give clues to its identity as well. Thus, although the N—H and O—H regions overlap,

$$O–H \qquad 3650–3200 \text{ cm}^{-1}$$

$$N–H \qquad 3500–3300 \text{ cm}^{-1}$$

FIGURE 2.4 The infrared spectrum of 3-methyl-2-butanone (neat liquid, KBr plates).

FIGURE 2.5 A comparison of the intensities of the C=O and C=C absorption bands.

FIGURE 2.6 A comparison of the shapes of the absorption bands for the O—H and N—H groups.

the N—H absorption usually has one or two *sharp* absorption bands of lower intensity, whereas O—H, when it is in the N—H region, usually gives a *broad* absorption peak. Also, primary amines give *two* absorptions in this region, whereas alcohols as pure liquids give only one (Fig. 2.6). Figure 2.6 also shows typical patterns for the C—H stretching frequencies at about 3000 cm^{-1}.

Therefore, while you study the sample spectra in the pages that follow, take notice of shapes and intensities. They are as important as the frequency at which an absorption occurs, and the eye must be trained to recognize these features. Often, when reading the literature of organic chemistry, you will find absorptions referred to as strong (s), medium (m), weak (w), broad, or sharp. The author is trying to convey some idea of what the peak looks like without actually drawing the spectrum.

2.8 CORRELATION CHARTS AND TABLES

To extract structural information from infrared spectra, you must be familiar with the frequencies at which various functional groups absorb. You may consult **infrared correlation tables,** which provide as much information as is known about where the various functional groups absorb. The references listed at the end of this chapter contain extensive series of correlation tables. Sometimes, the absorption information is presented in the form of a chart called a **correlation chart.** Table 2.3 is a simplified correlation table; a more detailed chart appears in Appendix 1.

The volume of data in Table 2.3 looks as though it may be difficult to assimilate. However, it is really quite easy if you start simply and then slowly increase your familiarity with and ability to interpret the finer details of an infrared spectrum. You can do this most easily by first establishing the broad visual patterns of Figure 2.2 quite firmly in mind. Then, as a second step, memorize a "typical absorption value"—a single number that can be used as a pivotal value— for each of the functional groups in this pattern. For example, start with a simple aliphatic ketone as a model for all typical carbonyl compounds. The typical aliphatic ketone has a carbonyl absorption of about 1715 ± 10 cm^{-1}. Without worrying about the variation, memorize 1715 cm^{-1} as the base value for carbonyl absorption. Then, more slowly, familiarize yourself with the extent of the carbonyl range and the visual pattern showing where the different kinds of carbonyl groups appear throughout this region. See, for instance, Section 2.14, which gives typical values for the various types of carbonyl compounds. Also, learn how factors such as ring strain and conjugation affect the base values (i.e., in which direction the values are shifted). Learn the trends, always keeping the memorized base value (1715 cm^{-1}) in mind. As a beginning, it might prove useful to memorize the base values for this approach given in Table 2.4. Notice that there are only eight of them.

TABLE 2.3
A SIMPLIFIED CORRELATION CHART

Type of Vibration			Frequency (cm^{-1})	Intensity	Page Reference
C−H	Alkanes	(stretch)	3000–2850	s	31
	−CH$_3$	(bend)	1450 and 1375	m	
	−CH$_2$−	(bend)	1465	m	
	Alkenes	(stretch)	3100–3000	m	33
		(out-of-plane bend)	1000–650	s	
	Aromatics	(stretch)	3150–3050	s	43
		(out-of-plane bend)	900–690	s	
	Alkyne	(stretch)	ca. 3300	s	35
	Aldehyde		2900–2800	w	56
			2800–2700	w	
C−C	Alkane		Not interpretatively useful		
C=C	Alkene		1680–1600	m–w	33
	Aromatic		1600 and 1475	m–w	43
C≡C	Alkyne		2250–2100	m–w	35
C=O	Aldehyde		1740–1720	s	56
	Ketone		1725–1705	s	58
	Carboxylic acid		1725–1700	s	62
	Ester		1750–1730	s	64
	Amide		1700–1640	s	70
	Anhydride		1810 and 1760	s	73
	Acid chloride		1800	s	72
C−O	Alcohols, ethers, esters, carboxylic acids, anhydrides		1300–1000	s	47, 50, 62, 64, and 73
O−H	Alcohols, phenols				
	Free		3650–3600	m	47
	H-bonded		3400–3200	m	47
	Carboxylic acids		3400–2400	m	62
N−H	Primary and secondary amines and amides				
	(stretch)		3500–3100	m	74
	(bend)		1640–1550	m–s	74
C−N	Amines		1350–1000	m–s	74
C=N	Imines and oximes		1690–1640	w–s	77
			2260–2240	m	77
			2270–1940	m–s	77
			1550 and 1350	s	79
			2550	w	81
			1050	s	81
	...ides		1375–1300 and 1350–1140	s	82
C−X	Fluoride		1400–1000	s	85
	Chloride		785–540	s	85
	Bromide, iodide		< 667	s	85

T A B L E 2.4
BASE VALUES FOR ABSORPTIONS OF BONDS

O—H	3400 cm^{-1}	C≡C	2150 cm^{-1}
N—H	3400	C=O	1715
C—H	3000	C=C	1650
C≡N	2250	C—O	1100

2.9 HOW TO APPROACH THE ANALYSIS OF A SPECTRUM (*OR* WHAT YOU CAN TELL AT A GLANCE)

When analyzing the spectrum of an unknown, concentrate your first efforts on determining the presence (or absence) of a few major functional groups. The C=O, O—H, N—H, C—O, C=C, C≡C, C≡N, and NO_2 peaks are the most conspicuous and give immediate structural information if they are present. Do not try to make a detailed analysis of the C—H absorptions near 3000 cm^{-1}; almost all compounds have these absorptions. Do not worry about subtleties of the exact environment in which the functional group is found. Following is a major checklist of the important gross features.

1. Is a carbonyl group present? The C=O group gives rise to a strong absorption in the region $1820-1660 \text{ cm}^{-1}$. The peak is often the strongest in the spectrum and of medium width. You can't miss it.

2. If C=O is present, check the following types (if it is absent, go to step 3):

   ACIDS   Is O—H also present?
   - *Broad* absorption near $3400-2400 \text{ cm}^{-1}$ (usually overlaps C—H).

   AMIDES   Is N—H also present?
   - Medium absorption near 3400 cm^{-1}; sometimes a double peak with equivalent halves.

   ESTERS   Is C—O also present?
   - Strong-intensity absorptions near $1300-1000 \text{ cm}^{-1}$.

   ANHYDRIDES   *Two* C=O absorptions near 1810 and 1760 cm^{-1}.

   ALDEHYDES   Is aldehyde C—H present?
   - Two weak absorptions near 2850 and 2750 cm^{-1} on right side of the aliphatic C—H absorptions.

   KETONES   The preceding five choices have been eliminated.

3. If C=O is absent:

   ALCOHOLS, PHENOLS   Check for O—H.
   - *Broad* absorption near $3400-3300 \text{ cm}^{-1}$.
   - Confirm this by finding C—O near $1300-1000 \text{ cm}^{-1}$.

   AMINES   Check for N—H.
   - Medium absorption(s) near 3400 cm^{-1}.

   ETHERS   Check for C—O near $1300-1000 \text{ cm}^{-1}$ (and absence of O—H near 3400 cm^{-1}).

4. Double bonds and/or aromatic rings

- C=C is a weak absorption near 1650 cm.
- Medium-to-strong absorptions in the region $1600-1450$ cm^{-1}; these often imply an aromatic ring.
- Confirm the double bond or aromatic ring by consulting the C$-$H region; aromatic and vinyl C$-$H occur to the left of 3000 cm^{-1} (aliphatic C$-$H occurs to the right of this value).

5. Triple bonds

- C≡N is a medium, sharp absorption near 2250 cm^{-1}.
- C≡C is a weak, sharp absorption near 2150 cm^{-1}.
- Check also for acetylenic C$-$H near 3300 cm^{-1}.

6. Nitro groups

- Two strong absorptions at $1600-1530$ cm^{-1} and $1390-1300$ cm^{-1}.

7. Hydrocarbons

- None of the preceding is found.
- Major absorptions are in C$-$H region near 3000^{-1}.
- Very simple spectrum; the only other absorptions appear near 1460 and 1375 cm^{-1}.

The beginning student should resist the idea of trying to assign or interpret *every* peak in the spectrum. You simply will not be able to do it. Concentrate first on learning these *major* peaks and recognizing their presence or absence. This is best done by carefully studying the illustrative spectra in the sections that follow.

A SURVEY OF THE IMPORTANT FUNCTIONAL GROUPS, WITH EXAMPLES

The following sections describe the behaviors of important functional groups toward infrared radiation. These sections are organized as follows:

1. The *basic* information about the functional group or type of vibration is abstracted and placed in a **Spectral Analysis Box,** where it may be consulted easily.

2. Examples of spectra follow the basic section. The *major* absorptions of diagnostic value are indicated on each spectrum.

3. Following the spectral examples, a discussion section provides details about the functional groups and other information that may be of use in identifying organic compounds.

2.10 HYDROCARBONS: ALKANES, ALKENES, AND ALKYNES

A. Alkanes

Alkanes show very few absorption bands in the infrared spectrum. They yield four or more C$-$H stretching peaks near 3000 cm^{-1} plus CH$_2$ and CH$_3$ bending peaks in the range $1475-1365$ cm^{-1}.

SPECTRAL ANALYSIS BOX

ALKANES

The spectrum is usually simple, with few peaks.

C—H   Stretch occurs around 3000 cm^{-1}.

In alkanes (except strained ring compounds), sp^3 C—H absorption always occurs at frequencies less than 3000 cm^{-1} (3000–2840 cm^{-1}).

If a compound has vinylic, aromatic, acetylenic, or cyclopropyl hydrogens, the C—H absorption is greater than 3000 cm^{-1}. These compounds have sp^2 and sp hybridizations (see Sections 2.10B and 2.10C).

CH$_2$ Methylene groups have a characteristic bending absorption of approximately 1465 cm^{-1}.

CH$_3$ Methyl groups have a characteristic bending absorption of approximately 1375 cm^{-1}.

CH$_2$ The bending (rocking) motion associated with four or more CH$_2$ groups in an open chain occurs at about 720 cm^{-1} (called a *long-chain band*).

C—C   Stretch not interpretatively useful; many weak peaks.

Examples: decane (Fig. 2.7), mineral oil (Fig. 2.8), and cyclohexane (Fig. 2.9).

FIGURE 2.7 The infrared spectrum of decane (neat liquid, KBr plates).

FIGURE 2.8 The infrared spectrum of mineral oil (neat liquid, KBr plates).

FIGURE 2.9 The infrared spectrum of cyclohexane (neat liquid, KBr plates).

B. Alkenes

Alkenes show many more peaks than alkanes. The principal peaks of diagnostic value are the C—H stretching peaks for the sp^2 carbon at values greater than 3000 cm^{-1}, along with C—H peaks for the sp^3 carbon atoms appearing below that value. Also prominent are the out-of-plane bending peaks that appear in the range 1000–650 cm^{-1}. For unsymmetrical compounds, you should expect to see the C=C stretching peak near 1650 cm^{-1}.

SPECTRAL ANALYSIS BOX

ALKENES

=C—H   Stretch for sp^2 C—H occurs at values greater than 3000 cm^{-1} (3095–3010 cm^{-1}).

=C—H   Out-of-plane (oop) bending occurs in the range 1000–650 cm^{-1}.

These bands can be used to determine the degree of substitution on the double bond (see discussion).

C=C   Stretch occurs at 1660–1600 cm^{-1}; conjugation moves C=C stretch to lower frequencies and increases the intensity.

Symmetrically substituted bonds (e.g., 2,3-dimethyl-2-butene) do not absorb in the infrared (no dipole change).

Symmetrically disubstituted (*trans*) double bonds are often vanishingly weak in absorption; *cis* are stronger.

Examples: 1-hexene (Fig. 2.10), cyclohexene (Fig. 2.11), *cis*-2-pentene (Fig. 2.12), and *trans*-2-pentene (Fig. 2.13).

FIGURE 2.10 The infrared spectrum of 1-hexene (neat liquid, KBr plates).

FIGURE 2.11 The infrared spectrum of cyclohexene (neat liquid, KBr plates).

FIGURE 2.12 The infrared spectrum of *cis*-2-pentene (neat liquid, KBr plates).

FIGURE 2.13 The infrared spectrum of *trans*-2-pentene (neat liquid, KBr plates).

C. Alkynes

Terminal alkynes will show a prominent peak at about 3300 cm^{-1} for the *sp*-hybridized C—H. A C≡C will also be a prominent feature in the spectrum for the terminal alkyne, appearing at about 2150 cm^{-1}. The alkyl chain will show C—H stretching frequencies for the *sp*^{3} carbon atoms. Other features include the bending bands for CH$_2$ and CH$_3$ groups. Nonterminal alkynes will not show the C—H band at 3300 cm^{-1}. The C≡C at 2150 cm^{-1} will be very weak or absent from the spectrum.

SPECTRAL ANALYSIS BOX

ALKYNES

≡C—H   Stretch for *sp* C—H usually occurs near 3300 cm^{-1}.

C≡C   Stretch occurs near 2150 cm^{-1}; conjugation moves stretch to lower frequency. Disubstituted or symmetrically substituted triple bonds give either no absorption or weak absorption.

Examples: 1-octyne (Fig. 2.14) and 4-octyne (Fig. 2.15).

FIGURE 2.14 The infrared spectrum of 1-octyne (neat liquid, KBr plates).

FIGURE 2.15 The infrared spectrum of 4-octyne (neat liquid, KBr plates).

DISCUSSION SECTION

C−H Stretch Region

The C−H stretching and bending regions are two of the most difficult regions to interpret in infrared spectra. The C−H stretching region, which ranges from 3300 to 2750 cm^{-1}, is generally the more useful of the two. As discussed in Section 2.4, the frequency of the absorption of C−H bonds is a function mostly of the type of hybridization that is attributed to the bond. The sp-$1s$ C−H bond present in acetylenic compounds is stronger than the sp^2-$1s$ bond present in C=C double-bond compounds (vinyl compounds). This strength results in a larger vibrational force constant and a higher frequency of vibration. Likewise, the sp^2-$1s$ C−H absorption in vinyl compounds occurs at a higher frequency than the sp^3-$1s$ C−H absorption in saturated aliphatic compounds. Table 2.5 gives some physical constants for various C−H bonds involving sp-, sp^2-, and sp^3-hybridized carbon.

As Table 2.5 demonstrates, the frequency at which the C−H absorption occurs indicates the type of carbon to which the hydrogen is attached. Figure 2.16 shows the entire C−H stretching region. Except for the aldehyde hydrogen, an absorption frequency of less than 3000 cm^{-1} usually implies a saturated compound (only sp^3-$1s$ hydrogens). An absorption frequency higher than 3000 cm^{-1} but not above about 3150 cm^{-1} usually implies aromatic or vinyl hydrogens. However, cyclopropyl C−H bonds, which have extra s character because of the need to put more p character into the ring C−C bonds to reduce angle distortion, also give rise to absorption in the region of 3100 cm^{-1}. Cyclopropyl hydrogens can easily be distinguished from aromatic hydrogens or vinyl hydrogens by cross-reference to the C=C and C−H out-of-plane regions. The aldehyde C−H stretch appears at lower frequencies than the saturated C−H absorptions and normally consists of two weak

TABLE 2.5

PHYSICAL CONSTANTS FOR sp-, sp^2-, AND sp^3-HYBRIDIZED CARBON AND THE RESULTING C−H ABSORPTION VALUES

Bond	≡C−H	=C−H	−C−H
Type	sp-$1s$	sp^2-$1s$	sp^3-$1s$
Length	1.08 Å	1.10 Å	1.12 Å
Strength	506 kJ	444 kJ	422 kJ
IR frequency	3300 cm^{-1}	~3100 cm^{-1}	~2900 cm^{-1}

FIGURE 2.16 The C—H stretch region.

absorptions at about 2850 and 2750 cm^{-1}. The 2850-cm^{-1} band usually appears as a shoulder on the saturated C—H absorption bands. The band at 2750 cm^{-1} is rather weak and may be missed in an examination of the spectrum. However, it appears at lower frequencies than aliphatic sp^3 C—H bands. If you are attempting to identify an aldehyde, look for this pair of weak but very diagnostic bands for the aldehyde C—H stretch.

Table 2.6 lists the sp^3-hybridized C—H stretching vibrations for methyl, methylene, and methine. The tertiary C—H (methine hydrogen) gives only one weak C—H stretch absorption, usually near 2890 cm^{-1}. Methylene hydrogens (—CH$_2$—), however, give rise to two C—H stretching bands, representing the symmetric (sym) and asymmetric (asym) stretching modes of the group. In effect, the 2890-cm^{-1} methine absorption is split into two bands at 2926 cm^{-1} (asym) and 2853 cm^{-1} (sym). The asymmetric mode generates a larger dipole moment and is of greater intensity than the symmetric mode. The splitting of the 2890-cm^{-1} methine absorption is larger in the case of a methyl group. Peaks appear at about 2962 and 2872 cm^{-1}. Section 2.3 showed the asymmetric and symmetric stretching modes for methylene and methyl.

Since several bands may appear in the C—H stretch region, it is probably a good idea to decide only whether the absorptions are acetylenic (3300 cm^{-1}), vinylic or aromatic (> 3000 cm^{-1}), aliphatic (< 3000 cm^{-1}), or aldehydic (2850 and 2750 cm^{-1}). Further interpretation of C—H stretching vibrations may not be worth extended effort. The C—H *bending vibrations* are often more useful for determining whether methyl or methylene groups are present in a molecule.

TABLE 2.6
STRETCHING VIBRATIONS FOR VARIOUS sp^3-HYBRIDIZED C—H BONDS

Group		Stretching Vibration (cm^{-1})	
		Asymmetric	Symmetric
Methyl	CH$_3$—	2962	2872
Methylene	—CH$_2$—	2926	2853
Methine	—C— (with | above and below, H below)	2890	Very weak

C—H Bending Vibrations for Methyl and Methylene

The presence of methyl and methylene groups, when not obscured by other absorptions, may be determined by analyzing the region from 1465 to 1370 cm^{-1}. As shown in Figure 2.17, the band due to CH_2 scissoring usually occurs at 1465 cm^{-1}. One of the bending modes for CH_3 usually absorbs strongly near 1375 cm^{-1}. These two bands can often be used to detect methylene and methyl groups, respectively. Furthermore, the 1375-cm^{-1} methyl band is usually split into *two* peaks of nearly equal intensity (symmetric and asymmetric modes) if a geminal dimethyl group is present. This doublet is often observed in compounds with isopropyl groups. A *tert*-butyl group results in an even wider splitting of the 1375-cm^{-1} band into two peaks. The 1370-cm^{-1} band is more intense than the 1390-cm^{-1} one. Figure 2.18 shows the expected patterns for the isopropyl and *tert*-butyl groups. Note that some variation from these idealized patterns may occur. Nuclear magnetic resonance spectroscopy may be used to confirm the presence of these groups. In cyclic hydrocarbons, which do not have attached methyl groups, the 1375-cm^{-1} band is missing, as can be seen in the spectrum of cyclohexane (Fig. 2.9). Finally, a rocking band (Section 2.3) appears near 720 cm^{-1} for long-chain alkanes of four carbons or more (see Fig. 2.7).

C=C Stretching Vibrations

Simple Alkyl-Substituted Alkenes. The C=C stretching frequency usually appears between 1670 and 1640 cm^{-1} for simple noncyclic (acyclic) alkenes. The C=C frequencies increase as alkyl groups are added to a double bond. For example, simple monosubstituted alkenes yield values near 1640 cm^{-1}, 1,1-disubstituted alkenes absorb at about 1650 cm^{-1}, and tri- and tetrasubstituted alkenes absorb near 1670 cm^{-1}. *Trans*-Disubstituted alkenes absorb at higher frequencies (1670 cm^{-1})

F I G U R E   2.17 The C—H bending vibrations for methyl and methylene groups.

F I G U R E   2.18 C—H bending patterns for the isopropyl and *tert*-butyl groups.

than *cis*-disubstituted alkenes (1658 cm^{-1}). Unfortunately, the C=C group has a rather weak intensity, certainly much weaker than a typical C=O group. In many cases, such as in tetrasubstituted alkenes, the double bond absorption may be so weak that it is not observed at all. Recall from Section 2.1 that if the attached groups are arranged symmetrically, no change in dipole moment occurs during stretching, and hence no infrared absorption is observed. *Cis*-Alkenes, which have less symmetry than *trans*-alkenes, generally absorb more strongly than the latter. Double bonds in rings, because they are often symmetric or nearly so, absorb more weakly than those not contained in rings. Terminal double bonds in monosubstituted alkenes generally have stronger absorption.

Conjugation Effects. Conjugation of a C=C double bond with either a carbonyl group or another double bond provides the multiple bond with more single-bond character (through resonance, as the following example shows), a lower force constant K, and thus a lower frequency of vibration. For example, the vinyl double bond in styrene gives an absorption at 1630 cm^{-1}.

With several double bonds, the number of C=C absorptions often corresponds to the number of conjugated double bonds. An example of this correspondence is found in 1,3-pentadiene, where absorptions are observed at 1600 and 1650 cm^{-1}. In the exception to the rule, butadiene gives only one band near 1600 cm^{-1}. If the double bond is conjugated with a carbonyl group, the C=C absorption shifts to a lower frequency and is also intensified by the strong dipole of the carbonyl group. Often, two closely spaced C=C absorption peaks are observed for these conjugated systems, resulting from two possible conformations.

Ring-Size Effects with Internal Double Bonds. The absorption frequency of *internal (endocyclic)* double bonds in cyclic compounds is very sensitive to ring size. As shown in Figure 2.19, the absorption frequency decreases as the internal angle decreases, until it reaches a minimum at 90° in cyclobutene. The frequency increases again for cyclopropene when the angle drops to 60°. This initially unexpected increase in frequency occurs because the C=C vibration in cyclopropene is strongly coupled to the attached C—C single-bond vibration. When the attached C—C bonds are perpendicular to the C=C axis, as in cyclobutene, their vibrational mode is orthogonal to that of the C=C bond (i.e., on a different axis) and does not couple. When the angle is greater than 90° (120° in the following example), the C—C single-bond stretching vibration can be resolved into two components, one of which is coincident with the direction of the C=C stretch. In the diagram, components **a** and **b** of the C—C stretching vector are shown. Since component **a** is in line with the C=C stretching vector, the C—C and C=C bonds are coupled, leading to a higher frequency of absorption. A similar pattern exists for cyclopropene, which has an angle less than 90°.

1650 cm⁻¹   1646   1611   1566   1656

Endo double bonds

(a) Strain moves the peak to the right.
Anomaly: Cyclopropene

(b) If an endo double bond is at a ring fusion, the absorption moves to the right an amount equivalent to the change that would occur if one carbon were removed from the ring.

e.g.:   ~1611 cm⁻¹

F I G U R E 2.19 C=C stretching vibrations in endocyclic systems.

Significant increases in the frequency of the absorption of a double bond contained in a ring are observed when one or two alkyl groups are attached directly to the double bond. The increases are most dramatic for small rings, especially cyclopropenes. For example, Figure 2.20 shows that the base value of 1656 cm⁻¹ for cyclopropene increases to about 1788 cm⁻¹ when one alkyl group is attached to the double bond; with two alkyl groups the value increases to about 1883 cm⁻¹.

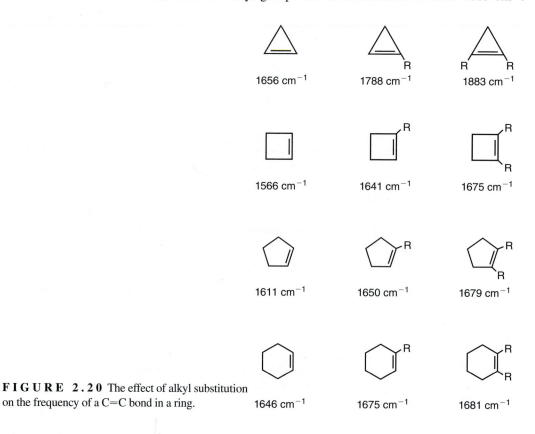

1656 cm⁻¹   1788 cm⁻¹   1883 cm⁻¹

1566 cm⁻¹   1641 cm⁻¹   1675 cm⁻¹

1611 cm⁻¹   1650 cm⁻¹   1679 cm⁻¹

F I G U R E 2.20 The effect of alkyl substitution on the frequency of a C=C bond in a ring.

1646 cm⁻¹   1675 cm⁻¹   1681 cm⁻¹

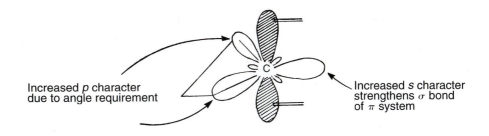

Increased *p* character
due to angle requirement

Increased *s* character
strengthens σ bond
of π system

| 1940 cm⁻¹ | 1780 | 1678 | 1657 | 1655 | 1651 |

$H_2C=C=CH_2$

Allene

Exo double bonds

(a) Strain moves the peak to the left.

(b) Ring fusion moves the absorption to the left.

FIGURE 2.21 C=C stretching vibrations in exocyclic systems.

The figure shows additional examples. It is important to realize that the ring size must be determined before the illustrated rules are applied. Notice, for example, that the double bonds in the 1,2-dialkylcyclopentene and 1,2-dialkylcyclohexene absorb at nearly the same value.

Ring-Size Effects with External Double Bonds. *External (exocyclic)* double bonds give an increase in absorption frequency with decreasing ring size, as shown in Figure 2.21. Allene is included in the figure because it is an extreme example of an *exo* double-bond absorption. Smaller rings require the use of more *p* character to make the C—C bonds form the requisite small angles (recall the trend: $sp = 180°$, $sp^2 = 120°$, $sp^3 = 109°$, $sp^{>3} = <109°$). This removes *p* character from the sigma bond of the double bond but gives it more *s* character, thus strengthening and stiffening the double bond. The force constant K is then increased, and the absorption frequency increases.

C—H Bending Vibrations for Alkenes

The C—H bonds in alkenes can vibrate by bending both in plane and out of plane when they absorb infrared radiation. The scissoring in-plane vibration for terminal alkenes occurs at about 1415 cm⁻¹. This band appears at this value as a medium-to-weak absorption for both monosubstituted and 1,1-disubstituted alkenes.

The most valuable information for alkenes is obtained from analysis of the C—H out-of-plane region of the spectrum, which extends from 1000 to 650 cm⁻¹. These bands are frequently the strongest peaks in the spectrum. The number of absorptions and their positions in the spectrum can be used to indicate the substitution pattern on the double bond.

C—H out-of-plane bending

Monosubstituted Double Bonds (Vinyl). This substitution pattern gives rise to two strong bands, one near 990 cm^{-1} and the other near 910 cm^{-1} for alkyl-substituted alkenes. An overtone of the 910-cm^{-1} band usually appears at 1820 cm^{-1} and helps confirm the presence of the vinyl group. The 910-cm^{-1} band is shifted to a lower frequency, as low as 810 cm^{-1}, when a group attached to the double bond can release electrons by a resonance effect (Cl, F, OR). The 910-cm^{-1} group shifts to a higher frequency, as high as 960 cm^{-1}, when the group withdraws electrons by a resonance effect (C=O, C≡N). The use of the out-of-plane vibrations to confirm the monosubstituted structure is considered very reliable. The absence of these bands almost certainly indicates that this structural feature is not present within the molecule.

cis- **and** *trans-1,2-Disubstituted Double Bonds.* A *cis* arrangement about a double bond gives one strong band near 700 cm^{-1}, while a *trans* double bond absorbs near 970 cm^{-1}. This kind of information can be valuable in the assignment of stereochemistry about the double bond (see Figs. 2.12 and 2.13).

1,1-Disubstituted Double Bonds. One strong band near 890 cm^{-1} is obtained for a *gem*-dialkyl-substituted double bond. When electron-releasing or electron-withdrawing groups are attached to the double bond, shifts similar to that just given for monosubstituted double bonds are observed.

Trisubstituted Double Bonds. One medium-intensity band near 815 cm^{-1} is obtained.

Tetrasubstituted Double Bonds. These alkenes do not give any absorption in this region because of the absence of a hydrogen atom on the double bond. In addition, the C=C stretching vibration is very weak (or absent) at about 1670 cm^{-1} in these highly substituted systems.

Figure 2.22 shows the C—H out-of-plane bending vibrations for substituted alkenes, together with the frequency ranges.

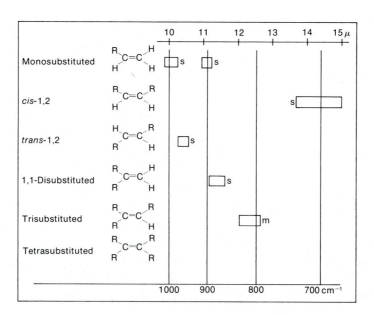

FIGURE 2.22 The C—H out-of-plane bending vibrations for substituted alkenes.

2.11 AROMATIC RINGS

Aromatic compounds show a number of absorption bands in the infrared spectrum, many of which are not of diagnostic value. The C—H stretching peaks for the sp^2 carbon appear at values greater than 3000 cm^{-1}. Since C—H stretching bands for alkenes appear in the same range, it may be difficult to use the C—H stretching bands to differentiate between alkenes and aromatic compounds. However, the C=C stretching bands for aromatic rings usually appear between 1600 and 1450 cm^{-1} outside the usual range where the C=C appears for alkenes (1650 cm^{-1}). Also prominent are the out-of-plane bending peaks that appear in the range 900–690 cm^{-1}, which, along with weak overtone bands at 2000–1667 cm^{-1}, can be used to assign substitution on the ring.

SPECTRAL ANALYSIS BOX

AROMATIC RINGS

=C—H   Stretch for sp^2 C—H occurs at values greater than 3000 cm^{-1} (3050–3010 cm^{-1}).

=C—H   Out-of-plane (oop) bending occurs at 900–690 cm^{-1}. These bands can be used with great utility to assign the ring substitution pattern (see discussion).

C=C   Ring stretch absorptions often occur in pairs at 1600 cm^{-1} and 1475 cm^{-1}.

Overtone/combination bands appear between 2000 and 1667 cm^{-1}. These *weak* absorptions can be used to assign the ring substitution pattern (see discussion).

Examples: toluene (Fig. 2.23), *ortho*-diethylbenzene (Fig. 2.24), *meta*-diethylbenzene (Fig. 2.25), *para*-diethylbenzene (Fig. 2.26), and styrene (Fig. 2.27).

FIGURE 2.23 The infrared spectrum of toluene (neat liquid, KBr plates).

F I G U R E   2 . 2 4 The infrared spectrum of *ortho*-diethylbenzene (neat liquid, KBr plates).

F I G U R E   2 . 2 5 The infrared spectrum of *meta*-diethylbenzene (neat liquid, KBr plates).

F I G U R E   2 . 2 6 The infrared spectrum of *para*-diethylbenzene (neat liquid, KBr plates).

FIGURE 2.27 The infrared spectrum of styrene (neat liquid, KBr plates).

DISCUSSION SECTION

C—H Bending Vibrations

The in-plane C—H bending vibrations occur between 1300 and 1000 cm^{-1}. However, these bands are rarely useful because they overlap other, stronger absorptions that occur in this region.

The out-of-plane C—H bending vibrations, which appear between 900 and 690 cm^{-1}, are far more useful than the in-plane bands. These extremely intense absorptions, resulting from strong coupling with adjacent hydrogen atoms, can be used to assign the positions of substituents on the aromatic ring. The assignment of structure based on these out-of-plane bending vibrations is most reliable for alkyl-, alkoxy-, halo-, amino-, or carbonyl-substituted aromatic compounds. Aromatic nitro compounds, derivatives of aromatic carboxylic acids, and derivatives of sulfonic acids sometimes lead to unsatisfactory interpretation.

Reliable interpretation   Unreliable interpretation

Monosubstituted Rings. This substitution pattern always gives a strong absorption near 690 cm^{-1}. If this band is absent, no monosubstituted ring is present. A second strong band usually appears near 750 cm^{-1}. When the spectrum is taken in a halocarbon solvent, the 690-cm^{-1} band may be obscured by the strong C—X stretch absorptions. The typical two-peak monosubstitution pattern appears in the spectra of toluene (Fig. 2.23) and styrene (Fig. 2.27). In addition, the spectrum of styrene shows a pair of bands for the vinyl out-of-plane bending modes.

ortho-Disubstituted Rings (1,2-Disubstituted Rings). One strong band near 750 cm^{-1} is obtained. This pattern is seen in the spectrum of *ortho*-diethylbenzene (Fig. 2.24).

meta-Disubstituted Rings (1,3-Disubstituted Rings). This substitution pattern gives the 690-cm^{-1} band plus one near 780 cm^{-1}. A third band of medium intensity is often found near 880 cm^{-1}. This pattern is seen in the spectrum of *meta*-diethylbenzene (Fig. 2.25).

para-*Disubstituted Rings (1,4-Disubstituted Rings).* One strong band appears in the region from 800 to 850 cm^{-1}. This pattern is seen in the spectrum of *para*-diethylbenzene (Fig. 2.26).

Figure 2.28a shows the C—H out-of-plane bending vibrations for the common substitution patterns already given plus some others, together with the frequency ranges. Note that the bands appearing in the 720- to 667-cm^{-1} region (shaded boxes) actually result from C=C out-of-plane ring bending vibrations rather than from C—H out-of-plane bending.

Combinations and Overtone Bands

Many *weak* combination and overtone absorptions appear between 2000 and 1667 cm^{-1}. The relative shapes and number of these peaks can be used to tell whether an aromatic ring is mono-, di-, tri-, tetra-, penta-, or hexasubstituted. Positional isomers can also be distinguished. Since the absorptions are weak, these bands are best observed by using neat liquids or concentrated solutions. If the compound has a high-frequency carbonyl group, this absorption will overlap the weak overtone bands so that no useful information can be obtained from the analysis of the region.

Figure 2.28b shows the various patterns obtained in this region. The monosubstitution pattern that appears in the spectra of toluene (Fig. 2.23) and styrene (Fig. 2.27) is particularly useful and

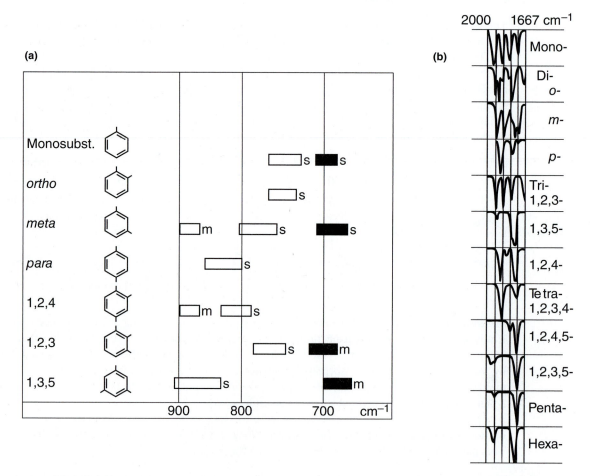

FIGURE 2.28 (a) The C—H out-of-plane bending vibrations for substituted benzenoid compounds. (*s = strong, m = medium*) (b) The 2000- to 1667-cm^{-1} region for substituted benzenoid compounds (from Dyer, John R., *Applications of Absorption Spectroscopy of Organic Compounds,* Prentice–Hall, Englewood Cliffs, N.J., 1965).

helps confirm the out-of-plane data given in the preceding section. Likewise, the *ortho-, meta-,* and *para*-disubstituted patterns may be consistent with the out-of-plane bending vibrations discussed earlier. The spectra of *ortho*-diethylbenzene (Fig. 2.24), *meta*-diethylbenzene (Fig. 2.25), and *para*-diethylbenzene (Fig. 2.26) each show bands in *both* the 2000- to 1667-cm^{-1} and 900- to 690-cm^{-1} regions, consistent with their structures. Note, however, that the out-of-plane vibrations are generally more useful for diagnostic purposes.

2.12 ALCOHOLS AND PHENOLS

Alcohols and phenols will show strong and broad hydrogen-bonded O—H stretching bands centered between 3400 and 3300 cm^{-1}. In solution, it will also be possible to observe a "free" O—H (non H—bonded) stretching band at about 3600 cm^{-1} (sharp and weaker) to the left of the hydrogen-bonded O—H peak. In addition, a C—O stretching band will appear in the spectrum at 1260–1000 cm^{-1}.

SPECTRAL ANALYSIS BOX

ALCOHOLS AND PHENOLS

O—H   The free O—H stretch is a *sharp* peak at 3650–3600 cm^{-1}. This band appears in combination with the hydrogen-bonded O—H peak when the alcohol is dissolved in a solvent (see discussion).

The hydrogen-bonded O—H band is a *broad* peak at 3400–3300 cm^{-1}. This band is usually the only one present in an alcohol that has not been dissolved in a solvent (neat liquid). When the alcohol is dissolved in a solvent, the free O—H and hydrogen-bonded O—H bands are present together, with the relatively weak free O—H on the left (see discussion).

C—O—H   Bending appears as a broad and weak peak at 1440–1220 cm^{-1}, often obscured by the CH$_3$ bendings.

C—O   Stretching vibration usually occurs in the range 1260–1000 cm^{-1}. This band can be used to assign a primary, secondary, or tertiary structure to an alcohol (see discussion).

Examples: The hydrogen-bonded O—H stretch is present in the pure liquid (neat) samples of 1-hexanol (Fig. 2.29), 2-butanol (Fig. 2.30), and *para*-cresol (Fig. 2.31).

FIGURE 2.29 The infrared spectrum of 1-hexanol (neat liquid, KBr plates).

FIGURE 2.30 The infrared spectrum of 2-butanol (neat liquid, KBr plates).

FIGURE 2.31 The infrared spectrum of *para*-cresol (neat liquid, KBr plates).

DISCUSSION SECTION

O—H Stretching Vibrations

When alcohols or phenols are determined as pure (neat) liquid films, as is common practice, a broad O—H stretching vibration is obtained for intermolecular hydrogen bonding in the range from 3400 to 3300 cm^{-1}. Figure 2.32a shows this band, which is observed in the spectra of 1-hexanol (Fig. 2.29) and 2-butanol (Fig. 2.30). Phenols also show the hydrogen-bonded O—H (Fig. 2.31). As the alcohol is diluted with carbon tetrachloride, a sharp "free" (non-hydrogen-bonded) O—H stretching band appears at about 3600 cm^{-1}, to the left of the broad band (Fig. 2.32b). When the solution is further diluted, the broad intermolecular hydrogen-bonded band is reduced considerably, leaving as the major band the free O—H stretching absorption (Fig. 2.32c). Intermolecular hydrogen bonding weakens the O—H bond, thereby shifting the band to lower frequency (lower energy).

Some workers have used the position of the free O—H stretching band to help assign a primary, secondary, or tertiary structure to an alcohol. For example, the free stretch occurs near 3640, 3630, 3620, and 3610 cm^{-1} for primary, secondary, and tertiary alcohols and for phenols, respectively. These absorptions can be analyzed only if the O—H region is expanded and carefully calibrated. Under the usual routine laboratory conditions, these fine distinctions are of little use. Far more useful information is obtained from the C—O stretching vibrations.

FIGURE 2.32 The O—H stretch region. (a) Hydrogen-bonded O—H only (neat liquid). (b) Free and hydrogen-bonded O—H (dilute solution). (c) Free and hydrogen-bonded O—H (very dilute solution).

Intramolecular hydrogen bonding, present in *ortho*-carbonyl-substituted phenols, usually shifts the broad O—H band to a lower frequency. For example, the O—H band is centered at about 3200 cm^{-1} in the neat spectrum of methyl salicylate, while O—H bands from normal phenols are centered at about 3350 cm^{-1}. The intramolecular hydrogen-bonded band does not change its position significantly even at high dilution because the internal bonding is not altered by a change in concentration.

Methyl salicylate

Although phenols often have broader O—H bands than alcohols, it is difficult to assign a structure based on this absorption; use the aromatic C=C region and the C—O stretching vibration (to be discussed shortly) to assign a phenolic structure. Finally, the O—H stretching vibrations in carboxylic acids also occur in this region. They may easily be distinguished from alcohols and phenols by the presence of a *very broad* band extending from 3400 to 2400 cm^{-1} *and* the presence of a carbonyl absorption (see Section 2.14D).

C—O—H Bending Vibrations

This bending vibration is coupled to H—C—H bending vibrations to yield some weak and broad peaks in the 1440 to 1220-cm^{-1} region. These broad peaks are difficult to observe because they are usually located under the more strongly absorbing CH$_3$ bending peaks at 1375 cm^{-1} (see Fig. 2.29).

C—O Stretching Vibrations

The strong C—O single-bond stretching vibrations are observed in the range from 1260 to 1000 cm^{-1}. Since the C—O absorptions are coupled with the adjacent C—C stretching vibrations, the position of the band may be used to assign a primary, secondary, or tertiary structure to an alcohol or to determine whether a phenolic compound is present. Table 2.7 gives the expected absorption bands for the C—O stretching vibrations in alcohols and phenols. For comparison, the O—H stretch values are also tabulated.

TABLE 2.7
C—O AND O—H STRETCHING VIBRATIONS IN ALCOHOLS AND PHENOLS

Compound	C—O Stretch (cm^{-1})	O—H Stretch (cm^{-1})
Phenols	1220	3610
3° Alcohols (saturated)	1150	3620
2° Alcohols (saturated)	1100	3630
1° Alcohols (saturated)	1050	3640

(C—O Stretch: Decrease ↓; O—H Stretch: Increase ↓)

Unsaturation on adjacent carbons or a cyclic structure lowers the frequency of C—O absorption.
2° examples:

1100 → 1070 cm^{-1} 1100 → 1070 cm^{-1} 1100 → 1060 cm^{-1}

1° examples:

1050 → 1017 cm^{-1} 1050 → 1030 cm^{-}

The spectrum of 1-hexanol, a primary alcohol, has its C—O absorption at 1058 cm^{-1} (Fig. 2.29), whereas that of 2-butanol, a secondary alcohol, has its C—O absorption at 1109 cm^{-1} (Fig. 2.30). Thus, both alcohols have their C—O bands near the expected values given in Table 2.7. Phenols give a C—O absorption at about 1220 cm^{-1} because of conjugation of the oxygen with the ring, which shifts the band to higher energy (more double-bond character). In addition to this band, an O—H in-plane bending absorption is usually found near 1360 cm^{-1} for neat samples of phenols. This latter band is also found in alcohols determined as neat (undiluted) liquids. It usually overlaps the C—H bending vibration for the methyl group at 1375 cm^{-1}.

The numbers in Table 2.7 should be considered *base* values. These C—O absorptions are shifted to lower frequencies when unsaturation is present on adjacent carbon atoms or when the O—H is attached to a ring. Shifts of 30 to 40 cm^{-1} from the base values are common, as seen in some selected examples in Table 2.7.

2.13 ETHERS

Ethers show at least one C—O band in the range 1300–1000 cm^{-1}. Simple aliphatic ethers can be distinguished from alkanes by the presence of the C—O band. In all other respects, the spectra of simple ethers look very similar to those of alkanes. Aromatic ethers, epoxides, and acetals are discussed in this section.

SPECTRAL ANALYSIS BOX

ETHERS

C—O   The most prominent band is that due to C—O stretch, 1300–1000 cm^{-1}. Absence of C=O and O—H is required to ensure that C—O stretch is not due to an ester or an alcohol. Phenyl alkyl ethers give two strong bands at about 1250 and 1040 cm^{-1}, while aliphatic ethers give one strong band at about 1120 cm^{-1}.

Examples: dibutyl ether (Fig. 2.33) and anisole (Fig. 2.34).

FIGURE 2.33 The infrared spectrum of dibutyl ether (neat liquid, KBr plates).

FIGURE 2.34 The infrared spectrum of anisole (neat liquid, KBr plates).

DISCUSSION SECTION

Ethers and related compounds such as epoxides, acetals, and ketals give rise to C—O—C stretching absorptions in the range from 1300 to 1000 cm^{-1}. Alcohols and esters also give strong C—O absorptions in this region, and these latter possibilities must be eliminated by observing the absence of bands in the O—H stretch region (Section 2.12) and in the C=O stretch region (Section 2.14), respectively. Ethers are generally encountered more often than epoxides, acetals, and ketals.

R—O—R   Ar—O—R CH_2=CH—O—R   RCH—CHR (epoxide)   R—C—H (R) with O—R above and O—R below

| Dialkyl ethers | Aryl ethers | Vinyl ethers | Epoxides | Acetals (ketals) |

Dialkyl Ethers. The asymmetric C—O—C stretching vibration leads to a single strong absorption that appears at about 1120 cm^{-1}, as seen in the spectrum of dibutyl ether (Fig. 2.33). The symmetric stretching band at about 850 cm^{-1} is usually very weak. The asymmetric C—O—C absorption also occurs at about 1120 cm^{-1} for a six-membered ring containing oxygen.

Aryl and Vinyl Ethers. Aryl alkyl ethers give rise to *two* strong bands: an asymmetric C—O—C stretch near 1250 cm^{-1} and a symmetric stretch near 1040 cm^{-1}, as seen in the spectrum of anisole (Fig. 2.34). Vinyl alkyl ethers also give two bands: one strong band assigned to an asymmetric stretching vibration at about 1220 cm^{-1} and one very weak band due to a symmetric stretch at about 850 cm^{-1}.

The shift in the asymmetric stretching frequencies in aryl and vinyl ethers to values higher than were found in dialkyl ethers can be explained through resonance. For example, the C—O band in vinyl alkyl ethers is shifted to a higher frequency (1220 cm^{-1}) because of the increased double-bond character, which strengthens the bond. In dialkyl ethers the absorption occurs at 1120 cm^{-1}. In addition, because resonance increases the polar character of the C=C double bond, the band at about 1640 cm^{-1} is considerably stronger than in normal C=C absorption (Section 2.10B).

$$\left[CH_2=CH-\ddot{O}-R \longleftrightarrow {}^{-}{:}CH_2-CH=\overset{+}{\underset{\cdot\cdot}{O}}-R \right] \qquad R-\ddot{\underset{\cdot\cdot}{O}}-R$$

| Resonance 1220 cm^{-1} | No resonance 1120 cm^{-1} |

Epoxides. These small-ring compounds give a *weak* ring-stretching band (breathing mode) in the range 1280–1230 cm^{-1}. Of more importance are the two *strong* ring deformation bands, one that appears between 950 and 815 cm^{-1} (asymmetric) and the other between 880 and 750 cm^{-1} (symmetric). For monosubstituted epoxides, this latter band appears in the upper end of the range, often near 835 cm^{-1}. Disubstituted epoxides have absorption in the lower end of the range, closer to 775 cm^{-1}.

Acetals and Ketals. Molecules that contain ketal or acetal linkages often give *four or five strong bands,* respectively, in the region from 1200 to 1020 cm^{-1}. These bands are often unresolved.

2.14 CARBONYL COMPOUNDS

The carbonyl group is present in aldehydes, ketones, acids, esters, amides, acid chlorides, and anhydrides. This group absorbs strongly in the range from 1850 to 1650 cm^{-1} because of its large change in dipole moment. Since the C=O stretching frequency is sensitive to attached atoms, the common functional groups already mentioned absorb at characteristic values. Figure 2.35 provides the normal base values for the C=O stretching vibrations of the various functional groups. The C=O frequency of a ketone, which is approximately in the middle of the range, is usually considered the reference point for comparisons of these values.

				cm^{-1}			
1810	1800	1760	1735	1725	1715	1710	1675
Anhydride (band 1)	Acid chloride	Anhydride (band 2)	Ester	Aldehyde	Ketone	Carboxylic acid	Amide

FIGURE 2.35 Normal base values for the C=O stretching vibrations for carbonyl groups.

The range of values given in Figure 2.35 may be explained through the use of electron-withdrawing effects (inductive effects), resonance effects, and hydrogen bonding. The first two effects operate in opposite ways to influence the C=O stretching frequency. First, an electronegative element may tend to draw in the electrons between the carbon and oxygen atoms through its electron-withdrawing effect, so that the C=O bond becomes somewhat stronger. A higher-frequency (higher-energy) absorption results. Since oxygen is more electronegative than carbon, this effect dominates in an ester to raise the C=O frequency above that of a ketone. Second, a resonance effect may be observed when the unpaired electrons on a nitrogen atom conjugate with the carbonyl group, resulting in increased single-bond character and a lowering of the C=O absorption frequency. This second effect is observed in an amide. Since nitrogen is less electronegative than an oxygen atom, it can more easily accommodate a positive charge. The resonance structure shown here introduces single-bond character into the C=O group and thereby lowers the absorption frequency below that of a ketone.

Ester   Amide

Electron-withdrawing effect raises   Resonance effect lowers C=O frequency
C=O frequency

In acid chlorides, the highly electronegative halogen atom strengthens the C=O bond through an enhanced inductive effect and shifts the frequency to values even higher than are found in esters. Anhydrides are likewise shifted to frequencies higher than are found in esters because of a concentration of electronegative oxygen atoms. In addition, anhydrides give two absorption bands that are due to symmetric and asymmetric stretching vibrations (Section 2.3).

A carboxylic acid exists in monomeric form *only* in very dilute solution, and it absorbs at about 1760 cm^{-1} because of the electron-withdrawing effect just discussed. However, acids in concentrated solution, in the form of neat liquid, or in the solid state (KBr pellet and Nujol) tend to dimerize via hydrogen bonding. This dimerization weakens the C=O bond and lowers the stretching force constant K, resulting in a lowering of the carbonyl frequency of saturated acids to about 1710 cm^{-1}.

Ketones absorb at a lower frequency than aldehydes because of their additional alkyl group, which is electron donating (compared to H) and supplies electrons to the C=O bond. This electron-releasing effect weakens the C=O bond in the ketone and lowers the force constant and the absorption frequency.

A.   Factors That Influence the C=O Stretching Vibration

Conjugation Effects. The introduction of a C=C bond adjacent to a carbonyl group results in delocalization of the π electrons in the C=O and C=C bonds. This conjugation increases the single-bond character of the C=O and C=C bonds in the resonance hybrid and hence lowers their force constants, resulting in a lowering of the frequencies of carbonyl and double-bond absorption. Conjugation with triple bonds also shows this effect.

Generally, the introduction of an α,β double bond in a carbonyl compound results in a 25- to 45-cm^{-1} lowering of the C=O frequency from the base value given in Figure 2.35. A similar lowering occurs when an adjacent aryl group is introduced. Further addition of unsaturation (γ,δ) results in a further shift to lower frequency, but only by about 15 cm^{-1} more. In addition, the C=C absorption shifts from its "normal" value, about 1650 cm^{-1}, to a lower-frequency value of about 1640 cm^{-1}, and the C=C absorption is greatly intensified. Often, two closely spaced C=O absorption peaks are observed for these conjugated systems, resulting from two possible conformations, the *s-cis* and *s-trans*. The *s-cis* conformation absorbs at a frequency higher than the *s-trans* conformation. In some cases, the C=O absorption is broadened rather than split into the doublet.

s-cis   *s-trans*

The following examples show the effects of conjugation on the C=O frequency.

α,β-Unsaturated ketone   Aryl-substituted aldehyde   Aryl-substituted acid
1715 $\longrightarrow$ 1690 cm^{-1} 1725 $\longrightarrow$ 1700 cm^{-1} 1710 $\longrightarrow$ 1680 cm^{-1}

Conjugation does not reduce the C=O frequency in amides. The introduction of α,β unsaturation causes an *increase in frequency* from the base value given in Figure 2.35. Apparently, the introduction of sp^2-hybridized carbon atoms removes electron density from the carbonyl group and strengthens the bond instead of interacting by resonance as in other carbonyl examples. Since the parent amide group is already highly stabilized (see p. 53), the introduction of the C=C unsaturation does not overcome this resonance.

Ring-Size Effects. Six-membered rings with carbonyl groups are unstrained and absorb at about the values given in Figure 2.35. Decreasing the ring size *increases the frequency* of the C=O absorption for the reasons discussed in Section 2.10 (C=C stretching vibrations and exocyclic double bonds). All of the functional groups listed in Figure 2.35, which can form rings, give increased frequencies of absorption with increased angle strain. For ketones and esters, there is often a 30-cm^{-1} increase in frequency for each carbon removed from the unstrained six-membered ring values. Some examples are

Cyclic ketone	Cyclic ketone	Cyclic ester	Cyclic amide
$1715 \rightarrow 1745$ cm^{-1}	$1715 \rightarrow 1780$ cm^{-1}	(lactone)	(lactam)
		$1735 \rightarrow 1770$ cm^{-1}	$1690 \rightarrow 1705$ cm^{-1}

In ketones, larger rings have frequencies that range from nearly the same value as in cyclohexanone (1715 cm^{-1}) to values slightly less than 1715 cm^{-1}. For example, cycloheptanone absorbs at about 1705 cm^{-1}.

α-Substitution Effects. When the carbon next to the carbonyl is substituted with a chlorine (or other halogen) atom, the carbonyl band shifts to a *higher frequency*. The electron-withdrawing effect removes electrons from the carbon of the C=O bond. This removal is compensated for by a tightening of the π bond (shortening), which increases the force constant and leads to an increase in the absorption frequency. This effect holds for all carbonyl compounds.

$$-\overset{\displaystyle }{\underset{\displaystyle \underset{\delta^-X}{\downarrow}}{C}} \leftarrow \overset{\displaystyle \overset{O}{\parallel}}{\underset{\delta^+}{C}}-$$

In ketones, two bands result from the substitution of an adjacent chlorine atom. One arises from the conformation in which the chlorine is rotated next to the carbonyl, and the other is due to the conformation in which the chlorine is away from the group. When the chlorine is next to the carbonyl, nonbonded electrons on the oxygen atom are repelled, resulting in a stronger bond and a higher absorption frequency. Information of this kind can be used to establish a structure in rigid ring systems, such as in the following examples:

Axial chlorine	Equatorial chlorine
~1725 cm^{-1}	~1750 cm^{-1}

Hydrogen-Bonding Effects. Hydrogen bonding to a carbonyl group lengthens the C=O bond and lowers the stretching force constant K, resulting in a *lowering* of the absorption frequency. Examples of this effect are the decrease in the C=O frequency of the carboxylic acid dimer and the lowering of the ester C=O frequency in methyl salicylate caused by intramolecular hydrogen bonding:

Methyl salicylate
1680 cm^{-1}

B.   Aldehydes

Aldehydes show a very strong band for the carbonyl group (C=O) that appears in the range of $1740–1725 \text{ cm}^{-1}$ for simple aliphatic aldehydes. This band is shifted to lower frequencies with conjugation to a C=C or phenyl group. A very important doublet can be observed in the C—H stretch region for the aldehyde C—H near 2850 and 2750 cm^{-1}. The presence of this doublet allows aldehydes to be distinguished from other carbonyl-containing compounds.

SPECTRAL ANALYSIS BOX

ALDEHYDES

C=O

R—C—H
  ‖
  O
C=O stretch appears in the range $1740–1725 \text{ cm}^{-1}$ for normal aliphatic aldehydes.

C=C—C—H (with structure)   Conjugation of C=O with α,β C=C; $1700–1680 \text{ cm}^{-1}$ for C=O and 1640 cm^{-1} for C=C.

Ar—C—H
  ‖
  O
Conjugation of C=O with phenyl; $1700–1660 \text{ cm}^{-1}$ for C=O and $1600–1450 \text{ cm}^{-1}$ for ring.

Ar—C=C—C—H
      ‖
      O
Longer conjugated system; 1680 cm^{-1} for C=O.

C—H   Stretch, aldehyde hydrogen (—CHO), consists of a pair of *weak* bands, one at $2860–2800 \text{ cm}^{-1}$ and the other at $2760–2700 \text{ cm}^{-1}$. It is easier to see the band at the lower frequency because it is not obscured by the usual C—H bands from the alkyl chain. The higher-frequency aldehyde C—H stretch is often buried in the aliphatic C—H bands.

Examples: nonanal (Fig. 2.36), crotonaldehyde (Fig. 2.37), and benzaldehyde (Fig. 2.38).

FIGURE 2.36 The infrared spectrum of nonanal (neat liquid, KBr plates).

FIGURE 2.37 The infrared spectrum of crotonaldehyde (neat liquid, KBr plates).

FIGURE 2.38 The infrared spectrum of benzaldehyde (neat liquid, KBr plates).

DISCUSSION SECTION

The spectrum of nonanal (Fig. 2.36) exhibits the normal aldehyde stretching frequency at 1725 cm^{-1}. Since the positions of these absorptions are not very different from those of ketones, it may not be easy to distinguish between aldehydes and ketones on this basis. Conjugation of the carbonyl group with an aryl or an α,β double bond shifts the normal C=O stretching band to a lower frequency (1700–1680 cm^{-1}), as predicted in Section 2.14A (Conjugation Effects). This effect is seen in crotonaldehyde (Fig. 2.37), which has α,β unsaturation, and in benzaldehyde (Fig. 2.38), in which an aryl group is attached directly to the carbonyl group. Halogenation on the α carbon leads to an increased frequency for the carbonyl group (p. 55).

The C—H stretching vibrations found in aldehydes (—CHO) at about 2750 and 2850 cm^{-1} are extremely important for distinguishing between ketones and aldehydes. Typical ranges for the pairs of C—H bands are 2860–2800 and 2760–2700 cm^{-1}. The band at 2750 cm^{-1} is probably the more useful of the pair because it appears in a region where other C—H absorptions (CH$_3$, CH$_2$, and so on) are absent. The 2850-cm^{-1} band often overlaps other C—H bands and is not as easy to see (see nonanal, Fig. 2.36). If the 2750-cm^{-1} band is present together with the proper C=O absorption value, an aldehyde functional group is almost certainly indicated.

The doublet that is observed in the range 2860–2700 cm^{-1} for an aldehyde is a result of *Fermi resonance* (p. 18). The second band appears when the aldehyde C—H *stretching* vibration is coupled with the first overtone of the medium-intensity aldehyde C—H *bending* vibration appearing in the range 1400–1350 cm^{-1}.

The medium-intensity absorption in nonanal (Fig. 2.36) at 1460 cm^{-1} is due to the scissoring (bending) vibration of the CH$_2$ group next to the carbonyl group. Methylene groups often absorb more strongly when they are attached directly to a carbonyl group.

C. Ketones

Ketones show a very strong band for the C=O group that appears in the range of 1720–1708 cm^{-1} for simple aliphatic ketones. This band is shifted to lower frequencies with conjugation to a C=C or phenyl group. An α-halogen atom will shift the C=O frequency to a higher value. Ring strain moves the absorption to a higher frequency in cyclic ketones.

SPECTRAL ANALYSIS BOX

KETONES

C=O

R—C—H
  ‖
  O

C=O stretch appears in the range 1720–1708 cm^{-1} for normal aliphatic ketones.

$\diagup$C=C—C—H
      | ‖
          O

Conjugation of C=O with α,β C=C; 1700–1675 cm^{-1} for C=O and 1644–1617 cm^{-1} for C=C.

Ar—C—H
   ‖
   O

Conjugation of C=O with phenyl; 1700–1680 cm^{-1} for C=O and 1600–1450 cm^{-1} for ring.

Ar—C—H   Conjugation with two aromatic rings; 1670–1600 cm^{-1} for C=O.
   ‖
   O

C=O   Cyclic ketones; C=O frequency increases with decreasing ring size.

C—C—C   Bending appears as a medium-intensity peak in the range
    ‖
    O   1300–1100 cm^{-1}.

Examples: 3-methyl-2-butanone (Fig. 2.4), mesityl oxide (Fig. 2.39), acetophenone (Fig. 2.40), cyclopentanone (Fig. 2.41), and 2,4-pentanedione (Fig. 2.42).

FIGURE 2.39 The infrared spectrum of mesityl oxide (neat liquid, KBr plates).

FIGURE 2.40 The infrared spectrum of acetophenone (neat liquid, KBr plates).

FIGURE 2.41 The infrared spectrum of cyclopentanone (neat liquid, KBr plates).

FIGURE 2.42 The infrared spectrum of 2,4-pentanedione (neat liquid, KBr plates).

DISCUSSION SECTION

Normal C=O Bands. The spectrum of 3-methyl-2-butanone (Fig. 2.4) exhibits a normal, or uncon-jugated, ketone stretching frequency at 1715 cm^{-1}. A very weak overtone band from the C=O (1715 cm^{-1}) appears at twice the frequency of the C=O absorption (3430 cm^{-1}). Small bands of this type should not be confused with O–H absorptions, which also appear near this value. The O–H stretching absorptions are *much more intense.*

Conjugation Effects. Conjugation of the carbonyl group with an aryl or an α,β double bond shifts the normal C=O stretching band (1715 cm^{-1}) to a lower frequency (1700–1675 cm^{-1}), as predicted in Section 2.14A. Rotational isomers may lead to a splitting or broadening of the carbonyl band. The effect of conjugation on the C=O band is seen in mesityl oxide (Fig. 2.39), which has α,β un-saturation, and in acetophenone (Fig. 2.40), in which an aryl group is attached to the carbonyl group. Both exhibit C=O shifts to lower frequencies. Figure 2.43 presents some typical C=O stretching vibrations, which demonstrate the influence of conjugation.

Cyclic Ketones (Ring Strain). Figure 2.44 provides some values for the C=O absorptions for cyclic ketones. Note that ring strain shifts the absorption values to a higher frequency, as was predicted in

| 5.83 | 5.83 | 5.90 | 5.93 | 5.95 | 6.01 | 6.10 μ |
| 1715 | 1715 | 1695 | 1685 | 1680 | 1665 | 1640 cm^{-1} |

FIGURE 2.43 The C=O stretching vibrations in conjugated ketones.

| 2140 | 1815 | 1780 | 1745 | 1715 | 1715 | 1705 cm^{-1} |

FIGURE 2.44 The C=O stretching vibrations for cyclic ketones and ketene.

Section 2.14A. Ketene is included in Figure 2.44 because it is an extreme example of an *exo* double-bond absorption. The *s* character in the C=O group increases as the ring size decreases, until it reaches a maximum value that is found in the *sp*-hybridized carbonyl carbon in ketene. The spectrum of cyclopentanone (Fig. 2.41) shows how ring strain increases the frequency of the carbonyl group.

α-Diketones (1,2-Diketones). Unconjugated diketones that have the two carbonyl groups adjacent to each other show one strong absorption peak at about 1716 cm^{-1}. If the two carbonyl groups are conjugated with aromatic rings, the absorption is shifted to a lower-frequency value, about 1680 cm^{-1}. In the latter case, a narrowly spaced doublet rather than a single peak may be observed due to symmetric and asymmetric absorptions.

β-Diketones (1,3-Diketones). Diketones with carbonyl groups located 1,3 with respect to each other may yield a more complicated pattern than those observed for most ketones (2,4-pentanedione, Fig. 2.42). These *β*-diketones often exhibit tautomerization, which yields an equilibrium mixture of enol and keto tautomers. Since many *β*-diketones contain large amounts of the enol form, you may observe carbonyl peaks for both the enol and keto tautomers.

Keto tautomer

C=O doublet
1723 cm^{-1} (symmetric stretch)
1706 cm^{-1} (asymmetric stretch)

Enol tautomer

C=O (hydrogen bonded), 1622 cm^{-1}
O—H (hydrogen bonded), 3200–2400 cm^{-1}

The carbonyl group in the enol form appearing at about 1622 cm^{-1} is substantially shifted and intensified in comparison to the normal ketone value, 1715 cm^{-1}. The shift is a result of internal hydrogen bonding, as discussed in Section 2.14A. Resonance, however, also contributes to the lowering of the carbonyl frequency in the enol form. This effect introduces single-bond character into the enol form.

A weak, broad O—H stretch is observed for the enol form at 3200–2400 cm^{-1}. Since the keto form is also present, a doublet for the asymmetric and symmetric stretching frequencies is observed for the two carbonyl groups (Fig. 2.42). The relative intensities of the enol and keto carbonyl absorptions depend on the percentages present at equilibrium. Hydrogen-bonded carbonyl groups in enol forms are often observed in the region 1640–1570 cm^{-1}. The keto forms generally appear as doublets in the range from 1730 to 1695 cm^{-1}.

α-Haloketones. Substitution of a halogen atom on the α carbon shifts the carbonyl absorption peak to a higher frequency, as discussed in Section 2.14A. Similar shifts occur with other electron-withdrawing groups, such as an alkoxy group (—O—CH$_3$). For example, the carbonyl group in chloroacetone appears at 1750 cm^{-1}, whereas that in methoxyacetone appears at 1731 cm^{-1}. When the more electronegative fluorine atom is attached, the frequency shifts to an even higher value, 1781 cm^{-1}, in fluoroacetone.

Bending Modes. A medium-to-strong absorption occurs in the range from 1300 to 1100 cm^{-1} for coupled stretching and bending vibrations in the C—CO—C group of ketones. Aliphatic ketones absorb to the right in this range (1220 to 1100 cm^{-1}), as seen in the spectrum of 3-methyl-2-butanone (Fig. 2.4), where a band appears at about 1180 cm^{-1}. Aromatic ketones absorb to the left in this range (1300 to 1220 cm^{-1}), as seen in the spectrum of acetophenone (Fig. 2.40), where a band appears at about 1260 cm^{-1}.

A medium-intensity band appears for a methyl group adjacent to a carbonyl at about 1370 cm^{-1} for the symmetric bending vibration. These methyl groups absorb with greater intensity than methyl groups found in hydrocarbons.

D.   Carboxylic Acids

Carboxylic acids show a very strong band for the C=O group that appears in the range of 1730–1700 cm^{-1} for simple aliphatic carboxylic acids in the *dimeric* form. This band is shifted to lower frequencies with conjugation to a C=C or phenyl group. The O—H stretch appears in the spectrum as a *very broad* band extending from 3400 to 2400 cm^{-1}. This broad band centers on about 3000 cm^{-1} and partially obscures the C—H stretching bands. If this very broad O—H stretch band is seen along with a C=O peak, it almost certainly indicates the compound is a carboxylic acid.

SPECTRAL ANALYSIS BOX

CARBOXYLIC ACIDS

O—H   Stretch, usually *very broad* (strongly H-bonded), occurs at 3400–2400 cm^{-1} and often overlaps the C—H absorptions.

C=O   Stretch, broad, occurs at 1730–1700 cm^{-1}. Conjugation moves the absorption to a lower frequency.

C—O   Stretch occurs in the range 1320–1210 cm^{-1}, medium intensity.

Examples: isobutyric acid (Fig. 2.45) and benzoic acid (Fig. 2.46).

FIGURE 2.45 The infrared spectrum of isobutyric acid (neat liquid, KBr plates).

FIGURE 2.46 The infrared spectrum of benzoic acid (Nujol mull, KBr plates). Dots indicate the Nujol (mineral oil) absorption bands (see Fig. 2.8).

DISCUSSION SECTION

The most characteristic feature in the spectrum of a carboxylic acid is the *extremely broad* O—H absorption occurring in the region from 3400 to 2400 cm^{-1}. This band is attributed to the strong hydrogen bonding present in the dimer, which was discussed in the introduction to Section 2.14. The absorption often obscures the C—H stretching vibrations that occur in the same region. If this broad hydrogen-bonded band is present *together with* the proper C=O absorption value, a carboxylic acid is almost certainly indicated. Figures 2.45 and 2.46 show the spectra of an aliphatic carboxylic acid and an aromatic carboxylic acid, respectively.

The carbonyl stretching absorption, which occurs at about 1730 to 1700 cm^{-1} for the dimer, is usually broader and more intense than that present in an aldehyde or a ketone. For most acids, when the acid is diluted with a solvent, the C=O absorption appears between 1760 and 1730 cm^{-1} for the monomer. However, the monomer is not often seen experimentally since it is usually easier to run the spectrum as a neat liquid. Under these conditions, as well as in a potassium bromide pellet or a Nujol mull, the dimer exists. It should be noted that some acids exist as dimers even at high dilution. Conjugation with a C=C or aryl group usually shifts the absorption band to a lower frequency, as predicted in Section 2.14A and as shown in the spectrum of benzoic acid (Fig. 2.46). Halogenation on the α carbon leads to an increase in the C=O frequency. Section 2.18 discusses salts of carboxylic acids.

The C—O stretching vibration for acids (dimer) appears near 1260 cm^{-1} as a medium-intensity band. A broad band, attributed to the hydrogen-bonded O—H out-of-plane bending vibration, appears at about 930 cm^{-1}. This latter band is usually of low-to-medium intensity.

E.   Esters

Esters show a very strong band for the C=O group that appears in the range of 1750–1735 cm^{-1} for simple aliphatic esters. The C=O band is shifted to lower frequencies when it is conjugated to a C=C or phenyl group. On the other hand, conjugation of a C=C or phenyl group with the *single-bonded oxygen* of an ester leads to an increased frequency from the range given above. Ring strain moves the C=O absorption to a higher frequency in cyclic esters (lactones).

SPECTRAL ANALYSIS BOX

ESTERS

C=O

R—C—O—R
  ‖
  O

C=O stretch appears in the range 1750–1735 cm^{-1} for normal aliphatic esters.

C=C—C—O—R
    | ‖
        O

Conjugation of C=O with α,β C=C; 1740–1715 cm^{-1} for C=O and 1640–1625 cm^{-1} for C=C (two bands for some C=C, *cis* and *trans*.

Ar—C—O—R
   ‖
   O

Conjugation of C=O with phenyl; 1740–1715 cm^{-1} for C=O and 1600–1450 cm^{-1} for ring.

Conjugation of a single-bonded oxygen atom with C=C or phenyl; 1765–1762 cm⁻¹ for C=O.

Cyclic esters (lactones); C=O frequency increases with decreasing ring size.

C—O   Stretch in two or more bands, one stronger and broader than the other, occurs in the range 1300–1000 cm⁻¹.

Examples: ethyl butyrate (Fig. 2.47), methyl methacrylate (Fig. 2.48), vinyl acetate (Fig. 2.49), methyl benzoate (Fig. 2.50), and methyl salicylate (Fig. 2.51).

FIGURE 2.47 The infrared spectrum of ethyl butyrate (neat liquid, KBr plates).

FIGURE 2.48 The infrared spectrum of methyl methacrylate (neat liquid, KBr plates).

FIGURE 2.49 The infrared spectrum of vinyl acetate (neat liquid, KBr plates).

FIGURE 2.50 The infrared spectrum of methyl benzoate (neat liquid, KBr plates).

FIGURE 2.51 The infrared spectrum of methyl salicylate (neat liquid, KBr plates).

DISCUSSION SECTION

General Features of Esters. The two most characteristic features in the spectrum of a normal ester are the strong C=O, which appears in the range from 1750 to 1735 cm^{-1}, and C—O stretching absorptions, which appear in the range from 1300 to 1000 cm^{-1}. Although some ester carbonyl groups may appear in the same general area as ketones, one can usually eliminate ketones from consideration by observing the *strong* and *broad* C—O stretching vibrations that appear in a region (1300 to 1000 cm^{-1}) where ketonic absorptions appear as weaker and narrower bands. For example, compare the spectrum of a ketone, mesityl oxide (Fig. 2.39) with that of an ester, ethyl butyrate (Fig. 2.47) in the 1300- to 1000-cm^{-1} region. Ethyl butyrate (Fig. 2.47) shows the typical C=O stretching vibration at about 1738 cm^{-1}.

Conjugation with a Carbonyl Group (α,β Unsaturation or Aryl Substitution). The C=O stretching vibrations are shifted by about 15 to 25 cm^{-1} to lower frequencies with α,β unsaturation or aryl substitution, as predicted in Section 2.14A (Conjugation Effects). The spectra of both methyl methacrylate (Fig. 2.48) and methyl benzoate (Fig. 2.50) show the C=O absorption shift from the position in a normal ester, ethyl butyrate (Fig. 2.47). Also notice that the C=C absorption band at 1630 cm^{-1} in methyl methacrylate has been intensified over what is obtained with a nonconjugated double bond (Section 2.10B).

CH$_3$CH$_2$CH$_2$COCH$_2$CH$_3$

Ethyl butyrate
1738 cm^{-1}

Methyl methacrylate
1725 cm^{-1}

Methyl benzoate
1724 cm^{-1}

Conjugation with the Ester Single-Bonded Oxygen. Conjugation involving the single-bonded oxygen shifts the C=O vibrations to higher frequencies. Apparently, the conjugation interferes with possible resonance with the carbonyl group, leading to an increase in the absorption frequency for the C=O band.

In the spectrum of vinyl acetate (Fig. 2.49), the C=O band appears at 1762 cm^{-1}, an increase of 25 cm^{-1} above a normal ester. Notice that the C=C absorption intensity is increased in a manner similar to the pattern obtained with vinyl ethers (Section 2.13). The substitution of an aryl group on the oxygen would exhibit a similar pattern.

Ethyl butyrate
1738 cm^{-1}

Vinyl acetate
1762 cm^{-1}

Phenyl acetate
1765 cm^{-1}

Figure 2.52 shows the general effect of α,β unsaturation or aryl substitution and conjugation with oxygen on the C=O vibrations.

Hydrogen-Bonding Effects. When intramolecular (internal) hydrogen bonding is present, the C=O is shifted to a lower frequency, as predicted in Section 2.14A and shown in the spectrum of methyl salicylate (Fig. 2.51).

Methyl salicylate
$1680\ cm^{-1}$

Cyclic Esters (Lactones). The C=O vibrations are shifted to higher frequencies with decreasing ring size, as predicted in Section 2.14A. The unstrained, six-membered cyclic ester δ-valerolactone absorbs at about the same value as a noncyclic ester ($1735\ cm^{-1}$). Because of increased angle strain, γ-butyrolactone absorbs at about $35\ cm^{-1}$ higher than δ-valerolactone.

δ-Valerolactone γ-Butyrolactone
$1735\ cm^{-1}$ $1770\ cm^{-1}$

FIGURE 2.52 The effect of α,β unsaturation or aryl substitution and conjugation with oxygen on the C=O vibrations in noncyclic (acyclic) esters.

Table 2.8 presents some typical lactones together with their C=O stretching absorption values. Inspection of these values reveals the influence of ring size, conjugation with a carbonyl group, and conjugation with the single-bond oxygen.

α-Halo Effects. Halogenation on the α carbon leads to an increase in the C=O frequency.

α-Keto Esters. In principle, one should see two carbonyl groups for a compound with "ketone" and "ester" functional groups. Usually, one sees a shoulder on the main absorption band near 1735 cm^{-1} or a single broadened absorption band.

$$R-\overset{\overset{\textstyle O}{\|}}{C}-\overset{\overset{\textstyle O}{\|}}{\underset{\alpha}{C}}-O-R$$

β-Keto Esters. Although this class of compounds exhibits tautomerization like that observed in β-diketones, less evidence exists for the enol form because β-keto esters do not enolize to as great an extent. β-Keto esters exhibit a *strong-intensity* doublet for the two carbonyl groups at about 1720 and 1740 cm^{-1} in the "keto" tautomer, presumably for the ketone and ester C=O groups. Evidence for the *weak-intensity* C=O band in the "enol" tautomer (often a doublet) appears at about 1650 cm^{-1}. Because of the low concentration of the enol tautomer, one generally cannot observe the broad O—H stretch that was observed in β-diketones.

TABLE 2.8
EFFECTS OF RING SIZE, α,β UNSATURATION, AND CONJUGATION WITH OXYGEN ON THE C=O VIBRATIONS IN LACTONES

Ring-Size Effects (cm^{-1})	α,β Conjugation (cm^{-1})	Conjugation with Oxygen (cm^{-1})
1735	1725	1760
1770	1750	1800
1820		

O=C(R—CH₂—C)(O—R) keto tautomer ⟷ enol tautomer

Keto tautomer   *Enol tautomer*

C—O Stretching Vibrations in Esters. Two (or more) bands appear for the C—O stretching vibrations in esters in the range from 1300 to 1000 cm⁻¹. Generally, the C—O stretch next to the carbonyl group (the "acid" side) of the ester is one of the strongest and broadest bands in the spectrum. This absorption appears between 1300 and 1150 cm⁻¹ for most common esters; esters of aromatic acids absorb nearer the higher-frequency end of this range, and esters of saturated acids absorb nearer the lower-frequency end. The C—O stretch for the "alcohol" part of the ester may appear as a weaker band in the range from 1150 to 1000 cm⁻¹. In analyzing the 1300- to 1000-cm⁻¹ region to confirm an ester functional group, do not worry about fine details. It is usually sufficient to find at least one very strong and broad absorption to help identify the compound as an ester.

F.   Amides

Amides show a very strong band for the C=O group that appears in the range of 1700–1640 cm⁻¹. The N—H stretch is observed in the range of 3475–3150 cm⁻¹. Unsubstituted (primary) amides, R—CO—NH₂, show two bands in the N—H region, while monosubstituted (secondary) amides, R—CO—NH—R, show only one band. The presence of N—H bands plus an unusually low value for the C=O would suggest the presence of an amide functional group. Disubstituted (tertiary) amides, R—CO—NR₂, will show the C=O in the range of 1680–1630 cm⁻¹, but will not show an N—H stretch.

SPECTRAL ANALYSIS BOX

AMIDES

C=O   Stretch occurs at approximately 1700–1640 cm⁻¹.

N—H   Stretch in primary amides (—NH₂) gives two bands near 3350 and 3180 cm⁻¹. Secondary amides have one band (—NH) at about 3300 cm⁻¹.

N—H   Bending occurs around 1640–1550 cm⁻¹ for primary and secondary amides.

Examples: propionamide (Fig. 2.53) and *N*-methylacetamide (Fig. 2.54).

FIGURE 2.53 The infrared spectrum of propionamide (Nujol mull, KBr plates). Dots indicate the Nujol (mineral oil) absorption bands (see Fig. 2.8).

FIGURE 2.54 The infrared spectrum of *N*-methylacetamide (neat liquid, KBr plates).

DISCUSSION SECTION

Carbonyl Absorption in Amides. Primary and secondary amides in the solid phase (potassium bromide pellet or Nujol) have broad C=O absorptions in the range from 1700–1640 cm^{-1}. The C=O band partially overlaps the N—H bending band which appears in the range 1680–1630 cm^{-1}, making the C=O band appear as a doublet. In very dilute solution, the band appears at about 1690 cm^{-1}. This effect is similar to that observed for carboxylic acids, in which hydrogen bonding reduces the frequency in the solid state or in concentrated solution. Tertiary amides, which cannot form hydrogen bonds, have C=O frequencies that are not influenced by the physical state and absorb in about the same range as do primary and secondary amides (1700–1640 cm^{-1}).

Primary amide   Secondary amide   Tertiary amide

Cyclic amides (lactams) give the expected increase in C=O frequency for decreasing ring size, as shown for lactones in Table 2.8.

~1660 cm^{-1} ~1705 cm^{-1} ~1745 cm^{-1}

N—H and C—N Stretching Bands. A pair of fairly strong N—H stretching bands appears at about 3350 cm^{-1} and 3180 cm^{-1} for a primary amide in the solid state (KBr or Nujol). The 3350- and 3180-cm^{-1} bands result from the asymmetric and symmetric vibrations, respectively (Section 2.3). Figure 2.53 shows an example, the spectrum of propionamide. In the solid state, secondary amides and lactams give one band at about 3300 cm^{-1}. A weaker band may appear at about 3100 cm^{-1} in secondary amides; it is attributed to a Fermi resonance overtone of the 1550-cm^{-1} band. A C—N stretching band appears at about 1400 cm^{-1} for primary amides.

N—H Bending Bands. In the solid state, primary amides give strong bending vibrational bands in the range from 1640 to 1620 cm^{-1}. They often nearly overlap the C=O stretching bands. Primary amides give other bending bands at about 1125 cm^{-1} and a very broad band in the range from 750 to 600 cm^{-1}. Secondary amides give relatively strong bending bands at about 1550 cm^{-1}; these are attributed to a combination of a C—N stretching band and an N—H bending band.

G. Acid Chlorides

Acid chlorides show a very strong band for the C=O group that appears in the range of 1810–1775 cm^{-1} for aliphatic acid chlorides. Acid chloride and anhydrides are the most common functional groups that have a C=O appearing at such a high frequency. Conjugation lowers the frequency.

SPECTRAL ANALYSIS BOX

ACID CHLORIDES

C=O   Stretch occurs in the range 1810–1775 cm^{-1} in unconjugated chlorides. Conjugation lowers the frequency to 1780–1760 cm^{-1}.

C—Cl   Stretch occurs in the range 730–550 cm^{-1}.

Examples: acetyl chloride (Fig. 2.55) and benzoyl chloride (Fig. 2.56).

FIGURE 2.55 The infrared spectrum of acetyl chloride (neat liquid, KBr plates).

FIGURE 2.56 The infrared spectrum of benzoyl chloride (neat liquid, KBr plates).

DISCUSSION SECTION

C=O Stretching Vibrations. By far the most common acid halides, and the only ones discussed in this book, are acid chlorides. The strong carbonyl absorption appears at a characteristically high frequency of about 1800 cm^{-1} for saturated acid chlorides. Figure 2.55 shows the spectrum of acetyl chloride. Conjugated acid chlorides absorb at a lower frequency (1780 to 1760 cm^{-1}), as predicted in Section 2.14A. Figure 2.56 shows an example of an aryl-substituted acid chloride, benzoyl chloride. In this spectrum, the main absorption occurs at 1774 cm^{-1}, but a weak shoulder appears on the higher-frequency side of the C=O band (about 1810 cm^{-1}). The shoulder is probably the result of an overtone of a strong band in the 1000- to 900-cm^{-1} range. A weak band is also seen at about 1900 cm^{-1} in the spectrum of acetyl chloride (Fig. 2.55). Sometimes, this overtone band is relatively strong.

In some aromatic acid chlorides, one may observe another rather strong band, often on the lower-frequency side of the C=O band, which makes the C=O appear as a doublet. This band, which appears in the spectrum of benzoyl chloride (Fig. 2.56) at about 1730 cm^{-1}, is probably a Fermi resonance band originating from an interaction of the C=O vibration, with an overtone of a strong band for aryl-C stretch often appearing in the range from 900 to 800 cm^{-1}. When a fundamental vibration couples with an overtone or combination band, the coupled vibration is called **Fermi resonance.** The Fermi resonance band may also appear on the higher-frequency side of the C=O in many aromatic acid chlorides. This type of interaction can lead to splitting in other carbonyl compounds as well.

C—Cl Stretching Vibrations. These bands, which appear in the range from 730 to 550 cm^{-1}, are best observed if KBr plates or cells are used. One strong C—Cl band appears in the spectrum of acetyl chloride. In other aliphatic acid chlorides, one may observe as many as four bands due to the many conformations that are possible.

H.   Anhydrides

Anhydrides show two strong bands for the C=O groups. Simple alkyl-substituted anhydrides generally give bands near 1820 and 1750 cm^{-1}. Anhydrides and acid chlorides are the most common functional groups that have a C=O peak appearing at such a high frequency. Conjugation shifts each of the bands to lower frequencies (about 30 cm^{-1} each). Simple five-membered ring anhydrides have bands at near 1860 and 1780 cm^{-1}.

SPECTRAL ANALYSIS BOX

ANHYDRIDES

C=O   Stretch always has two bands, 1830–1800 cm^{-1} and 1775–1740 cm^{-1}, with variable relative intensity. Conjugation moves the absorption to a lower frequency. Ring strain (cyclic anhydrides) moves the absorptions to a higher frequency.

C—O   Stretch (multiple bands) occurs in the range 1300–900 cm^{-1}.

Example: propionic anhydride (Fig. 2.57).

FIGURE 2.57 The infrared spectrum of propionic anhydride (neat liquid, KBr plates).

DISCUSSION SECTION

The characteristic pattern for noncyclic and saturated anhydrides is the appearance of *two strong bands,* not necessarily of equal intensities, in the regions from 1830 to 1800 cm^{-1} and from 1775 to 1740 cm^{-1}. The two bands result from asymmetric and symmetric stretch (Section 2.3). Conjugation shifts the absorption to a lower frequency, while cyclization (ring strain) shifts the absorption to a higher frequency. The *strong* and *broad* C—O stretching vibrations occur in the region from 1300 to 900 cm^{-1}. Figure 2.57 shows the spectrum of propionic anhydride.

2.15 AMINES

Primary amines, R—NH$_2$, show two N—H stretching bands in the range 3500–3300 cm^{-1}, whereas secondary amines, R$_2$N—H, show only one band in that region. Tertiary amines will not show an N—H stretch. Because of these features, it is easy to differentiate among primary, secondary, and tertiary amines by inspection of the N—H stretch region.

SPECTRAL ANALYSIS BOX

AMINES

N—H	Stretch occurs in the range 3500–3300 cm^{-1}. Primary amines have two bands. Secondary amines have one band: a vanishingly weak one for aliphatic compounds and a stronger one for aromatic secondary amines. Tertiary amines have no N—H stretch.
N—H	Bend in primary amines results in a broad band in the range 1640–1560 cm^{-1}. Secondary amines absorb near 1500 cm^{-1}.
N—H	Out-of-plane bending absorption can sometimes be observed near 800 cm^{-1}.
C—N	Stretch occurs in the range 1350–1000 cm^{-1}.

Examples: butylamine (Fig. 2.58), dibutylamine (Fig. 2.59), tributylamine (Fig. 2.60), and *N*-methylaniline (Fig. 2.61).

FIGURE 2.58 The infrared spectrum of butylamine (neat liquid, KBr plates).

FIGURE 2.59 The infrared spectrum of dibutylamine (neat liquid, KBr plates).

FIGURE 2.60 The infrared spectrum of tributylamine (neat liquid, KBr plates).

FIGURE 2.61 The infrared spectrum of *N*-methylaniline (neat liquid, KBr plates).

DISCUSSION SECTION

The N—H stretching vibrations occur in the range from 3500 to 3300 cm^{-1}. In neat liquid samples, the N—H bands are often weaker and sharper than an O—H band (see Fig. 2.6). Amines may sometimes be differentiated from alcohols on that basis. Primary amines, determined as neat liquids (hydrogen bonded), give *two bands* at about 3400 and 3300 cm^{-1}. The higher-frequency band in the pair is due to the asymmetric vibration, whereas the lower-frequency band results from a symmetric vibration (Section 2.3). In dilute solution, the two free N—H stretching vibrations are shifted to higher frequencies. Figure 2.58 shows the spectrum of an aliphatic primary amine. A low-intensity shoulder appears at about 3200 cm^{-1} on the low-frequency side of the symmetric N—H stretching band. This low-intensity band has been attributed to an overtone of the N—H *bending* vibration that appears near 1600 cm^{-1}. The 3200-cm^{-1} shoulder has been enhanced by a Fermi resonance interaction with the symmetric N—H stretching band near 3300 cm^{-1}. The overtone band is often even more pronounced in aromatic primary amines.

Aliphatic secondary amines determined as neat liquids give *one band* in the N—H stretching region at about 3300 cm^{-1}, but the band is often vanishingly weak. On the other hand, an aromatic secondary amine gives a stronger N—H band near 3400 cm^{-1}. Figures 2.59 and 2.61 are the spectra of an aliphatic secondary amine and an aromatic secondary amine, respectively. Tertiary amines do not absorb in this region, as shown in Figure 2.60.

In primary amines, the N—H bending mode (scissoring) appears as a medium- to strong-intensity (broad) band in the range from 1640 to 1560 cm^{-1}. In aromatic secondary amines, the band shifts to a lower frequency and appears near 1500 cm^{-1}. However, in aliphatic secondary amines the N—H bending vibration is very weak and usually is not observed. The N—H vibrations in aromatic compounds often overlap the aromatic C=C ring absorptions, which also appear in this region. An out-of-plane N—H bending vibration appears as a broad band near 800 cm^{-1} for primary and secondary amines. These bands appear in the spectra of compounds determined as neat liquids and are seen most easily in aliphatic amines (Figs. 2.58 and 2.59).

The C—N stretching absorption occurs in the region from 1350 to 1000 cm^{-1} as a medium to strong band for all amines. Aliphatic amines absorb from 1250 to 1000 cm^{-1}, whereas aromatic amines absorb from 1350 to 1250 cm^{-1}. The C—N absorption occurs at a higher frequency in aromatic amines because resonance increases the double-bond character between the ring and the attached nitrogen atom.

2.16 NITRILES, ISOCYANATES, ISOTHIOCYANATES, AND IMINES

Nitriles, isocyanates, and isothiocyanates all have *sp*-hybridized carbon atoms similar to the C≡C bond. They absorb in the region 2100–2270 cm^{-1}. The C=N bond of an imine has an *sp^2* carbon atom, however. Imines and similar compounds absorb near where double bonds appear, 1690–1640 cm^{-1}.

SPECTRAL ANALYSIS BOX

NITRILES   R—C≡N

—C≡N   Stretch is a medium-intensity, sharp absorption near 2250 cm^{-1}. Conjugation with double bonds or aromatic rings moves the absorption to a lower frequency.

Examples: butyronitrile (Fig. 2.62) and benzonitrile (Fig. 2.63).

ISOCYANATES   R—N=C=O

—N=C=O   Stretch in an isocyanate gives a broad, intense absorption near 2270 cm^{-1}.

Example: benzyl isocyanate (Fig. 2.64).

ISOTHIOCYANATES   R—N=C=S

—N=C=S   Stretch in an isothiocyanate gives one or two broad, intense absorptions centering near 2125 cm^{-1}.

IMINES   R$_2$C=N—R

—C=N—   Stretch in an imine, oxime, and so on gives a variable-intensity absorption in the range 1690–1640 cm^{-1}.

FIGURE 2.62 The infrared spectrum of butyronitrile (neat liquid, KBr plates).

FIGURE 2.63 The infrared spectrum of benzonitrile (neat liquid, KBr plates).

FIGURE 2.64 The infrared spectrum of benzyl isocyanate (neat liquid, KBr plates).

DISCUSSION SECTION

sp-*Hybridized Carbon.* The C≡N group in a nitrile gives a medium-intensity, sharp band in the triple-bond region of the spectrum (2270 to 2210 cm^{-1}). The C≡C bond, which absorbs near this region (2150 cm^{-1}), usually gives a weaker and broader band unless it is at the end of the chain. Aliphatic nitriles absorb at about 2250 cm^{-1}, whereas their aromatic counterparts absorb at lower frequencies, near 2230 cm^{-1}. Figures 2.62 and 2.63 are the spectra of an aliphatic nitrile and an aromatic nitrile, respectively. Aromatic nitriles absorb at lower frequencies with increased intensity because of conjugation of the triple bond with the ring. Isocyanates also contain an *sp*-hybridized carbon atom (R−N=C=O). This class of compounds gives a broad, intense band at about 2270 cm^{-1} (Fig. 2.64).

sp^2-*Hybridized Carbon.* The C=N bond absorbs in about the same range as a C=C bond. Although the C=N band varies in intensity from compound to compound, it usually is more intense than that obtained from the C=C bond. An oxime (R−CH=N−O−H) gives a C=N absorption in the range from 1690 to 1640 cm^{-1} and a broad O−H absorption between 3650 and 2600 cm^{-1}. An imine (R−CH=N−R) gives a C=N absorption in the range from 1690 to 1650 cm^{-1}.

2.17 NITRO COMPOUNDS

Nitro compounds show two strong bands in the infrared spectrum. One appears near 1550 cm^{-1} and the other near 1350 cm^{-1}. Although these two bands may partially overlap the aromatic ring region, 1600–1450 cm^{-1}, it is usually easy to see the NO$_2$ peaks.

SPECTRAL ANALYSIS BOX

NITRO COMPOUNDS

Aliphatic nitro compounds: asymmetric stretch (strong), 1600–1530 cm^{-1}; symmetric stretch (medium), 1390–1300 cm^{-1}.

Aromatic nitro compounds (conjugated): asymmetric stretch (strong), 1550–1490 cm^{-1}; symmetric stretch (strong), 1355–1315 cm^{-1}.

Examples: 1-nitrohexane (Fig. 2.65) and nitrobenzene (Fig. 2.66).

FIGURE 2.65 The infrared spectrum of 1-nitrohexane (neat liquid, KBr plates).

FIGURE 2.66 The infrared spectrum of nitrobenzene (neat liquid, KBr plates).

DISCUSSION SECTION

The nitro group (NO_2) gives two strong bands in the infrared spectrum. In aliphatic nitro compounds, the asymmetric stretching vibration occurs in the range from 1600 to 1530 cm^{-1}, and the symmetric stretching band appears between 1390 and 1300 cm^{-1}. An aliphatic nitro compound—for example, 1-nitrohexane (Fig. 2.65)—absorbs at about 1550 and 1380 cm^{-1}. Normally, its lower-frequency band is less intense than its higher-frequency band. In contrast with aliphatic nitro compounds, aromatic compounds give two bands of nearly equal intensity. Conjugation of a nitro group with an aromatic ring shifts the bands to lower frequencies: 1550–1490 cm^{-1} and 1355–1315 cm^{-1}. For example, nitrobenzene (Fig. 2.66) absorbs strongly at 1525 and 1350 cm^{-1}. The nitroso group (R—N=O) gives only one strong band, which appears in the range from 1600 to 1500 cm^{-1}.

2.18 CARBOXYLATE SALTS, AMINE SALTS, AND AMINO ACIDS

This section covers compounds with ionic bonds. Included here are carboxylate salts, amine salts, and amino acids. Amino acids are included in this section because of their zwitterionic nature.

SPECTRAL ANALYSIS BOX

CARBOXYLATE SALTS
$$R-\overset{\overset{\textstyle O}{\|}}{C}-O^- \quad Na^+$$

Asymmetric stretch (strong) occurs near 1600 cm^{-1}; symmetric stretch (strong) occurs near 1400 cm^{-1}.

Frequency of C=O absorption is lowered from the value found for the parent carboxylic acid because of resonance (more single-bond character).

AMINE SALTS NH_4^+ RNH_3^+ $R_2NH_2^+$ R_3NH^+

N—H   Stretch (broad) occurs at 3300–2600 cm^{-1}. The ammonium ion absorbs to the left in this range, while the tertiary amine salt absorbs to the right. Primary and secondary amine salts absorb in the middle of the range, 3100–2700 cm^{-1}. A broad band often appears near 2100 cm^{-1}.

N—H   Bend (strong) occurs at 1610–1500 cm^{-1}. Primary (two bands) is asymmetric at 1610 cm^{-1}, symmetric at 1500 cm^{-1}. Secondary absorbs in the range 1610–1550 cm^{-1}. Tertiary absorbs only weakly.

AMINO ACIDS

$$R-\underset{\underset{\textstyle NH_2}{|}}{CH}-\overset{\overset{\textstyle O}{\|}}{C}-OH \longrightarrow R-\underset{\underset{\textstyle ^+NH_3}{|}}{CH}-\overset{\overset{\textstyle O}{\|}}{C}-O^-$$

These compounds exist as zwitterions (internal salts) and exhibit spectra that are combinations of carboxylate and primary amine salts. Amino acids show NH_3^+ stretch (very broad), N—H bend (asymmetric/symmetric), and COO^- stretch (asymmetric/symmetric).

Example: leucine (Fig. 2.67).

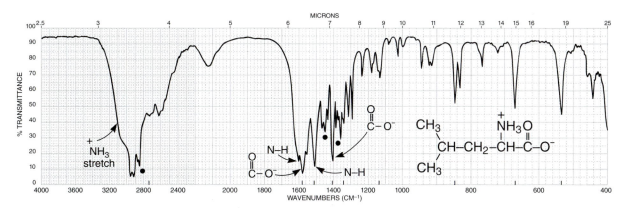

FIGURE 2.67 The infrared spectrum of leucine (Nujol mull, KBr plates). Dots indicate the Nujol (mineral oil) absorption bands (see Fig. 2.8).

2.19 SULFUR COMPOUNDS

Infrared spectral data for sulfur-containing compounds are covered in this section. Included here are single-bonded compounds (mercaptans or thiols and sulfides). Double-bonded S=O compounds are also included in this section.

SPECTRAL ANALYSIS BOX

MERCAPTANS (THIOLS) R—S—H

S—H   Stretch, one weak band, occurs near 2550 cm^{-1} and virtually confirms the presence of this group because few other absorptions appear here.

Example: benzenethiol (Fig. 2.68).

SULFIDES R—S—R

Little useful information is obtained from the infrared spectrum.

SULFOXIDES R—S—R
$$\overset{\|}{\underset{O}{}}$$

S=O   Stretch, one strong band, occurs near 1050 cm^{-1}.

FIGURE 2.68 The infrared spectrum of benzenethiol (neat liquid, KBr plates).

SULFONES

$$R-\overset{\overset{\displaystyle O}{\|}}{\underset{\underset{\displaystyle O}{\|}}{S}}-R$$

S=O   Asymmetric stretch (strong) occurs at 1300 cm⁻¹, symmetric stretch (strong) at 1150 cm⁻¹.

S=O   Asymmetric stretch (strong) occurs at 1300 cm^{-1}, symmetric stretch (strong) at 1150 cm^{-1}.

SULFONYL CHLORIDES

$$R-\overset{\overset{\displaystyle O}{\|}}{\underset{\underset{\displaystyle O}{\|}}{S}}-Cl$$

S=O   Asymmetric stretch (strong) occurs at 1375 cm^{-1}, symmetric stretch (strong) at 1185 cm^{-1}.

Example: benzenesulfonyl chloride (Fig. 2.69).

SULFONATES

$$R-\overset{\overset{\displaystyle O}{\|}}{\underset{\underset{\displaystyle O}{\|}}{S}}-O-R$$

S=O   Asymmetric stretch (strong) occurs at 1350 cm^{-1}, symmetric stretch (strong) at 1175 cm^{-1}.

S—O   Stretch, several strong bands, occurs in the range 1000–750 cm^{-1}.

Example: methyl *p*-toluenesulfonate (Fig. 2.70).

SULFONAMIDES
(Solid State)

S=O   Asymmetric stretch (strong) occurs at 1325 cm^{-1}, symmetric stretch (strong) at 1140 cm^{-1}.

N—H   Primary stretch occurs at 3350 and 3250 cm^{-1}; secondary stretch occurs at 3250 cm^{-1}; bend occurs at 1550 cm^{-1}.

Example: benzenesulfonamide (Fig. 2.71).

SULFONIC ACIDS
(Anhydrous)

$$R-\overset{\overset{\displaystyle O}{\|}}{\underset{\underset{\displaystyle O}{\|}}{S}}-O-H$$

S=O   Asymmetric stretch (strong) occurs at 1350 cm^{-1}, symmetric stretch (strong) at 1150 cm^{-1}.

S—O   Stretch (strong) occurs at 650 cm^{-1}.

FIGURE 2.69 The infrared spectrum of benzenesulfonyl chloride (neat liquid, KBr plates).

FIGURE 2.70 The infrared spectrum of methyl *p*-toluenesulfonate (neat liquid, KBr plates).

FIGURE 2.71 The infrared spectrum of benzenesulfonamide (Nujol mull, KBr plates). Dots indicate the Nujol (mineral oil) absorption bands (see Fig. 2.8).

☐ 2.20 PHOSPHORUS COMPOUNDS

Infrared spectral data for phosphorus-containing compounds are covered in this section. Included here are single-bonded compounds (P—H, P—R, and P—O—R). Double-bonded P=O compounds are also included in this section.

SPECTRAL ANALYSIS BOX

PHOSPHINES   RPH₂   R₂PH

P—H	Stretch, one strong, sharp band, at 2320–2270 cm⁻¹.
PH₂	Bend, medium bands, at 1090–1075 cm⁻¹ and 840–810 cm⁻¹.
P—H	Bend, medium band, at 990–885 cm⁻¹.
P—CH₃	Bend, medium bands, at 1450–1395 cm⁻¹ and 1346–1255 cm⁻¹.
P—CH₂—	Bend, medium band, at 1440–1400 cm⁻¹.

PHOSPHINE OXIDES   R₃P=O   Ar₃P=O

P=O	Stretch, one very strong band, at 1210–1140 cm⁻¹.

PHOSPHATE ESTERS   (RO)₃P=O

P=O	Stretch, one very strong band, at 1300–1240 cm⁻¹.
R—O	Stretch, one or two strong bands, at 1088–920 cm⁻¹.
P—O	Stretch, medium band, at 845–725 cm⁻¹.

☐ 2.21 ALKYL AND ARYL HALIDES

Infrared spectral data for halogen-containing compounds are covered in this section. Except for fluorine-containing samples, it is difficult to determine the presence or the absence of a halide in a compound via infrared spectroscopy. There are several reasons for this problem. First, the C–X absorption occurs at very low frequencies, to the extreme right of the spectrum, where a number of other bands appear (fingerprint). Second, the sodium chloride plates or cells that are often used obscure the region where halogens absorb (these plates are transparent only above 650 cm⁻¹).

The spectra of carbon tetrachloride and chloroform are shown in this section. These solvents are often used to dissolve solids for determining spectra in solution. Notice that these compounds have multiple chlorine atoms giving rise to strong C–Cl stretches at about 785 cm⁻¹ and 759 cm⁻¹, respectively. It is seldom useful to analyze compounds with a single chlorine atom because they are likely to be obscured by fingerprint peaks. Bromides and iodides absorb out of the range of normal NaCl plates. Mass spectral methods (Sections 3.7 and 4.9) provide more reliable information for halogen-containing compounds.

SPECTRAL ANALYSIS BOX

FLUORIDES   R—F

C—F   Stretch (strong) at 1400–1000 cm^{-1}. Monofluoroalkanes absorb at the lower-frequency end of this range, while polyfluoroalkanes give multiple strong bands in the range 1350–1100 cm^{-1}. Aryl fluorides absorb between 1250 and 1100 cm^{-1}.

CHLORIDES   R—Cl

C—Cl   Stretch (strong) in aliphatic chlorides occurs in the range 785–540 cm^{-1}. Primary chlorides absorb at the upper end of this range, while tertiary chlorides absorb near the lower end. Two or more bands may be observed due to the different conformations possible.

Multiple substitution on a single-carbon atom results in an intense absorption at the upper-frequency end of this range: CH_2Cl_2 (739 cm^{-1}), $HCCl_3$ (759 cm^{-1}), and CCl_4 (785 cm^{-1}). Aryl chlorides absorb between 1100 and 1035 cm^{-1}.

CH$_2$—Cl   Bend (wagging) at 1300–1230 cm^{-1}.

Examples: carbon tetrachloride (Fig. 2.72) and chloroform (Fig. 2.73).

BROMIDES   R—Br

C—Br   Stretch (strong) in aliphatic bromides occurs at 650–510 cm^{-1}, out of the range of routine spectroscopy using NaCl plates or cells. The trends indicated for aliphatic chlorides hold for bromides. Aryl bromides absorb between 1075 and 1030 cm^{-1}.

CH$_2$—Br   Bend (wagging) at 1250–1190 cm^{-1}.

IODIDES   R—I

C—I   Stretch (strong) in aliphatic iodides occurs at 600–485 cm^{-1}, out of the range of routine spectroscopy using NaCl plates or cells. The trends indicated for aliphatic chlorides hold for iodides.

CH$_2$—I   Bend (wagging) at 1200–1150 cm^{-1}.

FIGURE 2.72 The infrared spectrum of carbon tetrachloride (neat liquid, KBr plates).

FIGURE 2.73 The infrared spectrum of chloroform (neat liquid, KBr plates).

2.22 THE BACKGROUND SPECTRUM

In this final section, we take a look at a typical background spectrum. The infrared energy beam passes not only through the sample being measured but also through a length of air. Air contains two major infrared-active molecules: carbon dioxide and water vapor. Absorptions from these two molecules are contained in every spectrum. Since the FT-IR is a single-beam instrument (see Section 2.5B and Fig. 2.3b), it cannot remove these absorptions at the same time the sample spectrum is determined. That method is used by double-beam, dispersive instruments (Section 2.5A and Fig. 2.3a). Instead, the FT-IR determines the "background" spectrum (no sample in the path) and stores it in the computer memory. After a sample spectrum is determined, the computer subtracts the background spectrum from that of the sample, effectively removing the air peaks.

Figure 2.74 shows a typical background spectrum as determined by an FT-IR instrument. The two absorptions at 2350 cm^{-1} are due to the asymmetric stretching modes of carbon dioxide. The groups of peaks centered at 3750 cm^{-1} and 1600 cm^{-1} are due to the stretching and bending modes of atmospheric (gaseous) water molecules. The fine structure (spikes) in these absorptions are frequently seen in atmospheric water as well as other small *gas-phase* molecules, due to superimposed rotational energy level absorptions. In *liquids or solids,* the fine structure is usually blended together into a broad, smooth curve (see hydrogen bonding in alcohols, Section 2.12). Occasionally, other peaks may show up in the background, sometimes due to chemical coatings on the mirrors and sometimes due to degradation of the optics caused by adsorbed materials. Cleaning the optics can remedy the last situation.

The observed bell-curve shape of the background spectrum is due to differences in the output of the infrared source. The "lamp" has its highest output intensities at the wavelengths in the center of the spectrum and diminished intensities at wavelengths at either end of the spectrum. Because the source has unequal output intensity over the range of wavelengths measured, the FT-IR spectrum of the sample will also have a curvature. Most FT-IR instruments can correct this curvature using a software procedure called *autobaseline.* The autobaseline procedure corrects for imbalances in the source output and attempts to give the spectrum a horizontal baseline.

In solid samples (KBr pellets or dry-film preparations), additional imbalances in the baseline can be introduced due to "light-scattering" effects. Granular particles in a sample cause the source energy to be diffracted or scattered out of the main beam, causing loss of intensity. This scattering is usually greatest at the high-frequency (short-wavelength) end of the spectrum, the region from about 4000 cm^{-1} to 2500 cm^{-1}. This effect is often seen in spectra determined with KBr pellets in which the sample is either opaque or not ground to a sufficiently fine granule size; a rising baseline results as one moves to lower frequencies. The autobaseline procedure will also help combat this problem.

FIGURE 2.74 A background spectrum determined by an FT-IR instrument.

Finally, there will always be instances when the computer subtraction of the background will not be complete. This situation is readily recognized by the presence of the carbon dioxide "doublet" in the spectrum at 2350 cm^{-1}. Peaks at this wavenumber value are usually due to carbon dioxide and not to the sample being measured. A disconcerting, but not uncommon, situation occurs when the subtraction procedure favors the background. This causes the CO_2 doublet to go "negative" (upward from the baseline). Fortunately, few other functional groups absorb in the region near 2350 cm^{-1}, making identification of the CO_2 peaks relatively easy.

2.23 HOW TO SOLVE INFRARED SPECTRAL PROBLEMS

In this section, we will use the infrared spectrum to determine the structure of four different unknown compounds. Often, it is not possible to derive an exact structure using the infrared spectrum only. However, it is often possible to eliminate most of the potential structures. Section 2.9 provides important information on how to proceed in the analysis of a spectrum. In addition, you will find the following list of figures and tables to be particularly useful:

- Table 2.4: base values for functional groups

- Spectral analysis boxes for alcohols, amines, and nitriles

- Figure 2.35: types of C=O functional groups

- Factors that influence C=O stretching vibrations (Section 2.14A)

- Section 2.14C (ketones), including Figures 2.43 and 2.44

- Spectral analysis boxes for aldehydes, ketones, carboxylic acids, esters, and amides
- Figure 2.22: substitution patterns for groups attached to C=C
- Figure 2.28: substitution patterns for groups attached to a benzene ring

■ EXAMPLE 1

As a first example, determine the structure or structures possible for a compound with formula $C_5H_{12}O$ that is shown in Figure 2.75. First, determine the index of hydrogen deficiency (Section 1.4 in the previous chapter) for $C_5H_{12}O$. The "index" yields a value of 0. We conclude that they are no double bonds in the compound, ruling out either C=O or C=C which would have an "index" of one, each. Notice that the spectrum shows a strong and broad peak at about 3350 cm^{-1} that is assigned to an O—H group (Section 2.12). The structure may be further refined by investigating whether the compound is a primary, secondary or tertiary alcohol. To do this, look at the band near 1050 cm^{-1} for the C—O single bond stretch (Table 2.7). The best fit is to assign this band to a primary alcohol. Thus, we conclude that this compound is an unsaturated primary alcohol. Possible structures are shown below. We can choose between these primary alcohols by looking at the pair of peaks appearing at 1380 and 1370 cm^{-1} (Figure 2.17). The only compound that would fit is 3-methyl-1-butanol.

1-Pentanol   3-Methyl-1-butanol   2-Methyl-1-butanol

■ EXAMPLE 2

As a second example, determine the structure or structures possible for a compound with formula $C_6H_{10}O$ that is shown in Figure 2.76. The index of hydrogen deficiency is calculated to be 2 (Section 1.4 in the previous chapter). The strong absorption band at 1719 cm^{-1} is clearly assigned to a C=O peak, close to what we expect for a ketone. The peak shown at about 1640 cm^{-1} indicates the presence of a C=C bond. That would suggest that the compound is an unsaturated carbonyl compound which fits an "index" of 2. An sp^2 C—H pattern shows as a sharp peak appearing near 3100 cm^{-1}, which confirms the presence of a C=C functional group. A group of sp^3 C—H peaks appear that centers on 2940 cm^{-1}, which provides no help. Now consult Section 2.14 for the carbonyl group possibilities. Although we suspect that this is a ketone, it would be helpful to eliminate other C=O possibilities (see Figure 2.35). Comparison of the spectrum shown in Figure 2.76 with an aldehyde shown in Figure 2.36 eliminates this functional group possibility because we do not observe the doublet centering on 2800 cm^{-1} for the C—H group on an aldehyde. An ester or carboxylic acid are not possibilities because they have two oxygen atoms. Amides are not a possibility since the compound shown in Figure 2.76 does not contain a nitrogen atom. This confirms that the C=O group as the most likely associated with an unsaturated ketone.

We next need to determine where the C=C bond is located with respect to the C=O group. Is it is a conjugated ketone? This is not a likely prospect since the carbonyl group appears near the normal value for an unconjugated C=O group (see the "normal" values in Figure 2.43). This being the case, then we need to make sure that any structure considered must have the C=C out of conjugation with the C=O. So where is the C=C located? Figure 2.22 should help us determine where the

FIGURE 2.75 Example 1, an unknown compound with formula $C_5H_{12}O$.

C=C is located. The best fit would be to place the C=C bond at the end of the chain resulting in a monosubstituted pattern (H_2C=C). The peaks appearing at near 900 and 1000 cm^{-1} confirms this placement. In this case, we can assign a reasonable structure for $C_6H_{10}O$ shown in Figure 2.76. The best fit would be 5-hexen-2-one.

5-Hexen-2-one	4-Hexen-2-one	3-Hexen-2-one
Correct structure	Unconjugated, and would have a C=O near 1725, but the C=C would not fit Figure 2.22	Conjugated, would have C=O below 1700, eliminated as choice

■ **EXAMPLE 3**

As a third example, determine the structure or structures possible for a compound with formula C_7H_9N that is shown in Figure 2.77. The index of hydrogen deficiency is calculated to be 4 (Section 1.4 in the previous chapter). This very high value for the "index" may indicate that the compound has an

FIGURE 2.76 Example 2, an unknown compound with formula $C_6H_{10}O$.

aromatic ring. This is not certain, but it is a good first approximation. This suggestion is confirmed by the appearance of the aromatic ring patterns that are found between 1600 and 1450 cm^{-1} in Figure 2.77.

Assuming that the compound has an aromatic ring, we now need to determine the degree of substitution on the ring. An examination of Figure 2.28a for C—H out-of-plane bending suggests a *meta* substituted ring (1,3-disubstituted). The set of three peaks at about 700, 800, and 900 cm^{-1} fits nicely with the 1,3-disubstituted ring. The three-peak pattern of very weak bands between 1700 and 2000 cm^{-1} also is consistent with a 1,3-disubstituted ring (see Figure 2.28b). The single nitrogen atom suggests an amine functional group. We need to determine if this is a primary or secondary amine by consulting Section 2.15. A primary amine is a distinct possibility because of the strong NH$_2$ pattern with a two-peak pattern found between 3300 and 3500 cm^{-1} (see figure 2.58). A secondary amine would show a single N—H peak in that same range (see figure 2.59). To finish our analysis we need to decide where to place the remaining carbon atom. A methyl group would be suggested, and we would place this group on the third carbon atom on the aromatic ring. The N—H bend for the primary amine (broad band at about 1600 cm^{-1}) overlaps the aromatic ring patterns that occur between 1600 and 1450 cm^{-1}. The best fit to the spectrum is 3-methylaniline.

3-Methylaniline has the groups 1,3 (meta) with respect to each other

4-Methylaniline has groups para (1,4), so doesn't fit the spectrum.

2-Methylaniline has groups ortho (1,2) so doesn't fit the spectrum

N-methylaniline is a secondary amine, so won't fit the spectrum.

■ EXAMPLE 4

As a fourth example, determine the structure or structures possible for a compound with formula C$_5$H$_{10}$N$_2$ that is shown in Figure 2.78. The index of hydrogen deficiency is calculated to be 2 (Section 1.4 in the previous chapter). A glance at the spectrum shows a prominent peak near 2250 cm^{-1}. The only functional group appearing near this value is a nitrile (C≡N) or an alkyne (C≡C). Table 2.4 shows values for triple bonds. An alkyne would appear closer to 2150 cm^{-1} while a nitrile would appear at the value observed in the spectrum. Since the index of hydrogen deficiency has a value of 2, that fits a compound with a triple bond.

The remaining nitrogen atom would likely be an amine. Is it a primary, secondary, or tertiary amine? Section 2.15 shows spectra for these types of amines. Unfortunately, the region between 3600 and 3200 cm^{-1} does not reveal easily what type of amine may be present. The region is often obscured by the presence of water in the sample, or from weak overtone peaks from other parts of the molecule. However, a primary amine (RNH$_2$) should show a prominent doublet while the secondary amine (R$_2$NH) should show a singlet. Since no prominent peaks appear between 3600 and 3200 cm^{-1}, we can conclude that this compound is a nitrile (C≡N) with an attached tertiary amino group (R$_3$N). Nuclear magnetic resonance (NMR) would allow us to

FIGURE 2.77 Example 3, an unknown compound with formula C_7H_9N.

FIGURE 2.78 Example 4, an unknown compound with formula $C_5H_{10}N_2$

conclusively assign the structure. Without this information, several possibilities exist for compounds that would fit. We could compare the spectrum to an authentic sample, to determine that the compound is the following:

3-(*N,N*-Dimethylamino)propanenitrile

PROBLEMS

When a molecular formula is given, it is advisable to calculate an index of hydrogen deficiency (Section 1.4). The index often gives useful information about the functional group or groups that may be present in the molecule.

*1. In each of the following parts, a molecular formula is given. Deduce the structure that is consistent with the infrared spectrum. There may be more than one possible answer.

(a) C_3H_3Cl

(b) $C_{10}H_{14}$

(c) C_7H_8O

(d) $C_8H_{11}N$

(e) C_7H_7Cl

(f) $C_3H_5O_2Cl$

(g) $C_{10}H_{12}$ (two six-membered rings)

(h) C_4H_8O

*2. Ants emit tiny amounts of chemicals called alarm pheromones to warn other ants (of the same species) of the presence of an enemy. Several of the components of the pheromone in one species have been identified, and two of their structures follow. Which compound has the infrared spectrum shown?

Citral   Citronellal

*3. The main constituent of cinnamon oil has the formula C_9H_8O. From the following infrared spectrum, deduce the structure of this component.

*4. The infrared spectra of *cis-* and *trans*-3-hexen-1-ol follow. Assign a structure to each.

5. In each part, choose the structure that best fits the infrared spectrum shown.

*(a)

A

B

C

D

*(b)

*(d)

A

B

C

D

*(e)

A

B

C

D

(f)

(g)

(h)

(i)

(j) $CH_3-CH_2-CH_2-S-CH_2-CH_2-CH_3$ $CH_3(CH_2)_4CH_2-S-H$

A   **B**

$CH_3(CH_2)_4CH_2-O-H$ $CH_3-CH_2-CH_2-O-CH_2-CH_2-CH_3$

C   **D**

*6. The infrared spectra of some polymeric materials follow. Assign a structure to each of them, selected from the following choices: polyamide (nylon), poly(methyl methacrylate), polyethylene, polystyrene, and poly(acrylonitrile-styrene). You may need to look up the structures of these materials.

7. Assign a structure to each of the spectra shown. Choose from among the following 5-carbon esters:

$$CH_3-CH_2-\overset{\overset{\textstyle O}{\|}}{C}-O-CH=CH_2 \qquad\qquad CH_3-CH_2-CH_2-O-\overset{\overset{\textstyle O}{\|}}{C}-CH_3$$

$$CH_3-\overset{\overset{\textstyle O}{\|}}{C}-O-CH_2-CH=CH_2 \qquad\qquad CH_2=CH-\overset{\overset{\textstyle O}{\|}}{C}-O-CH_2-CH_3$$

8. Assign a structure to each of the following three spectra. The structures are shown here.

9. Assign a structure to each of the following three spectra. The structures are shown here.

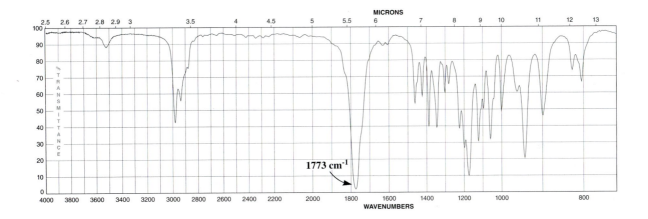

10. Assign a structure to each of the spectra shown. Choose from among the following 5-carbon alcohols:

$$CH_3$$
$$CH_3$$$C=CH—CH_2OH$$

$$CH_2=C$$$CH_3$$
$$CH_2CH_2OH$$

$$CH_3CH_2CH_2CH_2CH_2OH$$ $$CH_2=CH—CH_2CH_2CH_2OH$$

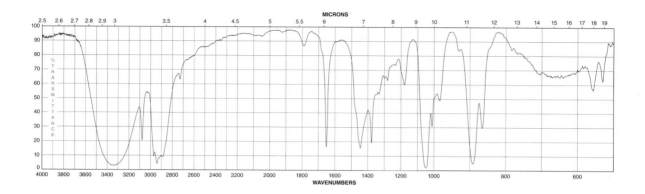

11. Substitution of an amino group on the *para* position of acetophenone shifts the C=O frequency from about 1685 to 1652 cm^{-1}, whereas a nitro group attached to the *para* position yields a C=O frequency of 1693 cm^{-1}. Explain the shift for each substituent from the 1685 cm^{-1} base value for acetophenone.

REFERENCES

Books and Compilations of Spectra

Bellamy, L. J., *The Infrared Spectra of Complex Molecules,* 3rd ed., John Wiley, New York, 1975.

Colthrup, N. B., L. H. Daly, and S. E. Wiberley, *Introduction to Infrared and Raman Spectroscopy,* 3rd ed., Academic Press, New York, 1990.

Lin-Vien, D., N. B. Colthrup, W. G. Fateley, and J. G. Grasselli, *The Handbook of Infrared and Raman Characteristic Frequencies of Organic Molecules,* Academic Press, New York, 1991.

Nakanishi, K., and P. H. Solomon, *Infrared Absorption Spectroscopy,* 2nd ed., Holden–Day, San Francisco, 1998.

Pouchert, C. J., *Aldrich Library of FT-IR Spectra,* Aldrich Chemical Co., Milwaukee, WI, 1985 (1st ed.) and 1997 (2nd ed.).

Pretsch, E., T. Clerc, J. Seibl, and W. Simon, *Tables of Spectral Data for Structure Determination of Organic Compounds,* 3rd ed., Springer-Verlag, Berlin, 1998, 1989. Translated from the German by K. Biemann.

Sadtler Standard Spectra, Sadtler Research Laboratories Division, Bio-Rad Laboratories, Inc., 3316 Spring Garden Street, Philadelphia, PA 19104-2596. Numerous FT-IR search libraries are available for computers.

Silverstein, R. M., F. X. Webster, and D. Kiemle, *Spectrometric Identification of Organic Compounds,* 7th ed., John Wiley, New York, 2005.

Selected Websites

http://www2.chemistry.msu.edu/faculty/reusch/VirtTxtJml/Spectrpy/spectro.htm

The excellent website presented by Professor William Reusch from the Department of Chemistry at Michigan State University includes background material for infrared spectroscopy, NMR spectroscopy, UV spectroscopy, and mass spectrometry. Some problems are included. The site includes links to other available sites for spectral problems.

http://sdbs.riodb.aist.go.jp/sdbs/cgi-bin/cre_index.cgi?lang=eng

This excellent free website presents an Integrated Spectral Data Base System for Organic Compounds (SDBS). It is organized and maintained by the National Institute of Advanced Industrial Science and Technology (AIST), Tsukuba, Ibaraki 305-8565, Japan. This database includes infrared, mass spectra, and NMR data (^{1}H and ^{13}C) for a number of compounds.

http://webbook.nist.gov/chemistry/

The National Institute of Standards and Technology (NIST) has developed the WebBook. This site includes gas-phase infrared spectra and mass spectral data for compounds. This site is not as useful as the SDBS website for infrared spectra, since most of the listed infrared spectra were determined in the gas phase rather than as a liquid phase. The mass spectral database is more useful.

http://www.chem.ucla.edu/~webspectra/

UCLA Department of Chemistry and Biochemistry in connection with Cambridge University Isotope Laboratories presents WebSpectra, an outstanding collection of unknowns for students to interpret. The website provides NMR spectra and some IR spectra. Solutions are provided. Formulas for each of the unknowns are provided. No mass spectral data is included. They provide links to other sites with problems for students to solve.

CHAPTER 3

MASS SPECTROMETRY
Part One: Basic Theory, Instrumentation and Sampling Techniques

The principles that underlie mass spectrometry pre-date all of the other instrumental techniques described in this book. The fundamental principles date to the late 1890s when J. J. Thomson determined the mass-to-charge ratio of the electron, and Wien studied magnetic deflection of anode rays and determined the rays were positively charged. Each man was honored with the Nobel Prize (Thomson in 1906 and Wien in 1911) for their efforts. In 1912–1913, J. J. Thomson studied the mass spectra of atmospheric gases and used a mass spectrum to demonstrate the existence of neon-22 in a sample of neon-20, thereby establishing that elements could have isotopes. The earliest mass spectrometer, as we know it today, was built by A. J. Dempster in 1918. However, the method of mass spectrometry did not come into common use until about 50 years ago, when inexpensive and reliable instruments became available.

Continued development of sample introduction and ionization techniques for high molecular weight (MW) compounds and biological samples in the 1980s and 1990s introduced mass spectrometry to a new community of researchers. The introduction of easily maintained lower-cost commercial instruments has made mass spectrometry an indispensable technique in numerous fields far removed from the laboratories of Thomson and Wien. Today, the biotechnology industry uses mass spectrometry to assay and sequence proteins, oligonucleotides, and polysaccharides. The pharmaceutical industry uses mass spectrometry in all phases of the drug development process, from lead compound discovery and structural analysis, to synthetic development and combinatorial chemistry, and to pharmacokinetics and drug metabolism. In health clinics around the world, mass spectrometry is used in testing blood and urine for everything from the presence and levels of certain compounds that are "markers" for disease states, including many cancers, to detecting the presence and quantitative analysis of illicit or performance-enhancing drugs. Environmental scientists rely on mass spectrometry to monitor water and air quality, and geologists use mass spectrometry to test the quality of petroleum reserves. Mass spectrometry is also used routinely in airport security screening and forensic investigations to detect traces of explosives.

To date, no fewer than five Nobel Prizes have been awarded for work directly related to mass spectrometry: J. J. Thomson (Physics, 1906) for "theoretical and experimental investigations on the conduction of electricity by gases"; F. W. Aston (Chemistry, 1922) for "discovery, by means of a mass spectrograph, of isotopes, in a large number of nonradioactive elements"; W. Paul (Physics, 1989) "for the development of the ion trap technique"; and most recently J. B. Fenn and K. Tanaka (Chemistry, 2002) "for the development of soft desorption ionization methods for mass spectrometric analyses of biological macromolecules."

3.1 THE MASS SPECTROMETER: OVERVIEW

In its simplest form, the mass spectrometer has five components (Fig. 3.1), and each will be discussed separately in this chapter. The first component of the mass spectrometer is the **sample inlet** (Section 3.2), which brings the sample from the laboratory environment (1 atm) to the lower

F I G U R E   3 . 1 The components of a mass spectrometer. (Adapted from Gross, J. H., *Mass Spectrometry: A Textbook*, Springer, Berlin, 2004. Reprinted by permission.)

pressure of the mass spectrometer. Pressures inside the mass spectrometer range from a few millimeters of mercury in a chemical ionization source to a few micrometers of mercury in the mass analyzer and detector regions of the instrument. The sample inlet leads to the **ion source** (Section 3.3), where the sample molecules are transformed into gas phase ions. Some instruments have been developed recently that combine the sample inlet and ion source under ambient conditions, thereby greatly simplifying sample preparation. The ions are then accelerated by an electromagnetic field. Next, the **mass analyzer** (Section 3.4) separates the sample ions based on their **mass-to-charge (*m/z*) ratio**. The ions then are counted by the **detector** (Section 3.5), and the signal is recorded and processed by the **data system,** typically a personal computer (PC). The output from the data system is the **mass spectrum**—a graph of the number of ions detected as a function of their *m/z* ratio.

3.2 SAMPLE INTRODUCTION

When we examine each of these essential mass spectrometer functions in detail, we see that the mass spectrometer is somewhat more complex than just described. Before the ions can be formed, a stream of molecules must be introduced into the **ion source** (ionization chamber) where the ionization takes place. A **sample inlet** system provides this stream of molecules.

A sample studied by mass spectrometry may be a gas, a liquid, or a solid. Enough of the sample must be converted to the vapor state to obtain the stream of molecules that must flow into the ionization chamber. With gases, of course, the substance is already vaporized, so a simple inlet system can be used. This inlet system is only partially evacuated so that the ionization chamber itself is at a lower pressure than the sample inlet system. The sample is introduced into a larger reservoir, from which the molecules of vapor can be drawn into the ionization chamber, which is at low pressure. To ensure that a steady stream of molecules is passing into the ionization chamber, the vapor travels through a small pinhole, called a **molecular leak,** before entering the chamber. The same system can be used for volatile liquids or solids. For less-volatile materials, the system can be designed to fit within an oven, which can heat the sample to increase the vapor pressure of the sample. Care must be taken not to heat any sample to a temperature at which it might decompose.

With nonvolatile samples, other sample inlet systems must be used. A common one is the **direct probe** method. The sample is placed on a thin wire loop or pin on the tip of the probe, which is then inserted through a vacuum lock into the ionization chamber. The sample probe is positioned close to the ion source. The probe can be heated, thus causing vapor from the sample to be evolved in proximity to the ionizing beam of electrons. A system such as this can be used to study samples of molecules with vapor pressures lower than 10^{-9} mmHg at room temperature.

In the last decade, a number of open air sample introduction methods have been developed that essentially eliminate sample preparation. In various atmospheric pressure chemical ionization (APCI) techniques, the sample is placed in a stream of ionized gas (Section 3.3B) or solvent aerosol (Section 3.3D) between the ion source and the inlet to the mass analyzer.

The most versatile sample inlet systems are constructed by connecting a chromatograph to the mass spectrometer. This sample introduction technique allows a complex mixture of components to be separated by the chromatograph, and the mass spectrum of each component may then be determined individually. A drawback of this method involves the need for rapid scanning by the mass spectrometer. The instrument must determine the mass spectrum of each component in the mixture *before* the next component exits from the chromatography column so that the first substance is not contaminated by the next before its spectrum has been obtained. Since high-efficiency columns are used in the chromatograph, in most cases compounds are completely separated before the eluent stream is analyzed. The instrument must have the capability of obtaining at least one scan per second over the *m/z* range of interest. Even more scans are necessary if a narrower range of masses is to be analyzed. The mass spectrometer that is coupled to the chromatograph should be relatively compact and capable of high resolution.

In **gas chromatography–mass spectrometry (GC-MS)**, the gas stream emerging from a gas chromatograph is admitted through a valve into a tube, where it passes over a molecular leak. Some of the gas stream is thus admitted into the ionization chamber of the mass spectrometer. In this way, it is possible to obtain the mass spectrum of every component in a mixture injected into the gas chromatograph. In effect, the mass spectrometer acts in the role of detector. Similarly, **high-performance liquid chromatography–mass spectrometry** (**HPLC-MS,** or more simply **LC-MS**) couples an HPLC instrument to a mass spectrometer through a special interface. The substances that elute from the HPLC column are detected by the mass spectrometer, and their mass spectra can be displayed, analyzed, and compared with standard spectra found in the computer library built into the instrument.

3.3 IONIZATION METHODS

A. Electron Ionization (EI)

Regardless of the method of sample introduction, once the stream of sample molecules has entered the mass spectrometer, the sample molecules must be converted to charged particles by the **ion source** before they can be analyzed and detected. The simplest and most common method for converting the sample to ions is **electron ionization (EI)**. In EI-MS, a beam of high-energy electrons is emitted from a **filament** that is heated to several thousand degrees Celsius. These high-energy electrons strike the stream of molecules that has been admitted from the sample inlet system. The electron–molecule collision strips an electron from the molecule, creating a cation. A **repeller plate,** which carries a positive electrical potential, directs the newly created ions toward a series of **accelerating plates.** A large potential difference, ranging from 1 to 10 kilovolts (kV), applied across these accelerating plates produces a beam of rapidly traveling positive ions. One or more **focusing slits** direct the ions into a uniform beam (Fig. 3.2).

Most of the sample molecules are not ionized at all but are continuously drawn off by vacuum pumps that are connected to the ionization chamber. Some of the molecules are converted to negative ions through the absorption of electrons. The repeller plate absorbs these negative ions. It is possible to reverse the polarity of the repeller and accelerating plates in some instruments, thereby allowing for mass analysis of negative ions (anions) that are created by electron capture when the sample molecules are hit by the electron beam. A small proportion of the positive ions that are formed may have a charge greater than one (a loss of more than one electron). These are accelerated in the same way as the singly charged positive ions.

The energy required to remove an electron from an atom or molecule is its **ionization potential** or **ionization energy.** Most organic compounds have ionization potentials ranging between 8 and 15 electron volts (eV). However, a beam of electrons does not create ions with high efficiency until it strikes the stream of molecules with a potential of 50 to 70 eV. To acquire reproducible spectral features, including fragmentation patterns, that can be readily compared with electronic databases, a standard 70-eV electron beam is used.

EI-MS has distinct advantages for routine mass spectrometry of small organic molecules. Electron ionization hardware is inexpensive and robust. The excess kinetic energy imparted to the sample during the EI process leads to significant fragmentation of the molecular ion (Chapter 4).

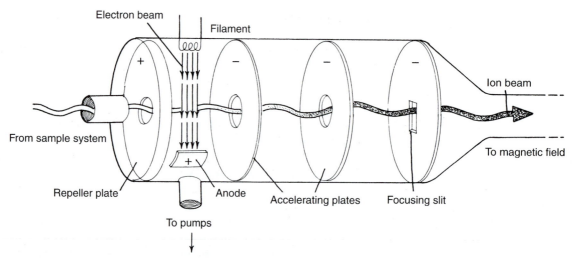

FIGURE 3.2 Electron ionization chamber.

The fragmentation pattern of a compound is reproducible, and many libraries of EI-MS data are available. This allows one to compare the mass spectrum of a sample compound against thousands of data sets in a spectral library in a few seconds using a PC, thus simplifying the process of determining or confirming a compound's identity.

The fragmentation of the molecular ion under EI conditions may also be considered a distinct disadvantage. Some compounds fragment so easily that the lifetime of the molecular ion is too short to be detected by the mass analyzer. Thus, one cannot determine a compound's molecular mass (Section 3.6) in such cases. Another drawback to EI-MS is that the sample must be relatively volatile so it can come into contact with the electron beam in the ionization chamber. This fact coupled with the fragmentation problem make it difficult to analyze high molecular weight (MW) compounds and most biomolecules using EI-MS.

B. Chemical Ionization (CI)

In **chemical ionization–mass spectrometry** (**CI-MS**), the sample molecules are combined with a stream of ionized reagent gas that is present in great excess relative to the sample. When the sample molecules collide with the preionized reagent gas, some of the sample molecules are ionized by various mechanisms, including proton transfer, electron transfer, and adduct formation. Almost any readily available gas or highly volatile liquid can be used as a reagent gas for CI-MS.

Common ionizing reagents for CI-MS include methane, ammonia, isobutane, and methanol. When methane is used as the CI reagent gas, the predominant ionization event is proton transfer from a CH_5^+ ion to the sample. Minor ions are formed by adduct formation between $C_2H_5^+$ and higher homologues with the sample. The methane is converted to ions as shown in Equations 3.1–3.4.

$$CH_4 \quad + \quad e^- \quad \rightarrow \quad CH_4^{\bullet+} \quad + \quad 2e^- \qquad \textbf{Equation 3.1}$$
$$CH_4^{\bullet+} \quad + \quad CH_4 \quad \rightarrow \quad CH_5^+ \quad + \quad \bullet CH_3 \qquad \textbf{Equation 3.2}$$

$$CH_4^{\bullet+} \quad \rightarrow \quad CH_3^+ \quad + \quad H\bullet \qquad \textbf{Equation 3.3}$$
$$CH_3^+ \quad + \quad CH_4 \quad \rightarrow \quad C_2H_5^+ \quad + \quad H_2 \qquad \textbf{Equation 3.4}$$

The sample molecule M is then ionized through the ion–molecule reactions in Equations 3.5 and 3.6:

$$M \quad + \quad CH_5^+ \quad \rightarrow \quad (M+H)^+ \quad + \quad CH_4 \qquad \textbf{Equation 3.5}$$
$$M \quad + \quad C_2H_5^+ \quad \rightarrow \quad (M+C_2H_5)^+ \qquad \textbf{Equation 3.6}$$

The situation is very similar for CI with ammonia as reagent gas (Equations 3.7–3.9):

NH_3	$+$	e^-	$\rightarrow$	$NH_3^{\bullet+}$	$+$	$2e^-$	**Equation 3.7**
$NH_3^{\bullet+}$	$+$	NH_3	$\rightarrow$	NH_4^+	$+$	$\bullet NH_2$	**Equation 3.8**
M	$+$	NH_4^+	$\rightarrow$	$(M+H)^+$	$+$	NH_3	**Equation 3.9**

Using isobutane as reagent gas produces *tert*-butyl cations (Equations 3.10 and 3.11), which readily protonate basic sites on the sample molecule (Equation 3.12). Adduct formation is also possible using isobutane in CI-MS (Equation 3.l3).

$(CH_3)_3CH$	$+$	e^-	$\rightarrow$	$(CH_3)_3CH^{\bullet+}$	$+$	$2e^-$	**Equation 3.10**
$(CH_3)_3CH^{\bullet+}$	$\rightarrow$	$(CH_3)_3C^+$	$+$	$H\bullet$			**Equation 3.11**
M	$+$	$(CH_3)_3C^+$	$\rightarrow$	$(M+H)^+$	$+$	$(CH_3)_2C{=}CH_2$	**Equation 3.12**
M	$+$	$(CH_3)_3C^+$	$\rightarrow$	$[M+C(CH_3)_3]^+$			**Equation 3.13**

Varying the reagent gas in CI-MS allows one to vary the selectivity of the ionization and degree of ion fragmentation. The choice of reagent gas should be made carefully to best match the **proton affinity** of the reagent gas with that of the sample to ensure efficient ionization of the sample without excessive fragmentation. The greater the difference between the proton affinity of the sample and that of the reagent gas, the more energy that is transferred to the sample during ionization. The excess energy produces an analyte ion in a highly excited vibrational state. If enough excess kinetic energy is transferred, the sample ion will fragment through the cleavage of covalent bonds. Therefore, using a reagent gas with a proton affinity matched closely to that of the sample will result in a greater number of intact molecular ions and smaller number of fragment ions. It is unlikely, of course, that one knows the precise proton affinity of the sample, but one can estimate the value by looking at tables of values determined for simple compounds with functional groups similar to the sample in question. A summary of common CI reagent gases and their ions/properties is presented in Table 3.1.

TABLE 3.1
SUMMARY OF CHEMICAL IONIZATION (CI) REAGENT GASES

Reagent Gas	Proton Affinity (kcal/mole)	Reagent Ion(s)	Analyte Ion(s)	Comments
H_2	101	H_3^+	$(M+H)^+$	Produces significant fragmentation
CH_4	132	$CH_5^+, C_2H_5^+$	$(M+H)^+, (M+C_2H_5)^+$	Less fragmentation than H_2, can form adducts
NH_3	204	NH_4^+	$(M+H)^+, (M+NH_4)^+$	Selective ionization, little fragmentation, some adduct formation
$(CH_3)_3CH$	196	$(CH_3)_3C^+$	$(M+H)^+,$ $[M+C(CH_3)_3)]^+$	Mild, selective protonation, little fragmentation
CH_3OH	182	$CH_3OH_2^+$	$(M+H)^+$	Degree of fragmentation observed between that of methane and isobutane
CH_3CN	188	CH_3CNH^+	$(M+H)^+$	Degree of fragmentation observed between that of methane and isobutane

F I G U R E   3 . 3 Comparison of CI-MS data of lavandulyl acetate using methane (top), isobutane (middle), and ammonia (bottom) as reagent gases. (From McLafferty, F. W., and F. Tureček, *Interpretation of Mass Spectra,* 4th ed., University Science Books, Mill Valley, CA, 1993. Reprinted with permission.)

As one can see from Figure 3.3, CI-MS of lavandulyl acetate (MW 196) gives mass spectra with very different appearances depending on the regent gas used to ionize the sample. In the top spectrum, the protonated molecular ion of lavandulyl acetate [(M + H)$^+$, m/z = 197] is barely visible, and the largest peak in the spectrum belongs to the fragment at m/z = 137. In the middle spectrum, acquired using isobutane as reagent gas, the protonated molecular ion at m/z = 197 is much more prominent, and there is less overall fragmentation. Fragmentation is still significant in this case, though, as the ion at m/z = 137 is still the most abundant in the spectrum. Finally, when lavandulyl acetate is ionized using ammonia, the protonated molecular ion is the most abundant ion (the base peak), and almost no fragmentation is observed. Note the presence of an adduct ion [(M + NH$_4$)$^+$, m/z = 214] present in this spectrum.

As a practical note, spectra acquired under CI conditions are usually acquired over a mass range above the m/z of the reagent gas ions. The ionized reagent gas is also detected by the spectrometer, and because the reagent gas is present in great excess relative to the sample, its ions would dominate the spectrum. Thus, CI (methane) spectra are typically acquired above $m/z = 50$ (CH_5^+ is $m/z = 17$, of course, but $C_2H_5^+$ [$m/z = 29$] and $C_3H_5^+$ [$m/z = 41$] are also present), and CI (isobutane) spectra are typically acquired above $m/z = 60$ or 70.

The main advantage of CI-MS is the selective production of intact quasi-molecular ions [$(M + H)^+$]. Figure 3.4 shows the mass spectrum of butyl methacrylate acquired under different ionization conditions. The molecular ion ($m/z = 142$) is barely visible in the EI-MS, but the $(M + H)^+$ ion ($m/z = 143$) is prominent in the CI-MS spectra. The CI-MS acquired using isobutane has

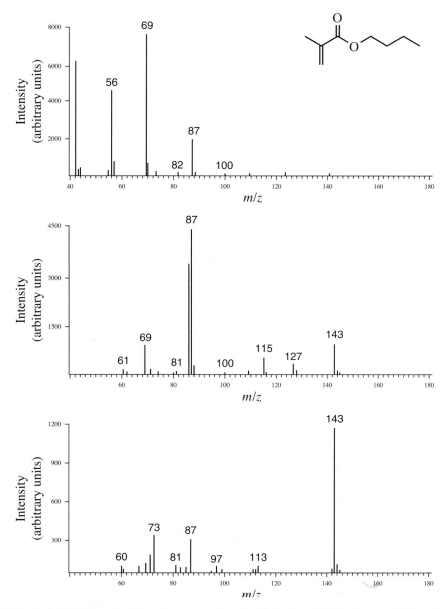

F I G U R E   3 . 4 MS of butyl methacrylate acquired under EI (top) and CI (methane, middle; isobutane, bottom) conditions. (From DeHoffmann, E., and V. Stroobant, *Mass Spectrometry: Principles and Applications,* 2nd ed., John Wiley and Sons, New York, 1999. Reprinted with permission.)

much less fragmentation than the CI-MS acquired using methane as the reagent gas. Other advantages to CI-MS include inexpensive and robust hardware. Like in EI-MS, however, the sample must be readily vaporized to be subjected to chemical ionization, which precludes the analysis of high molecular weight compounds and many biomolecules. CI ion sources are very similar in design to EI sources, and most modern mass spectrometers can switch from EI to CI mode in a matter of minutes.

While protonation is the most commonly encountered ionization method in CI-MS, other ionization processes may be exploited. For example, use of methyl nitrite/methane mixtures as reagent gas produces CH_3O^- that abstracts a proton from the sample, leading to a $(M – H)^-$ parent ion. Similarly, use of NF_3 as reagent gas produces F^- ion as a proton abstraction agent, also leading to $(M – H)^-$ ions. It is also possible to form negatively charged adducts under CI conditions.

Chemical ionization is also used in the open air technique known as direct analysis in real time (DART). In this ionization method (Figure 3.5) a reagent gas, usually He or Ar, is passed through a needle electrode with a 1–5 kV potential, creating excited gas atoms. The excited-state gas collides with atmospheric water to generate protonated water clusters (Equation 3.14). The sample is placed in the ion beam in any number of ways including a wire screen, glass capillary, or other solid surface. Collisions between the sample and these protonated water clusters transfer a proton to the sample (Equation 3.15). Analyte molecular ions then enter the mass analyzer through a small orifice.

$$He^* \quad + \quad n\,H_2O \quad \rightarrow \quad [(H_2O_{n-1})H]^+ \quad + \quad OH^- \quad + \quad He \qquad \textbf{Equation 3.14}$$
$$[(H_2O_{n-1})H]^+ \quad + \quad M \quad \rightarrow \quad (M + H)^+ \quad + \quad n\,H_2O \qquad \textbf{Equation 3.15}$$

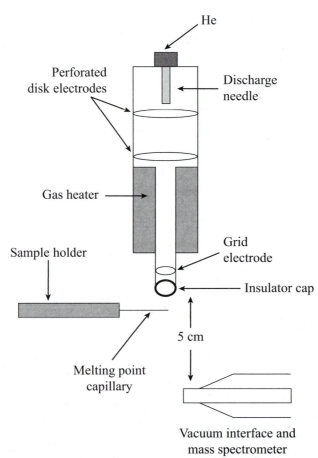

F I G U R E 3 . 5 Schematic representation of a direct analysis in real time (DART) source. (From Petucci, C. et al., *Analytical Chemistry 79*, (2007): 5064.)

C.   Desorption Ionization Techniques (SIMS, FAB, and MALDI)

Both EI and CI methods require a relatively volatile (low molecular weight) sample. More recently developed ionization techniques allow the analysis of large, nonvolatile molecules by mass spectrometry. Three of these methods, **secondary ion mass spectrometry (SIMS)**, **fast atom bombardment (FAB)**, and **matrix-assisted laser desorption ionization (MALDI)** are all **desorption ionization (DI)** techniques. In desorption ionization, the sample to be analyzed is dissolved or dispersed in a matrix and placed in the path of a high-energy (1- to 10-keV) beam of ions (SIMS), neutral atoms (FAB), or high-intensity photons (MALDI). Beams of Ar^+ or Cs^+ are often used in SIMS, and beams of neutral Ar or Xe atoms are common in FAB. Most MALDI spectrometers use a nitrogen laser that emits at 337 nm, but some applications use an infrared (IR) laser for direct analysis of samples contained in gels or thin-layer chromatography (TLC) plates. The collision of these ions/atoms/photons with the sample ionizes some of the sample molecules and ejects them from the surface (Fig. 3.6). The ejected ions are then accelerated toward the mass analyzer as with other ionization methods. Since FAB uses neutral atoms to ionize the sample, both positive-ion and negative-ion detection are possible. Molecular ions in SIMS and FAB are typically $(M + H)^+$ or $(M - H)^-$, but adventitious alkali metals can create $(M + Na)^+$ and $(M + K)^+$ ions also. SIMS and FAB ionization methods may be used on sample compounds with molecular weights up to about 20,000, such as polypeptides and oligonucleotides.

The matrix should be nonvolatile, relatively inert, and a reasonable electrolyte to allow ion formation. If the matrix compound is more acidic than the analyte, then predominantly $(M + H)^+$ ions will be formed, while mostly $(M - H)^-$ ions will result when the matrix is less acidic than the analyte. The matrix absorbs much of the excess energy imparted by the beam of ions/atoms and produces ions that contribute a large amount of background ions to the mass spectrum. In fact, chemical reactions within the matrix during ionization can contribute background ions in most mass

F I G U R E   3 . 6 Schematic representations of desorption ionization techniques.

FIGURE 3.7 Common matrices for SIMS and FAB mass spectrometry.

regions below about 600 *m/z*. Common matrix compounds for SIMS and FAB include glycerol, thioglycerol, 3-nitrobenzyl alcohol, di- and triethanolamine, and mixtures of dithiothreitol (DTT) and dithioerythritol (Fig. 3.7)

The matrix compounds used in MALDI are chosen for their ability to absorb the ultraviolet (UV) light from a laser pulse (337 nm for N_2 laser). Substituted nicotinic, picolinic, and cinnamic acid derivatives are often used in MALDI techniques (Fig. 3.8). The matrix absorbs most of the energy from the laser pulse, thus allowing for the creation of intact sample ions that are ejected from the matrix. MALDI mass spectrometry is useful for analytes spanning a wide range of molecular weights, from small polymers with average molecular weights of a few thousand atomic mass units (amu) to oligosaccharides, oligonucleotides and polypeptides, antibodies, and small proteins with molecular weights approaching 300,000 amu. Furthermore, MALDI requires only a few femtomoles (1×10^{-15} mole) of sample!

FIGURE 3.8 Common matrices for MALDI applications.

D. Electrospray Ionization (ESI)

An even more useful technique for studying high molecular weight biomolecules and other labile or nonvolatile compounds is **electrospray ionization** (**ESI**) and its cousin **thermospray ionization** (**TSI**). In ESI, a solution containing the sample molecules is sprayed out the end of a fine capillary into a heated chamber that is at nearly atmospheric pressure. The capillary through which the sample solution passes has a high voltage potential across its surface, and small, charged droplets are expelled into the ionization chamber. The charged droplets are subjected to a counterflow of a drying gas (usually nitrogen) that evaporates solvent molecules from the droplets. Thus, the charge density of each droplet increases until the electrostatic repulsive forces exceed the surface tension of the droplet (the Rayleigh limit), at which point the droplets break apart into smaller droplets. This process continues until solvent-free sample ions are left in the gas phase (Fig. 3.9). TSI occurs by a similar mechanism but relies on a heated capillary rather than one with an electrostatic potential to initially form the charged droplets. Negative ions may also be formed in ESI by loss of protons from the sample to basic species in solution. ESI has become much more common than TSI over the last decade or two, and because it relies on a sample in solution, ESI is the most logical method to be employed in LC-MS systems.

The charges of the ions generated using ESI do not necessarily reflect the charge state of the sample in solution. The charge transferred to the sample molecules (usually in the form of protons) arises from a combination of charge concentration in the droplets during evaporation of the aerosol and electrochemical processes stemming from the electrostatic potential of the capillary.

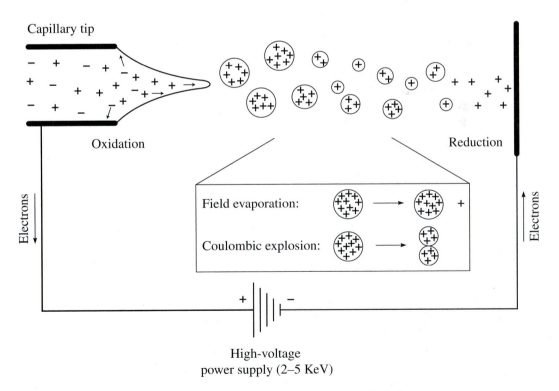

FIGURE 3.9 Schematic representation of electrospray ionization (ESI) showing both field evaporation and coulombic explosion. (From Gross, J. H., *Mass Spectrometry: A Textbook*, Springer, Berlin, 2004. Reprinted by permission.)

FIGURE 3.10 ESI-MS of proteins. Chicken egg white lysozyme in the absence (top) and presence (middle) of dithiothreitol. (From Gross, J. H., *Mass Spectrometry: A Textbook*, Spinger, Berlin, 2004. Reprinted with permission.)

The sample ions may bear a single charge or multiple charges. Figure 3.10 shows the ESI-MS of lysozyme from chicken egg white in the absence and presence of dithiothreitol. In the first spectrum, ions are observed representing protein molecules bearing 10^+, 11^+, 12^+, and 13^+ charges. The latter spectrum shows even more highly charged ions—including a peak from protein bearing a 20^+ charge. The formation of multiply charged ions is particularly useful in the MS analysis of proteins. Typical proteins can carry many protons due to the presence of basic amino acid side chains, resulting in peaks at $m/z = 600$–2000 for proteins with a molecular weight that approaches 200,000 amu.

The data shown in Figure 3.10 can be used to calculate the molecular mass for lysozyme. The mass is calculated by multiplying the charge on the lysozyme by the m/z value shown on the chromatogram. For example:

$$(10)(1432) = 14,320 \text{ AMU}$$
$$(12)(1193) = 14,316$$
$$(15)(955) = 14,325$$

Thus, the molecular mass of lysozyme is about 14,320 AMU.

ESI-MS is not limited to the study of large biomolecules, however. Many small molecules with molecular weight in the 100–1500 range can be studied by ESI-MS. Compounds that are too nonvolatile to be introduced by direct probe methods or are too polar or thermally labile to be introduced by GC-MS methods are ideal for study by LC-MS using ESI techniques.

Desorption electrospray ionization (DESI) combines the soft ionization of the electrospray technique with desorption of the sample ions from a surface. Unlike MALDI, however, no matrix is needed. The DESI technique uses electrosprayed aqueous aerosols to ionize and desorb analyte ions.

Developed around the same time as the DART technique (Section 3.3B), DESI interfaces with a mass analyzer using a heated ion transfer tube that is in some cases flexible and can be held in the researcher's hand directly above the sample surface.

Soft ionization techniques like DESI and DART coupled with the appropriate mass analyzer can produce accurate mass spectra to determine exact elemental composition (Section 3.7). The open ion source configuration allows the sampling from a number of matrices and surfaces as widely varied as intact plant material, cloth, concrete, and even human skin. Moving the sample in the ion beam provides spatial resolution, allowing one to observe differing compositions in different areas of the same sample. These techniques are useful for forensic and public safety applications with the ability to detect pictograms of material including biological molecules and explosive residues.

3.4 MASS ANALYSIS

Once the sample has been ionized, the beam of ions is accelerated by an electric field and then passes into the **mass analyzer,** the region of the mass spectrometer where the ions are separated according to their mass-to-charge (*m/z*) ratios. Just like there are many different ionization methods for different applications, there are also several types of mass analyzers. While some mass analyzers are more versatile than others, none of the options are one-size-fits-all.

A.   The Magnetic Sector Mass Analyzer

The kinetic energy of an accelerated ion is equal to

$$\frac{1}{2}mv^2 = zV \qquad \text{Equation 3.16}$$

where m is the mass of the ion, v is the velocity of the ion, z is the charge on the ion, and V is the potential difference of the ion-accelerating plates. In the **magnetic sector** mass analyzer (Fig. 3.11), the ions are passed between the poles of a magnet. In the presence of a magnetic field, a charged particle describes a curved flight path. The equation that yields the radius of curvature of this path is

$$r = \frac{mv}{zB} \qquad \text{Equation 3.17}$$

where r is the radius of curvature of the path, and B is the strength of the magnetic field. If these two equations are combined to eliminate the velocity term, the result is

$$\frac{m}{z} = \frac{B^2 r^2}{2V} \qquad \text{Equation 3.18}$$

As can be seen from Equation 3.18, the greater the value of *m/z*, the larger the radius of the curved path. The analyzer tube of the instrument is constructed to have a fixed radius of curvature. A particle with the correct *m/z* ratio can negotiate the curved analyzer tube and reach the detector. Particles with *m/z* ratios that are either too large or too small strike the sides of the analyzer tube and do not reach the detector. The method would not be very interesting if ions of only one mass could be detected. Therefore, the magnetic field strength is continuously varied (called a *magnetic field scan*) so that all of the ions produced in the ionization chamber can be detected. The record produced from the detector system is in the form of a plot of the numbers of ions versus their *m/z* values.

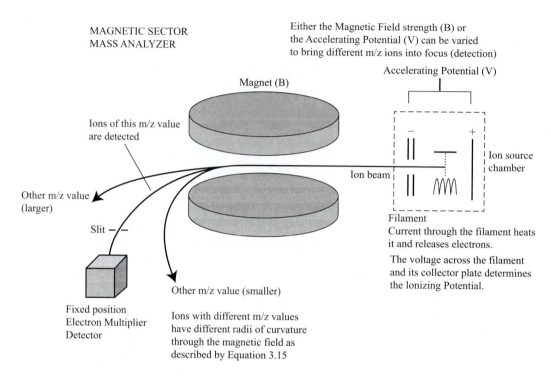

MAGNETIC SECTOR
MASS ANALYZER

Either the Magnetic Field strength (B) or
the Accelerating Potential (V) can be varied
to bring different m/z ions into focus (detection)

Accelerating Potential (V)

Magnet (B)

Ions of this m/z value
are detected

Ion source
chamber

Ion beam

Other m/z value
(larger)

Slit

Filament

Current through the filament heats
it and releases electrons.

The voltage across the filament
and its collector plate determines
the Ionizing Potential.

Fixed position
Electron Multiplier
Detector

Other m/z value (smaller)

Ions with different m/z values
have different radii of curvature
through the magnetic field as
described by Equation 3.15

FIGURE 3.11 Schematic of a magnetic sector mass analyzer.

An important consideration in mass spectrometry is **resolution,** defined according to the relationship

$$R = \frac{M}{\Delta M}$$

Equation 3.19

where R is the resolution, M is the mass of the particle, and ΔM is the difference in mass between a particle of mass M and the particle of next higher mass that can be resolved by the instrument. A magnetic sector analyzer can have R values approaching 10,000, depending on the radius of curvature and slit widths.

B. Double-Focusing Mass Analyzers

For many applications, much higher resolution is needed and can be achieved through modifications of this basic magnetic sector design. In fact, magnetic sector analyzers are used today only in **double-focusing mass spectrometers.** The particles leaving the ionization chamber do not all have precisely the same velocity, so the beam of ions passes through an electric field region before or after the magnetic sector (Fig. 3.12). In the presence of an electric field, the particles all travel at the same velocity. The particles describe a curved path in each of these regions, and the resolution of the mass analyzer improves—by a factor of 10 or more over the magnetic sector alone.

C. Quadrupole Mass Analyzers

In a **quadrupole mass analyzer** (Fig. 3.13), a set of four solid rods is arranged parallel to the direction of the ion beam. The rods should be hyperbolic in cross section, although cylindrical rods may be used. A direct-current (DC) voltage and a radiofrequency (RF) is applied to the

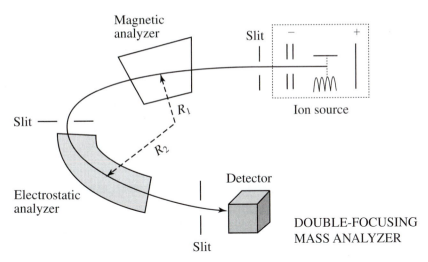

F I G U R E 3 . 1 2 Schematic of a double-focusing mass analyzer.

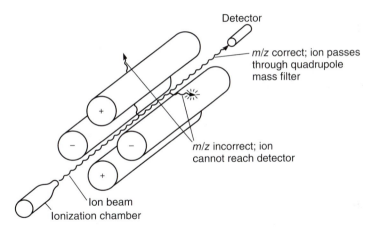

F I G U R E 3 . 1 3 Quadrupole mass analyzer.

rods, generating an oscillating electrostatic field in the region between the rods. Depending on the ratio of the RF amplitude to the DC voltage, ions acquire an oscillation in this electrostatic field. Ions of an incorrect *m/z* ratio (too small or too large) undergo an unstable oscillation. The amplitude of the oscillation continues to increase until the particle strikes one of the rods. Ions of the correct mass-to-charge ratio undergo a stable oscillation of constant amplitude and travel down the quadrupole axis with a "corkscrew"-type trajectory. These ions do not strike the quadrupole rods but pass through the analyzer to reach the detector. Like the magnetic sector analyzer, the quadrupole can be scanned from high to low values of *m/z*. A quadrupole mass analyzer is found in most "benchtop" GC-MS systems and typically has a *m/z* range from 0 to 1000, although quadrupole analyzers are available on some LC-MS systems with *m/z* ranges that approach 2000. Quadrupole mass spectrometers are low-resolution instruments (R~3000) incapable of providing exact elemental composition of the sample but their relatively low cost makes them popular for many applications. An additional drawback is their relatively slow acquisition rate due to their scanning nature.

The quadrupole **ion trap** mass analyzer operates by similar principles as the linear quadrupole described above and is a common mass analyzer found in GC-MS instruments. The ion trap consists of two hyperbolic endcap electrodes and a doughnut-shaped ring electrode (the endcap

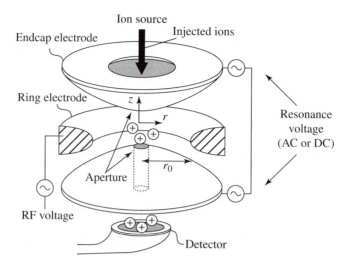

F I G U R E   3 . 1 4 Quadrupole ion trap mass analyzer. (From Gross, J. H., *Mass Spectrometry: A Textbook,* Springer, Berlin, 2004. Reprinted with permission.)

electrodes are connected). An alternating current (AC) (or DC) and an RF potential is applied between the endcaps and the ring electrode (Fig. 3.14). In the linear quadrupole analyzer, ions of different *m/z* values are allowed to pass in turn through the quadrupole by adjusting the RF and DC voltages. In the ion trap, ions of all *m/z* values are in the trap simultaneously, oscillating in concentric trajectories. Sweeping the RF potential results in the removal of ions with increasing *m/z* values by putting them in unstable trajectory that causes them to be ejected from the trap in the axial direction toward the detector. This process is called **resonant ejection.** Ion trap mass analyzers are somewhat more sensitive than linear quadrupole instruments, but they have similar resolution capabilities.

Because the ion trap contains ions of all values of *m/z* at the same time (as well as neutral molecules that were not ionized prior to entering the trap), ion trap mass analyzers are also sensitive to overload and ion–molecule collisions that complicate the resulting spectrum. Recall that not all of the sample molecules get ionized—many remain uncharged. These neutral species move in a random path in the ion trap, resulting in collisions with ions as the ions oscillate in their stable trajectories. These collisions result in chemical ionization-type ionization events (Equation 3.20). This is sometimes referred to as *self-CI.*

$$(R-H)^+ \quad + \quad M \quad \longrightarrow \quad R \quad + \quad (M+H)^+$$

(R−H)⁺	+	M	→	R	+	(M + H)⁺	
fragment ion		neutral		neutral		protonated	**Equation 3.20**
		sample molecule		or radical		molecular ion	

The result is an abnormally large (M + H)⁺ peak in the mass spectrum. This is observed in Figure 3.15, in which the base peak in the EI-MS of methyl dodecanoate under standard conditions has *m/z* = 215, representing an (M + H)⁺ ion produced in the ion trap from ion–molecule conditions. This self-CI process can be minimized by increasing ionization efficiency, reducing the number of ions in the trap (injecting less sample), or both. The bottom spectrum in Figure 3.15 was acquired under optimized ion trap conditions with a longer ion residence time. Now, the M⁺ ion is clearly visible, although the (M + 1) peak is still much larger than it should be based on isotopic contributions of ¹³C alone (see Section 3.7). Fortunately, the presence of the

FIGURE 3.15 EI-MS of methyl dodecanoate using a quadrupole ion trap mass analyzer. Standard conditions (top) and optimized conditions to minimize ion–molecule collisions and self-CI (bottom). (Reproduced from Varian, Inc.)

larger (M + 1) peak rarely has an adverse effect on spectral library searches done by a computer. The visual inspection of a sample spectrum to a printed standard spectrum is quite another matter. The self-CI peak becomes quite problematic when one is attempting to characterize unknowns if one does not know the molecular formula or functional groups present ahead of time.

D.   Time-of-Flight Mass Analyzers

The **time-of-flight (TOF)** mass analyzer is based on the simple idea that the velocities of two ions, created at the same instant with the same kinetic energy, will vary depending on the mass of the ions—the lighter ion will have a higher velocity. If these ions are traveling toward the mass spectrometer's detector, the faster (lighter) ion will strike the detector first. Examining this concept further, the kinetic energy of an ion accelerated through an electrical potential V will be

$$z\mathrm{V} = \frac{m\mathrm{v}^2}{2}$$ **Equation 3.21**

and the velocity of the ion is the length of the flight path L divided by the time *t* it takes the ion to travel over that distance:

$$\mathrm{v} = \frac{L}{t}$$ **Equation 3.22**

Replacing this expression for v in Equation 8.19 gives

$$z\mathrm{V} = \frac{m L^2}{2t^2}$$ **Equation 3.23**

Thus, it follows that

$$\frac{m}{z} = \frac{2\mathrm{V}t^2}{L^2}$$ **Equation 3.24**

The TOF mass analyzer (Fig. 3.16) requires very fast electronics to accurately measure ion flight times that may be submicrosecond. Furthermore, the ions in a TOF system must be created in short, well-defined pulses so that the ions all start their journey toward the detector at the same moment. The first requirement explains why TOF instrumentation (first developed in the 1940s

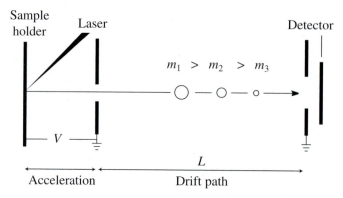

F I G U R E   3 . 1 6 Schematic representation of a MALDI/TOF mass spectrometer.

and 1950s) did not become widely used until the 1980s and 1990s, when suitable circuitry became cost-effective. The last requirement is perfectly suited for the MALDI ionization technique, and MALDI/TOF mass spectrometers have found wide use in the analysis of biomolecules and synthetic polymers. In theory, TOF mass analyzers have no upper limit to their effective mass range, and these mass analyzers have high sensitivity. Unlike magnetic sector or quadrupole spectrometers, in which some of the ions are "thrown away" during the experiment, TOF instruments are able to analyze (in principle) every ion created in the initial pulse. Mass data have been obtained using MALDI/TOF from samples with molecular weights of 300,000 amu and as little as a few hundred attomoles of material.

The major disadvantage of the TOF analyzer is its inherently low resolution. The mass resolution (**R**, Eq. 3.19) of the TOF instrument is proportional to the ion's flight time, so using longer drift tubes increases resolution. Flight tubes a few meters long are commonly used in high-end instruments. With shorter drift tubes, R of only 200–500 is possible. A modification to the TOF analyzer that increases resolution is the ion reflector. The reflector is an electric field behind the free drift region of the spectrometer that behaves as an ion mirror. The reflector is able to refocus ions of slightly different kinetic energies and, if set at a small angle, sends the ions on a path back toward the original ion source. This essentially doubles the ion flight path as well. In reflector TOF instruments, a mass resolution of several thousand is possible. Combining a quadrupole with a TOF analyzer (QTOF) will provide sufficient resolution in most cases for accurate mass determination (Section 3.6).

Time-of-flight mass spectrometers are relatively simple, which makes it possible to use them in the field. During the 1991 Gulf War, concern arose that Iraqi troops might be releasing chemical warfare agents against American troops. To guard against that possibility, the U.S. Army deployed a number of tracked vehicles, each equipped with a mass spectrometer. The mass spectrometer was used to sample the air and provide advance warning should any poisonous gases be released into the air. Basic TOF mass spectrometers are also used to detect residue from explosives and illegal drugs at security screening stations in airports. Because of their value for studying short-lived species, TOF mass spectrometers are particularly useful in kinetic studies, especially with applications to very fast reactions. Very rapid reactions such as combustion and explosions can be investigated with this technique.

3.5 DETECTION AND QUANTITATION: THE MASS SPECTRUM

The **detector** of a typical mass spectrometer consists of a counter that produces a current that is proportional to the number of ions that strike it. This sounds quite reasonable until one pauses to think about exactly how many ions will strike the detector in a typical experiment. Consider a typical application—analysis of a small organic molecule (MW = 250) by EI GC-MS. A 1.0-μL injection of a 1.0 mg/mL sample contains 3.6×10^{15} molecules. If the GC is running in split mode with a 1:100 ratio, only 3.6×10^{13} molecules enter the chromatographic column. A mass spectrum acquired at the top of the GC peak may only account for 10% of the material that elutes, and if only 1 in 1000 molecules is converted to an ion, just 3.6 billion ions are available. This still sounds like a lot of charged particles, but wait! In a scanning spectrometer, most of these ions never reach the detector; as the mass analyzer sweeps through the range of 35 to 300 m/z, most of the ions discharge on the quadrupole rods, for example. In a case like this, an ion of any given m/z value makes it through the analyzer only 1 time out of 300. Clearly, each peak in the mass spectrum represents a very small electrical signal, and the detector must be able to amplify this tiny current.

Through the use of **electron multiplier** circuits, this current can be measured so accurately that the current caused by just one ion striking the detector can be measured. These detectors are based on the simple concept of the Faraday cup, a metal cup that is in the path of ions emanating from the mass analyzer. When an ion strikes the surface of the electron multiplier two electrons are ejected. The approximately 2-kV potential difference between the opening and end of the detector

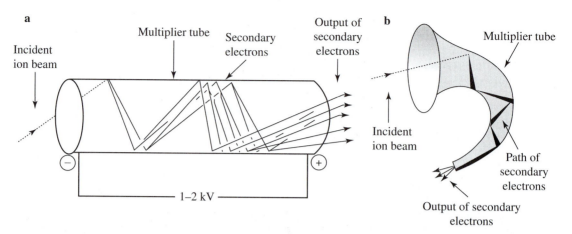

F I G U R E 3 . 1 7 Schematic representation of a linear channel electron multiplier (a) and a curved channel electron multiplier (b). (Based on, J. H., *Mass Spectrometry: A Textbook*, Spinger, Berlin, 2004.)

draws the electrons further into the electron multiplier, where each electron strikes the surface again, each causing the ejection of two more electrons. This process continues until the end of the electron multiplier is reached, and the electrical current is analyzed and recorded by the data system. The signal amplification just described will be 2^n, where n is the number of collisions with the electron multiplier surface. Typical electron multipliers provide a signal increase of 10^5–10^6. Two configurations of electron multipliers are shown in Figure 3.17. A curved electron multiplier shortens the ion path and results in a signal with less noise. Photomultiplier detectors operate on a similar principle as the electron multiplier, except ion collisions with the fluorescent screen in the photomultiplier result in photon emission proportional to the number of ion collisions. The intensity of the light (rather than electrical current) is then analyzed and recorded by the data system.

The signal from the detector is fed to a **recorder**, which produces the mass spectrum. In modern instruments, the output of the detector is fed through an interface to a computer. The computer can store the data, provide the output in both tabular and graphic forms, and compare the data to standard spectra, which are contained in spectra libraries that are also stored in the computer.

Figure 3.18 is a portion of a typical mass spectrum—that of dopamine, a substance that acts as a neurotransmitter in the central nervous system. The x-axis of the mass spectrum is the m/z ratio, and the y-axis is ion abundance. Mass spectral results may also be presented in tabular form, as in Table 3.2. The most abundant ion formed in the ionization chamber gives rise to the tallest peak in the mass spectrum, called the **base peak.** In the mass spectrum of dopamine, the base peak is indicated at an m/z value of 124. The spectral intensities are normalized by setting the base peak to relative abundance 100, and the rest of the ions are reported as percentages of the base peak intensity. The low end of the m/z range is typically 35 or 40 to eliminate the very large peaks from low-mass fragments from background ions arising from gases and small alkyl fragments. When acquiring data under CI conditions, the low end of the m/z range is set higher to eliminate the large peaks from the reagent gas ions.

As discussed earlier, in EI-MS, the beam of electrons in the ionization chamber converts some of the sample molecules to positive ions. The simple removal of an electron from a molecule yields an ion with weight that is the actual molecular weight of the original molecule. This is the **molecular ion,** which is usually represented by M^+ or $M^{\bullet+}$. Strictly speaking, the molecular ion is a **radical cation** since it contains an unpaired electron as well as a positive charge. The value of m/z at which the molecular ion appears on the mass spectrum, assuming that the ion has only one electron missing, gives the molecular weight of the original molecule. If you can identify the molecular ion peak

FIGURE 3.18 Partial EI-MS of dopamine.

in the mass spectrum, you will be able to use the spectrum to determine the molecular weight of an unknown substance. Ignoring heavy isotopes for the moment, the molecular ion peak is the peak in the mass spectrum with the largest m/z value; it is indicated in the graphic presentation in Figure 3.18 (m/z = 153).

Molecules in nature do not occur as isotopically pure species. Virtually all atoms have heavier isotopes that occur in characteristic natural abundances. Hydrogen occurs largely as ^{1}H, but about 0.02% of hydrogen atoms are the isotope ^{2}H. Carbon normally occurs as ^{12}C, but about 1.1% of carbon atoms are the heavier isotope ^{13}C. With the possible exception of fluorine and a few other elements, most elements have a certain percentage of naturally occurring heavier isotopes.

Peaks caused by ions bearing those heavier isotopes also appear in mass spectra. The relative abundances of such isotopic peaks are proportional to the abundances of the isotopes in nature. Most often, the isotopes occur one or two mass units above the mass of the "normal" atom. Therefore, besides looking for the molecular ion (M^+) peak, one would also attempt to locate **M + 1** and **M + 2** peaks. As Section 3.6 will demonstrate, the relative abundances of the $M + 1$ and $M + 2$ peaks can be used to determine the molecular formula of the substance being studied. In Figure 3.18, the isotopic peaks are the low-intensity peaks at m/z values (154 and 155) higher than that of the molecular ion peak (see also Table 3.2).

We have seen that the beam of electrons in the ionization chamber can produce the molecular ion. This beam is also sufficiently powerful to break some of the bonds in the molecule, producing a series of molecular fragments. The positively charged fragments are also accelerated in the ionization chamber, sent through the analyzer, detected, and recorded on the mass spectrum. These **fragment ions** appear at m/z values corresponding to their individual masses. Very often, a fragment ion, rather than the parent ion, is the most abundant ion produced in the mass spectrum. A second means of producing fragment ions exists if the molecular ion, once it is formed, is so unstable that it disintegrates before it can pass into the accelerating region of the ionization chamber. Lifetimes less than 10^{-6} sec are typical in this type of fragmentation. The fragments that are charged then appear as fragment ions in the mass spectrum. A great deal of structural information about a substance can be determined from an examination of the fragmentation pattern in the mass spectrum. Chapter 4 will examine some fragmentation patterns for common classes of compounds.

TABLE 3.2
EI-MS OF DOPAMINE. TABULAR REPRESENTATION OF THE DATA IN FIGURE 3.18

m/z	Relative Abundance	m/z	Relative Abundance	m/z	Relative Abundance
50	4.00	76	1.48	114	0.05
50.5	0.05	77	24.29	115	0.19
51	25.71	78	10.48	116	0.24
51.5	0.19	79	2.71	117	0.24
52	3.00	80	0.81	118	0.14
52.5	0.62	81	1.05	119	0.19
53	5.43	82	0.67	120	0.14
53.5	0.19	83	0.14	121	0.24
54	1.00	84	0.10	122	0.71
55	4.00	85	0.10	123	41.43
56	0.43	86	0.14	124	100.00 (base peak)
56.5	0.05 (metastable peak)	87	0.14	125	7.62
57	0.33	88	0.19	126	0.71
58	0.10	89	1.57	127	0.10
58.5	0.05	89.7	0.10 (metastable peak)	128	0.10
59	0.05	90	0.57	129	0.10
59.5	0.05	90.7	0.10 (metastable peak)	131	0.05
60	0.10	91	0.76	132	0.19
60.5	0.05	92	0.43	133	0.14
61	0.52	93	0.43	134	0.52
61.5	0.10	94	1.76	135	0.52
62	1.57	95	1.43	136	1.48
63	3.29	96	0.52	137	0.33
64	1.57	97	0.14	138	0.10
65	3.57	98	0.05	139	0.10
65.5	0.05	99	0.05	141	0.19
66	3.14	100.6	0.19 (metastable peak)	142	0.05
66.5	0.14	101	0.10	143	0.05
67	2.86	102	0.14	144	0.05
67.5	0.10	103	0.24	145	0.05
68	0.67	104	0.76	146	0.05
69	0.43	105	4.29	147	0.05
70	0.24	106	4.29	148	0.10
71	0.19	107	3.29	149	0.24
72	0.05	108	0.43	150	0.33
73	0.14	109	0.48	151	1.00
74	0.67	110	0.86	152	0.38
74.5	0.05	111	0.10	153	13.33 (molecular ion)
75	1.00	112	0.05	154	1.48
75.5	0.14	113	0.05	155	0.19

Ions with lifetimes on the order of 10^{-6} sec are accelerated in the ionization chamber before they have an opportunity to disintegrate. These ions may disintegrate into fragments *while they are passing into the analyzer region* of the mass spectrometer. The fragment ions formed at that point have considerably lower energy than normal ions since the uncharged portion of the original ion carries away some of the kinetic energy that the ion received as it was accelerated. As a result, the fragment ion produced in the analyzer follows an abnormal flight path on its way to the detector. This ion appears at an *m/z* ratio that depends on its own mass, as well as the mass of the original ion from which it formed. Such an ion gives rise to what is termed a **metastable ion peak** in the mass spectrum. Metastable ion peaks are usually broad peaks, and they frequently appear at nonintegral values of *m/z*. The equation that relates the position of the metastable ion peak in the mass spectrum to the mass of the original ion is

$$m_1^+ \longrightarrow m_2^+ \qquad\qquad \textbf{Equation 3.25}$$

and

$$m^* = \frac{(m_2)^2}{m_1} \qquad\qquad \textbf{Equation 3.26}$$

where m^* is the apparent mass of the metastable ion in the mass spectrum, m_1 is the mass of the original ion from which the fragment formed, and m_2 is the mass of the new fragment ion. A metastable ion peak is useful in some applications since its presence definitively links two ions together. Metastable ion peaks can be used to prove a proposed fragmentation pattern or to aid in the solution of structure proof problems.

3.6 DETERMINATION OF MOLECULAR WEIGHT

Section 3.3 showed that when a beam of high-energy electrons impinges on a stream of sample molecules, ionization of electrons from the molecules takes place. The resulting ions, called **molecular ions,** are then accelerated, sent through a magnetic field, and detected. If these molecular ions have lifetimes of at least 10^{-5} sec, they reach the detector without breaking into fragments. The user then observes the *m/z* ratio that corresponds to the molecular ion to determine the molecular weight of the sample molecule.

In practice, molecular weight determination is not quite as easy as the preceding paragraph suggests. First, you must understand that the value of the mass of any ion accelerated in a mass spectrometer is its true mass, the sum of the masses of each atom in that single ion, and not its molecular weight calculated from chemical atomic weights. The chemical scale of atomic weights is based on weighted averages of the weights of all of the isotopes of a given element. The mass spectrometer can distinguish between masses of particles bearing the most common isotopes of the elements and particles bearing heavier isotopes. Consequently, the masses that are observed for molecular ions are the masses of the molecules in which every atom is present as its most common isotope. In the second place, molecules subjected to bombardment by electrons may break apart into fragment ions. As a result of this fragmentation, mass spectra can be quite complex, with peaks appearing at a variety of *m/z* ratios. You must be quite careful to be certain that the suspected peak is indeed that of the molecular ion and not that of a fragment ion. This distinction becomes particularly crucial when the abundance of the molecular ion is low, as when the molecular ion is rather unstable and fragments easily. The masses of the ions detected in the mass spectrum must be measured accurately. An error of only one mass unit in the assignment of mass spectral peaks can render determination of structure impossible.

One method of confirming that a particular peak corresponds to a molecular ion is to vary the energy of the ionizing electron beam. If the energy of the beam is lowered, the tendency of the molecular ion to fragment lessens. As a result, the intensity of the molecular ion peak should increase with decreasing electron potential, while the intensities of the fragment ion peaks should decrease. Certain facts must apply to a molecular ion peak:

1. The peak must correspond to the ion of highest mass in the spectrum, excluding isotopic peaks that occur at higher masses. The isotopic peaks are usually of much lower intensity than the molecular ion peak. At the sample pressures used in most spectral studies, the probability that ions and molecules will collide to form heavier particles is quite low. Care must be taken, especially with GC-MS spectra, to recognize background ions that are a result of column bleed—small pieces of the silicone-based stationary phase of the capillary GC column.

2. The ion must have an odd number of electrons. When a molecule is ionized by an electron beam, it loses one electron to become a radical cation. The charge on such an ion is 1, thus making it an ion with an odd number of electrons.

3. The ion must be capable of forming the important fragment ions in the spectrum, particularly the fragments of relatively high mass, by loss of logical neutral fragments. Fragment ions in the range from $(M - 3)$ to $(M - 14)$ and $(M - 21)$ to $(M - 25)$ are not reasonable losses. Similarly, no fragment ion can contain a greater number of atoms of a particular element than the molecular ion. Chapter 4 will explain fragmentation processes in detail.

The observed abundance of the suspected molecular ion must correspond to expectations based on the assumed molecule structure. Highly branched substances undergo fragmentation very easily. Observation of an intense molecular ion peak for a highly branched molecule thus would be unlikely. The lifetimes of molecular ions vary according to the generalized sequence shown in below.

Another rule that is sometimes used to verify that a given peak corresponds to the molecular ion is the so-called **Nitrogen Rule.** This rule states that if a compound has an even number of nitrogen atoms (zero is an even number), its molecular ion will appear at an even mass value. On the other hand, a molecule with an odd number of nitrogen atoms will form a molecular ion with an odd mass. The Nitrogen Rule stems from the fact that nitrogen, although it has an even mass, has an odd-numbered valence. Consequently, an extra hydrogen atom is included as a part of the molecule,

	Aromatic compounds
	Conjugated alkenes
	Alicyclic compounds
	Organic sulfides
	Unbranched hydrocarbons
	Mercaptans
Molecular ion lifetime	Ketones
	Amines
	Esters
	Ethers
	Carboxylic acids
	Branched hydrocarbons
	Alcohols

giving it an odd mass. To picture this effect, consider ethylamine, $CH_3CH_2NH_2$. This substance has one nitrogen atom, and its mass is an odd number (45), whereas ethylenediamine, $H_2NCH_2CH_2NH_2$, has two nitrogen atoms, and its mass is an even number (60).

One must be careful when studying molecules containing chlorine or bromine atoms since these elements have two commonly occurring isotopes. Chlorine has isotopes of 35 (relative abundance = 75.77%) and 37 (relative abundance = 24.23%); bromine has isotopes of 79 (relative abundance = 50.5%) and 81 (relative abundance = 49.5%). When these elements are present, take special care not to confuse the molecular ion peak with a peak corresponding to the molecular ion with a heavier halogen isotope present. This is discussed further in Section 3.7B.

In many of the cases that you are likely to encounter in mass spectrometry, the molecular ion can be observed in the mass spectrum. Once you have identified that peak in the spectrum, the problem of molecular weight determination is solved. However, with molecules that form unstable molecular ions, you may not observe the molecular ion peak. Molecular ions with lifetimes less than 10^{-5} sec break up into fragments before they can be accelerated. The only peaks that are observed in such cases are those due to fragment ions. In many of these cases, using a mild CI method will allow for detection of the pseudomolecular ion $(M + H)^+$, and one can determine the molecular weight of the compound by simply subtracting one mass unit for the extra H atom present. If a molecular ion is not able to be detected by this method, then you will be obliged to deduce the molecular weight of the substance from the fragmentation pattern on the basis of known patterns of fragmentation for certain classes of compounds. For example, alcohols undergo dehydration very easily. Consequently, the initially formed molecular ion loses water (mass = 18) as a neutral fragment before it can be accelerated toward the mass analyzer. To determine the mass of an alcohol molecular ion, you must locate the heaviest fragment and keep in mind that it may be necessary to add 18 to its mass. Similarly, acetate esters undergo loss of acetic acid (mass = 60) easily. If acetic acid is lost, the weight of the molecular ion is 60 mass units higher than the mass of the heaviest fragment.

Since oxygen compounds form fairly stable oxonium ions and nitrogen compounds form ammonium ions, ion–molecule collisions form peaks in the mass spectrum that appear one mass unit *higher* than the mass of the molecular ion. This was referred to as self-CI in the discussion of the ion trap mass analyzer in Section 3.4. At times, the formation of ion–molecule products may be helpful in the determination of the molecular weight of an oxygen or nitrogen compound, but this self-CI can sometimes be confusing when one is trying to determine the true molecular ion in a spectrum of an unknown sample.

3.7 DETERMINATION OF MOLECULAR FORMULAS

A. Precise Mass Determination

Perhaps the most important application of high-resolution mass spectrometers is the determination of very precise molecular weights of substances. We are accustomed to thinking of atoms as having integral atomic masses—for example, H = 1, C = 12, and O = 16. However, if we determine atomic masses with sufficient precision, we find that this is not true. In 1923, Aston discovered that every isotopic mass is characterized by a small "mass defect." The mass of each atom actually differs from a whole mass number by an amount known as the *nuclear packing fraction*. Table 3.4 gives the actual masses of some atoms.

Depending on the atoms contained in a molecule, it is possible for particles of the same nominal mass to have slightly different measured masses when precise determinations are possible. To illustrate, a molecule with a molecular weight of 60.1 g/mole could be C_3H_8O, $C_2H_8N_2$, $C_2H_4O_2$, or CH_4N_2O (Table 3.3). Thus, a **low-resolution mass spectrum (LRMS)** will not be able to distinguish between these formulas. If one calculates the precise masses for each formula using the mass of the most common isotope for each element, however, mass differences between the formulas appear in the second and third decimal places. Observation of a molecular ion with a mass of 60.058 would

TABLE 3.3
SELECTED COMPARISONS OF MOLECULAR WEIGHTS AND PRECISE MASSES

Molecular Formula (MF)	Molecular Weight (MW) (g/mole)	Precise Mass
C_3H_8O	60.1	60.05754
$C_2H_8N_2$	60.1	60.06884
$C_2H_4O_2$	60.1	60.02112
CH_4N_2O	60.1	60.03242

establish that the unknown molecule is C_3H_8O. An instrument with a resolution of about 5320 would be required to distinguish among these peaks. That is well within the capability of modern mass spectrometers, which can attain resolutions greater than one part in 20,000. A **high-resolution mass spectrum (HRMS)**, then, not only determines the exact mass of the molecular ion, it allows one to know the exact molecular formula. Typical high-resolution instruments can determine an ion's m/z value to four or five decimal places. When the precise mass is measured to this degree of precision, only one formula (excluding isotopes) will fit the data. HRMS is extremely valuable to synthetic chemists as well as researchers doing natural product isolation/structure determination work or drug metabolism studies. It is interesting to compare the precision of molecular weight determinations by mass spectrometry with the chemical methods described in Chapter 1, Section 1.2. Chemical methods give results that are accurate to only two or three significant figures (±0.1% to 1%). Molecular weights determined by mass spectrometry have an accuracy of about ±0.005%. Clearly, mass spectrometry is much more precise than chemical methods of determining molecular weight. Precise mass values for some commonly encountered elements may be found in Table 3.4.

B. Isotope Ratio Data

The preceding section described a method of determining molecular formulas using data from high-resolution mass spectrometers. Another method of determining molecular formulas is to examine the relative intensities of the peaks due to the molecular ion and related ions that bear one or more heavy isotopes (the molecular ion cluster). This method would not be commonly used by researchers who have a high-resolution mass spectrometer at their disposal or are able to submit their samples to a service laboratory for exact mass analysis. Use of the molecular ion cluster can be useful, though, for a relatively quick determination of the molecular formula that does not require the much more expensive high-resolution instrument. This method is useless, of course, when the molecular ion peak is very weak or does not appear. Sometimes the isotopic peaks surrounding the molecular ion are difficult to locate in the mass spectrum, and the results obtained by this method may at times be rendered ambiguous.

The example of ethane (C_2H_6) can illustrate the determination of a molecular formula from a comparison of the intensities of mass spectral peaks of the molecular ion and the ions bearing heavier isotopes. Ethane has a molecular weight of 30 when it contains the most common isotopes of carbon and hydrogen. Its molecular ion peak should appear at a position in the spectrum corresponding to $m/z = 30$. Occasionally, however, a sample of ethane yields a molecule in which one of the carbon atoms is a heavy isotope of carbon, ^{13}C. This molecule would appear in the mass spectrum at $m/z = 31$. The relative abundance of ^{13}C in nature is 1.08% of the ^{12}C atoms. In the tremendous number of molecules in a sample of ethane gas, one of the carbon atoms of ethane will turn out to be a ^{13}C atom 1.08% of the time. Since there are two carbon atoms in the molecule, an ethane with mass 31 will

TABLE 3.4
PRECISE MASSES OF SOME COMMON ELEMENTS

Element	Atomic Weight	Nuclide	Mass
Hydrogen	1.00797	^1H	1.00783
		^2H	2.01410
Carbon	12.01115	^{12}C	12.0000
		^{13}C	13.00336
Nitrogen	14.0067	^{14}N	14.0031
		^{15}N	15.0001
Oxygen	15.9994	^{16}O	15.9949
		^{17}O	16.9991
		^{18}O	17.9992
Fluorine	18.9984	^{19}F	18.9984
Silicon	28.086	^{28}Si	27.9769
		^{29}Si	28.9765
		^{30}Si	29.9738
Phosphorus	30.974	^{31}P	30.9738
Sulfur	32.064	^{32}S	31.9721
		^{33}S	32.9715
		^{34}S	33.9679
Chlorine	35.453	^{35}Cl	34.9689
		^{37}Cl	36.9659
Bromine	79.909	^{79}Br	78.9183
		^{81}Br	80.9163
Iodine	126.904	^{127}I	126.9045

turn up (2×1.08) or 2.16% of the time. Thus, we would expect to observe a peak at $m/z = 31$ with an intensity of 2.16% of the molecular ion peak intensity at $m/z = 30$. This mass 31 peak is called the $M + 1$ peak since its mass is one unit higher than that of the molecular ion. You may notice that a particle of mass 31 could form in another manner. If a deuterium atom, ^2H, replaced one of the hydrogen atoms of ethane, the molecule would also have a mass of 31. The natural abundance of deuterium is only 0.016% of the abundance of ^1H atoms. The intensity of the $M + 1$ peak would be (6×0.016) or 0.096% of the intensity of the molecular ion peak if we consider only contributions due to deuterium. When we add these contributions to those of ^{13}C, we obtain the observed intensity of the $M + 1$ peak, which is 2.26% of the intensity of the molecular ion peak. An ion with $m/z = 32$ can form if *both* of the carbon atoms in an ethane molecule are ^{13}C. The probability that a molecule of formula $^{13}C_2H_6$ will appear in a natural sample of ethane is $(1.08 \times 1.08)/100$, or 0.01%.

A peak that appears two mass units higher than the mass of the molecular ion peak is called the $M + 2$ peak. The intensity of the $M + 2$ peak of ethane is only 0.01% of the intensity of the molecular ion peak. The contribution due to two deuterium atoms replacing hydrogen atoms would be $(0.016 \times 0.016)/100 = 0.00000256\%$, a negligible amount. To assist in the determination of the ratios of molecular ion, $M + 1$, and $M + 2$ peaks, Table 3.5 lists the natural abundances of some common elements and their isotopes. In this table, the relative abundances of the isotopes of each element are calculated by setting the abundances of the most common isotopes equal to 100.

To demonstrate how the intensities of the $M + 1$ and $M + 2$ peaks provide a unique value for a given molecular formula, consider two molecules of mass 42, propene (C_3H_6) and diazomethane

TABLE 3.5
NATURAL ABUNDANCES OF COMMON ELEMENTS AND THEIR ISOTOPES

Element						
			Relative Abundance			
Hydrogen	^1H	100	^2H	0.016		
Carbon	^{12}C	100	^{13}C	1.08		
Nitrogen	^{14}N	100	^{15}N	0.38		
Oxygen	^{16}O	100	^{17}O	0.04	^{18}O	0.20
Fluorine	^{19}F	100				
Silicon	^{28}Si	100	^{29}Si	5.10	^{30}Si	3.35
Phosphorus	^{31}P	100				
Sulfur	^{32}S	100	^{33}S	0.78	^{34}S	4.40
Chlorine	^{35}Cl	100			^{37}Cl	32.5
Bromine	^{79}Br	100			^{81}Br	98.0
Iodine	^{127}I	100				

(CH_2N_2). For propene, the intensity of the $M + 1$ peak should be $(3 \times 1.08) + (6 \times 0.016) = 3.34\%$, and the intensity of the $M + 2$ peak should be 0.05%. The natural abundance of ^{15}N isotopes of nitrogen is 0.38% of the abundance of ^{14}N atoms. In diazomethane, we expect the relative intensity of the $M + 1$ peak to be $1.08 + (2 \times 0.016) + (2 \times 0.38) = 1.87\%$ of the intensity of the molecular ion peak, and the intensity of the $M + 2$ peak would be 0.01% of the intensity of the molecular ion peak. Table 3.6 summarizes these intensity ratios. It shows that the two molecules have nearly the same molecular weight, but the relative intensities of the $M + 1$ and $M + 2$ peaks that they yield are quite different.

As an additional illustration, Table 3.7 compares the ratios of the molecular ion, $M + 1$, and $M + 2$ peaks for three substances of mass 28: carbon monoxide, nitrogen, and ethene. Again, notice that the relative intensities of the $M + 1$ and $M + 2$ peaks provide a means of distinguishing among these molecules.

As molecules become larger and more complex, the number of possible combinations that yield $M + 1$ and $M + 2$ peaks grows. For a particular combination of atoms, the intensities of these peaks relative to the intensity of the molecular ion peak are unique. Thus, the isotope ratio method can be used to establish the molecular formula of a compound. Examination of the intensity of the $M + 2$ peak is also useful for obtaining information about elements that may be present in the molecular formula. An unusually intense $M + 2$ peak can indicate that sulfur or silicon is present in the unknown

TABLE 3.6
ISOTOPE RATIOS FOR PROPENE AND DIAZOMETHANE

Compound	Molecular Mass	Relative Intensities		
		M	$M + 1$	$M + 2$
C_3H_6	42	100	3.34	0.05
CH_2N_2	42	100	1.87	0.01

TABLE 3.7
ISOTOPE RATIOS FOR CO, N₂, AND C₂H₄

Compound	Molecular Mass	Relative Intensities		
		M	$M+1$	$M+2$
CO	28	100	1.12	0.2
N_2	28	100	0.76	
C_2H_4	28	100	2.23	0.01

substance. The relative abundances of ^{33}S and ^{34}S are 0.78 and 4.40, respectively, and the relative abundance of ^{30}Si is 3.35. A trained chemist knows that a larger-than-normal $M+2$ peak can be the first hint that sulfur or silicon is present. Chlorine and bromine also have important $M+2$ isotopes, and they are discussed separately below.

Tables of possible combinations of carbon, hydrogen, oxygen, and nitrogen and intensity ratios for the $M+1$ and $M+2$ peaks for each combination have been developed. More extensive tables of intensity ratios for the $M+1$ and $M+2$ peaks may be found in specialized books on interpreting mass spectra. Accurate calculation of the relative intensities of isotope peaks in a molecular ion cluster for compounds containing several elements with isotopes is time consuming to do by hand as it requires polynomial expansions. Fortunately, many websites dealing with mass spectrometry have isotope calculators that make this a trivial task. Some of these sites may be found in the references at the end of this chapter.

For compounds containing only C, H, N, O, F, Si, P, and S, the relative intensities of $M+1$ and $M+2$ peaks can be estimated quickly using simplified calculations. The formula to calculate the $M+1$ peak intensity (relative to $M^+ = 100$) for a given formula is found in Equation 3.27. Similarly, the intensity of an $M+2$ peak intensity (relative to $M^+ = 100$) may be found by using Equation 3.28.

$$[M+1] = (\text{number of C} \times 1.1) + (\text{number of H} \times 0.015) + (\text{number of N} \times 0.37) \\ + (\text{number of O} \times 0.04) + (\text{number of S} \times 0.8) + (\text{number of Si} \times 5.1) \quad \textbf{Equation 3.27}$$

$$[M+2] = \frac{(\text{number of C} \times 1.1)^2}{200} + (\text{number of O} \times 0.2) + (\text{number of S} \times 4.4) + (\text{number of Si} \times 3.4)$$

Equation 3.28

When chlorine or bromine is present, the $M+2$ peak becomes very significant. The heavy isotope of each of these elements is two mass units heavier than the lighter isotope. The natural abundance of ^{37}Cl is 32.5% that of ^{35}Cl, and the natural abundance of ^{81}Br is 98.0% that of ^{79}Br. When either of these elements is present, the $M+2$ peak becomes quite intense. If a compound contains two chlorine or bromine atoms, a distinct $M+4$ peak, as well as an intense $M+2$ peak, should be observed. In such a case, it is important to exercise caution in identifying the molecular ion peak in the mass spectrum. Section 4.9 will discuss the mass spectral properties of the organic halogen compounds in greater detail. Table 3.8 gives the relative intensities of isotope peaks for various combinations of bromine and chlorine atoms, and Figure 3.19 illustrates them.

TABLE 3.8

RELATIVE INTENSITIES OF ISOTOPE PEAKS FOR VARIOUS COMBINATIONS OF BROMINE AND CHLORINE

Halogen	Relative Intensities			
	M	M + 2	M + 4	M + 6
Br	100	97.7		
Br₂	100	195.0	95.4	
Br₃	100	293.0	286.0	93.4
Cl	100	32.6		
Cl₂	100	65.3	10.6	
Cl₃	100	97.8	31.9	3.47
BrCl	100	130.0	31.9	
Br₂Cl	100	228.0	159.0	31.2
Cl₂Br	100	163.0	74.4	10.4

FIGURE 3.19 Mass spectra expected for various combinations of bromine and chlorine.

PROBLEMS

*1. A low-resolution mass spectrum of the alkaloid vobtusine showed the molecular weight to be 718. This molecular weight is correct for the molecular formulas $C_{43}H_{50}N_4O_6$ and $C_{42}H_{46}N_4O_7$. A high-resolution mass spectrum provided a molecular weight of 718.3743. Which of the possible molecular formulas is the correct one for vobtusine?

*2. A tetramethyltriacetyl derivative of oregonin, a diarylheptanoid xyloside found in red alder, was found by low-resolution mass spectrometry to have a molecular weight of 660. Possible molecular formulas include $C_{32}H_{36}O_{15}$, $C_{33}H_{40}O_{14}$, $C_{34}H_{44}O_{13}$, $C_{35}H_{48}O_{12}$, $C_{32}H_{52}O_{14}$, and $C_{33}H_{56}O_{13}$. High-resolution mass spectrometry indicated that the precise molecular weight was 660.278. What is the correct molecular formula for this derivative of oregonin?

*3. An unknown substance shows a molecular ion peak at $m/z = 170$ with a relative intensity of 100. The $M + 1$ peak has an intensity of 13.2, and the $M + 2$ peak has an intensity of 1.00. What is the molecular formula of the unknown?

*4. An unknown hydrocarbon has a molecular ion peak at $m/z = 84$, with a relative intensity of 31.3. The $M + 1$ peak has a relative intensity of 2.06, and the $M + 2$ peak has a relative intensity of 0.08. What is the molecular formula for this substance?

*5. An unknown substance has a molecular ion peak at $m/z = 107$, with a relative intensity of 100. The relative intensity of the $M + 1$ peak is 8.00, and the relative intensity of the $M + 2$ peak is 0.30. What is the molecular formula for this unknown?

*6. The mass spectrum of an unknown liquid shows a molecular ion peak at $m/z = 78$, with a relative intensity of 23.6. The relative intensities of the isotopic peaks are as follows:

$m/z = 79$	Relative intensity = 0.79
80	7.55
81	0.25

What is the molecular formula of this unknown?

REFERENCES

Chapman, J. R., *Practical Organic Mass Spectrometry: A Guide for Chemical and Biochemical Analysis*, 2nd ed., John Wiley, New York, 1995.

Crews, P., J. Rodriguez, and M. Jaspars, *Organic Structure Analysis*, 2nd ed., Oxford University Press, New York, 2010.

DeHoffmann, E. and V. Stroobant, *Mass Spectrometry: Principles and Applications*, 3rd ed., Wiley, New York, 2007.

Gross, J. H., and Roepstorff, *Mass Spectrometry: A Textbook*, 2nd ed., Springer, Berlin, 2011.

Lambert, J. B., S. Gronert, H. F. Shurvell, and D. A. Lightner, *Organic Structural Spectroscopy*, 2nd ed., Prentice Hall, Upper Saddle River, NJ, 2011.

March, R. E. and J. F. Todd, *Quadrupole Ion Trap Mass Spectrometry*, 2nd ed., Wiley-Interscience, New York, 2005.

McFadden, W. H., *Techniques of Combined Gas Chromatography/Mass Spectrometry: Applications in Organic Analysis*, Wiley-Interscience, New York, 1989.

McLafferty, F. W., and F. Tureāek, *Interpretation of Mass Spectra*, 4th ed., University Science Books, Mill Valley, Calif., 1993.

Pretsch, E., T. P. Buhlmann, and C. Affolter, *Structure Determination of Organic Compounds. Tables of Spectral Data*, 4th ed., Springer-Verlag, Berlin and New York, 2009.

Silverstein, R. M., F. X. Webster, and D. J. Kiemle, *Spectrometric Identification of Organic Compounds*, 7th ed., John Wiley, New York, 2005.

Smith, R. M., *Understanding Mass Spectra, A Basic Approach*, 2nd ed., Wiley, New York, 2004.

Watson, J. T. and O. D. Sparkman, *Introduction to Mass Spectrometry: Instrumentation, Applications, and Strategies for Data Interpretation*, 4th ed., Wiley, New York, 2007.

Articles

Cody, R. B., J. A. Laramée, and H. D. Durst, "Versatile New Ion Source for the Analysis of Materials in Open Air under Ambient Conditions," Analytical Chemistry 77 (2005): 2297-2302.

Takáts, Z., J. M. Wiseman, and R. G. Cooks, "Ambient Mass Spectrometry using Desporption Electrospray Ionization (DESI): Instrumentation, mechanisms and applications in forensics, chemistry, and biology," *Journal of Mass Spectrometry* 40 (2005): 1261-1275.

Selected Websites

National Institute of Materials and Chemical Research, Tsukuba, Ibaraki, Japan, *Integrated Spectra Data Base System for Organic Compounds (SDBS)*: http://sdbs.riodb.aist.go.jp/sdbs/cgi-bin/cre_index.cgi

National Institute of Standards and Technology, *NIST Chemistry WebBook*:

http://webbook.nist.gov/chemistry/

http://www.sisweb.com/mstools.htm

http://www.sisweb.com/index/referenc/academicsites.html

CHAPTER 4

MASS SPECTROMETRY
Part Two: Fragmentation and Structural Analysis

In electron ionization mass spectrometry (EI-MS), a molecule is bombarded by high-energy electrons in the ionization chamber (Section 3.3A). The collision between the sample molecules and the electrons initially results in the sample molecule losing one electron to form a radical cation. The molecule also absorbs a considerable amount of extra energy during its collision with the incident electrons. This extra energy places the molecular ion in a highly excited vibrational state. Under these conditions the radical cation may fragment (break apart) into smaller pieces (fragments) that can be detected in the mass spectrum. Analysis of these fragment ions provides structural information that can lead to the determination of molecular structure, often facilitated by supplementary spectroscopic information. In this chapter you will learn about the main types of bond cleavage that can be expected and show how an analysis of the fragments can lead to the structure of a molecule. Following a discussion of molecular ion structure and fundamental fragmentation mechanisms, a survey of the expected ion fragmentation patterns for the common functional group classes of compounds will help you understand the analysis procedures necessary to determine structure from mass spectral data.

4.1 THE INITIAL IONIZATION EVENT

As discussed previously, electron ionization results in highly energetic radical cations, many of which have sufficient energy to undergo one or more bond cleavage events leading to fragment ions. If the lifetime of the molecular ion is greater than 10^{-5} sec, a peak corresponding to the molecular ion will appear in the mass spectrum. However, molecular ions with lifetimes less than 10^{-5} sec break apart into fragments before they are accelerated within the ionization chamber and enter the mass analyzer. For a given compound, not all of the molecular ions formed by ionization have precisely the same lifetime; some have shorter lifetimes than others. As a result, in a typical EI mass spectrum one observes peaks corresponding to both the molecular ion and the fragment ions.

The ionization of the sample molecule forms a molecular ion that not only carries a positive charge but also has an unpaired electron. The molecular ion, then, is in fact a radical cation, and it contains an odd number of electrons. Odd electron ions ($OE^{\bullet+}$) have even mass (if no nitrogen is present in the compound; Section 3.6) and thus even electron ions (EE^+) will have odd mass. One cannot get very far in the discussion of ion fragmentation without first considering which electron is lost in the initial ionization event to form $M^{\bullet+}$. The electrons most likely to be ejected during the ionization event are the ones that are in the highest potential energy molecular orbitals, that is the electrons held least tightly by the molecule. Thus, it is easier to remove an electron from a non-bonding orbital, n, than it is to strip an electron from a π orbital. Similarly it is much easier to eject an electron from a π orbital in comparison to a σ orbital. The molecular ion can be represented with either a localized or a non-localized charge site. Some examples of loss of an electron and the notation for the molecular ion are shown below.

Loss of an electron from a non-bonding orbital:

Loss of an electron from a π orbital:

Loss of an electron from a σ orbital:

4.2 FUNDAMENTAL FRAGMENTATION PROCESSES

When drawing fragmentation mechanisms, it is essential to track the charge and radical sites carefully to prevent either misassignment of which fragment is the ion and which is neutral, or drawing highly improbable fragmentations. It is also important to keep in mind that the fragmentation is happening in the gas phase to an ion in a highly excited vibrational state. It is tempting to draw fragmentation mechanisms in the same way that one draws mechanisms for chemical reactions—with concerted bond-breaking or bond-making events. The vast majority of fragmentations in the mass spectrometer are likely stepwise in nature, although some processes like the retro Diels-Alder fragmentation are frequently drawn in a concerted fashion to emphasize the parallel to the chemical reaction more familiar to us. Finally, one needs to be consistent with the use of a single-headed arrow (fishhook, ⌒) for movement of a single electron and a double-headed arrow (⌒) for two-electron processes.

A. Stevenson's Rule

When fragment ions form in the mass spectrometer, they almost always do so by means of unimolecular processes. The low pressure of the ionization chamber makes it unlikely a significant number of bimolecular collisions could occur. The unimolecular processes that are energetically most favorable give rise to the most fragment ions. This is the idea behind **Stevenson's Rule:** The most probable fragmentation is the one that leaves the positive charge on the fragment with the lowest ionization energy. In other words, fragmentation processes that lead to the formation of more stable ions are favored over processes that lead to less stable ions. Where have we heard a statement like this before? This idea is grounded in the same concepts as Markovnikov's Rule, familiar to every organic chemist, which states that in the addition of a hydrogen halide to an alkene, the more stable carbocation forms the fastest and leads to the major product of the addition reaction. In fact, a great deal of the chemistry associated with ionic fragmentation can be explained in terms of what is known about carbocations in solution. For example, alkyl substitution stabilizes fragment ions (and promotes their formation) in much the same way that it stabilizes carbocations. Other familiar concepts will help one predict likely fragmentation processes: electronegativity, polarizability, resonance delocalization, the octet rule, and so on.

Often, fragmentation involves the loss of an electrically neutral fragment. This fragment does not appear in the mass spectrum, but its existence can be deduced by noting the difference in masses of the fragment ion and the original molecular ion. Again, processes that lead to the formation of a more stable neutral fragment are favored over those that lead to less stable neutral fragments.

An $OE^{\bullet+}$ can fragment in two ways: cleavage of a bond to create an EE^+ and a radical ($R^{\bullet}$) *or* by cleavage of bonds to create another $OE^{\bullet+}$ and a closed-shell neutral molecule (N). An EE^+, on the other hand, can only fragment in one way—cleavage of bonds to create another EE^+ and a closed-shell neutral molecule (N). This is the so-called **even-electron rule**. The most common mode of fragmentation involves the cleavage of one bond. In this process, the $OE^{\bullet+}$ yields a radical ($R^{\bullet}$) and an EE^+ fragment ion. Cleavages which lead to the formation of more stable carbocations are favored. When the loss of more than one possible radical is possible, a corollary to Stevenson's Rule is that the largest alkyl radical to be lost preferentially. Thus, ease of fragmentation to form ions increases in the order shown below.

$$H_3C^+ < RCH_2^+ < R_2CH^+ < R_3C^+ < H_2C{=}CHCH_2^+ \sim HC{\equiv}CCH_2^+ < C_6H_5CH_2^+$$

difficult   *easy*

B. Radical-Site-Initiated Cleavage: α-Cleavage

Before examining the characteristic fragmentation patterns of common organic functional groups, let us consider some of the most common modes of fragmentation. **Radical-site**-initiated fragmentation is one of the most common one-bond cleavages and is more commonly called an α-cleavage. The term α-cleavage is confusing to some because the bond that is broken is not directly attached to the radical site but is rather the bond to the next neighboring atom (the α-position). α-Cleavages may occur at saturated or unsaturated sites that may or may not involve a heteroatom (Y in Fig. 4.1).

C. Charge-Site-Initiated Cleavage: Inductive Cleavage

Another common one-bond cleavage is **charge-site**-initiated or **inductive cleavage,** often indicated in a fragmentation mechanism by the symbol *i*. Inductive cleavage involves the attraction of an electron pair by an electronegative heteroatom that ends up as a radical or as a closed-shell neutral molecule. While α-cleavage is a fragmentation of OE^+ only, inductive cleavage can operate on either an OE^+ or an EE^+, as seen in Figure 4.2.

FIGURE 4.1 Representative α-cleavage fragmentations (Y = heteroatom).

FIGURE 4.2 Representative inductive-cleavage fragmentations (Y = heteroatom).

D.   Two-Bond Cleavage

Some fragmentations involve cleavage of two bonds simultaneously. In this process, an elimination occurs, and the odd-electron molecular ion yields an OE$^+$ and an even-electron neutral fragment N, usually a stable small molecule of some type: H_2O, a hydrogen halide, or an alkene. Some examples of two-bond cleavages of this type are shown in Figure 4.3.

FIGURE 4.3 Common two-bond fragmentations (X = heteroatom).

E. Retro Diels-Alder Cleavage

Unsaturated six-membered rings can undergo a **retro Diels–Alder** fragmentation to produce the radical cation of a diene and a neutral alkene—the hypothetical precursors to the cyclohexene derivative if it had been prepared in the forward direction via the $[4\pi + 2\pi]$ diene + dienophile cycloaddition known to every organic chemist as the Diels–Alder reaction. A schematic representation of the retro Diels–Alder fragmentation is shown in Figure 4.4 Note that the unpaired electron and charge remain with the diene fragment according to Stevenson's Rule.

F. McLafferty Rearrangements

Another very common fragmentation that can occur with many substrates is the **McLafferty rearrangement** (Fig. 4.5). This fragmentation was first described by Fred McLafferty in 1956 and is one of the most predictable fragmentations, next to the simple α-cleavage. In the McLafferty rearrangement, a hydrogen atom on a carbon 3 atoms away from the radical cation of an alkene, arene, carbonyl, or imine (a so-called γ-hydrogen) is transferred to the charge site via a six-membered transition state, with concurrent cleavage of the sigma bond between the α and β positions of the tether.

FIGURE 4.4 A retro Diels–Alder fragmentation.

Z, Y = C, N, O

FIGURE 4.5 The McLafferty rearrangement.

This forms a new radical cation and an alkene with a π bond between what were the original β and γ carbons. For simplicity, the mechanism of the McLafferty rearrangement is usually drawn as a concerted process, as in Figure 4.5. There is experimental evidence, however, that the fragmentation is in fact stepwise, and as a general rule fragmentations that involve breaking more than one bond are probably stepwise. The McLafferty rearrangement is readily observed in the mass spectra of many organic functional groups, and several examples will be shown in the remaining sections of this chapter.

G.   Other Cleavage Types

In addition to these processes, fragmentation processes involving rearrangements, migrations of groups, and secondary fragmentations of fragment ions are also possible. These modes of fragmentation occur less often than the two cases already described, and additional discussion of them will be reserved for the compounds in which they are important.

4.3 FRAGMENTATION PATTERN OF HYDROCARBONS

A.   Alkanes

For saturated hydrocarbons and organic structures containing large saturated hydrocarbon skeletons, the methods of fragmentation are quite predictable. What is known about the stabilities of carbocations in solution can be used to help us understand the fragmentation patterns of alkanes. The mass spectra of alkanes are characterized by strong molecular ion peaks and a regular series of fragment ion peaks separated by 14 amu.

For a straight-chain, or "normal," alkane, a peak corresponding to the molecular ion can be observed as in the mass spectra of butane (Fig. 4.6) and octane (Fig. 4.7). As the carbon skeleton becomes more highly branched, the intensity of the molecular ion peak decreases. Straight-chain alkanes have fragments that are always primary carbocations. Since these ions are rather unstable, fragmentation is not favored. A significant number of the original molecules survive electron bombardment without fragmenting. Consequently, a molecular ion peak of significant intensity is observed. You will see this effect easily if you compare the mass spectrum of butane with that of isobutane (Fig. 4.8). The molecular ion peak in isobutane is much less intense than that in butane. Comparison of the mass

F I G U R E   4 . 6 Mass spectrum of butane.

SPECTRAL ANALYSIS BOX — Alkanes

MOLECULAR ION

Strong M⁺

FRAGMENT IONS

Loss of CH_2 units in a series: $M - 14$, $M - 28$, $M - 42$, etc.

FIGURE 4.7 EI mass spectrum of octane.

FIGURE 4.8 EI mass spectrum of isobutane.

F I G U R E   4 . 9 EI mass spectrum of 2,2,4-trimethylpentane (isooctane).

spectra of octane and 2,2,4-trimethylpentane (Fig. 4.9) provides a more dramatic illustration of the effect of chain branching on the intensity of the molecular ion peak. The molecular ion peak in 2,2,4-trimethylpentane is too weak to be observed, while the molecular ion peak in its straight-chain isomer is quite readily observed. The effect of chain branching on the intensity of the molecular ion peak can be understood by examining the method by which hydrocarbons undergo fragmentation.

Straight-chain hydrocarbons undergo fragmentation by breaking carbon–carbon bonds, resulting in a homologous series of fragmentation products. For example, in the case of butane, cleavage of the C1–C2 bond results in the loss of a methyl radical and the formation of the propyl carbocation ($m/z = 43$). Cleavage of the C2–C3 bond results in the loss of a ethyl radical and the formation of the ethyl carbocation ($m/z = 29$). In the case of octane, fragment peaks due to the hexyl ion ($m/z = 85$), the pentyl ion ($m/z = 71$), the butyl ion ($m/z = 57$), the propyl ion ($m/z = 43$), and the ethyl ion ($m/z = 29$) are observed. Notice that alkanes fragment to form clusters of peaks that are 14 mass units (corresponding to one CH_2 group) apart from each other. Other fragments within each cluster correspond to additional losses of one or two hydrogen atoms. As is evident in the mass spectrum of octane, the three-carbon ions appear to be the most abundant, with the intensities of each cluster uniformly decreasing with increasing fragment weight. Interestingly, for long-chain alkanes, the fragment corresponding to the loss of one carbon atom is generally absent. In the mass spectrum of octane, for example, a seven-carbon fragment should occur at a mass of 99, but it is not observed.

Cleavage of the carbon–carbon bonds of branched-chain alkanes can lead to secondary or tertiary carbocations. These ions are more stable than primary ions, of course, so fragmentation becomes a more favorable process. A greater proportion of the original molecules undergoes fragmentation, so the molecular ion peaks of branched-chain alkanes are considerably weaker or even absent. In isobutane, cleavage of a carbon–carbon bond yields an isopropyl carbocation, which is more stable than a normal propyl ion. Isobutane undergoes fragmentation more easily than butane because of the increased stability of its fragmentation products. With 2,2,4-trimethylpentane, the dominant cleavage event is the rupture of the C2-C3 bond, which leads to the formation of a *tert*-butyl carbocation. Since tertiary carbocations are the most stable of the saturated alkyl carbocations, this cleavage is particularly favorable and accounts for the intense fragment peak at $m/z = 57$.

B.   Cycloalkanes

Cycloalkanes generally form strong molecular ion peaks. Fragmentation via the loss of a molecule of ethene ($M - 28$) is common. The typical mass spectrum for a cycloalkane shows a relatively intense molecular ion peak. Fragmentation of ring compounds requires the cleavage of two carbon–carbon bonds, which is a more difficult process than cleavage of one such bond. Therefore, a larger proportion of cycloalkane molecules than of acyclic alkane molecules survives electron bombardment without undergoing fragmentation. In the mass spectra of cyclopentane (Fig. 4.10) and methylcyclopentane (Fig. 4.11), strong molecular ion peaks can be observed.

F I G U R E   4 . 1 0 EI mass spectrum of cyclopentane.

F I G U R E   4 . 1 1 EI mass spectrum of methylcyclopentane.

The fragmentation patterns of cycloalkanes may show mass clusters arranged in a homologous series, as in the alkanes. However, the most significant mode of cleavage of the cycloalkanes involves the loss of a molecule of ethene ($H_2C=CH_2$), either from the parent molecule or from intermediate OE^+. The peak at $m/z = 42$ in cyclopentane and the peak at $m/z = 56$ in methylcyclopentane result from the loss of ethene from the parent molecule. Each of these fragment peaks is the most intense in the mass spectrum. When the cycloalkane bears a side chain, loss of that side chain is a favorable mode of fragmentation. The fragment peak at $m/z = 69$ in the mass spectrum of methylcyclopentane is due to the loss of the CH_3 side chain, which results in a secondary carbocation.

SPECTRAL ANALYSIS BOX — Cycloalkanes

MOLECULAR ION	**FRAGMENT IONS**
Strong M^+	$M - 28$
	A series of peaks: $M - 15$, $M - 29$, $M - 43$, $M - 57$, etc.

Applying these pieces of information to the mass spectrum of bicyclo[2.2.1]heptane (Fig. 4.12), we can identify fragment peaks due to the loss of the side chain (the one-carbon bridge, plus an additional hydrogen atom) at $m/z = 81$ and the loss of ethene at $m/z = 68$. The fragment ion peak at $m/z = 67$ is due to the loss of ethene plus an additional hydrogen atom.

C. Alkenes

The mass spectra of most alkenes show distinct molecular ion peaks. Naturally, the mass of the molecular ion should correspond to a molecular formula with an index of hydrogen deficiency equal to at least *one* (see Chapter 1). Apparently, electron bombardment removes one of the electrons in the π bond, leaving the carbon skeleton relatively undisturbed. When alkenes undergo fragmentation processes, the resulting fragment ions have formulas corresponding to $C_nH_{2n}^+$ and $C_nH_{2n-1}^+$. It is sometimes difficult to locate double bonds in alkenes since they migrate readily. The similarity of

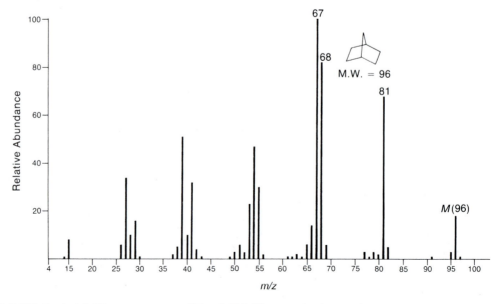

FIGURE 4.12 EI mass spectrum of bicyclo[2.2.1]heptane (norbornane).

the mass spectra of alkene isomers is readily seen in the mass spectra of three isomers of the formula C_5H_{10} (Figs. 4.13, 4.14, and 4.15). The mass spectra are very nearly identical, with the only difference being a large fragment at $m/z = 42$ in the spectrum of 1-pentene. This ion likely forms via loss of ethylene through a McLafferty-type rearrangement of the molecular ion. The allyl carbocation ($m/z = 41$) is an important fragment in the mass spectra of terminal alkenes and forms via an allylic α-cleavage as shown in Figure 4.1. The fragment at $m/z = 55$ is from loss of a methyl radical. This fragment is the base peak in the spectra of the diastereomeric pentene isomers since loss of the methyl group distal to the alkene creates an allylic cation that is resonance stabilized. Clearly, MS analysis alone is not enough to distinguish E- and Z-2-pentene. Use of IR analysis, particularly the C–H oop bending region, will allow one to definitively determine the configuration of the alkene.

FIGURE 4.13 EI-MS spectrum of 1-pentene.

FIGURE 4.14 EI-MS spectrum of Z-2-pentene.

S P E C T R A L   A N A L Y S I S   B O X —Alkenes

MOLECULAR ION	FRAGMENT IONS
Strong M^+	$m/z = 41$
	A series of peaks: $M - 15$, $M - 29$, $M - 43$, $M - 57$, etc.

F I G U R E   4 . 1 5 EI-MS spectrum of *E*-2-pentene.

The mass spectra of cycloalkenes show quite distinct molecular ion peaks. For many cycloalkenes, migration of bonds gives virtually identical mass spectra. Consequently, it may be impossible to locate the position of the double bond in a cycloalkene, particularly a cyclopentene or a cycloheptene. Cyclohexenes do have a characteristic fragmentation pattern that corresponds to a retro Diels–Alder reaction (Fig. 4.4). In the mass spectrum of the monoterpene limonene (Fig. 4.16), the intense peak at $m/z = 68$ corresponds to the diene fragment arising from the retro Diels–Alder fragmentation.

The mere presence of a cyclohexene moiety does not guarantee that a retro Diels–Alder fragmentation will be observed in the mass spectrum. Consider the mass spectra of α- and β-ionone (Fig. 4.17). The spectrum of α-ionone shows much more fragmentation in general and a peak at $m/z = 136$ in particular that is created by a retro Diels–Alder fragmentation of the cyclohexene ring and loss of isobutene. Retro Diels–Alder fragmentation of β-ionone should give a peak at $m/z = 164$ from loss of ethene, but the peak at that position is miniscule. In the case of β-ionone, loss of a methyl radical via α-cleavage adjacent to the ring double bond yields a relatively stable tertiary allylic cation. This fragmentation is not available to α-ionone.

D.   Alkynes

The mass spectra of alkynes are very similar to those of alkenes. The molecular ion peaks tend to be rather intense, and fragmentation patterns parallel those of the alkenes. As can be seen from the mass spectrum of l-pentyne (Fig. 4.18), an important fragmentation is the loss of an ethyl radical via an α-cleavage to produce the propargyl ion ($m/z = 39$). Similarly, loss of methyl radical in an α-cleavage

FIGURE 4.16 EI-MS spectrum of limonene.

of 2-pentyne produces a resonance-stabilized propargylic cation at ($m/z = 53$) (Fig. 4.19). Another important mode of fragmentation for terminal alkynes is the loss of the terminal hydrogen, yielding a strong $M - 1$ peak. This peak appears as the base peak ($m/z = 67$) in the spectrum of 1-pentyne.

SPECTRAL ANALYSIS BOX — Alkynes

MOLECULAR ION	**FRAGMENT IONS**
Strong M^+	$m/z = 39$
	Strong $M - 1$ peak

E.   Aromatic Hydrocarbons

The mass spectra of most aromatic hydrocarbons show intense molecular ion peaks. As is evident from the mass spectrum of benzene (Fig. 4.20), fragmentation of the benzene ring requires a great deal of energy. Such fragmentation is not observed to any significant extent. In the mass spectrum of toluene (Fig. 4.21), loss of a hydrogen atom from the molecular ion gives a strong peak at $m/z = 91$. Although it might be expected that this fragment ion peak is due to the benzyl carbocation ($C_6H_5CH_2^+$), isotope-labeling experiments suggest that the benzyl carbocation actually rearranges to form the aromatic delocalized **tropylium ion** ($C_7H_7^+$, Figure 4.25). When a benzene ring contains larger side chains, a favored mode of fragmentation is cleavage of the side chain to form initially a **benzyl cation,** which spontaneously rearranges to the tropylium ion. When the side chain attached to a benzene ring contains three or more carbons, ions formed by a McLafferty rearrangement can be observed.

The mass spectra of the xylene isomers (Figs. 4.22 and 4.23 for example) show a medium peak at $m/z = 105$, which is due to the loss of a hydrogen atom and the formation of the methyltropylium ion. More importantly, xylene loses one methyl group to form the tropylium ($m/z = 91$). The mass spectra of *ortho-*, *meta-*, and *para*-disubstituted aromatic rings are essentially identical. As a result, the substitution pattern of polyalkylated benzenes cannot be determined by mass spectrometry.

FIGURE 4.17 EI-MS spectra of α-ionone (top) and β-ionone (bottom).

F I G U R E 4 . 1 8 EI-MS spectrum of 1-pentyne.

F I G U R E 4 . 1 9 EI-MS spectrum of 2-pentyne.

The formation of a substituted tropylium ion is typical for alkyl-substituted benzenes. In the mass spectrum of isopropylbenzene (Fig. 4.24), a strong peak appears at $m/z = 105$. This peak corresponds to loss of a methyl group to form a methyl-substituted tropylium ion. The tropylium ion has characteristic fragmentations of its own. The tropylium ion can fragment to form the aromatic cyclopentadienyl cation ($m/z = 65$) plus ethyne (acetylene). The cyclopentadienyl cation in turn can fragment to form another equivalent of ethyne and the aromatic cyclopropenyl cation ($m/z = 39$) (Fig. 4.25).

In the mass spectrum of butylbenzene (Fig. 4.26), a strong peak due to the tropylium ion appears at $m/z = 91$. When the alkyl group attached to the benzene ring is a propyl group or larger, a McLafferty rearrangement is likely to occur, producing a peak at $m/z = 92$. Indeed, all alkylbenzenes bearing a side chain of three or more carbons and at least one hydrogen on the γ-carbon will exhibit a peak at

F I G U R E 4 . 2 0 EI-MS spectrum of benzene.

F I G U R E 4 . 2 1 EI-MS spectrum of toluene.

$m/z = 92$ in their mass spectra from the McLafferty rearrangement. Using butylbenzene as an example, this rearrangement is depicted below.

FIGURE 4.22 EI-MS spectrum of *ortho*-xylene.

FIGURE 4.23 EI-MS spectrum of *meta*-xylene.

FIGURE 4.24 EI-MS of isopropylbenzene (cumene).

FIGURE 4.25 Formation and fragmentation of the tropylium ion.

4.4 FRAGMENTATION PATTERNS OF ALCOHOLS, PHENOLS, AND THIOLS

The intensity of the molecular ion peak in the mass spectrum of a primary or secondary alcohol is usually rather low, and the molecular ion peak is often entirely absent in the mass spectrum of a tertiary alcohol. Common fragmentations of alcohols are α-cleavage adjacent to the hydroxyl group and dehydration.

SPECTRAL ANALYSIS BOX — Alcohols

MOLECULAR ION	FRAGMENT IONS
M$^+$ weak or absent	Loss of alkyl group
	$M - 18$

FIGURE 4.26 EI-MS of butylbenzene.

FIGURE 4.27 EI-MS of 1-pentanol.

The mass spectrum of straight-chain pentanol isomers, 1-pentanol (Fig. 4.27), 2-pentanol (Figure 4.28), and 3-pentanol (Fig. 4.29) all exhibit very weak molecular ion peaks at $m/z = 88$, while the molecular ion in the mass spectrum of the tertiary alcohol 2-methyl-2-butanol (Fig. 4.30) is entirely absent. The most important fragmentation reaction for alcohols is the loss of an alkyl group via α-cleavage. As discussed earlier, the largest alkyl group is most readily lost. In the spectrum of 1-pentanol (Fig. 4.27), the peak at $m/z = 31$ is due to the loss of a butyl group to form an $H_2C=OH^+$ ion. 2-Pentanol (Fig. 4.28) loses either a propyl group to form the $CH_3CH=OH^+$ fragment at $m/z = 45$ or a methyl radical to form the relatively small peak at $m/z = 73$ corresponding to $CH_3CH_2CH_2CH=OH^+$.

FIGURE 4.28 EI-MS of 2-pentanol.

FIGURE 4.29 EI-MS of 3-pentanol.

FIGURE 4.30 EI-MS of 2-methyl-2-butanol.

3-Pentanol loses an ethyl radical to form the $CH_3CH_2CH=OH^+$ ion at $m/z = 59$. The symmetry of 3-pentanol means there are two identical α-cleavage paths, making the peak corresponding to that ion even more prevalent (Fig. 4.29). 2-Methyl-2-butanol (Fig. 4.30) undergoes α-cleavage to lose a methyl radical two different ways, creating a considerable size peak at $m/z = 73$ in addition to the peak at $m/z = 59$ corresponding to the $(CH_3)_2C=OH^+$ ion formed by loss of an ethyl radical.

A second common mode of fragmentation involves dehydration. The importance of dehydration increases as the chain length of the alcohol increases. While the fragment ion peak resulting from dehydration ($m/z = 70$) is very intense in the mass spectrum of 1-pentanol, it is quite weak in the other pentanol isomers. Dehydration may occur by either **thermal dehydration** prior to ionization or by fragmentation of the molecular ion. Thermal dehydration is especially troublesome for alcohol samples analyzed by GC-MS. The injection port of the gas chromatograph is usually maintained at more than 200°C, and many alcohols, especially tertiary or allylic/benzylic, will dehydrate before the sample molecules even reach the GC column and certainly before the molecules reach the ion source of the mass spectrometer. Thermal dehydration is a **1,2-elimination** of water. If the alcohol molecules reach the ion source intact, however, dehydration of the molecular ion can still occur, but in this case it is a **1,4-elimination** of water via a cyclic mechanism:

Alcohols containing four or more carbons may undergo the *simultaneous* loss of both water and ethylene. This type of fragmentation is not prominent for 1-butanol but is responsible for the base peak at $m/z = 42$ in the mass spectrum of 1-pentanol (Fig. 4.27).

FIGURE 4.31 Fragmentation pathways for cyclohexanol.

Cyclic alcohols may undergo fragmentation by at least three different pathways, and these are illustrated for the case of cyclohexanol in Figure 4.31. The first fragmentation is simply an α-cleavage and loss of a hydrogen atom to yield an $M - 1$ fragment ion. The second fragmentation path begins with an initial α-cleavage of a ring bond adjacent to the hydroxyl-bearing carbon, followed by a 1,5-hydrogen migration. This moves the radical site back to a resonance-stabilized position adjacent to the oxonium ion. A second α-cleavage results in the loss of a propyl radical and formation of a protonated acrolein ion with $m/z = 57$. This fragmentation path is nearly identical to one that operates on cyclohexanone derivatives (Section 4.6B). The third fragmentation path of cyclic alcohols is dehydration via abstraction of a hydrogen atom from three or four carbons away (the hydrogen atom is transferred in a five- or six-membered cyclic transition state) to produce a

FIGURE 4.32 EI-MS of cyclohexanol.

FIGURE 4.33 EI-MS of benzyl alcohol.

bicyclic radical cation with $m/z = 82$. A peak corresponding to each of these fragment ions can be observed in the mass spectrum of cyclohexanol (Fig. 4.32).

Benzylic alcohols usually exhibit strong molecular ion peaks. The following sequence of reactions illustrates their principal modes of fragmentation. Loss of a hydrogen atom from the molecular ion leads to a hydroxytropylium ion ($m/z = 107$). The hydroxytropylium ion can lose carbon monoxide to form a resonance-delocalized cyclohexadienyl cation ($m/z = 79$). This ion can eliminate molecular hydrogen to create a phenyl cation, $C_6H_5^+$, $m/z = 77$. Peaks arising from these fragment ions can be observed in the mass spectrum of benzyl alcohol (Fig. 4.33).

The mass spectra of phenols usually show strong molecular ion peaks. In fact, the molecular ion at $m/z = 94$ is the base peak in the EI-MS of phenol (Fig. 4.34). Favored modes of fragmentation involve loss of a hydrogen atom to create an $M - 1$ peak (a small peak at $m/z = 93$), loss of carbon monoxide (CO) to produce a peak at $M - 28$ ($m/z = 66$), and loss of a formyl radical (HCO·) to give a peak at $M - 29$. In the case of phenol itself, this creates the aromatic cyclopentadienyl cation at $m/z = 65$. In some cases, the loss of 29 mass units may be sequential: initial loss of carbon monoxide followed by loss of a hydrogen atom. The mass spectrum of *ortho*-cresol (2-methylphenol) exhibits a much larger peak at $M - 1$ (Fig. 4.35) than does unsubstituted phenol. Note also the peaks at $m/z = 80$ and $m/z = 79$ in the *o*-cresol spectrum from loss of CO and formyl radical, respectively.

SPECTRAL ANALYSIS BOX — Phenols

MOLECULAR ION	FRAGMENT IONS
M⁺ strong	$M - 1$
	$M - 28$
	$M - 29$

Thiols show molecular ion peaks that are more intense than those of the corresponding alcohols. A characteristic feature of the mass spectra of sulfur compounds is the presence of a

F I G U R E   4 . 3 4 EI-MS of phenol.

FIGURE 4.35 EI-MS of 2-methylphenol (*ortho*-cresol).

significant $M + 2$ peak. This peak arises from the presence of the heavy isotope, ^{34}S, which has a natural abundance of 4.4%.

The fragmentation patterns of the thiols are very similar to those of the alcohols. As alcohols tend to undergo dehydration under some conditions, thiols tend to lose the elements of hydrogen sulfide, giving rise to an $M - 34$ peak.

4.5 FRAGMENTATION PATTERNS OF ETHERS AND SULFIDES

Aliphatic ethers tend to exhibit molecular ion peaks that are stronger than those of alcohols with the same molecular weights. Nevertheless, the molecular ion peaks of ethers are still rather weak. Principal modes of fragmentation include α-cleavage, formation of carbocation fragments through inductive cleavage (β-cleavage), and loss of alkoxy radicals.

SPECTRAL ANALYSIS BOX — Ethers

MOLECULAR ION	FRAGMENT IONS
M^+ weak, but observable	α-Cleavage
	$m/z = 43, 59, 73$, etc.
	$M - 31, M - 45, M - 59$, etc.

The fragmentation of the ethers is somewhat similar to that of the alcohols. In the mass spectrum of diisopropyl ether (Fig. 4.36), an α-cleavage gives rise to a peak at $m/z = 87$ due to the loss of a methyl radical. A second mode of fragmentation involves cleavage of the carbon–oxygen bond of an ether to yield an isopropoxyl radical and a isopropyl carbocation. Cleavage of this type in diisopropyl ether is responsible for the $C_3H_7^+$ fragment at $m/z = 43$. A third type of fragmentation occurs as a rearrangement reaction of one of the fragment ions rather than on the molecular ion itself. The rearrangement involves transfer of a hydrogen β to the oxonium ion with concurrent formation of an alkene. This type

FIGURE 4.36 EI-MS of diisopropyl ether.

of rearrangement is particularly favored when the α carbon of the ether is branched. In the case of di-isopropyl ether, this rearrangement gives rise to a $(HO=CHCH_3)^+$ fragment at $m/z = 45$.

$$R—CH=\overset{+}{O}—CH—\overset{H}{CH_2} \longrightarrow R—CH=\overset{+}{O}H + CH=CH_2$$

The mass spectrum of di-*sec*-butyl ether (Fig. 4.37) shows the same fragmentations. There are two possible α-cleavages in this compound, however. Loss of a methyl radical gives the very small

FIGURE 4.37 EI-MS of di-*sec*-butyl ether.

FIGURE 4.38 EI-MS of 2-ethyl-2-methyl-1,3-dioxolane.

$M - 15$ peak at $m/z = 115$, but loss of the larger ethyl radical gives the substantially larger peak at $m/z = 101$. Inductive cleavage of the C—O bond creates a *sec*-butyl cation at $m/z = 57$. Further rearrangement of the α-cleavage products produce ions at $m/z = 45$ and 59, corresponding to $(HO=CHCH_3)^+$ and $(HO=CHCH_2CH_3)^+$, respectively.

Acetals and ketals behave very similarly to ethers. However, fragmentation is even more favorable in acetals and ketals than in ethers, so the molecular ion peak of an acetal or ketal may be either extremely weak or totally absent. For example, in the mass spectrum of 2-ethyl-2-methyl-1,3-dioxolane (the ethylene ketal of methyl ethyl ketone), the molecular ion is not visible (Fig. 4.38). The highest mass peak is at $m/z = 101$ from loss of a methyl radical via α-cleavage, and an alternative α-cleavage produces the large peak at $m/z = 87$ formed by loss of an ethyl radical. The base peak in the spectrum is found at $m/z = 43$, typical of 2-methyl-1,3-dioxolanes.

Aromatic ethers may undergo cleavage reactions that involve loss of the alkyl group to form $C_6H_5O^+$ ions. These fragment ions then lose carbon monoxide to form cyclopentadienyl cations $(C_5H_5^+)$. In addition, an aromatic ether may lose the entire alkoxy group to yield phenyl cations $(C_6H_5^+)$. The mass spectrum of ethyl 4-methylphenyl ether (*p*-methylphenetole) exhibits a strong molecular ion at $m/z = 136$ as well as a fragment at $m/z = 107$ from loss of an ethyl radical (Fig. 4.39). The base peak at $m/z = 108$ arises from loss of ethene via a McLafferty rearrangement.

Sulfides (thioethers) show mass spectral patterns that are very similar to those of the ethers. As in the case of the thiols, thioethers show molecular ion peaks that tend to be more intense than those of the corresponding ethers, owing to the increased stability of the sulfur-centered radical cation. The α-cleavage of dimethyl disulfide derivatives of alkenes is a useful way to locate the position of an alkene within a long alkyl chain. Long chain alkenes are common structural features of insect pheromones and some marine natural products, and although two-dimensional NMR experiments can be used to locate alkene positions within a long chain, simple EI-MS analysis of the product of the alkene and dimethyldisulfide accomplishes the same task. The reaction requires only micrograms of material and usually allows one to observe both diagnostic ions produced by the cleavage of the C—C bond between the vicinal sulfide groups as shown below for a bioactive compound isolated from the currant stem girdler (Fig. 4.40). Cleavage of the molecular ion between the thioether groups produces fragments at $m/z = 173$ and 201, thereby locating the original alkene between carbons 9 and 10 of the fatty acid chain.

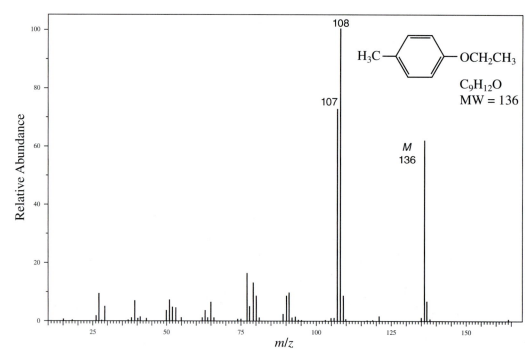

FIGURE 4.39 EI-MS of 4-methylphenetole.

◻ 4.6 FRAGMENTATION PATTERNS OF CARBONYL-CONTAINING COMPOUNDS

As we saw in the analysis of IR spectra, the spectroscopic signature of the carbonyl group provides a wealth of information. The same is true in mass spectrometry of compounds containing the C=O moiety. All carbonyl-containing compounds have some fragmentations in common: α-cleavage, inductive (β-) cleavage, and McLafferty rearrangements are ubiquitous. Some carbonyl-containing functional groups have more unique fragmentation behavior that provides structural clues as well. In this section, we will survey the common fragmentation patterns of C=O containing molecules.

A. Aldehydes

The molecular ion peak of an aliphatic aldehyde is usually observable, although at times it may be fairly weak. Principal modes of fragmentation include α-cleavage and β-cleavage. If the carbon chain attached to the carbonyl group contains at least three carbons, McLafferty rearrangement is also commonly observed.

SPECTRAL ANALYSIS BOX — Aldehydes

MOLECULAR ION	FRAGMENT IONS
M^+ weak, but observable (aliphatic)	Aliphatic:
M^+ strong (aromatic)	$m/z = 29$, $M - 29$,
	$M - 43$, $m/z = 44$
	Aromatic:
	$M - 1$, $M - 29$

FIGURE 4.40 EI-MS spectrum of (Z)-9-octadecen-4-olide and its dimethyldisulfide derivative showing. Cleavage between thioether groups (from Cosse, A. A., et al, *Journal of Chemical Ecology 27* (2001): 1841–1853.)

The appearance of an $M - 1$ peak due to the loss of one hydrogen atom is very characteristic of aldehydes. This peak is observed at $m/z = 85$ in the mass spectrum of valeraldehyde (Fig. 4.41). The peak due to the formation of HCO^+ can be observed at $m/z = 29$; this is also a very characteristic peak in the mass spectra of aldehydes. The second important mode of fragmentation for aldehydes is known as **β-cleavage** (inductive cleavage). In the case of valeraldehyde, β-cleavage creates a propyl cation ($m/z = 43$).

FIGURE 4.41 EI-MS of valeraldehyde.

The third major fragmentation pathway for aldehydes is the McLafferty rearrangement. The fragment ion formed in this rearrangement has $m/z = 44$ and is the base peak in the spectrum of valeraldehyde. The $m/z = 44$ peak is considered to be quite characteristic for aldehydes. As with all McLafferty rearrangements, of course, this rearrangement occurs only if the chain attached to the carbonyl group has three or more carbons.

Aromatic aldehydes also exhibit intense molecular ion peaks, and the loss of one hydrogen atom via α-cleavage is a very favorable process. The resulting $M - 1$ peak may in some cases be more intense than the molecular ion peak. In the mass spectrum of benzaldehyde (Fig. 4.42), the $M - 1$ peak appears at $m/z = 105$. Note also the peak at $m/z = 77$, which corresponds to the phenyl cation formed by loss of the formyl radical.

FIGURE 4.42 EI-MS of benzaldehyde.

B.   Ketones

The mass spectra of ketones show an intense molecular ion peak. Loss of the alkyl groups attached to the carbonyl group is one of the most important fragmentation processes. The pattern of fragmentation is similar to that of aldehydes. Loss of alkyl groups by means of α-cleavage is an important mode of fragmentation, and the larger of the two alkyl groups attached to the carbonyl group appears more likely to be lost, in keeping with Stevenson's Rule. The ion formed from this type of α-cleavage in ketones (and aldehydes) is the acylium ion ($RC{\equiv}O^{+}$). In the mass spectrum of 2-butanone (Fig. 4.43), the peak at $m/z = 43$ is more intense than the peak at $m/z = 57$, which is due to the loss of the methyl group. Similarly, in the mass spectrum of 2-octanone (Fig. 4.44) loss of the hexyl group, giving a peak at $m/z = 43$, is more likely than loss of the methyl group, which gives the weak peak at $m/z = 113$.

SPECTRAL ANALYSIS BOX — Ketones

MOLECULAR ION	FRAGMENT IONS
M⁺ strong	Aliphatic:
	$M - 15, M - 29, M - 43$, etc.
	$m/z = 43$
	$m/z = 58, 72, 86$, etc.
	$m/z = 42, 83$
	Aromatic:
	$m/z = 105, 120$

When the carbonyl group of a ketone has attached to it at least one alkyl group that is three or more carbon atoms in length, a McLafferty rearrangement is possible. The peak at $m/z = 58$ in the mass spectrum of 2-octanone is due to the fragment ion that results from this rearrangement.

FIGURE 4.43 EI-MS of 2-butanone.

F I G U R E 4 . 4 4 EI-MS of 2-octanone.

Cyclic ketones may undergo a variety of fragmentation and rearrangement processes. Outlines of these processes for the case of cyclohexanone follow. A fragment ion peak corresponding to each process appears in the mass spectrum of cyclohexanone (Fig. 4.45).

FIGURE 4.45 EI-MS of cyclohexanone.

Aromatic ketones undergo α-cleavage to lose the alkyl group and form the phenylacylium ($C_6H_5CO^+$, $m/z = 105$) ion. This ion can undergo secondary fragmentation to lose carbon monoxide, forming the $C_6H_5^+$ ion ($m/z = 77$). These peaks appear prominently in the mass spectrum of acetophenone (Fig. 4.46). With larger alkyl groups attached to the carbonyl group of an aromatic ketone, a rearrangement of the McLafferty type is likely, and the rearrangement can occur to the carbonyl and to the aromatic ring. In the case of butyrophenone, McLafferty rearrangement to the aromatic ring yields the fragment seen at $m/z = 106$, and the rearrangement to the carbonyl gives the fragment at $m/z = 120$ (Fig. 4.47). The $m/z = 120$ fragment ion may undergo additional α-cleavage to yield the $C_6H_5CO^+$ ion at $m/z = 105$.

McLafferty rearrangement

$m/z = 106$

McLafferty rearrangement

$m/z = 120$

FIGURE 4.46 EI-MS of acetophenone.

C. Esters

Fragmentation of esters is especially facile, but it is usually possible to observe weak molecular ion peaks in the mass spectra of methyl esters. The esters of higher alcohols form much weaker molecular ion peaks, and esters of alcohols larger than four carbons often form molecular ions that fragment too quickly to be observed. The most important fragmentation of esters is an α-cleavage that involves the loss of the alkoxy group to form the corresponding acylium ion, $RC\equiv O^+$ The acylium ion peak appears at $m/z = 71$ in the mass spectrum of methyl butyrate

FIGURE 4.47 EI-MS of butyrophenone.

FIGURE 4.48 EI-MS of methyl butyrate.

(Fig. 4.48). A second useful peak results from the loss of the alkyl group from the acyl side of the ester, leaving a fragment $H_3C-O-C=O^+$ that appears at $m/z = 59$ in the mass spectrum of methyl butyrate. Other fragment ion peaks include the $^+OCH_3$ fragment ($m/z = 31$) and the R^+ fragment from the acyl portion of the ester molecule, $CH_3CH_2CH_2^+$ in the case of methyl butyrate, at $m/z = 43$.

SPECTRAL ANALYSIS BOX — Esters

MOLECULAR ION

M^+ weak, but generally observable

FRAGMENT IONS

Methyl esters:

  $M - 31, m/z = 59, 74$

Higher esters:

  $M - 45, M - 59, M - 73$

  $m/z = 73, 87, 101$

  $m/z = 88, 102, 116$

  $m/z = 61, 75, 89$

  $m/z = 77, 105, 108$

  $M - 32, M - 46, M - 60$

Another important fragmentation of esters is the McLafferty rearrangement that produces the peak at $m/z = 74$ (for methyl esters). Ethyl, propyl, butyl, and higher alkyl esters also undergo α-cleavage and McLafferty rearrangements typical of the methyl esters. In addition, however, these esters may undergo an additional rearrangement of the alkoxy portion of the ester that results in fragments that appear in the series $m/z = 61, 75, 89$, and so on. This process is illustrated for butyl butyrate and is commonly referred to as the **McLafferty + 1 rearrangement** or the McLafferty

FIGURE 4.49 EI-MS of butyl butyrate.

rearrangement with double-hydrogen transfer (Fig. 4.49) Several other peaks in the mass spectrum of butyl butyrate are readily assigned by considering the common fragmentations. Loss of a propyl radical through α-cleavage forms the butoxyacylium ion at $m/z = 101$, McLafferty rearrangement on the acyl side of the ester creates the ion observed at $m/z = 73$, and loss of butoxy radical from the molecular ion yields the acylium ion seen at $m/z = 71$.

Benzyl esters undergo rearrangement to eliminate a neutral ketene molecule and the radical cation of benzyl alcohol at $m/z = 108$. The resulting ion is often the most intense peak in the mass spectrum of such a compound. This fragmentation is dominant in the mass spectrum of benzyl laurate, along with the benzyl cation/tropylium ion at $m/z = 91$ (Fig. 4.50). Other high-mass fragments in the benzyl laurate spectrum include a peak at $m/z = 199$ from loss of a benzyl radical and the peak at $m/z = 183$ from loss of benzyloxy radical via α-cleavage.

FIGURE 4.50 EI-MS of benzyl laurate.

Alkyl benzoate esters prefer to lose the alkoxy group to form the $C_6H_5C\equiv O^+$ ion ($m/z = 105$). This ion may lose carbon monoxide to form the phenyl cation ($C_6H_5^+$) at $m/z = 77$. Each of these peaks appears in the mass spectrum of methyl benzoate (Fig. 4.51). Alkyl substitution on benzoate esters appears to have little effect on the mass spectral results unless the alkyl group is in the *ortho* position with respect to the ester functional group. In this case, the alkyl group can interact with the ester function, with the elimination of a molecule of alcohol. This is observed in the mass spectrum of isobutyl salicylate (Fig. 4.52). The base peak at $m/z = 120$ arises from elimination of isobutyl alcohol via this *ortho* effect. The fragment at $m/z = 121$ comes from loss of isobutoxyl radical via standard α-cleavage, and the peak at $m/z = 138$ likely arises by elimination of isobutene from the molecular ion.

D. Carboxylic Acids

Aliphatic carboxylic acids generally show weak, but observable, molecular ion peaks. Aromatic carboxylic acids, on the other hand, show strong molecular ion peaks. The principal modes of fragmentation resemble those of the methyl esters.

With short-chain acids, the loss of OH and COOH through α-cleavage on either side of the C=O group may be observed. In the mass spectrum of butyric acid (Fig. 4.53) loss of ·OH gives rise to a

---

SPECTRAL ANALYSIS BOX — Carboxylic Acids

MOLECULAR ION	**FRAGMENT IONS**
Aliphatic carboxylic acids:	Aliphatic carboxylic acids:
M^+ weak, but observable	$M - 17$, $M - 45$
	$m/z = 45, 60$
Aromatic carboxylic acids:	Aromatic carboxylic acids:
M^+ strong	$M - 17$, $M - 45$
	$M - 18$

$m/z = 120$

$m/z = 138$

FIGURE 4.51 EI-MS of methyl benzoate.

FIGURE 4.52   EI-MS of isobutyl salicylate.

small peak at $m/z = 71$. Loss of COOH gives rise to a peak at $m/z = 45$. Loss of the alkyl group as a free radical, leaving the $COOH^+$ ion ($m/z = 45$), also appears in the mass spectrum and is characteristic of the mass spectra of carboxylic acids. With acids containing γ-hydrogens, the principal pathway for fragmentation is the McLafferty rearrangement. In the case of carboxylic acids, this rearrangement produces a prominent peak at $m/z = 60$.

FIGURE 4.53   EI-MS of butyric acid.

FIGURE 4.54 EI-MS of *para*-anisic acid.

Aromatic carboxylic acids produce intense molecular ion peaks. The most important fragmentation pathway involves loss of ·OH to form the $C_6H_5C \equiv O^+$ ($m/z = 105$), followed by loss of CO to form the $C_6H_5^+$ ion ($m/z = 77$). In the mass spectrum of *para*-anisic acid (Fig. 4.54), loss of ·OH gives rise to a peak at $m/z = 135$. Further loss of CO from this ion gives rise to a peak at $m/z = 107$. Benzoic acids bearing *ortho* alkyl, hydroxy, or amino substituents undergo loss of water through a rearrangement reaction similar to that observed for *ortho*-substituted benzoate esters, as illustrated at the end of Section 4.6C.

4.7 FRAGMENTATION PATTERNS OF AMINES

The value of the mass of the molecular ion can be of great help in identifying a substance as an amine. As stated in Section 3.6, a compound with an odd number of nitrogen atoms must have an odd molecular weight. On this basis, it is possible to quickly determine whether a substance could be an amine. Unfortunately, in the case of aliphatic amines, the molecular ion peak may be very weak or even absent.

SPECTRAL ANALYSIS BOX—Amines

MOLECULAR ION	FRAGMENT IONS
M^+ weak or absent	α-Cleavage
Nitrogen Rule obeyed	$m/z = 30$

The most intense peak in the mass spectrum of an aliphatic amine arises from α-cleavage:

When there is a choice of R groups to be lost through this process, the largest R group is lost preferentially. For primary amines that are not branched at the carbon next to the nitrogen, the most intense peak in the spectrum occurs at $m/z = 30$. It arises from α-cleavage:

$$\left[R\!-\!CH_2\!-\!NH_2\right]^{\ddot{+}} \longrightarrow R\cdot + CH_2\!=\!\overset{+}{N}H_2$$
$$m/z = 30$$

The presence of this peak is strong, although not conclusive, evidence that the test substance is a primary amine. The peak may arise from secondary fragmentation of ions formed from the fragmentation of secondary or tertiary amines as well. In the mass spectrum of ethylamine (Fig. 4.55), the $m/z = 30$ peak can be seen clearly.

The same β-cleavage peak can also occur for long-chain primary amines. Further fragmentation of the R group of the amine leads to clusters of fragments 14 mass units apart due to sequential loss of CH_2 units from the R group. Long-chain primary amines can also undergo fragmentation via the process

$$\left[R\!-\!CH_2 \underset{(CH_2)_n}{\diagdown} NH_2\right]^{\ddot{+}} \longrightarrow R\cdot + \underset{(CH_2)_n}{\overset{CH_2\!-\!\overset{+}{N}H_2}{\diagdown\diagup}}$$

This is particularly favorable when $n = 4$ since a stable six-membered ring results. In this case, the fragment ion appears at $m/z = 86$.

Secondary and tertiary amines also undergo fragmentation processes as described earlier. The most important fragmentation is β-cleavage. In the mass spectrum of diethylamine (Fig. 4.56),

CH$_3$—CH$_2$—NH$_2$
M.W. = 45

FIGURE 4.55 Mass spectrum of ethylamine.

F I G U R E   4 . 5 6   Mass spectrum of diethylamine.

the intense peak at $m/z = 58$ is due to loss of a methyl group. Again, in the mass spectrum of triethylamine (Fig. 4.57), loss of methyl produces the most intense peak in the spectrum, at $m/z = 86$. In each case, further fragmentation of this initially formed fragment ion produces a peak at $m/z = 30$.

F I G U R E   4 . 5 7   Mass spectrum of triethylamine.

Cyclic aliphatic amines usually produce intense molecular ion peaks. Their principal modes of fragmentation are as follows:

$$\left[\underset{\underset{CH_3}{|}}{\overset{\cdot+}{N}}\right] \longrightarrow \underset{\underset{CH_3}{|}}{\overset{+}{N}} + H\cdot \longrightarrow CH_3-\overset{+}{N}{\equiv}CH + \underset{H_2C-CH_2}{\overset{CH_2}{\diagup\diagdown}}$$

$$m/z = 85 \qquad m/z = 84 \qquad\qquad m/z = 42$$

$$\left[\underset{\underset{CH_3}{|}}{\overset{\cdot+}{N}}\right] \longrightarrow \underset{\underset{CH_3}{|}}{\overset{CH_2}{\overset{+}{N}}}{\overset{\cdot CH_2}{\diagdown}} + CH_2{=}CH_2 \longrightarrow CH_2{=}\overset{+}{N}{=}CH_2 + CH_3\cdot$$

$$m/z = 57 \qquad\qquad m/z = 42$$

Aromatic amines show intense molecular ion peaks. A moderately intense peak may appear at an m/z value one mass unit less than that of the molecular ion due to loss of a hydrogen atom. The fragmentation of aromatic amines can be illustrated for the case of aniline:

$$\left[\text{⟨⟩}{-}NH_2\right]^{\cdot+} \longrightarrow \left[\text{⟨⟩}{-}NH\right]^{+} + H\cdot \longrightarrow \left[\text{⟨⟩}\right]^{\cdot+} + HCN$$

$$m/z = 93 \qquad\qquad m/z = 92 \qquad\qquad\qquad m/z = 66$$

$$\left[\text{⟨⟩}\right]^{\cdot+} + H\cdot$$

$$m/z = 65$$

Very intense molecular ion peaks characterize substituted pyridines. Frequently, loss of a hydrogen atom to produce a peak at an m/z value one mass unit less than the molecular ion is also observed.

The most important fragmentation process for the pyridine ring is loss of the elements of hydrogen cyanide. This produces a fragment ion that is 27 mass units lighter than the molecular ion. In the mass spectrum of 3-methylpyridine (Fig. 4.58), you can see the peak due to loss of hydrogen ($m/z = 92$) and the peak due to loss of hydrogen cyanide ($m/z = 66$).

When the alkyl side chain attached to a pyridine ring contains three or more carbons arranged linearly, fragmentation via the McLafferty rearrangement can also occur.

$$\left[\underset{H}{\overset{N}{\text{⟨⟩}}}R\right]^{\cdot+} \longrightarrow \left[\underset{H}{\overset{N}{\text{⟨⟩}}}{=}\right]^{\cdot+} + \overset{R}{\diagup}$$

$$m/z = 93$$

FIGURE 4.58 Mass spectrum of 3-methylpyridine.

This mode of cleavage is most important for substituents attached to the number 2 position of the ring.

4.8 FRAGMENTATION PATTERNS OF OTHER NITROGEN COMPOUNDS

As is true of amines, nitrogen-bearing compounds such as amides, nitriles, and nitro compounds must follow the Nitrogen Rule (explained more completely in Section 3.6): If they contain an odd number of nitrogen atoms, they must have an odd-numbered molecular weight.

Amides

The mass spectra of amides usually show observable molecular ion peaks. The fragmentation patterns of amides are quite similar to those of the corresponding esters and acids. The presence of a strong fragment ion peak at $m/z = 44$ is usually indicative of a primary amide. This peak arises from α-cleavage of the following sort.

$$\left[R-\overset{\overset{\displaystyle O}{\|}}{C}-NH_2 \right]^{+\cdot} \longrightarrow R\cdot \; + \; [O{=}C{=}NH_2]^+$$
$$m/z = 44$$

Once the carbon chain in the acyl moiety of an amide becomes long enough to permit the transfer of a hydrogen attached to the γ position, McLafferty rearrangements become possible. For primary amides, the McLafferty rearrangement gives rise to a fragment ion peak at $m/z = 59$. For N-alkylamides, analogous peaks at m/z values of 73, 87, 101, and so on often appear.

$$\left[\begin{array}{c} \underset{H_2N}{\overset{O}{\diagup}} \overset{H}{\diagdown} \overset{R}{\diagup} \end{array} \right]^{+\cdot} \longrightarrow \left[\underset{H_2N}{\overset{O^{\diagdown H}}{\diagup}} \right]^{+\cdot} + \overset{R}{\|}$$
$$m/z = 59$$

FIGURE 4.59 Mass spectrum of hexanenitrile.

Nitriles

Aliphatic nitriles usually undergo fragmentation so readily that the molecular ion peak is too weak to be observed. However, most nitriles form a peak due to the loss of one hydrogen atom, producing an ion of the type $R-CH=C=N^+$. Although this peak may be weak, it is a useful diagnostic peak in characterizing nitriles. In the mass spectrum of hexanenitrile (Fig. 4.59), this peak appears at $m/z = 96$.

When the alkyl group attached to the nitrile functional group is a propyl group or some longer hydrocarbon group, the most intense peak in the mass spectrum results from a McLafferty rearrangement:

$$m/z = 41$$

This peak, which appears in the mass spectrum of hexanenitrile, can be quite useful in characterizing an aliphatic nitrile. Unfortunately, as the alkyl group of a nitrile becomes longer, the probability of formation of the $C_3H_5^+$ ion, which also appears at $m/z = 41$, increases. With high molecular weight nitriles, most of the fragment ions of mass 41 are $C_3H_5^+$ ions rather than ions formed as a result of a McLafferty rearrangement.

The strongest peak in the mass spectrum of an aromatic nitrile is the molecular ion peak. Loss of cyanide occurs, giving, in the case of benzonitrile (Fig. 4.60), the $C_6H_5^+$ ion at $m/z = 77$. More important fragmentation involves loss of the elements of hydrogen cyanide. In benzonitrile, this gives rise to a peak at $m/z = 76$.

Nitro Compounds

The molecular ion peak for an aliphatic nitro compound is seldom observed. The mass spectrum is the result of fragmentation of the hydrocarbon part of the molecule. However, the mass spectra of

FIGURE 4.60 Mass spectrum of benzonitrile.

nitro compounds may show a moderate peak at $m/z = 30$, corresponding to the NO^+ ion, and a weaker peak at $m/z = 46$, corresponding to the NO_2^+ ion. These peaks appear in the mass spectrum of 1-nitropropane (Fig. 4.61). The intense peak at $m/z = 43$ is due to the $C_3H_7^+$ ion.

Aromatic nitro compounds show intense molecular ion peaks. The characteristic NO^+ ($m/z = 30$) and NO_2^+ ($m/z = 46$) peaks appear in the mass spectrum. The principal fragmentation pattern, however, involves loss of all or part of the nitro group. Using nitrobenzene (Fig. 4.62) as an example, this fragmentation pattern may be described as follows:

4.9 FRAGMENTATION PATTERNS OF ALKYL CHLORIDES AND ALKYL BROMIDES

The most dramatic feature of the mass spectra of alkyl chlorides and alkyl bromides is the presence of an important $M + 2$ peak. This peak arises because both chlorine and bromide are present in nature in two isotopic forms, each with a significant natural abundance.

SPECTRAL ANALYSIS BOX — Alkyl Halides

MOLECULAR ION

Strong $M + 2$ peak
  (for Cl, $M/M + 2 = 3:1$;
  for Br, $M/M + 2 = 1:1$)

FRAGMENT IONS

Loss of Cl or Br
Loss of HCl
α-Cleavage

FIGURE 4.61 Mass spectrum of 1-nitropropane.

FIGURE 4.62 Mass spectrum of nitrobenzene.

FIGURE 4.63 Mass spectrum of 1-bromohexane.

For aliphatic halogen compounds, the molecular ion peak is strongest with alkyl iodides, less strong for alkyl bromides, weaker for alkyl chlorides, and weakest for alkyl fluorides. Furthermore, as the alkyl group increases in size or as the amount of branching at the α position increases, the intensity of the molecular ion peak decreases.

There are several important fragmentation mechanisms for the alkyl halides. Perhaps the most important is the simple loss of the halogen atom, leaving a carbocation. This fragmentation is most important when the halogen is a good leaving group. Therefore, this type of fragmentation is most prominent in the mass spectra of the alkyl iodides and the alkyl bromides. In the mass spectrum of 1-bromohexane (Fig. 4.63), the peak at $m/z = 85$ is due to the formation of the hexyl ion. This ion undergoes further fragmentation to form a $C_3H_7^+$ ion at $m/z = 43$. The corresponding heptyl ion peak at $m/z = 99$ in the mass spectrum of 2-chloroheptane (Fig. 4.64) is quite weak.

FIGURE 4.64 Mass spectrum of 2-chloroheptane.

Alkyl halides may also lose a molecule of hydrogen halide according to the process

$$[R-CH_2-CH_2-X]^{\ddot{+}} \longrightarrow [R-CH=CH_2]^{\ddot{+}} + HX$$

This mode of fragmentation is most important for alkyl fluorides and chlorides and is less important for alkyl bromides and iodides. In the mass spectrum of 1-bromohexane, the peak corresponding to the loss of hydrogen bromide at $m/z = 84$ is very weak. However, for 2-chloroheptane, the peak corresponding to the loss of hydrogen chloride at $m/z = 98$ is quite intense.

A less important mode of fragmentation is α-cleavage, for which a fragmentation mechanism might be

$$\left[R-CH_2-X \right]^{\ddot{+}} \longrightarrow R\cdot + CH_2=X^+$$

When the α position is branched, the heaviest alkyl group attached to the α carbon is lost with greatest facility. The peaks arising from α-cleavage are usually rather weak.

A fourth fragmentation mechanism involves rearrangement and loss of an alkyl radical:

The corresponding cyclic ion can be observed at $m/z = 135$ and 137 in the mass spectrum of 1-bromohexane and at $m/z = 105$ and 107 in the mass spectrum of 2-chloroheptane. Such fragmentation is important only in the mass spectra of long-chain alkyl chlorides and bromides.

The molecular ion peaks in the mass spectra of benzyl halides are usually of sufficient intensity to be observed. The most important fragmentation involves loss of halogen to form the $C_7H_7^+$ ion. When the aromatic ring of a benzyl halide carries substituents, a substituted phenyl cation may also appear.

The molecular ion peak of an aromatic halide is usually quite intense. The most important mode of fragmentation involves loss of halogen to form the $C_6H_5^+$ ion.

Although the fragmentation patterns we have described are well characterized, the most interesting feature of the mass spectra of chlorine- and bromine-containing compounds is the presence of two molecular ion peaks. As Section 3.6 indicated, chlorine occurs naturally in two isotopic forms. The natural abundance of chlorine of mass 37 is 32.5% that of chlorine of mass 35. The natural abundance of bromine of mass 81 is 98.0% that of ^{79}Br. Therefore, the intensity of the $M + 2$ peak in a chlorine-containing compound should be 32.5% of the intensity of the molecular ion peak, and the intensity of the $M + 2$ peak in a bromine-containing compound should be almost equal to the intensity of the molecular ion peak. These pairs of molecular ion peaks (sometimes called doublets) appear in the mass spectra of ethyl chloride (Fig. 4.65) and ethyl bromide (Fig. 4.66).

Table 3.8 in Section 3.7 can be used to determine what the ratio of the intensities of the molecular ion and isotopic peaks should be when more than one chlorine or bromine is present in the same molecule. The mass spectra of dichloromethane (Fig. 4.67), dibromomethane (Fig. 4.68), and 1-bromo-2-chloroethane (Fig. 4.69) are included here to illustrate some of the combinations of halogens listed in Figure 3.19.

Unfortunately, it is not always possible to take advantage of these characteristic patterns to identify halogen compounds. Frequently, the molecular ion peaks are too weak to permit accurate measurement of the ratio of the intensities of the molecular ion and isotopic peaks. However, it is often

FIGURE 4.65 Mass spectrum of ethyl chloride.

FIGURE 4.66 Mass spectrum of ethyl bromide.

possible to make such a comparison on certain fragment ion peaks in the mass spectrum of a halogen compound. The mass spectrum of 1-bromohexane (Fig. 4.63) may be used to illustrate this method. The presence of bromine can be determined using the fragment ion peaks at m/z values of 135 and 137.

Since iodine and fluorine exist in nature in the form of only one isotope each, their mass spectra do not show isotopic peaks. The presence of halogen must be deduced either by noting the unusually weak $M + 1$ peak or by observing the mass difference between the fragment ions and the molecular ion.

FIGURE 4.67 Mass spectrum of dichloromethane.

FIGURE 4.68 Mass spectrum of dibromomethane.

▢ 4.10 COMPUTERIZED MATCHING OF SPECTRA WITH SPECTRAL LIBRARIES

Once a digitized mass spectrum is in hand, a basic PC can compare that data set to a library of tens of thousands of mass spectra within seconds and provide a list of potential matches. Each peak in a spectrum is categorized by the search program by uniqueness and relative abundance. Higher mass peaks are usually more characteristic of the compound in question than are commonly encountered low mass peaks, so the peaks with larger m/z may be weighted more heavily in the search algorithm. The output of a library search is a table that lists the names of possible

FIGURE 4.69 Mass spectrum of 1-bromo-2-chloroethane.

compounds, their molecular formulas, and an indicator of the probability that the spectrum of the test compound matches the spectrum in the database. The probability is determined by the number of peaks (and their intensities) that can be matched. This type of table is often called a *hit list*. Figure 4.70 is the mass spectrum of an unknown liquid substance with an observed boiling point of

FIGURE 4.70 EI-MS of an unknown liquid.

TABLE 4.1
RESULT OF LIBRARY SEARCH FOR UNKNOWN LIQUID

Name	Molecular Weight	Formula	Probability	CAS No.
1. Benzene, 1-chloro-2-methyl-	126	C_7H_7Cl	94	000095-49-8
2. Benzene, 1-chloro-3-methyl-	126	C_7H_7Cl	70	000108-41-8
3. Benzene, 1-chloro-4-methyl-	126	C_7H_7Cl	60	000106-43-4
4. Benzene, (chloromethyl)-	126	C_7H_7Cl	47	000100-44-7
5. 1,3,5-Cycloheptatriene, 1-chloro-	126	C_7H_7Cl	23	032743-66-1

158°C to 159°C. Table 4.1 reproduces the type of information that the computer would display as a hit list. Notice that the information includes the name of each compound that the computer has used for matching, its molecular weight and molecular formula, and its Chemical Abstracts Service (CAS) Registry number.

The information in Table 4.1 indicates that the unknown liquid is most likely **1-chloro-2-methylbenzene** since the probability of a correct match is placed at 94%. It is interesting to note that the *meta* and *para* isomers show probabilities of 70% and 60%, respectively. It is tempting to simply accept the results of the computer-based library search as correct, but the method is not an absolute guarantee that the identity of a sample has been correctly determined. A visual inspection of the experimental and library spectra must be included as part of the process. A computer can compare a mass spectrum it has determined with the spectra in these databases.

4.11 STRATEGIC APPROACH TO ANALYZING MASS SPECTRA AND SOLVING PROBLEMS

Like any other problem involving the correlation of spectral data with structure, having a well-defined strategy for analyzing mass spectra is the key to success. It is also true that chemical intuition plays an important role as well, and of course there is no substitute for practical experience. Before diving into the mass spectrum itself, take an inventory of what is known about the sample. Is the elemental composition known? Has the molecular formula been determined by exact mass analysis? What functional groups are present in the compound? What is the sample's "chemical history"? For example, how has the sample been handled? From what sort of chemical reaction was the compound isolated? And the questions can continue.

The first step in analyzing the mass spectrum itself is identifying the molecular ion. See Section 3.6 to review the requirements for a molecular ion. Once the molecular ion is identified, note its nominal mass and examine the isotope cluster (if the formula is not already known) for the presence of Cl, Br, and other $M + 2$ elements. Depending on whether the m/z value of the molecular ion is odd or even, the nitrogen rule will tell you how many nitrogens, if any, to incorporate into your analysis. If the molecular ion is not visible, consider running the sample under CI conditions to determine the molecular mass of the sample. If acquiring more data is not an option, consider what logical losses could have created the high mass peaks in the spectrum you have (loss of water from an alcohol, for example).

After analysis of the molecular ion cluster, examine the high mass peaks in your spectrum and determine whether the mass losses are odd or even. If an even number of nitrogens are present (zero is even), odd mass losses correspond to simple homolytic cleavages, and even mass losses are from rearrangements (this is reversed if there are an odd number of nitrogens present). Try to assign these mass losses to a radical fragment or neutral molecule. Next, look for readily identifiable fragments: phenylacylium ions, tropylium ions, phenyl cations, cyclopentadienyl cations, and so on.

Finally, use the fragmentation information to piece together a proposed structure. More than one potential structure may be reasonable pending further analysis. In some cases, it may only be possible to come up with a partial structure. Although tempting at times, remember that it is very risky to propose structures (or eliminate possible structures) on the *absence* of data: "That structure should give a peak at $m/z = Q$ from a McLafferty rearrangement, but there is no peak there—therefore that structure is wrong." When you have assembled a potential structure, reanalyze the fragmentation of that structure and see if it agrees with the experimental data. Comparison of your data to reference spectra from compounds with similar structures and functional groups can be very informative, and conducting a mass spectral library search of your spectrum against a database will likely provide some clues to the compound's identity, if not an exact match.

4.12 HOW TO SOLVE MASS SPECTRAL PROBLEMS

In this section, we will use EI-MS data to determine the structure of two unknown compounds. Even if the compound you are analyzing is not in searchable MS databases available to you, it is still possible to determine the structure of the compound with a few key pieces of data. If the molecular formula is available, either from an exact mass determination (Section 3.6) or Rule of Thirteen analysis (Section 1.5) on the molecular ion, the process is much simpler. Furthermore, knowing the main functional group(s) in the compound will assist in analyzing the fragmentation pattern. Information from an infrared spectrum and/or NMR spectra are useful in this regard.

■ SOLVED EXAMPLE 1

An unknown compound has the mass spectrum shown. The infrared spectrum of the unknown shows significant peaks at

3102 cm^{-1}	3087	3062	3030	1688
1598	1583	1460	1449	1353
1221	952	746	691	

There is also a band from aliphatic C—H stretching from 2879 to 2979 cm^{-1}.

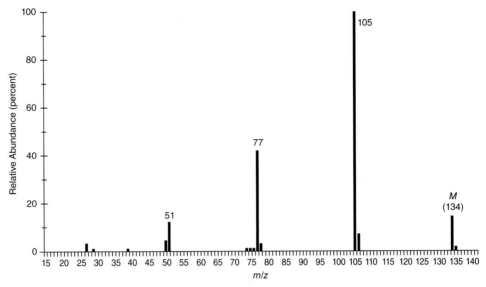

The molecular ion appears at an m/z value of 134. Applying the Rule of Thirteen gives the following possible molecular formulas:

$$C_{10}H_{14} \qquad U = 4$$
$$C_9H_{10}O \qquad U = 5$$

The infrared spectrum shows a $C = O$ peak at 1688 cm^{-1}. The position of this peak, along with the C$-$H stretching peaks in the 3030–3102 cm^{-1} range and $C = C$ stretching peaks in the 1449–1598 cm^{-1} range, suggests a ketone in which the carbonyl group is conjugated with a benzene ring. Such a structure would be consistent with the second molecular formula and with the index of hydrogen deficiency.

The base peak in the mass spectrum appears at $m/z = 105$. This peak is likely due to the formation of a benzoyl cation.

Subtracting the mass of the benzoyl ion from the mass of the molecular ion gives a difference of 29, suggesting that an ethyl group is attached to the carbonyl carbon. The peak appearing at $m/z = 77$ arises from the phenyl cation.

If we assemble all of the "pieces" suggested in the data, as described above, we conclude that the unknown compound must have been **propiophenone (1-phenyl-1-propanone).**

■ **SOLVED EXAMPLE 2**

While cleaning out old samples from your lab, you come across a vial labeled simply "decanone." You obtain an EI GC-MS of the material in the vial and obtain the mass spectrum shown in the figure. Use the fragmentation pattern to determine which isomer of decanone is in the vial.

There are four isomers of "decanone," and if we analyze the common fragmentations of ketones for each one, we should find some differences we can match to the given spectrum. The most common fragmentations for ketones are α-cleavage and the McLafferty rearrangement. Let's begin there. If we make a table of *m/z* values for the fragments expected for each isomer (remember there will be two different α-cleavages and perhaps two McLafferty rearrangements for each compound) and look for them in the given spectrum, we get the following. **BOLD = present** in spectrum; *Italic = NOT* present in spectrum.

Decanone isomer	α-cleavage fragments (*m/z*)	McLafferty fragments (*m/z*)
2-decanone	**43** and *141*	**58**
3-decanone	**57** and *127*	*72*
4-decanone	**71** and **113**	*128* and **86**
5-decanone	**85** and **99**	*100* and *114*

Now, 3-decanone and 5-decanone can be ruled out since there is no fragment from McLafferty rearrangement in either case. One would expect either the ion at *m/z* = 43 or 58 to be the base peak for 2-decanone. However, the ions at *m/z* = 71, 86, and 113 are almost as large as the ion at *m/z* = 58. Furthermore, it would be surprising if 2-decanone did not produce both possible α-cleavage ions, and there is no peak at *m/z* = 141. Therefore, we can conclude that the spectrum above belongs to 4-decanone.

1. Assign a structure that would be expected to give rise to each of the following mass spectra. *Note*: Some of these problems may have more than one reasonable answer. In some cases, infrared spectral data have been included in order to make the solution to the problems more reasonable. We recommend that you review the index of hydrogen deficiency (Section 1.4) and the Rule of Thirteen (Section 1.5) and apply those methods to each of the following problems.

*(a) The infrared spectrum has no interesting features except aliphatic C−H stretching and bending.

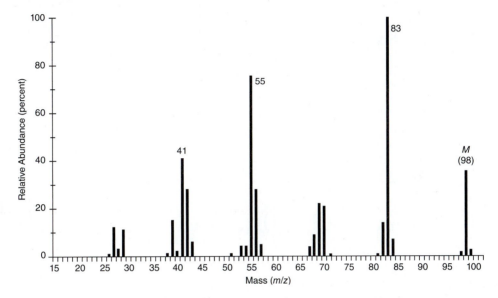

*(b) The infrared spectrum has a medium-intensity peak at about 1650 cm^{-1}. There is also a C−H out-of-plane bending peak near 880 cm^{-1}.

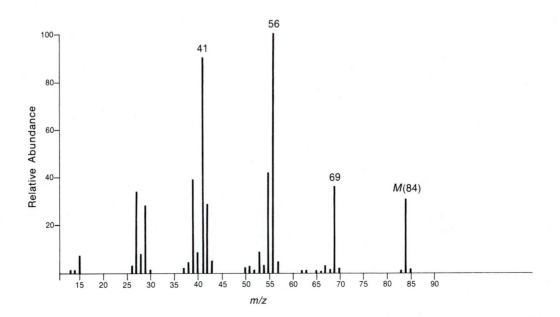

*(c) The infrared spectrum of this unknown has a prominent, broad peak at 3370 cm^{-1}. There is also a strong peak at 1159 cm^{-1}. The mass spectrum of this unknown does not show a molecular ion peak. You will have to deduce the molecular weight of this unknown from the heaviest fragment ion peak, which arises from the loss of a methyl group from the molecular ion.

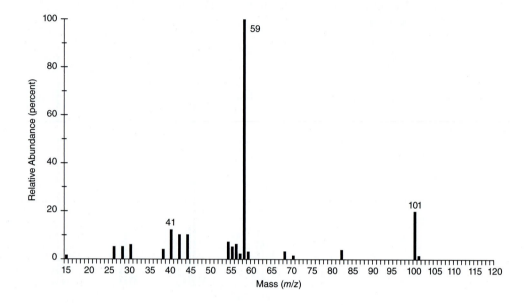

*(d) This unknown contains oxygen, but it does not show any significant infrared absorption peaks above 3000 cm^{-1}.

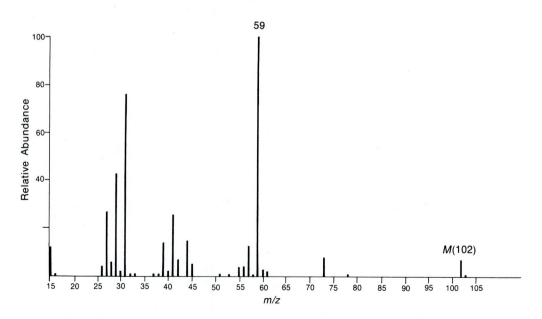

*(e) The infrared spectrum of this unknown shows a strong peak near 1725 cm^{-1}.

*(f) The infrared spectrum of this unknown shows a strong peak near 1715 cm^{-1}.

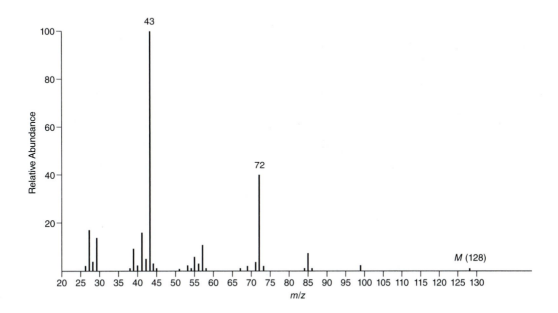

*(g) The infrared spectrum of this compound lacks any significant absorption above 3000 cm^{-1}. There is a prominent peak near 1740 cm^{-1} and another strong peak near 1200 cm^{-1}.

*(h) The infrared spectrum of this substance shows a very strong, broad peak in the range of 2500–3000 cm^{-1}, as well as a strong, somewhat broadened peak at about 1710 cm^{-1}.

*(i) The ^{13}C NMR spectrum of this unknown shows only four peaks in the region 125–145 ppm. The infrared spectrum shows a very strong, broad peak extending from 2500 to 3500 cm^{-1}, as well a strong and somewhat broadened peak at 1680 cm^{-1}.

*(j) Note the molecular ion on this substance has odd mass.

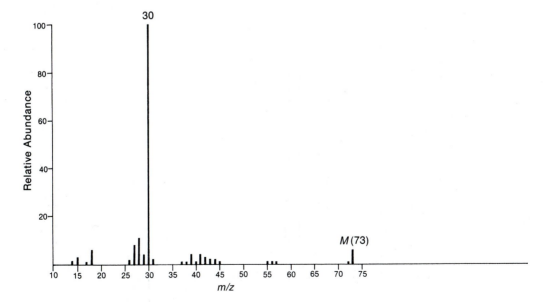

*(k) Notice the *M* **+ 2** peak in the mass spectrum.

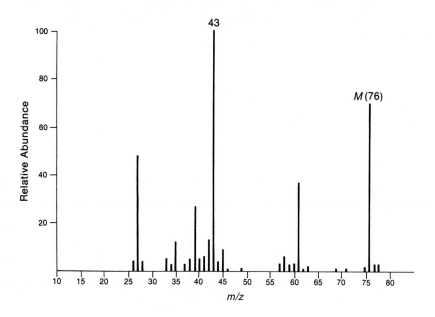

*(l) The infrared spectrum of this unknown shows two strong peaks, one near 1350 cm^{-1} and the other near 1550 cm^{-1}. Notice that the mass of the molecular ion is *odd*.

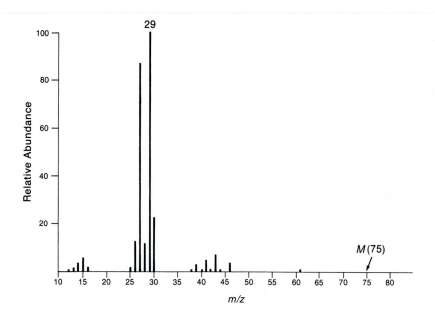

*(m) There is a sharp peak of medium intensity near 2250 cm^{-1} in the infrared spectrum of this compound.

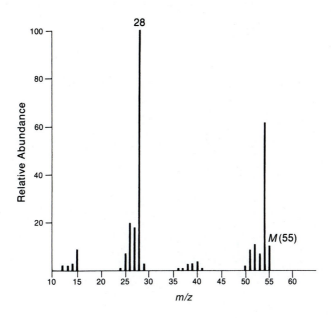

*(n) Consider the fragment ions at $m/z = 127$ and 128. From what ions might these peaks arise?

*(o)

*(p)

*(q)

*(r)

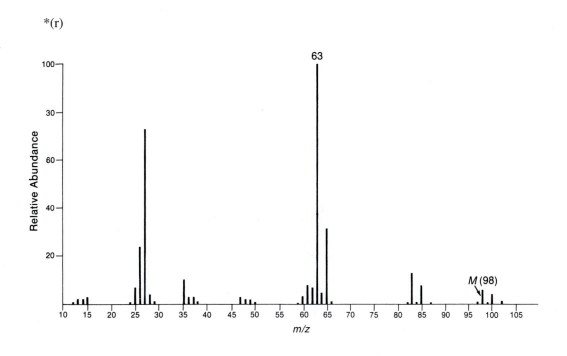

*(s) The infrared spectrum of this unknown shows a sharp peak at 3087 cm^{-1} and a sharp peak at 1612 cm^{-1} in addition to other absorptions. The unknown contains chlorine atoms, but some of the isotopic peaks ($M + n$) are too weak to be seen.

2. The mass spectrum of 3-butyn-2-ol shows a large peak at $m/z = 55$. Draw the structure of the fragment and explain why it is particularly stable.

3. How could the following pairs of isomeric compounds be differentiated by mass spectrometry?

(a)

(b)

(c)

(d)

$$CH_3-CH_2-CH_2-\underset{\underset{\textstyle OH}{|}}{CH}-CH_3 \quad \text{and} \quad CH_3-\underset{\underset{\textstyle OH}{|}}{\overset{\overset{\textstyle CH_3}{|}}{C}}-CH_2-CH_3$$

(e)

$$CH_3-CH_2-CH_2-CH_2-CH_2-Br \quad \text{and} \quad CH_3-CH_2-\underset{\underset{\textstyle Br}{|}}{CH}-CH_2-CH_3$$

(f)

$$CH_3-\underset{\underset{\textstyle CH_3}{|}}{CH}-CH_2-\overset{\overset{\textstyle O}{\|}}{C}-O-CH_3 \quad \text{and} \quad CH_3-CH_2-CH_2-\overset{\overset{\textstyle O}{\|}}{C}-O-CH_2-CH_3$$

(g)

$$CH_3-CH_2-\overset{\overset{\textstyle O}{\|}}{C}-\underset{\underset{\textstyle CH_3}{|}}{N}\overset{\textstyle CH_3}{} \quad \text{and} \quad CH_3-CH_2-CH_2-\overset{\overset{\textstyle O}{\|}}{C}-\underset{\underset{\textstyle H}{|}}{N}\overset{\textstyle CH_3}{}$$

(h) $$CH_3-CH_2-CH_2-CH_2-NH_2 \quad \text{and} \quad CH_3-CH_2-CH_2-NH-CH_3$$

(i)

(j)

4. Use the mass spectrum and either the NMR spectrum (Chapter 5) or the infrared spectrum (Chapter 2) to deduce the structure of each of the following compounds.

(a) $C_4H_7BrO_2$ ^1H NMR, 300 MHz, 2.9 ppm (triplet, 2H), 3.6 ppm (triplet, 2H) and 3.8 ppm (singlet, 3H)

(b) $C_4H_7ClO_2$ ^1H NMR, 300 MHz, 1.7 ppm (doublet, 3H), 3.8 ppm (singlet, 3H), and 4.4 ppm (quartet, 1H)

(c) $C_8H_6O_3$ ^{1}H NMR, 300 MHz, 6.1 ppm (singlet, 2H), 6.9 ppm (doublet, 1H), and 7.3 ppm (singlet, 1H), 7.4 ppm (doublet, 1H), 9.8 ppm (singlet, 1H); significant IR absorbances at 1687, 1602, 1449, 1264, 1038, 929, and 815 cm^{-1}.

(d) The infrared spectrum lacks any significant peaks above 3000 cm^{-1}.

(e) The infrared spectrum contains a single, strong peak at 3280 cm^{-1}.

(f) The infrared spectrum contains a single, strong peak at 1723 cm^{-1}.

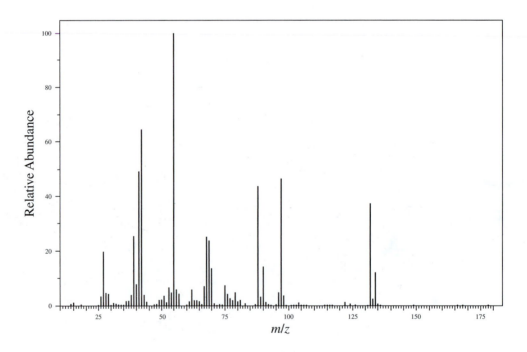

5. For each structure shown below

- Identify the site of initial ionization under EI conditions.
- Determine the structure of the ion indicated by the *m/z* value(s).
- Draw a fragmentation mechanism that accounts for the formation of the fragment ions.

    (a) Fragment ion at *m/z* = 98 (base peak in spectrum)

    (b) Fragment ion at *m/z* = 95 (base peak in spectrum)

    (c) Fragment ions at *m/z* = 103 and 61 (base peak)

    (d) Fragment ions at *m/z* = 95 (base peak) and 43

    (e) Fragment ion at *m/z* = 58 (base peak)

(f) Fragment ion at $m/z = 120$ (base peak)

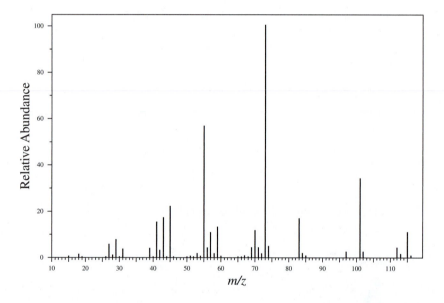

(g) Fragment ions at $m/z = 100$ (base peak), 91, 72, and 44

6. For each mass spectrum below, determine the structure of the prominent fragment ions and draw a fragmentation mechanism to explain their formation.

(a) 3-Methyl-3-heptanol

(b) Dicyclohexylamine

(c) 3,3,5-Trimethylcyclohexanone

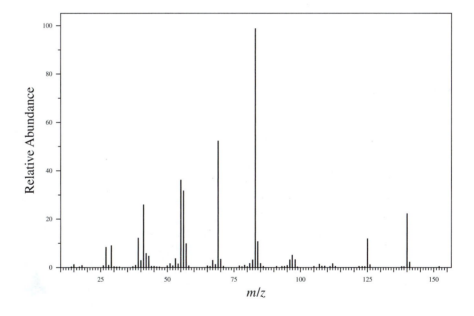

7. All dialkyl phthalate esters exhibit a base peak at $m/z = 149$. What is the structure of this fragment ion? Draw a mechanism that accounts for its formation from diethyl phthalate.

8. (a) The EI-MS of *ortho*-nitrotoluene (MW = 137) shows a large fragment ion at $m/z = 120$. The EI-MS of α,α,α-trideutero-*ortho*-nitrotoluene does **not** have a significant fragment ion at $m/z = 120$ but does have a peak at $m/z = 122$. Show the fragmentation process that explains these observations.

    (b) The EI mass spectra for methyl 2-methylbenzoate and methyl 3-methylbenzoate are reproduced below. Determine which spectrum belongs to which isomer and explain your answer.

Spectrum 1

Spectrum 2

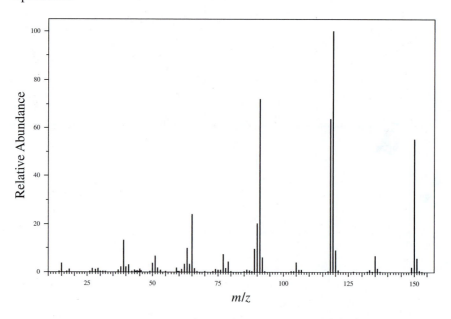

9. The methanol extract of the sponge *Haliclona vansoesti* is toxic to brine shrimp. The extract was chromatographed and the main component isolated. The compound was determined to have the two-dimensional structure shown (R = H) after analysis of the IR and 1-D NMR data. The compound was converted to the triacetate (R = Ac) followed by reaction of a small amount (0.3 mg) of the triacetate with dimethyl disulfide and iodine. GC-MS analysis of the resulting adduct showed a molecular ion at *m/z* = 463 and fragment ions at *m/z* = 145 and 318. Determine the exact position of the double bond in this natural product.

$$n + m = 8$$

10. The structure of pyrinodemin A, isolated from the sponge *Amphimedon* sp., was isolated in 1999 and assigned the structure below, with the Z-alkene between carbons 16′ and 17′. Owing to its structural novelty and reported biological activity, a number of groups undertook its synthesis. It was soon determined that the original structure was not correct and that the natural material was likely a positional isomer with respect to the alkene. Other proposed structures have the alkene between C15′–C16′ and between C14′ and C15′. Show how chemical derivatization of the natural material and MS analysis could be used to solve the mystery of pyrinodemin A's structure.

11. The structure of passifloricin A, an antiproazoal natural product isolated from the stinking passion flower *Passiflora foetida*, was assigned the structure below. A number of stereoisomers of this compound were prepared, but none of them produced spectral data that matched that of the natural material. Reexamination of the spectral data reported for natural passifloricin A turned up two issues: the natural product has only two CH(OR)CH$_2$CH(OR) units (the proposed structure has three) and the natural material has a prominent peak at *m/z* = 225 in the EI-MS. None of the stereoisomers of the synthetic compounds produce an ion at that *m/z* value, but all of them have a prominent ion at *m/z* = 193. Using this information, propose a more likely structure for passifloricin A.

12. The leaves of the Brazilian tree *Senna multijuga* contain a number of pyridine alkaloids that inhibit acetylcholinesterase. Two recently isolated compounds have the structure shown.

$$n + m = 10$$

Isomer A: EI-MS, *m/z* (rel. int.): 222(20), 150(10), 136(25), 123(100).
Isomer B: EI-MS, *m/z* (rel. int.): 236(20), 150(10), 136(25), 123(100).

(a) Use the mass spectral data provided to determine the precise location of the hydroxyl group in each isomer.

(b) What is the structure of the ion at *m/z* = 123? Draw a mechanism for its formation.

REFERENCES

Books

Chapman, J. R., *Practical Organic Mass Spectrometry: A Guide for Chemical and Biochemical Analysis*, 2nd ed., John Wiley, New York, 1995.

Crews, P., J. Rodriguez, and M. Jaspars, *Organic Structure Analysis*, 2nd ed., Oxford University Press, New York, 2010.

DeHoffmann, E., and V. Stroobant, *Mass Spectrometry: Principles and Applications*, 3rd ed., Wiley, New York, 2007.

Lambert, J. B., S. Gronert, H. F. Shurvell, D. A. Lightner, *Organic Structural Spectroscopy*, 2nd ed., Prentice Hall, Upper Saddle River, NJ, 2011.

McFadden, W. H., *Techniques of Combined Gas Chromatography/Mass Spectrometry: Applications in Organic Analysis*, Wiley-Interscience, New York, 1989.

McLafferty, F. W., and F. Tureček, *Interpretation of Mass Spectra*, 4th ed., University Science Books, Mill Valley, CA, 1993.

Pretsch, E., T. P. Buhlmann, and C. Affolter, *Structure Determination of Organic Compounds. Tables of Spectral Data*, 4th ed., Springer-Verlag, Berlin and New York, 2009.

Silverstein, R. M., F. X. Webster, and D. J. Kiemle, *Spectrometric Identification of Organic Compounds*, 7th ed., John Wiley, New York, 2005.

Smith, R. M., *Understanding Mass Spectra, A Basic Approach*, 2nd ed., Wiley, New York, 2004.

Watson, J. T., and O. D. Sparkman, *Introduction to Mass Spectrometry: Instrumentation, Applications, and Strategies for Data Interpretation*, 4th ed., Wiley, New York, 2007.

Articles

Carlson, D. A., C.-S. Roan, R. A. Yost, and J. Hector, "Dimethyl Disulfide Derivatives of Long Chain Alkenes, Alkadienes, and Alkatrienes for Gas Chromatography/Mass Spectrometry," *Analytical Chemistry 61* (1989):1564-1571.

Cosse, A. A., R. J. Bartelt, D. G. James, and R. J. Petroski, "Identification of a Female-Specific, Antennally Active Volatile Compound of the Currant Stem Girdler," *Journal of Chemical Ecology 27* (2001):1841-1853.

McFadden, W. H., L. E. Boggs, and R. G. Buttery "Specific Rearrangements in the Mass Spectra of Butyl Hexanoates and Similar Aliphatic Esters," *Journal of Physical Chemistry 70*, (1966): 3516-3523.

Yuan, G., and J. Yan "A Method for the Identification of the Double-Bond Position Isomeric Linear Tetradecenols and Related Compounds Based on Mass Spectra of Dimethyl Disulfide Derivatives," *Rapid Communications in Mass Spectrometry 16* (2002) 11-14.

Selected Websites

National Institute of Materials and Chemical Research, Tsukuba, Ibaraki, Japan, *Integrated Spectra Data Base System for Organic Compounds (SDBS)*: http://sdbs.riodb.aist.go.jp/sdbs/cgi-bin/cre_index.cgi

National Institute of Standards and Technology, *NIST Chemistry WebBook*: http://webbook.nist.gov/chemistry/
http://www.sisweb.com/mstools.htm
http://www.sisweb.com/index/referenc/academicsites.html
http://www.sisweb.com/mstools.htm

NUCLEAR MAGNETIC RESONANCE SPECTROSCOPY

Part One: Basic Concepts

Nuclear magnetic resonance (NMR) is a spectroscopic method that is even more important to the organic chemist than infrared spectroscopy. Many nuclei may be studied by NMR techniques, but hydrogen and carbon are most commonly available. Whereas infrared (IR) spectroscopy reveals the types of functional groups present in a molecule, NMR gives information about the number of magnetically distinct atoms of the type being studied. When hydrogen nuclei (protons) are studied, for instance, one can determine the number of each of the distinct types of hydrogen nuclei as well as obtain information regarding the nature of the immediate environment of each type. Similar information can be determined for the carbon nuclei. The combination of IR and NMR data is often sufficient to determine completely the structure of an unknown molecule.

5.1 NUCLEAR SPIN STATES

Many atomic nuclei have a property called **spin:** the nuclei behave as if they were spinning. In fact, any atomic nucleus that possesses either *odd* mass, *odd* atomic number, or both has a quantized **spin angular momentum** and a magnetic moment. The more common nuclei that possess spin include $_1^1H$, $_1^2H$, $_6^{13}C$, $_7^{14}N$, $_8^{17}O$, and $_9^{19}F$. Notice that the nuclei of the ordinary (most abundant) isotopes of carbon and oxygen, $_6^{12}C$ and $_8^{16}O$, are not included among those with the spin property. However, the nucleus of the ordinary hydrogen atom, the proton, does have spin. For each nucleus with spin, the number of allowed spin states it may adopt is quantized and is determined by its nuclear spin quantum number I. For each nucleus, the number I is a physical constant, and there are $2I + 1$ allowed spin states with integral differences ranging from $+I$ to $-I$. The individual spin states fit into the sequence

$$+I, (I - 1), \ldots, (-I + 1), -I \qquad \text{Equation 5.1}$$

For instance, a proton (hydrogen nucleus) has the spin quantum number $I = \frac{1}{2}$ and has two allowed spin states $[2(\frac{1}{2}) + 1 = 2]$ for its nucleus: $-\frac{1}{2}$ and $+\frac{1}{2}$. For the chlorine nucleus, $I = \frac{3}{2}$ and there are four allowed spin states $[2(\frac{3}{2}) + 1 = 4]$: $-\frac{3}{2}, -\frac{1}{2}, +\frac{1}{2}$, and $+\frac{3}{2}$. Table 5.1 gives the spin quantum numbers of several nuclei.

TABLE 5.1
SPIN QUANTUM NUMBERS OF SOME COMMON NUCLEI

Element	$_1^1H$	$_1^2H$	$_6^{12}C$	$_6^{13}C$	$_7^{14}N$	$_8^{16}O$	$_8^{17}O$	$_9^{19}F$	$_{15}^{31}P$	$_{17}^{35}Cl$
Nuclear spin quantum number	$\frac{1}{2}$	1	0	$\frac{1}{2}$	1	0	$\frac{5}{2}$	$\frac{1}{2}$	$\frac{1}{2}$	$\frac{3}{2}$
Number of spin states	2	3	0	2	3	0	6	2	2	4

In the absence of an applied magnetic field, all the spin states of a given nucleus are of equivalent energy (degenerate), and in a collection of atoms, all of the spin states should be almost equally populated, with the same number of atoms having each of the allowed spins.

5.2 NUCLEAR MAGNETIC MOMENTS

Spin states are not of equivalent energy in an applied magnetic field because the nucleus is a charged particle, and any moving charge generates a magnetic field of its own. Thus, the nucleus has a magnetic moment μ generated by its charge and spin. A hydrogen nucleus may have a clockwise $(+\frac{1}{2})$ or counterclockwise $(-\frac{1}{2})$ spin, and the nuclear magnetic moments (μ) in the two cases are pointed in opposite directions. In an applied magnetic field, all protons have their magnetic moments either aligned with the field or opposed to it. Figure 5.1 illustrates these two situations.

Hydrogen nuclei can adopt only one or the other of these orientations with respect to the applied field. The spin state $+\frac{1}{2}$ is of lower energy since it is aligned with the field, while the spin state $-\frac{1}{2}$ is of higher energy since it is opposed to the applied field. This should be intuitively obvious to anyone who thinks a little about the two situations depicted in Figure 5.2, involving magnets. The aligned configuration of magnets is stable (low energy). However, where the magnets are opposed (not aligned), the center magnet is repelled out of its current (high-energy) orientation. If the central

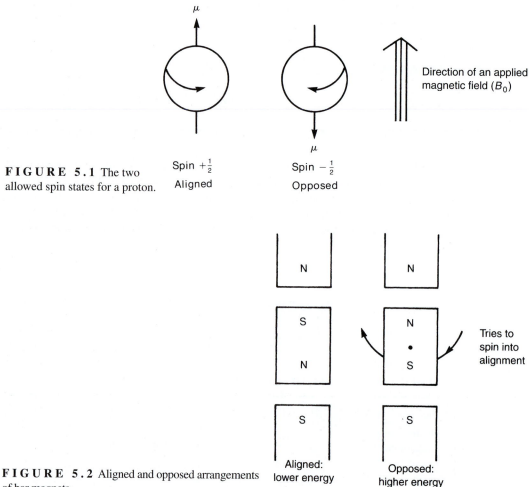

FIGURE 5.1 The two allowed spin states for a proton.

Spin $+\frac{1}{2}$
Aligned

Spin $-\frac{1}{2}$
Opposed

Direction of an applied magnetic field (B_0)

FIGURE 5.2 Aligned and opposed arrangements of bar magnets.

Aligned: lower energy

Opposed: higher energy

Tries to spin into alignment

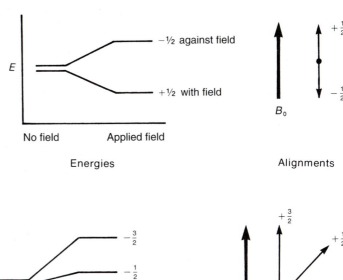

FIGURE 5.3 The spin states of a proton in the absence and in the presence of an applied magnetic field.

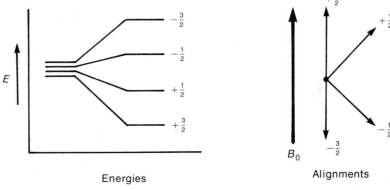

FIGURE 5.4 The spin states of a chlorine atom both in the presence and in the absence of an applied magnetic field.

magnet were placed on a pivot, it would spontaneously spin around the pivot into alignment (low energy). Hence, as an external magnetic field is applied, the degenerate spin states split into two states of unequal energy, as shown in Figure 5.3.

In the case of a chlorine nucleus, there are four energy levels, as shown in Figure 5.4. The $+\frac{3}{2}$ and $-\frac{3}{2}$ spin states are aligned with the applied field and opposed to the applied field, respectively. The $+\frac{1}{2}$ and $-\frac{1}{2}$ spin states have intermediate orientations, as indicated by the vector diagram on the right in Figure 5.4.

◻ 5.3 ABSORPTION OF ENERGY

The nuclear magnetic resonance phenomenon occurs when nuclei aligned with an applied field are induced to absorb energy and change their spin orientation with respect to the applied field. Figure 5.5 illustrates this process for a hydrogen nucleus.

The energy absorption is a quantized process, and the energy absorbed must equal the energy difference between the two states involved.

$$E_{\text{absorbed}} = (E_{-\frac{1}{2}\text{ state}} - E_{+\frac{1}{2}\text{ state}}) = h\upsilon \qquad \textbf{Equation 5.2}$$

In practice, this energy difference is a function of the strength of the applied magnetic field B_0, as illustrated in Figure 5.6.

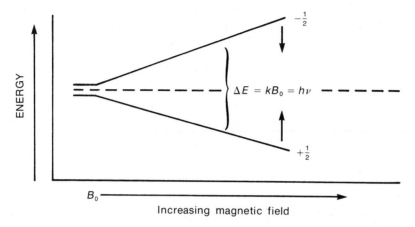

F I G U R E 5.5 The NMR absorption process for a proton.

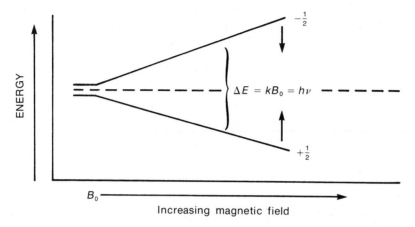

F I G U R E 5.6 The spin-state energy separation as a function of the strength of the applied magnetic field B_0.

The stronger the applied magnetic field, the greater the energy difference between the possible spin states:

$$\Delta E = f(B_0) \qquad\qquad \textbf{Equation 5.3}$$

The magnitude of the energy-level separation also depends on the particular nucleus involved. Each nucleus (hydrogen, chlorine, and so on) has a different ratio of magnetic moment to angular momentum since each has different charge and mass. This ratio, called the **magnetogyric ratio γ**, is a constant for each nucleus and determines the energy dependence on the magnetic field:

$$\Delta E = f(\gamma B_0) = h\nu \qquad\qquad \textbf{Equation 5.4}$$

Since the angular momentum of the nucleus is quantized in units of $h/2\pi$, the final equation takes the form

$$\Delta E = \gamma\left(\frac{h}{2\pi}\right) B_0 = h\nu \qquad\qquad \textbf{Equation 5.5}$$

Solving for the frequency of the absorbed energy,

$$\nu = \left(\frac{\gamma}{2\pi}\right) B_0 \qquad\qquad \textbf{Equation 5.6}$$

If the correct value of γ for the proton is substituted, one finds that an unshielded proton should absorb radiation of frequency 42.6 MHz in a field of strength 1 Tesla (10,000 Gauss) or radiation of frequency 60.0 MHz in a field of strength 1.41 Tesla (14,100 Gauss). Table 5.2 shows the field strengths and frequencies at which several nuclei have resonance (i.e., absorb energy and make spin transitions).

TABLE 5.2
FREQUENCIES AND FIELD STRENGTHS AT WHICH SELECTED NUCLEI HAVE THEIR NUCLEAR RESONANCES

Isotope	Natural Abundance (%)	Field Strength, B_0 (Tesla[a])	Frequency, ν (MHz)	Magnetogyric Ratio, γ (radians/Tesla)
^{1}H	99.98	1.00	42.6	267.53
		1.41	60.0	
		2.35	100.0	
		4.70	200.0	
		7.05	300.0	
^{2}H	0.0156	1.00	6.5	41.1
^{13}C	1.108	1.00	10.7	67.28
		1.41	15.1	
		2.35	25.0	
		4.70	50.0	
		7.05	75.0	
^{19}F	100.0	1.00	40.0	251.7
^{31}P	100.0	1.00	17.2	108.3

[a] 1 Tesla = 10,000 Gauss.

Although many nuclei are capable of exhibiting magnetic resonance, the organic chemist is mainly interested in hydrogen and carbon resonances. This chapter emphasizes hydrogen. Chapter 6 will discuss nuclei other than hydrogen—for example, carbon-13, fluorine-19, phosphorus-31, and deuterium (hydrogen-2).

For a proton (the nucleus of a hydrogen atom), if the applied magnetic field has a strength of approximately 1.41 Tesla, the difference in energy between the two spin states of the proton is about 2.39×10^{-5} kJ/mole. Radiation with a frequency of about 60 MHz (60,000,000 Hz), which lies in the radiofrequency (RF) region of the electromagnetic spectrum, corresponds to this energy difference. Other nuclei have both larger and smaller energy differences between spin states than do hydrogen nuclei. The earliest nuclear magnetic resonance spectrometers applied a variable magnetic field with a range of strengths near 1.41 Tesla and supplied a constant radiofrequency radiation of 60 MHz. They effectively induced transitions only among proton (hydrogen) spin states in a molecule and were not useful for other nuclei. Separate instruments were required to observe transitions in the nuclei of other elements, such as carbon and phosphorus. Fourier transform instruments (Section 5.7B), which are in common use today, are equipped to observe the nuclei of several different elements in a single instrument. Instruments operating at frequencies of 300 and 400 MHz are now quite common, and instruments with frequencies above 600 MHz are found in the larger research universities.

5.4 THE MECHANISM OF ABSORPTION (RESONANCE)

To understand the nature of a nuclear spin transition, the analogy of a child's spinning top is useful. Protons absorb energy because they begin to precess in an applied magnetic field. The phenomenon of precession is similar to that of a spinning top. Owing to the influence of the earth's gravitational field, the top begins to "wobble," or precess, about its axis (Fig. 5.7a). A spinning nucleus behaves in a similar fashion under the influence of an applied magnetic field (Fig. 5.7b).

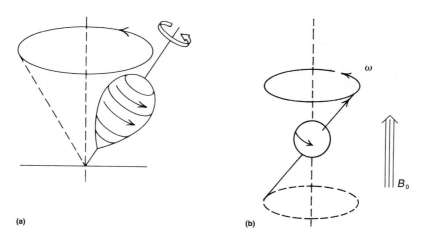

FIGURE 5.7 (a) A top precessing in the earth's gravitational field; (b) the precession of a spinning nucleus resulting from the influence of an applied magnetic field.

When the magnetic field is applied, the nucleus begins to precess about its own axis of spin with angular frequency ω, which is sometimes called its **Larmor frequency.** The frequency at which a proton precesses is directly proportional to the strength of the applied magnetic field; the stronger the applied field, the higher the rate (angular frequency ω) of precession. For a proton, if the applied field is 1.41 Tesla (14,100 Gauss), the frequency of precession is approximately 60 MHz.

Since the nucleus has a charge, the precession generates an oscillating electric field of the same frequency. If radiofrequency waves of this frequency are supplied to the precessing proton, the energy can be absorbed. That is, when the frequency of the oscillating electric field component of the incoming radiation just matches the frequency of the electric field generated by the precessing nucleus, the two fields can couple, and energy can be transferred from the incoming radiation to the nucleus, thus causing a spin change. This condition is called **resonance,** and the nucleus is said to have resonance with the incoming electromagnetic wave. Figure 5.8 schematically illustrates the resonance process.

FIGURE 5.8 The nuclear magnetic resonance process; absorption occurs when $\upsilon = \omega$.

5.5 POPULATION DENSITIES OF NUCLEAR SPIN STATES

For a proton, if the applied magnetic field has a strength of approximately 1.41 Tesla, resonance occurs at about 60 MHz, and using $\Delta E = h\nu$, we can calculate that the difference in energy between the two spin states of the proton is about 2.39×10^{-5} kJ/mole. Thermal energy resulting from room temperature is sufficient to populate both of these energy levels since the energy separation between the two levels is small. There is, however, a slight excess of nuclei in the lower-energy spin state. The magnitude of this difference can be calculated using the Boltzmann distribution equations. Equation 5.7 gives the Boltzmann ratio of nuclear spins in the upper and lower levels.

$$\frac{N_{upper}}{N_{lower}} = e^{-\Delta E/kT} = e^{-h\nu/kT}$$

Equation 5.7

$$h = 6.624 \times 10^{-34} \text{ J·sec}$$

$$k = 1.380 \times 10^{-23} \text{ J/K · molecule}$$

$$T = \text{absolute temperature (K)}$$

where ΔE is the energy difference between the upper and lower energy states, and k is the molecular (not molar) gas constant. Since $\Delta E = h\nu$, the second form of the equation is derived, where ν is the operating frequency of the instrument and h is Planck's constant.

Using Equation 5.7, one can calculate that at 298 K (25°C), for an instrument operating at 60 MHz there are 1,000,009 nuclei in the lower (favored) spin state for every 1,000,000 that occupy the upper spin state:

$$\frac{N_{upper}}{N_{lower}} = 0.999991 = \frac{1,000,000}{1,000,009}$$

In other words, in approximately 2 million nuclei, there are only 9 more nuclei in the lower spin state. Let us call this number (9) the **excess population** (Fig. 5.9).

The excess nuclei are the ones that allow us to observe resonance. When the 60-MHz radiation is applied, it not only induces transitions upward but also stimulates transitions downward. If the populations of the upper and lower states become exactly equal, we observe no net signal. This situation is called **saturation.** One must be careful to avoid saturation when performing an NMR experiment. Saturation is achieved quickly if the power of the radiofrequency signal is too high. Therefore, the very small excess of nuclei in the lower spin state is quite important to NMR spectroscopy, and we can see that very sensitive NMR instrumentation is required to detect the signal.

If we increase the operating frequency of the NMR instrument, the energy difference between the two states increases (see Fig. 5.6), which causes an increase in this excess. Table 5.3 shows how the excess increases with operating frequency. It also clearly shows why modern instrumentation has been designed with increasingly higher operating frequencies. The sensitivity of the instrument is increased, and the resonance signals are stronger, because more nuclei can undergo transition at higher frequency. Before the advent of higher-field instruments, it was very difficult to observe less-sensitive nuclei such as carbon-13, which is not very abundant (1.1%) and has a detection frequency much lower than that of hydrogen (see Table 5.2).

Population

_____ N

$N = 1,000,000$
Excess $= 9$

_____ $N + 9$

FIGURE 5.9 The excess population of nuclei in the lower spin state at 60 MHz.

TABLE 5.3
VARIATION OF ^{1}H EXCESS NUCLEI
WITH OPERATING FREQUENCY

Frequency (MHz)	Excess Nuclei
20	3
40	6
60	9
80	12
100	16
200	32
300	48
600	96

5.6 THE CHEMICAL SHIFT AND SHIELDING

Nuclear magnetic resonance has great utility because not all protons in a molecule have resonance at exactly the same frequency. This variability is due to the fact that the protons in a molecule are surrounded by electrons and exist in slightly different electronic (magnetic) environments from one another. The valence-shell electron densities vary from one proton to another. The protons are **shielded** by the electrons that surround them. In an applied magnetic field, the valence electrons of the protons are caused to circulate. This circulation, called a **local diamagnetic current,** generates a counter magnetic field that opposes the applied magnetic field. Figure 5.10 illustrates this effect, which is called **diamagnetic shielding** or **diamagnetic anisotropy.**

Circulation of electrons around a nucleus can be viewed as being similar to the flow of an electric current in an electric wire. From physics, we know that the flow of a current through a wire induces a magnetic field. In an atom, the local diamagnetic current generates a secondary, induced magnetic field that has a direction opposite that of the applied magnetic field.

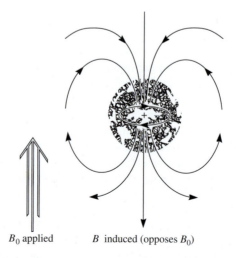

B_0 applied B induced (opposes B_0)

FIGURE 5.10 Diamagnetic anisotropy—the diamagnetic shielding of a nucleus caused by the circulation of valence electrons.

As a result of diamagnetic anisotropy, each proton in a molecule is shielded from the applied magnetic field to an extent that depends on the electron density surrounding it. The greater the electron density around a nucleus, the greater the induced counter field that opposes the applied field. The counter field that shields a nucleus diminishes the net applied magnetic field that the nucleus experiences. As a result, the nucleus precesses at a lower frequency. This means that it also absorbs radiofrequency radiation at this lower frequency. Each proton in a molecule is in a slightly different chemical environment and consequently has a slightly different amount of electronic shielding, which results in a slightly different resonance frequency.

These differences in resonance frequency are very small. For instance, the difference between the resonance frequencies of the protons in chloromethane and those in fluoromethane is only 72 Hz when the applied field is 1.41 Tesla. Since the radiation used to induce proton spin transitions at that magnetic field strength is of a frequency near 60 MHz, the difference between chloromethane and fluoromethane represents a change in frequency of only slightly more than one part per million! It is very difficult to measure exact frequencies to that precision; hence, no attempt is made to measure the exact resonance frequency of any proton. Instead, a reference compound is placed in the solution of the substance to be measured, and the resonance frequency of each proton in the sample is measured relative to the resonance frequency of the protons of the reference substance. In other words, the frequency *difference* is measured directly. The standard reference substance that is used universally is **tetramethylsilane, $(CH_3)_4Si$,** also called **TMS.** This compound was chosen initially because the protons of its methyl groups are more shielded than those of most other known compounds. At that time, no compounds that had better-shielded hydrogens than TMS were known, and it was assumed that TMS would be a good reference substance since it would mark one end of the range. Thus, when another compound is measured, the resonances of its protons are reported in terms of how far (in Hertz) they are shifted from those of TMS.

The shift from TMS for a given proton depends on the strength of the applied magnetic field. In an applied field of 1.41 Tesla the resonance of a proton is approximately 60 MHz, whereas in an applied field of 2.35 Tesla (23,500 Gauss) the resonance appears at approximately 100 MHz. The ratio of the resonance frequencies is the same as the ratio of the two field strengths:

$$\frac{100 \text{ MHz}}{60 \text{ MHz}} = \frac{2.35 \text{ Tesla}}{1.41 \text{ Tesla}} = \frac{23,500 \text{ Gauss}}{14,100 \text{ Gauss}} = \frac{5}{3}$$

Hence, for a given proton, the shift (in Hertz) from TMS is $\frac{5}{3}$ larger in the 100-MHz range ($B_0 = 2.35$ Tesla) than in the 60-MHz range ($B_0 = 1.41$ Tesla). This can be confusing for workers trying to compare data if they have spectrometers that differ in the strength of the applied magnetic field. The confusion is easily overcome if one defines a new parameter that is independent of field strength—for instance, by dividing the shift in Hertz of a given proton by the frequency in megahertz of the spectrometer with which the shift value was obtained. In this manner, a field-independent measure called the **chemical shift** (δ) is obtained

$$\delta = \frac{\text{(shift in Hz)}}{\text{(spectrometer frequency in MHz)}} \qquad \textbf{Equation 5.8}$$

The chemical shift in δ units expresses the amount by which a proton resonance is shifted from TMS, in parts per million (ppm), of the spectrometer's basic operating frequency. Values of δ for a given proton are always the same irrespective of whether the measurement was made at 60 MHz ($B_0 = 1.41$ Tesla) or at 100 MHz ($B_0 = 2.35$ Tesla). For instance, at 60 MHz the shift of the protons in CH_3Br is 162 Hz from TMS, while at 100 MHz the shift is 270 Hz. However, both of these correspond to the same value of δ (2.70 ppm):

$$\delta = \frac{162 \text{ Hz}}{60 \text{ MHz}} = \frac{270 \text{ Hz}}{100 \text{ MHz}} = 2.70 \text{ ppm}$$

By agreement, most workers report chemical shifts in **delta (δ) units,** or **parts per million (ppm),** of the main spectrometer frequency. On this scale, the resonance of the protons in TMS comes at exactly 0.00 ppm (by definition).

The NMR spectrometer actually scans from high δ values to low ones (as will be discussed in Section 5.7). Following is a typical chemical shift scale with the sequence of δ values that would be found on a typical NMR spectrum chart.

Direction of scan $\Rightarrow$

												TMS			
12	11	10	9	8	7	6	5	4	3	2	1	0	−1	−2	δ scale (ppm)

5.7 THE NUCLEAR MAGNETIC RESONANCE SPECTROMETER

A. The Continuous-Wave (CW) Instrument

Figure 5.11 schematically illustrates the basic elements of a classical 60-MHz NMR spectrometer. The sample is dissolved in a solvent containing no interfering protons (usually $CDCl_3$), and a small amount of TMS is added to serve as an internal reference. The sample cell is a small cylindrical glass tube that is suspended in the gap between the faces of the pole pieces of the magnet. The sample is spun around its axis to ensure that all parts of the solution experience a relatively uniform magnetic field.

Also in the magnet gap is a coil attached to a 60-MHz radiofrequency (RF) generator. This coil supplies the electromagnetic energy used to change the spin orientations of the protons. Perpendicular to the RF oscillator coil is a detector coil. When no absorption of energy is taking place, the detector coil picks up none of the energy given off by the RF oscillator coil. When the sample absorbs energy, however, the reorientation of the nuclear spins induces a RF signal in the plane of the detector coil, and the instrument responds by recording this as a **resonance signal,** or **peak.**

At a constant field strength, the distinct types of protons in a molecule precess at slightly different frequencies. Rather than changing the frequency of the RF oscillator to allow each of the

FIGURE 5.11 The basic elements of the classical nuclear magnetic resonance spectrometer.

protons in a molecule to come into resonance, the CW NMR spectrometer uses a constant-frequency RF signal and varies the magnetic field strength. As the magnetic field strength is increased, the precessional frequencies of all the protons increase. When the precessional frequency of a given type of proton reaches 60 MHz, it has resonance. The magnet that is varied is, in fact, a two-part device. There is a main magnet, with a strength of about 1.41 Tesla, which is capped by electromagnet pole pieces. By varying the current through the pole pieces, the worker can increase the main field strength by as much as 20 parts per million (ppm). Changing the field in this way systematically brings all of the different types of protons in the sample into resonance.

As the field strength is increased linearly, a pen travels across a recording chart. A typical spectrum is recorded as in Figure 5.12. As the pen travels from left to right, the magnetic field is increasing. As each chemically distinct type of proton comes into resonance, it is recorded as a peak on the chart. The peak at $\delta = 0$ ppm is due to the internal reference compound TMS. Since highly shielded protons precess more slowly than relatively unshielded protons, it is necessary to increase the field to induce them to precess at 60 MHz. Hence, highly **shielded** protons appear to the right of this chart, and less shielded, or **deshielded,** protons appear to the left. The region of the chart to the left is sometimes said to be **downfield** (or at **low field**), and that to the right, **upfield** (or at **high field**). Varying the magnetic field as is done in the usual spectrometer is exactly equivalent to varying the radio frequency, RF and a change of 1 ppm in the magnetic field strength (increase) has the same effect as a 1-ppm change (decrease) in the RF frequency (see Eq. 5.6). Hence, changing the field strength instead of the RF frequency is only a matter of instrumental design. Instruments that vary the magnetic field in a continuous fashion, scanning from the downfield end to the upfield end of the spectrum, are called **continuous-wave (CW) instruments.** Because the chemical shifts of

FIGURE 5.12 The 60-MHz ^{1}H nuclear magnetic resonance spectrum of phenylacetone (the absorption peak at the far right is caused by the added reference substance TMS).

the peaks in this spectrum are calculated from *frequency* differences from TMS, this type of spectrum (Fig. 5.12) is said to be a **frequency-domain spectrum.**

A distinctive characteristic enables one to recognize a CW spectrum. Peaks generated by a CW instrument have **ringing,** a decreasing series of oscillations that occurs after the instrument has scanned through the peak (Fig. 5.13). Ringing occurs because the excited nuclei do not have time to relax back to their equilibrium state before the field, and pen, of the instrument have advanced to a new position. The excited nuclei have a relaxation rate that is slower than the rate of scan. As a result, they are still emitting an oscillating, rapidly decaying signal, which is recorded as ringing. Ringing is desirable in a CW instrument and is considered to indicate that the field homogeneity is well adjusted. Ringing is most noticeable when a peak is a sharp singlet (a single, isolated peak).

B.   The Pulsed Fourier Transform (FT) Instrument

The CW type of NMR spectrometer, which was described in Section 5.7A, operates by exciting the nuclei of the isotope under observation one type at a time. In the case of ^1H nuclei, each distinct type of proton (phenyl, vinyl, methyl, and so on) is excited individually, and its resonance peak is observed and recorded independently of all the others. As we scan, we look at first one type of hydrogen and then another, scanning until all of the types have come into resonance.

An alternative approach, common to modern, sophisticated instruments, is to use a powerful but short burst of energy, called a **pulse,** that excites all of the magnetic nuclei in the molecule simultaneously. In an organic molecule, for instance, all of the ^1H nuclei are induced to undergo resonance at the same time. An instrument with a 2.1-Tesla magnetic field uses a short (1- to 10-μsec) burst of 90-MHz energy to accomplish this. The source is turned on and off very quickly, generating a pulse similar to that shown in Figure 5.14a. According to a variation of the Heisenberg Uncertainty Principle, even though the frequency of the oscillator generating this pulse is set to 90 MHz, if the duration of the pulse is very short, the frequency content of the pulse is uncertain because the oscillator was not on long enough to establish a solid fundamental frequency. Therefore, the pulse actually contains a *range of frequencies* centered about the fundamental, as shown in Figure 5.14b. This range of frequencies is great enough to excite all of the distinct types of hydrogens in the molecule at once with this single burst of energy.

When the pulse is discontinued, the excited nuclei begin to lose their excitation energy and return to their original spin state, or **relax.** As each excited nucleus relaxes, it emits electromagnetic radiation.

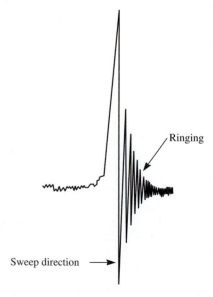

FIGURE 5.13 A CW peak that shows ringing.

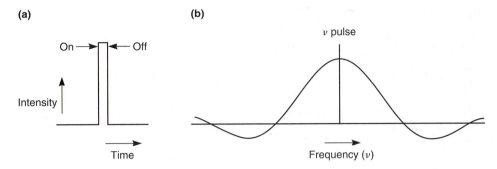

FIGURE 5.14 A short pulse. (a) The original pulse; (b) the frequency content of the same pulse.

Since the molecule contains many different nuclei, many different frequencies of electromagnetic radiation are emitted simultaneously. This emission is called a **free-induction decay (FID)** signal (Fig. 5.15). Notice that the intensity of the FID decays with time as all of the nuclei eventually lose their excitation. The FID is a superimposed combination of all the frequencies emitted and can be quite complex. We usually extract the individual frequencies due to different nuclei by using a computer and a mathematical method called a Fourier transform (FT) analysis, which is described later in this section.

If we look at a very simple molecule such as acetone, we can avoid the inherent complexities of the Fourier transform and gain a clearer understanding of the method. Figure 5.16a shows the FID for the hydrogens in acetone. This FID was determined in an instrument with a 7.05-Tesla magnet operating at 300 MHz.

Since acetone has only one type of hydrogen (all six hydrogens are equivalent), the FID curve is composed of a single sinusoidal wave. The signal decays exponentially with time as the nuclei relax and their signal diminishes. Since the horizontal axis on this signal is time, the FID is sometimes called a **time-domain signal.** If the signal did not decay in intensity, it would appear as a sine (or cosine) wave of constant intensity, as shown in Figure 5.16b. One can calculate the frequency of this wave from its measured wavelength λ (difference between the maxima).

The determined frequency is not the exact frequency emitted by the methyl hydrogens. Due to the design of the instrument, the basic frequency of the pulse is not the same as the frequency of the acetone resonance. The observed FID is actually an interference signal between the radiofrequency source (300 MHz in this case) and the frequency emitted by the excited nucleus, where the wavelength is given by

$$\lambda = \frac{1}{v_{acetone} - v_{pulse}}$$ **Equation 5.9**

In other words, this signal represents the difference in the two frequencies. Since the frequency of the pulse is known, we could readily determine the exact frequency. However, we do not need to know it since we are interested in the chemical shift of those protons, which is given by

$$\delta'_{acetone} = \frac{v_{acetone} - v_{pulse}}{v_{pulse}}$$ **Equation 5.10**

which can be reduced to the unit analysis

$$ppm = \frac{(Hz)}{MHz}$$

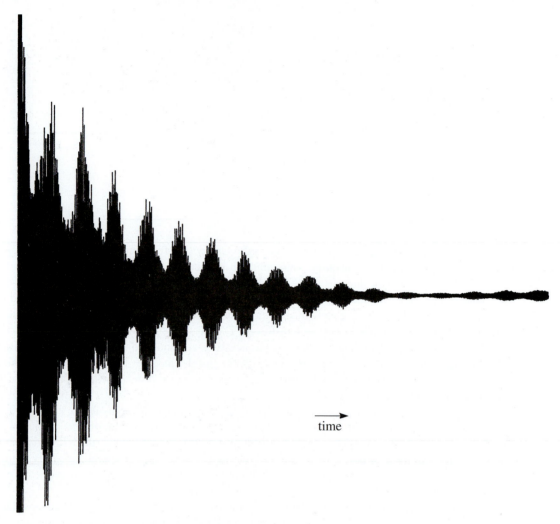

time

F I G U R E 5 . 1 5 The ^{1}H free-induction decay (FID) signal of ethyl phenylacetate (300 MHz).

showing that δ'_{acetone} is the chemical shift of the protons of acetone from the position of the pulse, not from TMS. If we know δ_{TMS}, the position of TMS from the pulse, the actual chemical shift of this peak can be calculated by the adjustment

$$\delta_{\text{actual}} = (\delta'_{\text{acetone}} - \delta'_{\text{TMS}})$$ **Equation 5.11**

We can now plot this peak as a chemical shift on a standard NMR spectrum chart (Fig. 5.16c). The peak for acetone appears at about 2.1 ppm. We have converted the time-domain signal to a **frequency-domain signal,** which is the standard format for a spectrum obtained by a CW instrument.

Now consider the ^{1}H FID from ethyl phenylacetate (Fig. 5.15). This complex molecule has many types of hydrogens, and the FID is the superimposition of many *different* frequencies, each of which could have a *different* decay rate! A mathematical method called a **Fourier transform,** however, will separate each of the individual components of this signal and convert them to frequencies. The Fourier transform breaks the FID into its separate sine or cosine wave components. This procedure is too complex to be carried out by eye or by hand; it requires a computer. Modern pulsed

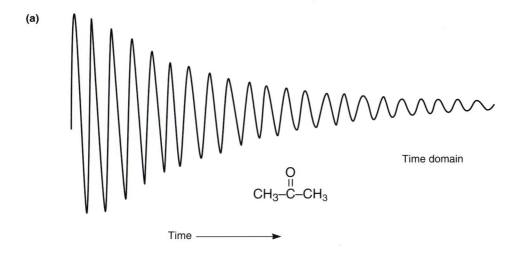

(a)

Time domain

$$CH_3-\overset{\overset{O}{\|}}{C}-CH_3$$

Time ⟶

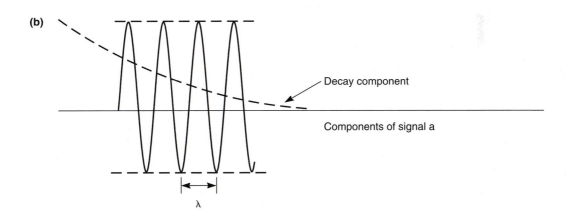

(b)

Decay component

Components of signal a

λ

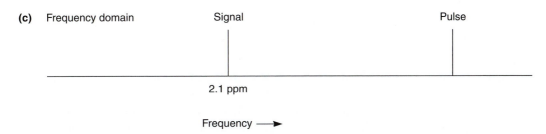

(c)   Frequency domain   Signal   Pulse

2.1 ppm

Frequency ⟶

FIGURE 5.16 (a) An FID curve for the hydrogens in acetone (time domain); (b) the appearance of the FID when the decay is removed; (c) the frequency of this sine wave plotted on a frequency chart (frequency domain).

FT-NMR spectrometers have computers built into them that not only can work up the data by this method but also can control all of the settings of the instrument.

The pulsed FT method described here has several advantages over the CW method. It is more sensitive, and it can measure weaker signals. Five to 10 minutes are required to scan and record a

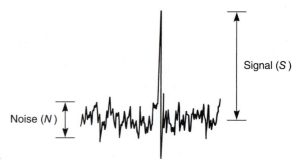

F I G U R E 5 . 1 7 The signal-to-noise ratio.

CW spectrum; a pulsed experiment is much faster, and a measurement of an FID can be performed in a few seconds. With a computer and fast measurement, it is possible to repeat and average the measurement of the FID signal. This is a real advantage when the sample is small, in which case the FID is weak in intensity and has a great amount of noise associated with it. **Noise** is random electronic signals that are usually visible as fluctuations of the baseline in the signal (Fig. 5.17). Since noise is random, its intensity does not increase as many iterations of the spectrum are added together. Using this procedure, one can show that the signal-to-noise ratio improves as a function of the square root of the number of scans *n*:

$$\frac{S}{N} = f\sqrt{\mathrm{n}}$$

Pulsed FT-NMR is therefore especially suitable for the examination of nuclei that are not very abundant in nature, nuclei that are not strongly magnetic, or very dilute samples.

Modern NMR spectrometers use superconducting magnets, which can have field strengths as high as 21 Tesla and operate at 900 MHz. A superconducting magnet is made of special alloys and must be cooled to liquid helium temperatures. The magnet is usually surrounded by a Dewar flask (an insulated chamber) containing liquid helium; in turn, this chamber is surrounded by another one containing liquid nitrogen. Instruments operating at frequencies above 100 MHz have superconducting magnets. NMR spectrometers with frequencies of 300 MHz, 400 MHz, and 500 MHz are now common in chemistry; instruments with frequencies of 900 MHz are used for special research projects.

5.8 CHEMICAL EQUIVALENCE—A BRIEF OVERVIEW

All of the protons found in chemically identical environments within a molecule are **chemically equivalent,** and they often exhibit the same chemical shift. Thus, all the protons in tetramethylsilane (TMS) or all the protons in benzene, cyclopentane, or acetone—which are molecules that have protons that are equivalent by symmetry considerations—have resonance at a single value of δ (but a different value from that of each of the other molecules in the same group). Each such compound gives rise to a single absorption peak in its NMR spectrum. The protons are said to be chemically equivalent. On the other hand, a molecule that has sets of protons that are chemically distinct from one another may give rise to a different absorption peak from each set, in which case the sets of protons are chemically nonequivalent. The following examples should help clarify these relationships:

Molecules giving rise to one NMR absorption peak—all protons chemically equivalent

(benzene)   (cyclopentane, CH₂ ring)

$CH_3-\overset{\overset{\displaystyle O}{\|}}{C}-CH_3$ $(CH_3)_4Si$

Molecules giving rise to two NMR absorption peaks—two different sets of chemically equivalent protons

$CH_2-\overset{\overset{\displaystyle O}{\|}}{C}-OCH_3$
$|$
$CH_2-\overset{\overset{\displaystyle O}{\|}}{C}-OCH_3$

(p-xylene: ring with CH₃ at top and bottom, H at positions)

$CH_3-\overset{\overset{\displaystyle O}{\|}}{C}-OCH_3$ CH_3-O-CH_2Cl

Molecules giving rise to three NMR absorption peaks—three different sets of chemically equivalent protons

$CH_3-O-CH_2-\overset{\overset{\displaystyle CH_3}{|}}{\underset{\underset{\displaystyle CH_3}{|}}{C}}-CH_3$

$CH_3-\overset{\overset{\displaystyle O}{\|}}{C}-CH_2-O-CH_3$

You can see that an NMR spectrum furnishes a valuable type of information on the basis of the number of different peaks observed; that is, the number of peaks corresponds to the number of chemically distinct types of protons in the molecule. Often, protons that are chemically equivalent are also **magnetically equivalent.** Note, however, that *in some instances, protons that are chemically equivalent are not magnetically equivalent.* We will explore this circumstance in Chapter 7, which examines chemical and magnetic equivalence in more detail.

5.9 INTEGRALS AND INTEGRATION

The NMR spectrum not only distinguishes how many different types of protons a molecule has, but also reveals how many of each type are contained within the molecule. In the NMR spectrum, the area under each peak is proportional to the number of hydrogens generating that peak. Hence, in phenylacetone (see Fig. 5.12), the area ratio of the three peaks is 5:2:3, the same as the ratio of the numbers of the three types of hydrogens. The NMR spectrometer has the capability to electronically **integrate** the area under each peak. It does this by tracing over each peak a vertically rising line, called the **integral,** which rises in height by an amount proportional to the area under the peak. Figure 5.18 is a 60-MHz NMR spectrum of benzyl acetate, showing each of the peaks integrated in this way.

F I G U R E   5 . 1 8 Determination of the integral ratios for benzyl acetate (60 MHz).

Note that the height of the integral line does not give the absolute number of hydrogens. It gives the relative number of each type of hydrogen. For a given integral to be of any use, there must be a second integral to which it may be referred. Benzyl acetate provides a good example of this. The first integral rises for 55.5 divisions on the chart paper; the second, 22.0 divisions; and the third, 32.5 divisions. These numbers are relative. One can find ratios of the types of protons by dividing each of the larger numbers by the smallest number:

$$\frac{55.5 \text{ div}}{22.0 \text{ div}} = 2.52 \qquad \frac{22.0 \text{ div}}{22.0 \text{ div}} = 1.00 \qquad \frac{32.5 \text{ div}}{22.0 \text{ div}} = 1.48$$

Thus, the number ratio of the protons of all the types is 2.52:1.00:1.48. If we assume that the peak at 5.1 ppm is really due to two hydrogens, and if we assume that the integrals are slightly (as much as 10%) in error, then we arrive at the true ratio by multiplying each figure by 2 and rounding off to 5:2:3. Clearly, the peak at 7.3 ppm, which integrates for five protons, arises from the resonance of the aromatic ring protons, whereas that at 2.0 ppm, which integrates for three protons, is due to the methyl protons. The two-proton resonance at 5.1 ppm arises from the benzyl protons. Notice that the integrals give the simplest ratio, but not necessarily the true ratio, of numbers of protons of each type.

The spectrum of benzyl acetate shown in Figure 5.19 was obtained on a modern FT-NMR instrument operating at 300 MHz. The spectrum is similar to that obtained at 60 MHz. Integral lines are shown as before, but in addition, you will observe that digitized integral values for the integrals are printed below the peaks. The areas under the curve are relative and not absolute. The integral values are proportional to the actual number of protons represented by the peak. You will need to "massage" the numbers shown in Figure 5.19 to obtain the actual number of protons represented by a particular peak. You will find that it is much easier to do the math when digitized values are provided rather than by measuring the change in heights of the integral line. Notice

FIGURE 5.19 An integrated spectrum of benzyl acetate determined on a 300-MHz FT-NMR instrument.

that benzyl acetate has 10 total protons, so you need to massage the numbers to obtain 10 protons. Proceed as follows:

Divide by the Smallest Integral Value	Multiply by 2	Round Off
4.58/1.92 = 2.39	(2.39)(2) = 4.78	5H
1.92/1.92 = 1.0	(1.0)(2) = 2.0	2H
2.80/1.92 = 1.46	(1.46)(2) = 2.92	3H
		10H

5.10 CHEMICAL ENVIRONMENT AND CHEMICAL SHIFT

If the resonance frequencies of all protons in a molecule were the same, NMR would be of little use to the organic chemist. Not only do different types of protons have different chemical shifts, but each also has a characteristic value of chemical shift. Every type of proton has only a limited range of δ values over which it gives resonance. Hence, the numerical value (in δ units or ppm) of the chemical shift for a proton gives a clue regarding the type of proton originating the signal, just as an infrared frequency gives a clue regarding the type of bond or functional group.

For instance, notice that the aromatic protons of both phenylacetone (Fig. 5.12) and benzyl acetate (Fig. 5.18) have resonance near 7.3 ppm, and that both of the methyl groups attached directly to a carbonyl have resonance at about 2.1 ppm. Aromatic protons characteristically have resonance near 7 to 8 ppm, whereas acetyl groups (methyl groups of this type) have their resonance near 2 ppm. These values of chemical shift are diagnostic. Notice also how the resonance of the benzyl ($-CH_2-$) protons comes at a higher value of chemical shift (5.1 ppm) in benzyl acetate than in phenylacetone (3.6 ppm). Being attached to the electronegative element oxygen, these protons are more deshielded (see Section 5.11) than those in phenylacetone. A trained chemist would readily recognize the probable presence of the oxygen from the value of chemical shift shown by these protons.

F I G U R E   5 . 2 0 A simplified correlation chart for proton chemical shift values.

It is important to learn the ranges of chemical shifts over which the most common types of protons have resonance. Figure 5.20 is a correlation chart that contains the most essential and frequently encountered types of protons. Table 5.4 lists the chemical shift ranges for selected types of protons. For the beginner, it is often difficult to memorize a large body of numbers relating to chemical shifts and proton types. One actually need do this only crudely. It is more important to "get a feel" for the regions and the types of protons than to know a string of actual numbers. To do this, study Figure 5.20 carefully. Table 5.4 and Appendices 2 and 3 give more detailed listings of chemical shifts.

▨ 5.11 LOCAL DIAMAGNETIC SHIELDING

A. Electronegativity Effects

The trend of chemical shifts that is easiest to explain is that involving electronegative elements substituted on the same carbon to which the protons of interest are attached. The chemical shift simply increases as the electronegativity of the attached element increases. Table 5.5 illustrates this relationship for several compounds of the type CH_3X.

Multiple substituents have a stronger effect than a single substituent. The influence of the substituent drops off rapidly with distance, an electronegative element having little effect on protons that are more than three carbons distant. Table 5.6 illustrates these effects for the underlined protons.

Section 5.6 briefly discussed the origin of the electronegativity effect. Electronegative substituents attached to a carbon atom, because of their electron-withdrawing effects, reduce the valence electron density around the protons attached to that carbon. These electrons, it will be recalled, shield the proton from the applied magnetic field. Figure 5.10 illustrates this effect, called *local diamagnetic shielding*. Electronegative substituents on carbon reduce the local diamagnetic shielding in the vicinity of the attached protons because they reduce the electron density around those protons. Substituents that have this type of effect are said to deshield the proton. The greater the electronegativity of the substituent, the more it deshields protons and hence the greater is the chemical shift of those protons.

TABLE 5.4
APPROXIMATE CHEMICAL SHIFT RANGES (PPM) FOR SELECTED TYPES OF PROTONS[a]

$R-CH_3$		0.7 – 1.3	$R-N-\overset{\shortmid}{\underset{\shortmid}{C}}-H$		2.2 – 2.9
$R-CH_2-R$		1.2 – 1.4			
R_3CH		1.4 – 1.7	$R-S-\overset{\shortmid}{\underset{\shortmid}{C}}-H$		2.0 – 3.0
$R-\overset{\shortmid}{C}=\overset{\shortmid}{C}-\overset{\shortmid}{\underset{\shortmid}{C}}-H$		1.6 – 2.6	$I-\overset{\shortmid}{\underset{\shortmid}{C}}-H$		2.0 – 4.0
$R-\overset{O}{\overset{\|}{C}}-\overset{\shortmid}{\underset{\shortmid}{C}}-H, H-\overset{O}{\overset{\|}{C}}-\overset{\shortmid}{\underset{\shortmid}{C}}-H$		2.1 – 2.4	$Br-\overset{\shortmid}{\underset{\shortmid}{C}}-H$		2.7 – 4.1
			$Cl-\overset{\shortmid}{\underset{\shortmid}{C}}-H$		3.1 – 4.1
$RO-\overset{O}{\overset{\|}{C}}-\overset{\shortmid}{\underset{\shortmid}{C}}-H, HO-\overset{O}{\overset{\|}{C}}-\overset{\shortmid}{\underset{\shortmid}{C}}-H$		2.1 – 2.5	$R-\overset{O}{\underset{O}{\overset{\|}{\underset{\|}{S}}}}-O-\overset{\shortmid}{\underset{\shortmid}{C}}-H$		ca. 3.0
$N≡C-\overset{\shortmid}{\underset{\shortmid}{C}}-H$		2.1 – 3.0	$RO-\overset{\shortmid}{\underset{\shortmid}{C}}-H, HO-\overset{\shortmid}{\underset{\shortmid}{C}}-H$		3.2 – 3.8
⬡$-\overset{\shortmid}{\underset{\shortmid}{C}}-H$		2.3 – 2.7	$R-\overset{O}{\overset{\|}{C}}-O-\overset{\shortmid}{\underset{\shortmid}{C}}-H$		3.5 – 4.8
$R-C≡C-H$		1.7 – 2.7	$O_2N-\overset{\shortmid}{\underset{\shortmid}{C}}-H$		4.1 – 4.3
$R-S-H$	var	1.0 – 4.0[b]	$F-\overset{\shortmid}{\underset{\shortmid}{C}}-H$		4.2 – 4.8
$R-\overset{\shortmid}{N}-H$	var	0.5 – 4.0[b]			
$R-O-H$	var	0.5 – 5.0[b]	$R-\overset{\shortmid}{C}=\overset{\shortmid}{C}-H$		4.5 – 6.5
⬡$-O-H$	var	4.0 – 7.0[b]	⬡$-H$		6.5 – 8.0
⬡$-\overset{\shortmid}{N}-H$	var	3.0 – 5.0[b]	$R-\overset{O}{\overset{\|}{C}}-H$		9.0 – 10.0
$R-\overset{O}{\overset{\|}{C}}-\overset{\shortmid}{N}-H$	var	5.0 – 9.0[b]	$R-\overset{O}{\overset{\|}{C}}-OH$		11.0 – 12.0

[a]For those hydrogens shown as $-\overset{\shortmid}{C}-H$, if that hydrogen is part of a methyl group (CH_3) the shift is generally at the low end of the range given, if the hydrogen is in a methylene group ($-CH_2-$) the shift is intermediate, and if the hydrogen is in a methine group ($-CH-$), the shift is typically at the high end of the range given.

[b]The chemical shift of these groups is variable, depending not only on the chemical environment in the molecule, but also on concentration, temperature, and solvent.

TABLE 5.5
DEPENDENCE OF THE CHEMICAL SHIFT OF CH_3X ON THE ELEMENT X

Compound CH_3X	CH_3F	CH_3OH	CH_3Cl	CH_3Br	CH_3I	CH_4	$(CH_3)_4Si$
Element X	F	O	Cl	Br	I	H	Si
Electronegativity of X	4.0	3.5	3.1	2.8	2.5	2.1	1.8
Chemical shift δ	4.26	3.40	3.05	2.68	2.16	0.23	0

TABLE 5.6
SUBSTITUTION EFFECTS

$CH Cl_3$	CH_2Cl_2	CH_3Cl	$-CH_2Br$	$-CH_2-CH_2Br$	$-CH_2-CH_2CH_2Br$
7.27	5.30	3.05	3.30	1.69	1.25

B. Hybridization Effects

The second important set of trends is that due to differences in the hybridization of the atom to which hydrogen is attached.

sp^3 Hydrogens

Referring to Figure 5.20 and Table 5.4, notice that all hydrogens attached to purely sp^3 carbon atoms ($C-CH_3$, $C-CH_2-C$, $C-CH-C$, cycloalkanes) have resonance in the limited range from
0 to 2 ppm, provided that no electronegative elements or π-bonded groups are nearby. At the extreme right of this range are TMS (0 ppm) and hydrogens attached to carbons in highly strained rings (0–1 ppm)—as occurs, for example, with cyclopropyl hydrogens. Most methyl groups occur near 1 ppm if they are attached to other sp^3 carbons. Methylene-group hydrogens (attached to sp^3 carbons) appear at greater chemical shifts (near 1.2 to 1.4 ppm) than do methyl-group hydrogens. Tertiary methine hydrogens occur at higher chemical shift than secondary hydrogens, which in turn have a greater chemical shift than do primary or methyl hydrogens. The following diagram illustrates these relationships:

Of course, hydrogens on an sp^3 carbon that is attached to a heteroatom ($-O-CH_2-$, and so on) or to an unsaturated carbon ($-C=C-CH_2-$) do not fall in this region but have greater chemical shifts.

sp^2 Hydrogens

Simple vinyl hydrogens ($-C=C-H$) have resonance in the range from 4.5 to 7 ppm. In an sp^2-1s $C-H$ bond, the carbon atom has more s character (33% s), which effectively renders it "more electronegative" than an sp^3 carbon (25% s). Remember that s orbitals hold electrons closer to the nucleus than do the carbon p orbitals. If the sp^2 carbon atom holds its electrons more tightly, this results in less shielding for the H nucleus than in an sp^3-1s bond. Thus, vinyl hydrogens have a greater chemical shift (5 to 6 ppm) than aliphatic hydrogens on sp^3 carbons (1 to 4 ppm). Aromatic hydrogens appear in a range farther downfield (7 to 8 ppm). The downfield positions of vinyl and aromatic resonances are, however, greater than one would expect based on these hybridization differences. Another effect, called **anisotropy,** is responsible for the largest part of these shifts (and will be discussed in Section 5.12). Aldehyde protons (also attached to sp^2 carbon) appear even farther downfield (9 to 10 ppm) than aromatic protons since the inductive effect of the electronegative oxygen atom further decreases the electron density on the attached proton. Aldehyde protons, like aromatic and alkene protons, exhibit an anomalously large chemical shift due to anisotropy (Section 5.12).

An aldehyde

sp Hydrogens

Acetylenic hydrogens ($C-H$, sp-1s) appear anomalously at 2 to 3 ppm owing to anisotropy (to be discussed in Section 5.12). On the basis of hybridization alone, as already discussed, one would expect the acetylenic proton to have a chemical shift greater than that of the vinyl proton. An sp carbon should behave as if it were more electronegative than an sp^2 carbon. This is the opposite of what is actually observed.

C.   Acidic and Exchangeable Protons; Hydrogen Bonding

Acidic Hydrogens

Some of the least-shielded protons are those attached to carboxylic acids. These protons have their resonances at 10 to 12 ppm.

Both resonance and the electronegativity effect of oxygen withdraw electrons from the acid proton.

Hydrogen Bonding and Exchangeable Hydrogens

Protons that can exhibit hydrogen bonding (e.g., hydroxyl or amino protons) exhibit extremely variable absorption positions over a wide range. They are usually found attached to a heteroatom. Table 5.7 lists the ranges over which some of these types of protons are found. The more hydrogen bonding that takes place, the more deshielded a proton becomes. The amount of hydrogen bonding

TABLE 5.7
TYPICAL RANGES FOR PROTONS WITH VARIABLE CHEMICAL SHIFT

Acids	RCOOH	10.5–12.0 ppm
Phenols	ArOH	4.0–7.0
Alcohols	ROH	0.5–5.0
Amines	RNH_2	0.5–5.0
Amides	$RCONH_2$	5.0–8.0
Enols	CH=CH—OH	>15

is often a function of concentration and temperature. The more concentrated the solution, the more molecules can come into contact with each other and hydrogen bond. At high dilution (no H bonding), hydroxyl protons absorb near 0.5–1.0 ppm; in concentrated solution, their absorption is closer to 4–5 ppm. Protons on other heteroatoms show similar tendencies.

Free (dilute solution)   Hydrogen bonded (concentrated solution)

Hydrogens that can exchange either with the solvent medium or with one another also tend to be variable in their absorption positions. The following equations illustrate possible situations:

$$R{-}O{-}H_a + R'{-}O{-}H_b \rightleftharpoons R{-}O{-}H_b + R'{-}O{-}H_a$$

$$R{-}O{-}H + H{:}SOLV \rightleftharpoons R{-}\overset{+}{\underset{\underset{H}{|}}{O}}{-}H + {:}SOLV^{-}$$

$$R{-}O{-}H + {:}SOLV \rightleftharpoons H{:}SOLV^{+} + R{-}O{:}^{-}$$

Chapter 8 will discuss all of these situations in more detail.

5.12 MAGNETIC ANISOTROPY

Figure 5.20 clearly shows that there are some types of protons with chemical shifts that are not easily explained by simple considerations of the electronegativity of the attached groups. For instance, consider the protons of benzene and other aromatic systems. Aryl protons generally have a chemical shift as large as that of the proton of chloroform! Alkenes, alkynes, and aldehydes also have protons with resonance values that are not in line with the expected magnitudes of any electron-withdrawing or hybridization effects. In each of these cases, the anomalous shift is due to the presence of an unsaturated system (one with π electrons) in the vicinity of the proton in question.

Take benzene, for example. When it is placed in a magnetic field, the π electrons in the aromatic ring system are induced to circulate around the ring. This circulation is called a **ring current.** The moving electrons generate a magnetic field much like that generated in a loop of wire through which a current is induced to flow. The magnetic field covers a spatial volume large enough that it influences the shielding of the benzene hydrogens. Figure 5.21 illustrates this phenomenon.

The benzene hydrogens are said to be deshielded by the diamagnetic anisotropy of the ring. In electromagnetic terminology, an isotropic field is one of either uniform density or spherically symmetric distribution; an anisotropic field is not isotropic; that is, it is nonuniform. An applied magnetic field is anisotropic in the vicinity of a benzene molecule because the labile electrons in the ring interact with the applied field. This creates a nonhomogeneity in the immediate vicinity of the molecule. Thus, a proton attached to a benzene ring is influenced by three magnetic fields: the strong magnetic field applied by the electromagnets of the NMR spectrometer and two weaker fields, one due to the usual shielding by the valence electrons around the proton, and the other due to the anisotropy generated by the ring-system π electrons. It is the anisotropic effect that gives the benzene protons a chemical shift that is greater than expected. These protons just happen to lie in a deshielding region of the anisotropic field. If a proton were placed in the center of the ring rather than on its periphery, it would be found to be shielded since the field lines there would have the opposite direction from those at the periphery.

All groups in a molecule that have π electrons generate secondary anisotropic fields. In acetylene, the magnetic field generated by induced circulation of the π electrons has a geometry such that the acetylenic hydrogens are shielded (Fig. 5.22). Hence, acetylenic hydrogens have

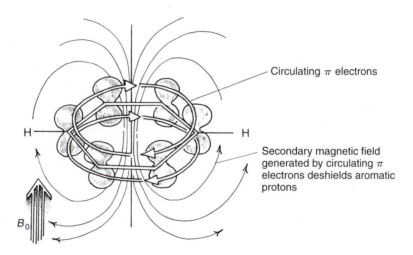

FIGURE 5.21 Diamagnetic anisotropy in benzene.

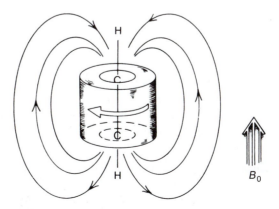

FIGURE 5.22 Diamagnetic anisotropy in acetylene.

resonance at higher field than expected. The shielding and deshielding regions due to the various π electron functional groups have characteristic shapes and directions, and Figure 5.23 illustrates these for a number of groups. Protons falling within the conical areas are shielded, and those falling outside the conical areas are deshielded. The magnitude of the anisotropic field diminishes with distance, and beyond a certain distance there is essentially no anisotropic effect. Figure 5.24 shows the effects of anisotropy in several actual molecules.

FIGURE 5.23 Anisotropy caused by the presence of π electrons in some common multiple-bond systems.

FIGURE 5.24 The effects of anisotropy in some actual molecules.

☐ 5.13 SPIN–SPIN SPLITTING (*n* + 1) RULE

We have discussed the manner in which the chemical shift and the integral (peak area) can give information about the number and types of hydrogens contained in a molecule. A third type of information to be found in the NMR spectrum is that derived from the spin–spin splitting phenomenon. Even in simple molecules, one finds that each type of proton rarely gives a single resonance peak. For instance, in 1,1,2-trichloroethane there are two chemically distinct types of hydrogens:

$$Cl-\underset{\underset{Cl}{|}}{\overset{\overset{\textcircled{H}}{|}}{C}}-\boxed{CH_2}-Cl$$

On the basis of the information given thus far, one would predict two resonance peaks in the NMR spectrum of 1,1,2-trichloroethane, with an area ratio (integral ratio) of 2:1. In reality, the high-resolution NMR spectrum of this compound has five peaks: a group of three peaks (called a **triplet**) at 5.77 ppm and a group of two peaks (called a **doublet**) at 3.95 ppm. Figure 5.25 shows this spectrum. The methine (CH) resonance (5.77 ppm) is said to be split into a triplet, and the methylene resonance (3.95 ppm) is split into a doublet. The area under the three triplet peaks is 1, relative to an area of 2 under the two doublet peaks.

This phenomenon, called **spin–spin splitting,** can be explained empirically by the so-called ***n* + 1 Rule.** Each type of proton "senses" the number of equivalent protons (*n*) on the carbon atom(s) next to the one to which it is bonded, and its resonance peak is split into (*n* + 1) components.

Examine the case at hand, 1,1,2-trichloroethane, utilizing the *n* + 1 Rule. First, the lone methine hydrogen is situated next to a carbon bearing two methylene protons. According to the rule, it has

FIGURE 5.25 The ¹H NMR spectrum of 1,1,2-trichloroethane (60 MHz).

two equivalent neighbors ($n = 2$) and is split into $n + 1 = 3$ peaks (a triplet). The methylene protons are situated next to a carbon bearing only one methine hydrogen. According to the rule, these protons have one neighbor ($n = 1$) and are split into $n + 1 = 2$ peaks (a doublet).

Two neighbors give a triplet
($n + 1 = 3$) (area = 1)

One neighbor gives a doublet
($n + 1 = 2$) (area = 2)

Equivalent protons
behave as a group

Before proceeding to explain the origin of this effect, let us examine two simpler cases predicted by the $n + 1$ Rule. Figure 5.26 is the spectrum of ethyl iodide (CH_3CH_2I). Notice that the methylene protons are split into a quartet (four peaks), and the methyl group is split into a triplet (three peaks). This is explained as follows:

Three equivalent neighbors give a quartet
($n + 1 = 4$) (area = 2)

Two equivalent neighbors give a triplet
($n + 1 = 3$) (area = 3)

FIGURE 5.26 The ^{1}H NMR spectrum of ethyl iodide (60 MHz).

Finally, consider 2-nitropropane, which has the spectrum given in Figure 5.27.

One neighbor gives a doublet   Six equivalent neighbors give a septet
($n + 1 = 2$) (area = 6)   ($n + 1 = 7$) (area = 1)

Notice that in the case of 2-nitropropane there are two adjacent carbons that bear hydrogens (two carbons, each with three hydrogens), and that all six hydrogens as a group split the methine hydrogen into a **septet.**

Also notice that the chemical shifts of the various groups of protons make sense according to the discussions in Sections 5.10 and 5.11. Thus, in 1,1,2-trichloroethane, the methine hydrogen (on a carbon bearing two Cl atoms) has a larger chemical shift than the methylene protons (on a carbon bearing only one Cl atom). In ethyl iodide, the hydrogens on the carbon-bearing iodine have a larger chemical shift than those of the methyl group. In 2-nitropropane, the methine proton (on the carbon bearing the nitro group) has a larger chemical shift than the hydrogens of the two methyl groups.

Finally, note that the spin–spin splitting gives a new type of structural information. It reveals how many hydrogens are adjacent to each type of hydrogen that is giving an absorption peak or, as in these cases, an absorption multiplet. For reference, some commonly encountered spin–spin splitting patterns are collected in Table 5.8.

F I G U R E 5 . 2 7 The ^{1}H NMR spectrum of 2-nitropropane (60 MHz).

T A B L E 5 . 8
SOME EXAMPLES OF COMMONLY OBSERVED SPLITTING PATTERNS IN COMPOUNDS

1H		Cl—C(Cl)(H)—C(Br)(H)—Br		1H
1H		Cl—C(Cl)(H)—CH$_2$—Cl		2H
2H		Cl—CH$_2$—CH$_2$—Br		2H
1H		Cl—C(Cl)(H)—CH$_3$		3H
2H		Cl—CH$_2$—CH$_3$		3H
1H		Br—C(H)(CH$_3$)—CH$_3$		6H

Downfield   Upfield

5.14 THE ORIGIN OF SPIN–SPIN SPLITTING

Spin–spin splitting arises because hydrogens on adjacent carbon atoms can "sense" one another. The hydrogen on carbon A can sense the spin direction of the hydrogen on carbon B. In some molecules of the solution, the hydrogen on carbon B has spin $+\frac{1}{2}$ (X-type molecules); in other molecules of the solution, the hydrogen on carbon B has spin $-\frac{1}{2}$ (Y-type molecules). Figure 5.28 illustrates these two types of molecules.

The chemical shift of proton A is influenced by the direction of the spin in proton B. Proton A is said to be **coupled** to proton B. Its magnetic environment is affected by whether proton B has a $+\frac{1}{2}$ or a $-\frac{1}{2}$ spin state. Thus, proton A absorbs at a slightly different chemical shift value in type X molecules than in type Y molecules. In fact, in X-type molecules, proton A is slightly deshielded because the field of proton B is aligned with the applied field, and its magnetic moment adds to the applied field. In Y-type molecules, proton A is slightly shielded with respect to what its chemical shift would be in the absence of coupling. In this latter case, the field of proton B diminishes the effect of the applied field on proton A.

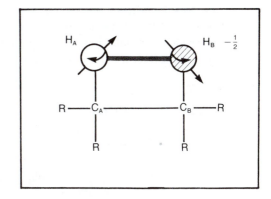

X-type molecule

Y-type molecule

FIGURE 5.28 Two different molecules in a solution with differing spin relationships between protons H_A and H_B.

Since in a given solution there are approximately equal numbers of X- and Y-type molecules at any given time, two absorptions of nearly equal intensity are observed for proton A. The resonance of proton A is said to have been split by proton B, and the general phenomenon is called **spin–spin splitting.** Figure 5.29 summarizes the spin–spin splitting situation for proton A.

Of course, proton A also "splits" proton B since proton A can adopt two spin states as well. The final spectrum for this situation consists of two doublets:

Two doublets will be observed in any situation of this type except one in which protons A and B are identical by symmetry, as in the case of the first of the following molecules:

The first molecule would give only a single NMR peak since protons A and B have the same chemical shift value and are, in fact, identical. The second molecule would probably exhibit the two-doublet spectrum since protons A and B are not identical and would surely have different chemical shifts.

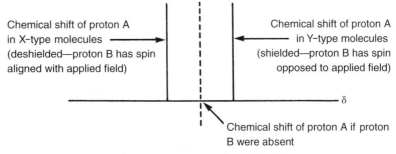

FIGURE 5.29 The origin of spin–spin splitting in proton A's NMR spectrum.

Note that except in unusual cases, coupling (spin–spin splitting) occurs only between hydrogens on adjacent carbons. Hydrogens on nonadjacent carbon atoms generally do not couple strongly enough to produce observable splitting, although there are some important exceptions to this generalization, which Chapter 7 will discuss.

☐ 5.15 THE ETHYL GROUP (CH₃CH₂–)

Now consider ethyl iodide, which has the spectrum shown in Figures 5.26 and 5.30. The methyl protons give rise to a triplet centered at 1.83 ppm, and the methylene protons give a quartet centered at 3.20 ppm. This pattern and the relative intensities of the component peaks can be explained with the use of the model for the two-proton case outlined in Section 5.13. First, look at the methylene protons and their pattern, which is a quartet. The methylene protons are split by the methyl protons, and to understand the splitting pattern, you must examine the various possible spin arrangements of the protons for the methyl group, which are shown in Figure 5.31.

Some of the eight possible spin arrangements are identical to each other since one methyl proton is indistinguishable from another and since there is free rotation in a methyl group. Taking this into consideration, there are only four different types of arrangements. There are, however, three possible ways to obtain the arrangements with net spins of $+\frac{1}{2}$ and $-\frac{1}{2}$. Hence, these arrangements are three times more probable statistically than are the $+\frac{3}{2}$ and $-\frac{3}{2}$ spin arrangements. Thus, one notes in the splitting pattern of the methylene protons that the center two peaks are more intense than the outer ones. In fact, the intensity ratios are 1:3:3:1. Each of these different spin arrangements of the methyl protons (except the sets of degenerate ones, which are effectively identical) gives the methylene protons in that molecule a different chemical shift value. Each of the spins in the $+\frac{3}{2}$ arrangement tends to deshield the methylene proton with respect to its position in the absence of coupling. The $+\frac{1}{2}$ arrangement also deshields the methylene proton, but only slightly, since the two opposite spins cancel each other's effects. The $-\frac{1}{2}$ arrangement shields the methylene proton slightly, whereas the $-\frac{3}{2}$ arrangement shields the methylene proton more strongly.

FIGURE 5.30 The ethyl splitting pattern.

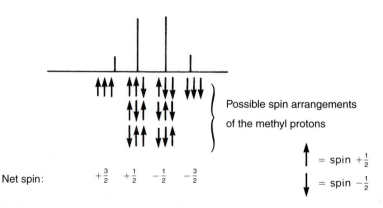

FIGURE 5.31 The splitting pattern of methylene protons due to the presence of an adjacent methyl group.

Net spin: +1   0   −1

Possible spin arrangements
of the methyl protons

FIGURE 5.32 The splitting pattern of methyl protons due to the presence of an adjacent methylene group.

Keep in mind that there are, in fact, four different "types" of molecules in the solution, each type having a different methyl spin arrangement. Each spin arrangement causes the methylene protons in that molecule to have a chemical shift different from those in a molecule with another methyl spin arrangement (except, of course, when the spin arrangements are indistinguishable, or degenerate). Molecules having the $+\frac{1}{2}$ and $-\frac{1}{2}$ spin arrangements are three times more numerous in solution than those with the $+\frac{3}{2}$ and $-\frac{3}{2}$ spin arrangements.

Figure 5.32 provides a similar analysis of the methyl splitting pattern, showing the four possible spin arrangements of the methylene protons. Examination of this figure makes it easy to explain the origin of the triplet for the methyl group and the intensity ratios of 1:2:1.

Now one can see the origin of the ethyl pattern and the explanation of its intensity ratios. The occurrence of spin–spin splitting is very important for the organic chemist as it gives additional structural information about molecules. Namely, it reveals the number of nearest proton neighbors each type of proton has. From the chemical shift one can determine what type of proton is being split, and from the integral (the area under the peaks) one can determine the relative numbers of the types of hydrogen. This is a great amount of structural information, and it is invaluable to the chemist attempting to identify a particular compound.

5.16 PASCAL'S TRIANGLE

We can easily verify that the intensity ratios of multiplets derived from the $n + 1$ Rule follow the entries in the mathematical mnemonic device called **Pascal's triangle** (Fig. 5.33). Each entry in the triangle is the sum of the two entries above it and to its immediate left and right. Notice that the intensities of the outer peaks of a multiplet such as a septet are so small compared to the inner peaks that they are often obscured in the baseline of the spectrum. Figure 5.27 is an example of this phenomenon.

Singlet							1						
Doublet						1		1					
Triplet					1		2		1				
Quartet				1		3		3		1			
Quintet			1		4		6		4		1		
Sextet		1		5		10		10		5		1	
Septet	1		6		15		20		15		6		1

FIGURE 5.33 Pascal's triangle.

5.17 THE COUPLING CONSTANT

Section 5.15 discussed the splitting pattern of the ethyl group and the intensity ratios of the multiplet components but did not address the quantitative amount by which the peaks were split. The distance between the peaks in a simple multiplet is called the **coupling constant** *J*. The coupling constant is a measure of how strongly a nucleus is affected by the spin states of its neighbor. The spacing between the multiplet peaks is measured on the same scale as the chemical shift, and the coupling constant is always expressed in Hertz (Hz). In ethyl iodide, for instance, the coupling constant *J* is 7.5 Hz. To see how this value was determined, consult Figures 5.26 and 5.34.

The spectrum in Figure 5.26 was determined at 60 MHz; thus, each ppm of chemical shift (δ unit) represents 60 Hz. Inasmuch as there are 12 grid lines per ppm, each grid line represents (60 Hz)/12 = 5 Hz. Notice the top of the spectrum. It is calibrated in cycles per second (cps), which are the same as Hertz, and since there are 20 chart divisions per 100 cps, one division equals (100 cps)/20 = 5 cps = 5 Hz. Now examine the multiplets. The spacing between the component peaks is approximately 1.5 chart divisions, so

$$J = 1.5 \text{ div} \times \frac{5 \text{ Hz}}{1 \text{ div}} = 7.5 \text{ Hz}$$

That is, the coupling constant between the methyl and methylene protons is 7.5 Hz. When the protons interact, the magnitude (in ethyl iodide) is always of this same value, 7.5 Hz. The amount of coupling is *constant*, and hence *J* can be called a coupling constant.

The invariant nature of the coupling constant can be observed when the NMR spectrum of ethyl iodide is determined at both 60 MHz and 100 MHz. A comparison of the two spectra indicates that the 100-MHz spectrum is greatly expanded over the 60-MHz spectrum. The chemical shift in Hertz for the CH_3 and CH_2 protons is much larger in the 100-MHz spectrum, although the chemical shifts in δ units (ppm) for these protons remain identical to those in the 60-MHz spectrum. Despite the expansion of the spectrum determined at the higher spectrometer frequency, careful examination of the spectra indicates that the coupling constant between the CH_3 and CH_2 protons is 7.5 Hz in both spectra! The spacings of the lines of the triplet and the lines of the quartet do not expand when the spectrum of ethyl iodide is determined at 100 MHz. The extent of coupling between these two sets of protons remains constant irrespective of the spectrometer frequency at which the spectrum was determined (Fig. 5.35).

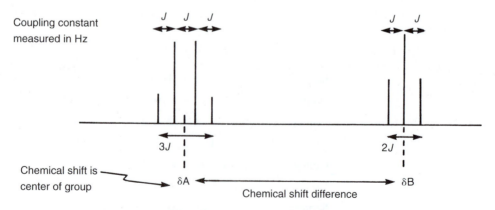

FIGURE 5.34 The definition of the coupling constants in the ethyl splitting pattern.

FIGURE 5.35 An illustration of the relationship between the chemical shift and the coupling constant.

For the interaction of most aliphatic protons in acyclic systems, the magnitudes of coupling constants are always near 7.5 Hz. Compare, for example, 1,1,2-trichloroethane (Fig. 5.25), for which $J = 6$ Hz, and 2-nitropropane (Fig. 5.27), for which $J = 7$ Hz. These coupling constants are typical for the interaction of two hydrogens on adjacent sp^3-hybridized carbon atoms. Two hydrogen atoms on adjacent carbon atoms can be described as a three-bond interaction and abbreviated as ^3J. Typical values for this most commonly observed coupling is approximately 6 to 8 Hz. The bold lines in the diagram show how the hydrogen atoms are three bonds away from each other.

Coupling constants on modern FT-NMR spectrometers are more easily determined by printing Hertz values directly on the peaks. It is a simple matter of subtracting these values to determine the coupling constants in Hertz. See, for example, the spectra shown in Figures 5.40 and 5.46, in which peaks have been labeled in Hertz. Section 7.2 in Chapter 7 describes the various types of coupling constants associated with two-bond (^2J), three-bond (^3J), and four-bond (^4J) interactions.

^3J *cis* = 10 Hz ^3J *trans* = 16 Hz

In alkenes, the ^3J coupling constants for hydrogen atoms that are *cis* to each other have values near 10 Hz, while the ^3J coupling constants for hydrogen atoms that are *trans* are larger, 16 Hz. A study of the magnitude of the coupling constant can give important structural information (see Section 7.8 in Chapter 7).

Table 5.9 gives the approximate values of some representative ^3J coupling constants. A more extensive list of coupling constants appears in Chapter 7, Section 7.2, and in Appendix 5.

Before closing this section, we should take note of an axiom: *the coupling constants of the groups of protons that split one another must be identical* within experimental error. This axiom is extremely useful in interpreting a spectrum that may have several multiplets, each with a different coupling constant.

$J = 7$ Hz	$J = 5$ Hz	$J = 7$ Hz	$J = 5$ Hz
A	B	C	D

Take, for example, the preceding spectrum, which shows three triplets and one quartet. Which triplet is associated with the quartet? It is, of course, the one that has the same *J* values as are

TABLE 5.9

SOME REPRESENTATIVE ^3J COUPLING CONSTANTS AND THEIR APPROXIMATE VALUES (HZ)

H H │ │ C—C 6 to 8	*ortho* 6 to 10	a,a 8 to 14 a,e 0 to 7 e,e 0 to 5
11 to 18	8 to 11	*cis* 6 to 12 *trans* 4 to 8
H H 6 to 15		*cis* 2 to 5 *trans* 1 to 3
H 4 to 10 CH		5 to 7

found in the quartet. The protons in each group interact to the same extent. In this example, with the J values given, clearly quartet A ($J = 7$ Hz) is associated with triplet C ($J = 7$ Hz) and not with triplet B or D ($J = 5$ Hz). It is also clear that triplets B and D are related to each other in the interaction scheme.

Multiplet skewing ("leaning") is another effect that can sometimes be used to link interacting multiplets. There is a tendency for the outermost lines of a multiplet to have nonequivalent heights. For instance, in a triplet, line 3 may be slightly taller than line 1, causing the multiplet to "lean." When this happens, the taller peak is usually in the direction of the proton or group of protons causing the splitting. This second group of protons leans toward the first one in the same fashion. If arrows are drawn on both multiplets in the directions of their respective skewing, these arrows will point at each other. See Figures 5.25 and 5.26 for examples.

5.18 A COMPARISON OF NMR SPECTRA AT LOW– AND HIGH–FIELD STRENGTHS

Section 5.17 showed that, for a given proton, the frequency shift (in Hertz) from TMS is larger when the spectrum is determined at a higher field; however, all coupling constants remain the same as they were at low field (see Fig. 5.35). Even though the shifts in Hertz increase, the chemical shifts (in ppm) of a given proton at low field and high field are the same because we divide by a different operating frequency in each case to determine the chemical shift (Eq. 5.8). If we compare the spectra of a compound determined at both low field and high field, however, the gross appearances of the spectra will differ because, although the coupling constant has the same magnitude in Hertz regardless of operating frequency, the number of Hertz per ppm unit changes. At 60 MHz, for instance, a ppm unit equals 60 Hz, whereas at 300 MHz a ppm unit equals 300 Hz. The coupling constant does not change, but it becomes a smaller fraction of a ppm unit!

When we plot the two spectra on paper to the same parts-per-million scale (same spacing in length for each ppm), the splittings in the high-field spectrum appear compressed, as in Figure 5.36, which shows the 60-MHz and 300-MHz spectra of 1-nitropropane. The coupling has not changed in size; it has simply become a smaller fraction of a ppm unit. At higher field, it becomes necessary to use an expanded parts-per-million scale (more space per ppm) to observe the splittings. The 300-MHz multiplets are identical to those observed at 60 MHz. This can be seen in Figure 5.36b, which shows expansions of the multiplets in the 300-MHz spectrum.

With 300-MHz spectra, therefore, it is frequently necessary to show expansions if one wishes to see the details of the multiplets. In some of the examples in this chapter, we have used 60-MHz spectra—not because we are old-fashioned, but because these spectra show the multiplets more clearly without the need for expansions.

In most cases, the expanded multiplets from a high-field instrument are identical to those observed with a low-field instrument. However, there are also cases in which complex multiplets

FIGURE 5.36 The NMR spectrum of 1-nitropropane. (a) Spectrum determined at 60 MHz; (b) spectrum determined at 300 MHz (with expansions).

become simplified when higher field is used to determine the spectrum. This simplification occurs because the multiplets are moved farther apart, and a type of interaction called a second-order interaction is reduced or even completely removed. Chapter 7 will discuss second-order interactions.

5.19 SURVEY OF TYPICAL ¹H NMR ABSORPTIONS BY TYPE OF COMPOUND

In this section, we will review the typical NMR absorptions that may be expected for compounds in each of the most common classes of organic compounds. These guidelines can be consulted whenever you are trying to establish the class of an unknown compound. Coupling behaviors commonly observed in these compounds are also included in the tables. This coupling information was not covered in this chapter, but it is discussed in Chapters 7 and 8. It is included here so that it will be useful if you wish to use this survey later.

A. Alkanes

Alkanes can have three different types of hydrogens (methyl, methylene, and methine), each of which appears in its own region of the NMR spectrum.

SPECTRAL ANALYSIS BOX—Alkanes

CHEMICAL SHIFTS

R—CH$_3$	0.7–1.3 ppm	Methyl groups are often recognizable as a tall singlet, doublet, or triplet even when overlapping other CH absorptions.
R—CH$_2$—R	1.2–1.4 ppm	In long chains, all of the methylene (CH$_2$) absorptions may be overlapped in an unresolvable group.
R$_3$CH	1.4–1.7 ppm	Note that methine hydrogens (CH) have a larger chemical shift than those in methylene or methyl groups.

COUPLING BEHAVIOR

—CH—CH—	$^3J \approx 7\text{–}8$ Hz	In hydrocarbon chains, adjacent hydrogens will generally couple, with the spin–spin splitting following the *n*+1 Rule.

In alkanes (aliphatic or saturated hydrocarbons), all of the **CH** hydrogen absorptions are typically found from about 0.7 to 1.7 ppm. Hydrogens in methyl groups are the most highly shielded type of proton and are found at chemical shift values (0.7–1.3 ppm) lower than methylene (1.2–1.4 ppm) or methine hydrogens (1.4–1.7 ppm).

In long hydrocarbon chains, or in larger rings, all of the **CH** and **CH$_2$** absorptions may overlap in an unresolvable group. Methyl group peaks are usually separated from other types of hydrogens, being found at lower chemical shifts (higher field). However, even when methyl hydrogens are located within an unresolved cluster of peaks, the methyl peaks can often be recognized as tall singlets, doublets, or triplets clearly emerging from the absorptions of the other types of protons. Methine protons are usually separated from the other protons, being shifted further downfield.

Figure 5.37 shows the spectrum of the hydrocarbon octane. Note that the integral can be used to estimate the total number of hydrogens (the ratio of CH$_3$ to CH$_2$-type carbons) because all of

FIGURE 5.37 ^{1}H spectrum of octane (300 MHz).

the CH_2 hydrogens are in one group and the CH_3 hydrogens are in the other. The NMR spectrum shows the lowest whole-number ratios. You need to multiply by 2 to give the actual number of protons.

B.   Alkenes

Alkenes have two types of hydrogens: vinyl (those attached directly to the double bond) and allylic hydrogens (those attached to the α *carbon,* the carbon atom attached to the double bond). Each type has a characteristic chemical shift region.

SPECTRAL ANALYSIS BOX—Alkenes

CHEMICAL SHIFTS

C=C−H	4.5–6.5 ppm	Hydrogens attached to a double bond (vinyl hydrogens) are deshielded by the anisotropy of the adjacent double bond.
C=C−C−H	1.6–2.6 ppm	Hydrogens attached to a carbon adjacent to a double bond (allyic hydrogens) are also deshielded by the anisotropy of the double bond, but because the double bond is more distant, the effect is smaller.

COUPLING BEHAVIOR

H−C=C−H	$^3J_{trans} \approx 11$–18 Hz $^3J_{cis} \approx 6$–15 Hz	The splitting patterns of vinyl protons may be complicated by the fact that they may not be equivalent even when located on the same carbon of the double bond (Section 7.6).
−C=C−H (with H below)	$^2J \approx 0$ –3 Hz	
−C=C−C−H (with H below)	$^4J \approx 0$ –3 Hz	When allylic hydrogens are present in an alkene, they may show long-range allylic coupling (Section 7.7) to hydrogens on the far double-bond carbon as well as the usual splitting due to the hydrogen on the adjacent (nearest) carbon.

Two types of NMR absorptions are typically found in alkenes: **vinyl absorptions** due to protons directly attached to the double bond (4.5–6.5 ppm) and **allylic absorptions** due to protons located on a carbon atom adjacent to the double bond (1.6–2.6 ppm). Both types of hydrogens are deshielded due to the anisotropic field of the π electrons in the double bond. The effect is smaller for the allylic hydrogens because they are more distant from the double bond. A spectrum of 2-methyl-1-pentene is shown in Figure 5.38. Note the vinyl hydrogens at 4.7 ppm and the allylic methyl group at 1.7 ppm.

The splitting patterns of both vinyl and allylic hydrogens can be quite complex due to the fact that the hydrogens attached to a double bond are rarely equivalent and to the additional complication that allylic hydrogens can couple to all of the hydrogens on a double bond, causing additional splittings. These situations are discussed in Chapter 7, Sections 7.8 and 7.9.

FIGURE 5.38 ^{1}H spectrum of 2-methyl-1-pentene (300 MHz).

C. Aromatic Compounds

Aromatic compounds have two characteristic types of hydrogens: aromatic ring hydrogens (benzene ring hydrogens) and benzylic hydrogens (those attached to an adjacent carbon atom).

SPECTRAL ANALYSIS BOX—Aromatic Compounds

CHEMICAL SHIFTS

6.5–8.0 ppm

Hydrogens attached to an aromatic (benzenoid) ring have a large chemical shift, usually near 7.0 ppm. They are deshielded by the large anisotropic field generated by the electrons in the ring's π system.

2.3–2.7 ppm

Benzylic hydrogens are also deshielded by the anisotropic field of the ring, but they are more distant from the ring, and the effect is smaller.

COUPLING BEHAVIOR

$^3J_{ortho} \approx$ 7–10 Hz
$^4J_{meta} \approx$ 2–3 Hz
$^5J_{para} \approx$ 0–1 Hz

Splitting patterns for the protons on a benzene ring are discussed in Section 7.10. It is often possible to determine the positions of the substituents on the ring from these splitting patterns and the magnitudes of the coupling constants.

The hydrogens attached to aromatic rings are easily identified. They are found in a region of their own (6.5–8.0 ppm) in which few other types of hydrogens show absorption. Occasionally, a highly deshielded vinyl hydrogen will have its absorption in this range, but this is not frequent. The hydrogens on an aromatic ring are more highly deshielded than those attached to double bonds due to the large anisotropic field that is generated by the circulation of the π electrons in the ring (ring current). See Section 5.12 for a review of this specialized behavior of aromatic rings.

The largest chemical shifts are found for ring hydrogens when electron-withdrawing groups such as $-NO_2$ are attached to the ring. These groups deshield the attached hydrogens by withdrawing electron density from the ring through resonance interaction. Conversely, electron-donating groups like methoxy ($-OCH_3$) increase the shielding of these hydrogens, causing them to move upfield.

Nonequivalent hydrogens attached to a benzene ring will interact with one another to produce spin–spin splitting patterns. The amount of interaction between hydrogens on the ring is dependent on the number of intervening bonds or the distance between them. *Ortho* hydrogens ($^3J \approx 7$–10 Hz) couple more strongly than *meta* hydrogens ($^4J \approx 2$–3 Hz), which in turn couple more strongly than *para* hydrogens ($^5J \approx 0$–1 Hz). It is frequently possible to determine the substitution pattern of the ring by the observed splitting patterns of the ring hydrogens (Section 7.10). One pattern that is easily recognized is that of a *para*-disubstituted benzene ring (Fig. 7.64). The spectrum of α-chloro-*p*-xylene is shown in Figure 5.39. The highly deshielded ring hydrogens appear at 7.2 ppm and clearly show a *para*-disubstitution pattern. The chemical shift of the methyl protons at 2.3 ppm shows a smaller deshielding effect. The large shift of the methylene hydrogens is due to the electronegativity of the attached chlorine.

D.   Alkynes

Terminal alkynes (those with a triple bond at the end of a chain) will show an acetylenic hydrogen, as well as the α hydrogens found on carbon atoms next to the triple bond. The acetylenic hydrogen will be absent if the triple bond is in the middle of a chain.

F I G U R E   5 . 3 9 ^{1}H spectrum of α-chloro-*p*-xylene (300 MHz).

SPECTRAL ANALYSIS BOX—Alkynes

CHEMICAL SHIFTS

C≡C—H	1.7–2.7 ppm	The terminal or acetylenic hydrogen has a chemical shift near 1.9 ppm due to anisotropic shielding by the adjacent π bonds.
C≡C—CH—	1.6–2.6 ppm	Protons on a carbon next to the triple bond are also affected by the π system.

COUPLING BEHAVIOR

H—C≡C—C—H	$^4J = 2$–3 Hz	"Allylic coupling" is often observed in alkynes, but is relatively small.

In terminal alkynes (compounds in which the triple bond is in the 1-position), the acetylenic proton appears near 1.9 ppm. It is shifted upfield because of the shielding provided by the π electrons (Fig. 5.22). A spectrum of 1-pentyne is shown in Figure 5.40, where the insets show expansions of the 1.94 and 2.17-ppm regions of the spectrum for protons **c** and **d**, respectively. The peaks in the expansions have been labeled with Hertz (Hz) values so that coupling constants can be calculated. Note that the acetylenic proton (**c**) at 1.94 ppm appears as a triplet with a coupling constant of between 2.6 and 3.0 Hz. This coupling constant is calculated by subtraction: 585.8 – 583.2 = 2.6 Hz or

FIGURE 5.40 ^{1}H spectrum of 1-pentyne (300 MHz).

583.2 – 580.2 = 3.0 Hz, and they will vary somewhat because of experimental error. Values less than 7.0 Hz (^3J) are often attributed to a long-range coupling found in terminal alkynes, in which four bond (^4J) coupling can occur. Sections 7.2 and 7.10 in Chapter 7 provide more information about long-range coupling.

^4J = 2.6 Hz

There are two H atoms four bonds away, $n = 2 + 1$ = triplet

^3J = 7 Hz

There are two H atoms three bonds away, $n = 2 + 1$ = triplet

Proton **d** is split into a triplet by the two neighboring protons (^3J), and then the triplet is split again into doublets (see inset for proton **d** in Fig. 5.40). The type of pattern is referred to as a *triplet of doublets*. The ^3J coupling constant is calculated by subtraction, for example, counting from left to right, peak 6 from peak 4 (648.3 – 641.3 = 7.0 Hz). The ^4J coupling constant can also be calculated from the triplet of doublets, for example, peak 6 from peak 5 (643.9 – 641.3 = 2.6 Hz).

The sextet for the CH_2 group (**b**) at 1.55 ppm in Figure 3.40 results from coupling with a total of five adjacent (^3J) hydrogen atoms on carbons **d** and **a**. Finally, the triplet for the CH_3 group (**a**) at 1.0 ppm results from coupling with two adjacent (^3J) hydrogen atoms on carbon **b**.

E. Alkyl Halides

In alkyl halides, the α hydrogen (the one attached to the same carbon as the halogen) will be deshielded.

SPECTRAL ANALYSIS BOX—Alkyl Halides

CHEMICAL SHIFTS

−CH−I	2.0– 4.0 ppm	The chemical shift of a hydrogen atom attached to the same carbon as a halide atom will increase (move further downfield).
−CH−Br	2.7– 4.1 ppm	This deshielding effect is due to the electronegativity of the attached halogen atom. The extent of the shift is increased as the electronegativity of the attached atom increases, with the largest shift found in compounds containing fluorine.
−CH−Cl	3.1– 4.1 ppm	
−CH−F	4.2– 4.8 ppm	

COUPLING BEHAVIOR

−CH−F	$^2J \approx 50$ Hz	Compounds containing fluorine will show spin–spin splitting due to coupling between the fluorine and the hydrogens on either the same or the adjacent carbon atom. ^{19}F has a spin of $\frac{1}{2}$. The other halogens (I, Cl, Br) do not show any coupling.
−CH−CF−	$^3J \approx 20$ Hz	

Hydrogens attached to the same carbon as a halogen are deshielded (local diamagnetic shielding) due to the electronegativity of the attached halogen (Section 5.11A). The amount of deshielding

FIGURE 5.41 ¹H spectrum of 1-chlorobutane (300 MHz).

increases as the electronegativity of the halogen increases, and it is further increased when multiple halogens are present.

Compounds containing fluorine will show coupling between the fluorine and the hydrogens both on the same carbon (−**CHF**) and those hydrogens on the adjacent carbon (**CH**−**CF**−). Since the spin of fluorine (¹⁹F) is $\frac{1}{2}$, the $n + 1$ Rule can be used to predict the multiplicities of the attached hydrogens. Other halogens do not cause spin–spin splitting of hydrogen peaks.

The spectrum of 1-chlorobutane is shown in Figure 5.41. Note the large downfield shift (deshielding) of the hydrogens on carbon 1 due to the attached chlorine.

F.   Alcohols

In alcohols, both the hydroxyl proton and the α hydrogens (those on the same carbon as the hydroxyl group) have characteristic chemical shifts.

SPECTRAL ANALYSIS BOX—Alcohols

CHEMICAL SHIFTS

C−OH	0.5–5.0 ppm	The chemical shift of the −OH hydrogen is highly variable, its position depending on concentration, solvent, and temperature. The peak may be broadened at its base by the same set of factors.
CH−O−H	3.2–3.8 ppm	Protons on the α carbon are deshielded by the electronegative oxygen atom and are shifted downfield in the spectrum.

COUPLING BEHAVIOR

CH−OH	*No coupling* (*usually*), or $^3J = 5$ Hz	Because of the rapid chemical exchange of the −OH proton in many solutions, coupling is not usually observed between the −OH proton and those hydrogens attached to the α carbon.

The chemical shift of the −**OH** hydrogen is variable, its position depending on concentration, solvent, temperature, and presence of water or of acidic or basic impurities. This peak can be found anywhere in the range of 0.5–5.0 ppm. The variability of this absorption is dependent on the rates of −**OH** proton exchange and the amount of hydrogen bonding in the solution (Section 8.1).

The −**OH** hydrogen is usually not split by hydrogens on the adjacent carbon (−**CH**−**OH**) because rapid exchange decouples this interaction (Section 8.1).

$$-CH-OH + HA \rightleftharpoons -CH-OH + HA$$ **no coupling if exchange is rapid**

Exchange is promoted by increased temperature, small amounts of acid impurities, and the presence of water in the solution. In ultrapure alcohol samples, −**CH**−**OH** coupling is observed. A freshly purified and distilled sample, or a previously unopened commercial bottle, may show this coupling.

On occasion, one may use the rapid exchange of an alcohol as a method for identifying the −**OH** absorption. In this method, a drop of D_2O is placed in the NMR tube containing the alcohol solution. After shaking the sample and sitting for a few minutes, the −**OH** hydrogen is replaced by deuterium, causing it to disappear from the spectrum (or to have its intensity reduced).

$$-CH-OH + D_2O \rightleftharpoons -CH-OD + HOD$$ **deuterium exchange**

The hydrogen on the adjacent carbon (−**CH**−**OH**) appears in the range 3.2–3.8 ppm, being deshielded by the attached oxygen. If exchange of the **OH** is taking place, this hydrogen will not show any coupling with the −**OH** hydrogen, but will show coupling to any hydrogens on the adjacent carbon located further along the carbon chain. If exchange is not occurring, the pattern of this hydrogen may be complicated by differently sized coupling constants for the −**CH**−**OH** and −**CH**−**CH**−**O**− couplings (Section 8.1).

A spectrum of 2-methyl-1-propanol is shown in Figure 5.42. Note the large downfield shift (3.4 ppm) of the hydrogens attached to the same carbon as the oxygen of the hydroxyl group. The hydroxyl group appears at 2.4 ppm, and in this sample it shows some coupling to the hydrogens on the adjacent carbon. The methine proton at 1.75 ppm has been expanded and inset on the spectrum. There are nine

F I G U R E   5 . 4 2 ¹H spectrum of 2-methyl-1-propanol (300 MHz).

peaks (nonet) in that pattern, suggesting coupling with the two methyl groups and one methylene group, $n = (3 + 3 + 2) + 1 = 9$.

G.   Ethers

In ethers, the α hydrogens (those attached to the α *carbon*, which is the carbon atom attached to the oxygen) are highly deshielded.

SPECTRAL ANALYSIS BOX—Ethers

CHEMICAL SHIFTS

R—O—CH—   3.2–3.8 ppm   The hydrogens on the carbons attached to the oxygen are deshielded due to the electronegativity of the oxygen.

In ethers, the hydrogens on the carbon next to oxygen are deshielded due to the electronegativity of the attached oxygen, and they appear in the range 3.2–3.8 ppm. Methoxy groups are especially easy to identify as they appear as a tall singlet in this area. Ethoxy groups are also easy to identify, having both an upfield triplet and a distinct quartet in the region of 3.2–3.8 ppm. An exception is found in epoxides, in which, due to ring strain, the deshielding is not as great, and the hydrogens on the ring appear in the range 2.5–3.5 ppm.

The spectrum of butyl methyl ether is shown in Figure 5.43. The absorption of the methyl and methylene hydrogens next to the oxygen are both seen at about 3.4 ppm. The methoxy peak is

FIGURE 5.43 ^{1}H spectrum of butyl methyl ether (300 MHz).

unsplit and stands out as a tall, sharp singlet. The methylene hydrogens are split into a triplet by the hydrogens on the adjacent carbon of the chain.

H. Amines

Two characteristic types of hydrogens are found in amines: those attached to nitrogen (the hydrogens of the amino group) and those attached to the α carbon (the same carbon to which the amino group is attached).

SPECTRAL ANALYSIS BOX—Amines

CHEMICAL SHIFTS

R—N—H	0.5–4.0 ppm	Hydrogens attached to a nitrogen have a variable chemical shift depending on the temperature, acidity, amount of hydrogen bonding, and solvent.
—CH—N—	2.2–2.9 ppm	The α hydrogen is slightly deshielded due to the electronegativity of the attached nitrogen.
(ring)—N—H	3.0–5.0 ppm	This hydrogen is deshielded due to the anisotropy of the ring and the resonance that removes electron density from nitrogen and changes its hybridization.

COUPLING BEHAVIOR

—N—H	$^1J \approx 50$ Hz	Direct coupling between a nitrogen and an attached hydrogen is not usually observed but is quite large when it occurs. More commonly, this coupling is obscured by quadrupole broadening by nitrogen or by proton exchange. See Sections 8.4 and 8.5.
—N—CH	$^2J \approx 0$ Hz	This coupling is usually not observed.
C—N—H (H)	$^3J \approx 0$ Hz	Due to chemical exchange, this coupling is usually not observed.

Location of the —NH absorptions is not a reliable method for the identification of amines. These peaks are extremely variable, appearing over a wide range of 0.5–4.0 ppm, and the range is extended in aromatic amines. The position of the resonance is affected by temperature, acidity, amount of hydrogen bonding, and solvent. In addition to this variability in position, the —NH peaks are often very broad and weak without any distinct coupling to hydrogens on an adjacent carbon atom. This condition can be caused by chemical exchange of the —NH proton or by a property of nitrogen atoms called *quadrupole broadening* (see Section 8.5). The amino hydrogens will exchange with D_2O, as already described for alcohols, causing the peak to disappear.

$$-N-H + D_2O \rightleftharpoons -N-D + DOH$$

The —NH peaks are strongest in aromatic amines (anilines), in which resonance appears to strengthen the NH bond by changing the hybridization. Although nitrogen is a spin-active element ($I = 1$), coupling is usually not observed between either attached or adjacent hydrogen atoms, but it can appear in certain specific cases. Reliable prediction is difficult.

FIGURE 5.44 ^1H spectrum of propylamine (300 MHz).

The hydrogens α to the amino group are slightly deshielded by the presence of the electronegative nitrogen atom, and they appear in the range 2.2–2.9 ppm. A spectrum of propylamine is shown in Figure 5.44. Notice the weak, broad NH absorptions at 1.8 ppm and that there appears to be a lack of coupling between the hydrogens on the nitrogen and those on the adjacent carbon atom.

I.   Nitriles

In nitriles, only the α hydrogens (those attached to the same carbon as the cyano group) have a characteristic chemical shift.

SPECTRAL ANALYSIS BOX—Nitriles

CHEMICAL SHIFTS

$-CH-C\equiv N$ 2.1–3.0 ppm   The α hydrogens are slightly deshielded by the cyano group.

Hydrogens on the adjacent carbon of a nitrile are slightly deshielded by the anisotropic field of the π-bonded electrons appearing in the range 2.1–3.0 ppm. A spectrum of valeronitrile is shown in Figure 5.45. The hydrogens next to the cyano group appear near 2.35 ppm.

FIGURE 5.45 ¹H spectrum of valeronitrile (300 MHz).

J.   Aldehydes

Two types of hydrogens are found in aldehydes: the aldehyde hydrogen and the α hydrogens (those hydrogens attached to the same carbon as the aldehyde group).

SPECTRAL ANALYSIS BOX—Aldehydes

CHEMICAL SHIFTS

R—CHO	9.0–10.0 ppm	The aldehyde hydrogen is shifted far downfield due to the anisotropy of the carbonyl group (C=O).
R—CH—CH=O	2.1–2.4 ppm	Hydrogens on the carbon adjacent to the C=O group are also deshielded due to the carbonyl group, but they are more distant, and the effect is smaller.

COUPLING BEHAVIOR

—CH—CHO	$^{3}J \approx 1$–3 Hz	Coupling occurs between the aldehyde hydrogen and hydrogens on the adjacent carbon, but ^{3}J is small.

The chemical shift of the proton in the aldehyde group (−**CHO**) is found in the range of 9–10 ppm. Protons appearing in this region are very indicative of an aldehyde group since no other protons appear in this region. The aldehyde proton at 9.64 ppm appears as a doublet in the inset of Figure 5.46, with a $^3J = 1.5$ Hz, for 2-methylpropanal (isobutyraldehyde). NMR is far more reliable than infrared spectroscopy for confirming the presence of an aldehyde group. The other regions have also been expanded and shown as insets on the spectrum and are summarized as follows:

Proton **a** 1.13 ppm (doublet, $^3J = 342.7 − 335.7 = 7.0$ Hz)
Proton **b** 2.44 ppm (septet of doublets, $^3J = 738.0 − 731.0 = 7.0$ Hz and $^4J = 725.5 − 724.0 = 1.5$ Hz)
Proton **c** 9.64 ppm (doublet, $^3J = 2894.6 − 2893.1 = 1.5$Hz)

The CH group (**b**) adjacent to the carbonyl group appears in the range of 2.1 to 2.4 ppm, which is typical for protons on the α carbon. In the present case, the pattern at 2.44 appears as a septet of doublets resulting from coupling with the adjacent two CH$_3$ groups ($n = 6 + 1 = 7$) and coupling with the aldehyde proton resulting in a septet of doublets ($n = 1 + 1 = 2$).

Notice that the two methyl groups (**a**) appear as a doublet, integrating for 6 H with a $^3J = 7.0$ Hz. The $n + 1$ Rule predicts a doublet because of the presence of one adjacent proton on carbon **b**.

K. Ketones

Ketones have only one distinct type of hydrogen atom—those attached to the α carbon.

F I G U R E   5 . 4 6 ^{1}H spectrum of 2-methylpropanal (isobutyraldehyde) at 300 MHz.

SPECTRAL ANALYSIS BOX—Ketones

CHEMICAL SHIFTS

R—CH—C=O   2.1–2.4 ppm   The α hydrogens in ketones are deshielded by the anisotropy
　　| of the adjacent C=O group.
　　R

In a ketone, the hydrogens on the carbon next to the carbonyl group appear in the range 2.1–2.4 ppm. If these hydrogens are part of a longer chain, they will be split by any hydrogens on the adjacent carbon, which is further along the chain. Methyl ketones are quite easy to distinguish since they show a sharp three-proton singlet near 2.1 ppm. Be aware that all hydrogens on a carbon next to a carbonyl group give absorptions within the range of 2.1–2.4 ppm. Therefore, ketones, aldehydes, esters, amides, and carboxylic acids would all give rise to NMR absorptions in this same area. It is necessary to look for the absence of other absorptions (−CHO, −OH, −NH$_2$, −OCH$_2$R, etc.) to confirm the compound as a ketone. Infrared spectroscopy would also be of great assistance in differentiating these types of compounds. Absence of the aldehyde, hydroxyl, amino, or ether stretching absorptions would help confirm the compound as a ketone.

A spectrum of 5-methyl-2-hexanone is shown in Figure 5.47. Notice the tall singlet at 2.2 ppm for the methyl group (**d**) next to the carbonyl group. This is quite characteristic of a methyl ketone. Since there are no adjacent protons, one observes a singlet integrating for 3 H. Typically, carbon atoms with more attached protons are more shielded. Thus, the methyl group appears further upfield than the methylene group (**e**), which has fewer attached protons. The quartet for the methylene group **b** is clearly visible at about 1.45 ppm, but it partly overlaps the multiplet for the single proton **c** appearing at about 1.5 ppm. The doublet for the two methyl groups at about 0.9 ppm integrates for about 6 H. Remember that the doublet results from the two equivalent methyl groups seeing one adjacent proton (^3J).

F I G U R E 5 . 4 7 ^{1}H spectrum of 5-methyl-2-hexanone (300 MHz).

L. Esters

Two distinct types of hydrogen are found in esters: those on the carbon atom attached to the oxygen atom in the *alcohol part* of the ester and those on the α carbon in the *acid part* of the ester (that is, those attached to the carbon next to the C=O group).

SPECTRAL ANALYSIS BOX—Esters

CHEMICAL SHIFTS

$$\alpha \quad\quad O$$
$$-CH_2-C$$
$$\quad\quad\quad\quad O-CH_2-$$
$$2.1\text{–}2.5 \text{ ppm} \quad\quad 3.5\text{–}4.8 \text{ ppm}$$

The α hydrogens in esters are deshielded by the anisotropy of the adjacent (C=O) group.

Hydrogens on the carbon attached to the single-bonded oxygen are deshielded due to the electronegativity of oxygen.

All hydrogens on a carbon next to a carbonyl group give absorptions in the same area (2.1–2.5 ppm). The anisotropic field of the carbonyl group deshields these hydrogens. Therefore, ketones, aldehydes, esters, amides, and carboxylic acids would all give rise to NMR absorptions in this same area. The peak in the 3.5- to 4.8-ppm region is the key to identifying an ester. The large chemical shift of these hydrogens is due to the deshielding effect of the electronegative oxygen atom, which is attached to the same carbon. Either of the two types of hydrogens mentioned may be split into multiplets if they are part of a longer chain.

A spectrum for isobutyl acetate is shown in Figure 5.48. Note that the tall singlet (**c**) at 2.1 ppm integrating for 3 H is the methyl group attached to the C=O group. If the methyl group had been attached to the singly bonded oxygen atom, it would have appeared near 3.5 to 4.0 ppm. Chemical shift information tells you to which side of the $-CO_2-$ group the methyl group is attached. The $-CH_2-$ group (**d**) attached to the oxygen atom is shifted downfield to about 3.85 ppm because of

FIGURE 5.48 ^{1}H spectrum of isobutyl acetate (300 MHz).

the electronegativity of the oxygen atom. That group integrates for 2 H and appears as a doublet because of the one neighboring proton (**b**) on the methine carbon atom. That single proton on the methine carbon appears as a multiplet that is split by the neighboring two methyl groups (**a**) and the methylene group (**d**) into a nonet (nine peaks, at 1.95 ppm). Finally, the two methyl groups appear as a doublet at 0.9 ppm, integrating for 6 H.

M. Carboxylic Acids

Carboxylic acids have the acid proton (the one attached to the $-COOH$ group) and the α hydrogens (those attached to the same carbon as the carboxyl group).

SPECTRAL ANALYSIS BOX—Carboxylic Acids

CHEMICAL SHIFTS

R$-$COOH	11.0–12.0 ppm	This hydrogen is deshielded by the attached oxygen, and it is highly acidic. This (usually broad) signal is a very characteristic peak for carboxylic acids. Note in some cases, however, due to factors discussed below, that this peak may be so broadened as to disappear into the baseline.
$-$CH$-$COOH	2.1–2.5 ppm	Hydrogens adjacent to the carbonyl group are slightly deshielded.

In carboxylic acids, the hydrogen of the carboxyl group ($-COOH$) has resonance in the range 11.0–12.0 ppm. With the exception of the special case of a hydrogen in an enolic OH group that has strong internal hydrogen bonding, no other common type of hydrogen appears in this region. A peak in this region is a strong indication of a carboxylic acid. Since the carboxyl hydrogen has no neighbors, it is usually unsplit; however, hydrogen bonding and exchange may cause the peak become *broadened* (become very wide at the base of the peak) and show very low intensity. Sometimes the acid peak is so broad that it disappears into the baseline. In that case, the acidic proton may not be observed. Infrared spectroscopy is very reliable for determining the presence of a carboxylic acid. As with alcohols, this hydrogen will exchange with water and D_2O. In D_2O, proton exchange will convert the group to $-COOD$, and the $-COOH$ absorption near 12.0 ppm will disappear.

$$R-COOH + D_2O \rightleftharpoons R-COOD + DOH \qquad \textbf{exchange in } \mathbf{D_2O}$$

Carboxylic acids are often insoluble in $CDCl_3$, and it is common practice to determine their spectra in D_2O to which a small amount of sodium metal is added. This basic solution (NaOD, D_2O) will remove the proton, making a soluble sodium salt of the acid. However, when this is done the $-COOH$ absorption will disappear from the spectrum.

$$R-COOH + NaOD \rightleftharpoons R-COO^-Na^+ + DOH$$
$$\textbf{insoluble} \qquad\qquad \textbf{soluble}$$

A spectrum of ethylmalonic acid is shown in Figure 5.49. The $-COOH$ absorption integrating for 2 H is shown as an inset on the spectrum. Notice that this peak is very broad due to hydrogen bonding and exchange. Also notice that proton **c** is shifted downfield to 3.1 ppm, resulting from the effect of two neighboring carbonyl groups. The normal range for a proton next to just one carbonyl group would be expected to appear in the range 2.1 to 2.5 ppm.

FIGURE 5.49 ^{1}H spectrum of ethylmalonic acid (300 MHz).

N. Amides

SPECTRAL ANALYSIS BOX—Amides

CHEMICAL SHIFTS

R(CO)−N−**H**	5.0–9.0 ppm	Hydrogens attached to an amide nitrogen are variable in chemical shift, the value being dependent on the temperature, concentration, and solvent.
−CH−CONH−	2.1–2.5 ppm	The α hydrogens in amides absorb in the same range as other acyl (next to C=O) hydrogens. They are slightly deshielded by the carbonyl group.
R(CO)−N−**CH**	2.2–2.9 ppm	Hydrogens on the carbon next to the nitrogen of an amide are slightly deshielded by the electronegativity of the attached nitrogen.

COUPLING BEHAVIOR

−N−H	$^1J \approx 50$ Hz	In cases in which this coupling is seen (rare), it is quite large, typically 50 Hz or more. In most cases, either the quadrupole moment of the nitrogen atom or chemical exchange decouples this interaction.
−N−CH−	$^2J \approx 0$ Hz	Usually not seen for the same reasons stated above.
−N−CH− H	$^3J \approx 0$–7 Hz	Exchange of the amide NH is slower than in amines, and splitting of the adjacent **CH** is observed even if the **NH** is broadened.

Amides have three distinct types of hydrogens: those attached to nitrogen, α hydrogens attached to the carbon atom on the carbonyl side of the amide group, and hydrogens attached to a carbon atom that is also attached to the nitrogen atom.

The −NH absorptions of an amide group are highly variable, depending not only on their environment in the molecule, but also on temperature and the solvent used. Because of resonance between the unshared pairs on nitrogen and the carbonyl group, rotation is restricted in most amides. Without rotational freedom, the two hydrogens attached to the nitrogen in an unsubstituted amide are not equivalent, and *two different absorption peaks* will be observed, one for each hydrogen (Section 8.6). Nitrogen atoms also have a quadrupole moment (Section 8.5), its magnitude depending on the particular molecular environment. If the nitrogen atom has a large quadrupole moment, the attached hydrogens will show peak broadening (a widening of the peak at its base) and an overall reduction of its intensity.

Hydrogens adjacent to a carbonyl group (regardless of type) all absorb in the same region of the NMR spectrum: 2.1–2.5 ppm.

The spectrum of butyramide is shown in Figure 5.50. Notice the separate absorptions for the two −NH hydrogens (6.6 and 7.2 ppm). This occurs due to restricted rotation in this compound. The hydrogens next to the C=O group appear characteristically at 2.1 ppm.

O. Nitroalkanes

In nitroalkanes, α hydrogens, those hydrogen atoms that are attached to the same carbon atom to which the nitro group is attached, have a characteristically large chemical shift.

S P E C T R A L   A N A L Y S I S   B O X—Nitroalkanes

−CH−NO_2	4.1–4.4 ppm	Deshielded by the nitro group.

F I G U R E   5 . 5 0 ^{1}H spectrum of butyramide (300 MHz).

FIGURE 5.51 ¹H spectrum of 1-nitrobutane (300 MHz).

Hydrogens on a carbon next to a nitro group are highly deshielded and appear in the range 4.1–4.4 ppm. The electronegativity of the attached nitrogen and the positive formal charge assigned to that nitrogen clearly indicate the deshielding nature of this group.

A spectrum of 1-nitrobutane is shown in Figure 5.51. Note the large chemical shift (4.4 ppm) of the hydrogens on the carbon adjacent to the nitro group.

5.20 HOW TO SOLVE NMR SPECTRA PROBLEMS

In this section, the general approach to solving NMR problems will be discussed and two specific sample problems will be solved in detail. The following figures and tables are the most important sources of information for a beginning student to consult.

Figure 5.20   The chemical shifts of important types of protons

Table 5.4   Proton chemical shifts (more detail)

Table 5.9   Basic coupling constants

Section 5.19   A survey of what to expect for each type of functional group

When first approaching an unknown, if the class of compound is not known, start with the infrared spectrum (when given) and the molecular formula. The infrared spectrum will help determine the main functional groups that are present. However, the number of heteroatoms and the index of unsaturation can also help. A compound with one oxygen and an index of zero is likely an alcohol or an ether—then look for the highly deshielded protons next to oxygen (4−5 ppm) in the ¹H-NMR spectrum. One oxygen and an index of one is likely an aldehyde or a ketone, although an unsaturated alcohol or ether are also possible—look in the ¹H-NMR in the region where protons next to a carbonyl group are found (2–3 ppm). If these peaks are absent, look for protons on a double bond (5–6 ppm). Two oxygens and an index of one is likely an ester or an acid – look in the ^{1}H-NMR for an acid proton (10–12 ppm), or for the highly deshielded protons next to an ester oxygen (4–5 ppm). The same approach can be used with nitrogen heteroatoms—amines, nitriles, and so on. If the index of unsaturation is greater than four, look for a benzene ring (peak or peaks near 7 ppm). Try to account for every double bond or ring that the index of unsaturation suggests is present. A cursory examination of the ^{1}H−NMR chemical shifts of the various groups of peaks will often help you make a good preliminary guess at the correct functional group even if you do not have a formula or an infrared spectrum.

Once you have determined the probable functional group, you can consult Section 5.19 to see if the entire spectrum fits the expected patterns for the suspected functional group. Section 5.19 is a useful tool that should not be overlooked.

Once you have determined the types of hydrogens present in the compound, it is time to assign a group of hydrogens to each carbon. Each splitting pattern in the spectrum represents the hydrogens on a single carbon or a group of equivalent carbons. Starting with the molecular formula, you should first subtract the atoms present in the functional group, and then determine the number of hydrogens in each group by using the integrals and the splitting patterns. You should be able to account for every carbon in the molecule, assign the number of hydrogens on each of these carbons, and determine the number of neighboring hydrogens for the splitting group belonging to each carbon. Once this is completed, you can begin to assemble a structure that fits the spectrum. We will do these final steps in the following two examples.

If you are totally stuck, an approach that usually works is to actually draw all the structures you might think are possible, predict what their NMR spectra should look like, and compare them to the unknown compound. Usually only one structure will match the NMR spectrum of the unknown.

Also do not hesitate to list the pieces you know are present, and try to assemble then into a structure that fits the spectrum. For instance, you may know that you have a ester group −O−(C=O)−, a CH_2 group integrating for two, and two methyl groups integrating for three protons each (CH_3 and CH_3), one unsplit with a large chemical shift and the other at a chemical shift representative of protons next to a carbonyl group. The group of protons on the carbon next to the carbonyl is a quartet (three neighbors), and the methyl group is a triplet (two neighbors). This compound can only assemble as $CH_3−O−(C=O)−CH_2CH_3$.

Be sure to look carefully at how the spectrum is presented. Not all spectra require that they be displayed on a scale that covers the entire chemical shift range. Notice, for instance, that the spectra given in Example 1 only cover the range from 0 to 5 ppm since there are no peaks with greater chemical shifts than 5.0 ppm. If there is a big gap between peaks, the part of the spectrum with larger chemical shifts may be offset and redrawn above the main baseline. It is also possible that some peaks will be expanded and offset in order to better see the splitting or determine the coupling constants.

■ EXAMPLE 1

The following compounds are isomeric esters derived from acetic acid, each with formula $C_5H_{10}O_2$. Each of the spectra has been expanded so that you will be able see the splitting patterns.

Since we are told that these compounds are esters, we already know the functional group and can subtract one carbon and two oxygens (CO_2) from the molecular formula, leaving four carbons and ten hydrogens to elucidate. You can familiarize yourself with ester spectra by going to Section 5.19L.

With the first spectrum (1a) as an example, you can use the integral curve traced on the spectrum to calculate the number of hydrogen atoms represented in each multiplet (Section 5.9). Alternatively, you can avoid the laborious task of counting squares or using a ruler to measure the height of

¹H NMR
300 MHz   C₅H₁₀O₂

1.76   2.64   1.78   2.59

(a)

¹H NMR
300 MHz   C₅H₁₀O₂

0.86   2.55   5.22

(b)

each integral! It is far easier to determine the integral values by using the integral numbers listed just below the peaks. These numbers are the integrated values of the area under the peaks. They are proportional to the actual number of protons, within experimental error. The process: Divide each of the integral values by the smallest integral value to get the values shown in the second column (1.76/1.76 = 1.0; 2.64/1.76 = 1.5; 1.77/1.76 = 1.01; 2.59/1.76 = 1.47). The values shown in the third column are obtained by multiplying by 2 and rounding off the resulting values. If everything works out, you should find that the total number of protons should equal the number of protons in the formula, in this case 10 protons.

Often, one can inspect the spectrum and visually approximate the relative numbers of protons, thus avoiding the mathematical approach shown in the table. Using this eyeball approach, you can

1.76	1.0	2 H
2.64	1.5	3 H
1.77	1.01	2 H
2.59	1.47	3 H
		10 protons

determine that the second spectrum (1b) yields a ratio of 1:3:6 =10 H. The septet at about 4.9 ppm in this spectrum has been expanded and redrawn above the baseline so that you can see it better.

Returning to the first ester, spectrum 1a, we see a sharp singlet, integrating for three hydrogens at about 2.1 ppm. This is an unsplit methyl group (CH_3) with no neighbors. The chemical shift is typical for hydrogens on a carbon next to a carbonyl group ($CH_3-C=O$). In contrast, the other peak that integrates for three hydrogens is a triplet near 0.9 ppm. This methyl group (CH_3) has an ordinary aliphatic (saturated) carbon nearby with two neighbors, most likely a CH_2 group. The triplet with large chemical shift at about 4.1 ppm is most likely from a CH_2 group attached to an oxygen ($-O-CH_2-$) which deshields the hydrogens. The final set of peaks at about 1.65 ppm arises from a CH_2 group (integral =2) with five neighbors. Since CH_5 groups are impossible, this set of hydrogens must have two neighboring carbons, a CH_2 and a CH_3 group. This means we have an n-propyl group attached to oxygen ($-O-CH_2-CH_2-CH_3$). The central carbon in this chain gives the sextet. Adding all of this together we get propyl ethanoate ($CH_3(C=O)O-CH_2CH_2CH_3$) as the answer for 1a. This ester can also be named propyl acetate, as it is the propyl ester of acetic acid.

In spectrum 1b, the singlet integrating for three hydrogens at 2.1 ppm, just as in Problem 1a above, is an unsplit methyl group next to a carbonyl group ($CH_3-C=O$). The septet at 5.0 ppm is a single hydrogen (CH) with six neighbors. Since CH_6 is impossible, those neighbors must be two methyl groups ($CH_3 + CH_3$). Those two methyl groups are both attached directly to the CH group. Since they are equivalent methyl groups (by rotation), they both have the same chemical shift. However, since they each have a single neighbor (CH) they are each split into a pair of precisely overlapping doublets. Note that the two peaks here *do not* represent two different methyl group singlets, each with a different chemical shift. They represent two overlapping methyl groups (chemically equivalent) that coincide in chemical shift and are each split into the same two overlapping doublet peaks by the adjacent CH group. That means that the alkyl portion of this ester is an isopropyl group attached to oxygen ($-O-CH(CH_3)_2$). The answer for compound 2b is isopropyl ethanoate (also called isopropyl acetate) with the final structure $CH_3(C=O)O-CH(CH_3)_2$.

n-propyl acetate

isopropyl acetate

■ EXAMPLE 2

The two isomeric compounds with the formula $C_3H_5ClO_2$ have NMR spectra shown in Example 2a and 2b. The downfield protons appearing in the NMR spectra at about 12.1 and 11.5 ppm, respectively, are shown as insets.

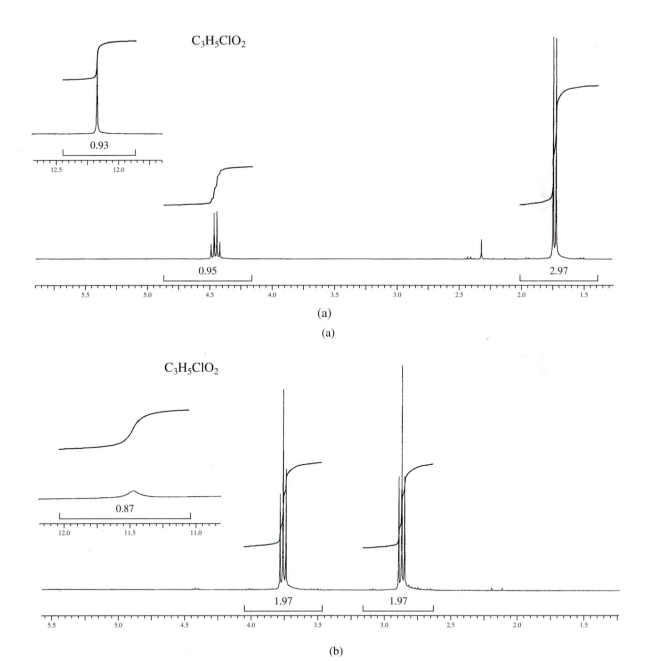

(a)

(b)

In both spectra, the peaks at 12.2 and 11.5 ppm (both are offset above the baseline) are in the region expected for the proton attached to a carboxylic acid group. In spectrum 1a, the acid peak (integrating for one hydrogen) is a sharp singlet, while in spectrum 1b it is substantially broadened, probably due to a moderate exchange rate. See Section 3.19M for a full discussion of the NMR spectra of carboxylic acids and a short explanation of peak broadening.

The molecular formula has an index of unsaturation of one (due to the acid carbonyl) and shows the required two oxygen atoms. Therefore, we can subtract $-CO_2H$ from the formula, leaving C_2H_4Cl to elucidate. The most direct approach here is to see that there are only two possible structures for that remaining two-carbon portion of the compound: $Cl-CH_2CH_2-$ and $CH_3CH(Cl)-$. From these two arrangements, you should see that the first will give two triplets, one with a larger chemical shift due to chlorine, and that the second will give a doublet and a downfield (deshielded) quartet for the hydrogen on the same carbon as the chlorine. Note the greater chemical shift for the hydrogen in 2a on the same carbon as the acid group (4.45 ppm) compared to that of the two hydrogens in 2b that are simply next to the acid group (3.75 ppm). The compounds are 2-chloropropanoic acid (2a) and 3-chloropropanoic acid (2b).

2-chloroproanoic acid   **3-chloropropanoic acid**

PROBLEMS

*1. What are the allowed nuclear spin states for the following atoms?

(a) ^{14}N (b) ^{13}C (c) ^{17}O (d) ^{19}F

*2. Calculate the chemical shift in parts per million (δ) for a proton that has resonance 128 Hz downfield from TMS on a spectrometer that operates at 60 MHz.

*3. A proton has resonance 90 Hz downfield from TMS when the field strength is 1.41 Tesla (14,100 Gauss) and the oscillator frequency is 60 MHz.

(a) What will be its shift in Hertz if the field strength is increased to 2.82 Tesla and the oscillator frequency to 120 MHz?

(b) What will be its chemical shift in parts per million (δ)?

*4. Acetonitrile (CH_3CN) has resonance at 1.97 ppm, whereas methyl chloride (CH_3Cl) has resonance at 3.05 ppm, even though the dipole moment of acetonitrile is 3.92 D and that of methyl chloride is only 1.85 D. The larger dipole moment for the cyano group suggests that the electronegativity of this group is greater than that of the chlorine atom. Explain why the methyl hydrogens on acetonitrile are actually more shielded than those in methyl chloride, in contrast with the results expected on the basis of electronegativity. (*Hint:* What kind of spatial pattern would you expect for the magnetic anisotropy of the cyano group, CN?)

*5. The position of the OH resonance of phenol varies with concentration in solution, as the following table shows. On the other hand, the hydroxyl proton of *ortho*-hydroxyacetophenone appears at 12.05 ppm and does not show any great shift upon dilution. Explain.

Concentration w/v in CCl$_4$	δ (ppm)
100%	7.45
20%	6.75
10%	6.45
5%	5.95
2%	4.88
1%	4.37

Phenol

12.05 ppm

o-Hydroxyacetophenone

*6. The chemical shifts of the methyl groups of three related molecules, pinane, α-pinene, and β-pinene, follow.

CH$_3$ 1.17 ppm   CH$_3$ CH$_3$ 0.99 ppm   1.01 ppm

Pinane

CH$_3$ 1.27 ppm   (CH$_3$)   CH$_3$ 1.63 ppm   0.85 ppm

α-Pinene

CH$_3$ 1.23 ppm   (CH$_3$)   CH$_2$ 0.72 ppm

β-Pinene

Build models of these three compounds and then explain why the two circled methyl groups have such small chemical shifts.

*7. In benzaldehyde, two of the ring protons have resonance at 7.87 ppm, and the other three have resonance in the range from 7.5 to 7.6 ppm. Explain.

*8. Make a three-dimensional drawing illustrating the magnetic anisotropy in 15,16-dihydro-15, 16-dimethylpyrene, and explain why the methyl groups are observed at −4.2 ppm in the ^{1}H NMR spectrum.

CH$_3$ CH$_3$

15,16-Dihydro-15,16-dimethylpyrene

*9. Work out the spin arrangements and splitting patterns for the following spin system:

$$\begin{array}{ccc} & H_A & H_B \\ & | & | \\ Cl-&C-&C-Br \\ & | & | \\ & Cl & H_B \end{array}$$

*10. Explain the patterns and intensities of the isopropyl group in isopropyl iodide.

$$\begin{array}{c} CH_3 \\ \diagdown \\ CH-I \\ \diagup \\ CH_3 \end{array}$$

*11. What spectrum would you expect for the following molecule?

$$\begin{array}{cccc} & Cl & H & Cl \\ & | & | & | \\ H-&C-&C-&C-H \\ & | & | & | \\ & Cl & Cl & Cl \end{array}$$

*12. What arrangement of protons would give two triplets of equal area?

*13. Predict the appearance of the NMR spectrum of propyl bromide.

*14. The following compound, with the formula $C_4H_8O_2$, is an ester. Give its structure and assign the chemical shift values.

*15. The following compound is a monosubstituted aromatic hydrocarbon with the formula C_9H_{12}. Give its structure and assign the chemical shift values.

*16. The following compound is a carboxylic acid that contains a bromine atom: $C_4H_7O_2Br$. The peak at 10.97 ppm was moved onto the chart (which runs only from 0 to 8 ppm) for clarity. What is the structure of the compound?

*17. The compound that gives the following NMR spectrum has the formula $C_3H_6Br_2$. Draw the structure.

*18. Draw the structure of an ether with formula $C_5H_{12}O_2$ that fits the following NMR spectrum:

*19. Following are the NMR spectra of three isomeric esters with the formula $C_7H_{14}O_2$, all derived from propanoic acid. Provide a structure for each.

^1H NMR
300 MHz

$C_7H_{14}O_2$

(a)

$C_7H_{14}O_2$

^1H NMR
300 MHz

(b)

^{1}H NMR
300 MHz

$C_7H_{14}O_2$

(c)

*20. The two isomeric compounds with the formula $C_{10}H_{14}$ have NMR spectra shown below. Make no attempt to interpret the aromatic proton area between 7.1 and 7.3 ppm except to determine the number of protons attached to the aromatic ring. Draw the structures of the isomers.

$C_{10}H_{14}$

(a)

$C_{10}H_{14}$

(b)

*21. The compound with the formula $C_8H_{11}N$ has the NMR spectra shown. The infrared spectrum shows a doublet at about 3350 cm^{-1}. Make no attempt to interpret the aromatic proton area between 7.1 and 7.3 ppm except to determine the number of protons attached to the aromatic ring. Draw the structure of the compound.

$C_8H_{11}N$

22. The NMR spectra are shown for two isomeric compounds with formula $C_{10}H_{12}O$. Their infrared spectra show strong bands near 1715 cm^{-1}. Make no attempt to interpret the aromatic proton area between 7.1 and 7.4 ppm except to determine the number of protons attached to the aromatic ring. Draw the structure of the compounds.

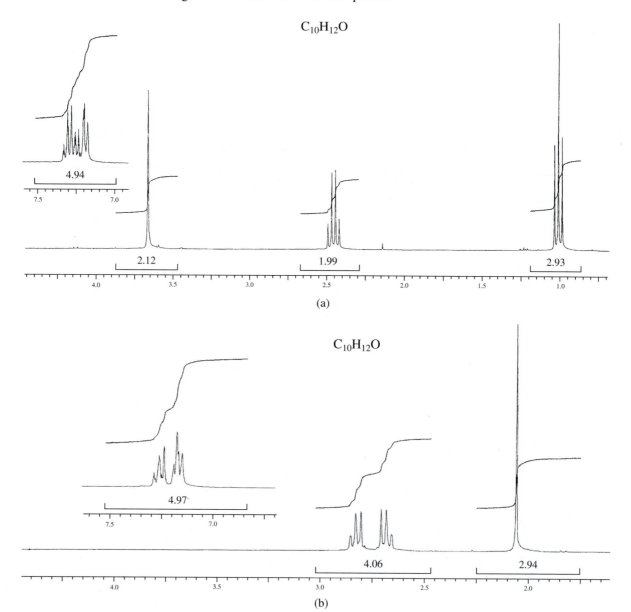

$C_{10}H_{12}O$

(a)

$C_{10}H_{12}O$

(b)

23. The NMR spectra are shown in parts a, b, c, and d for four isomeric compounds with formula $C_{10}H_{12}O_2$. Their infrared spectra show strong bands near 1735 cm^{-1}. Make no attempt to interpret the aromatic proton area between 7.0 and 7.5 ppm except to determine the number of protons attached to the aromatic ring. Draw the structures of the compounds.

$C_{10}H_{12}O_2$

(a)

$C_{10}H_{12}O_2$

(b)

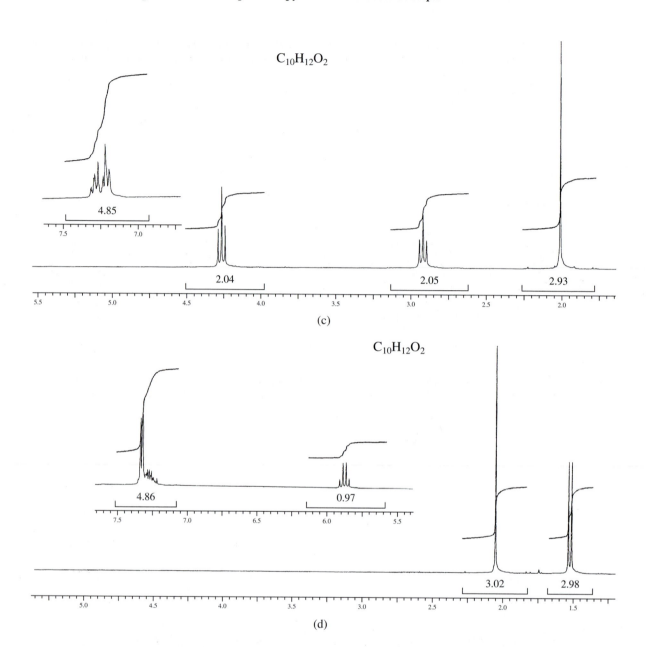

24. Along with the following NMR spectrum, this compound, with formula $C_5H_{10}O_2$, shows bands at 3450 cm^{-1} (broad) and 1713 cm^{-1} (strong) in the infrared spectrum. Draw its structure.

$C_5H_{10}O_2$

25. The NMR spectrum for an ester with formula $C_5H_6O_2$ is shown below. The infrared spectrum shows medium-intensity bands at 3270 and 2118 cm^{-1}. Draw the structure of the compound.

$C_5H_6O_2$

26. The NMR spectrum is shown for a compound with formula $C_7H_{12}O_4$. The infrared spectrum has strong absorption at 1740 cm^{-1} and has several strong bands in the range 1333 to 1035 cm^{-1}. Draw the structure of this compound.

$C_7H_{12}O_4$

REFERENCES

Textbooks

Berger, S., and S. Braun, *200 and More NMR Experiments*, Wiley-VCH, Weinheim, 2004.

Crews, P., J. Rodriguez, and M. Jaspars, *Organic Structure Analysis*, 2nd ed., Oxford University Press, New York, 2009.

Field, L. D., S. Sternhell and J. R. Kalman, Organic Structures from Spectra, 5th ed., Wiley, 2013.

Friebolin, H., *Basic One- and Two-Dimensional NMR Spectroscopy*, 5th ed., Wiley-VCH Publishers, New York, 2010.

Gunther, H., *NMR Spectroscopy*, 2nd ed., John Wiley and Sons, New York, 1995.

Jackman, L. M., and S. Sternhell, *Nuclear Magnetic Resonance Spectroscopy in Organic Chemistry*, 2nd ed., Pergamon Press, New York, 1969.

Lambert, J. B., S. Groenert, H. F. Shurvell, D. A. Lightner, and R. G. Cooks, *Organic Structural Spectroscopy*, Prentice Hall, Upper Saddle River, NJ, 2010.

Macomber, R. S., *NMR Spectroscopy: Essential Theory and Practice*, College Outline Series, Harcourt, Brace Jovanovich, New York, 1988.

Macomber, R. S., *A Complete Introduction to Modern NMR Spectroscopy*, John Wiley and Sons, New York, 1997.

Sanders, J. K. M., and B. K. Hunter, *Modern NMR Spectroscopy—A Guide for Chemists*, 2nd ed., Oxford University Press, Oxford, 1993.

Silverstein, R. M., F. X. Webster and D. J. Kiemle, *Spectrometric Identification of Organic Compounds*, 7th ed., John Wiley and Sons, 2005.

Williams, D. H., and I. Fleming, *Spectroscopic Methods in Organic Chemistry*, 6th ed., McGraw-Hill Book Co. Ltd., London, 2008.

Selected Websites

http://www2.chemistry.msu.edu/faculty/reusch/VirtTxtJml/Spectrpy/spectro.htm

This excellent website presented by Professor William Reusch from the Department of Chemistry at the Michigan State University includes background material for infrared spectroscopy, NMR spectroscopy, UV spectroscopy and mass spectrometry. Some problems are included. The site includes links to other available sites for spectral problems.

http://sdbs.riodb.aist.go.jp

This excellent free website presents an integrated Integrated Spectral DataBase System (SDBS) for Organic Compounds, National Institute of Advanced Industrial Science and Technology (AIST), Tsukuba, Ibaraki 305-8565, Japan. This database includes infrared, mass spectra, and NMR data (proton and carbon-13) for a large number of compounds.

http://www.chem.ucla.edu/~webspectra/

UCLA Department of Chemistry and Biochemistry in connection with Cambridge University Isotope Laboratories, presents WebSpectra an outstanding collection of unknowns for students to interpret. The website provides NMR spectra and some IR spectra. Solutions are provided. No mass spectral data are included. They provide links to other sites with problems for students to solve.

http://www.nd.edu/~smithgrp/structure/workbook.html

Combined structure problems provided by the Smith group at Notre Dame University.

Compilations of Spectra

Ault, A., and M. R. Ault, *A Handy and Systematic Catalog of NMR Spectra, 60 MHz with Some 270 MHz,* University Science Books, Mill Valley, CA, 1980.

Pouchert, C. J., *The Aldrich Library of NMR Spectra, 60 MHz,* 2nd ed., Aldrich Chemical Company, Milwaukee, WI, 1983.

Pouchert, C. J., and J. Behnke, *The Aldrich Library of ^{13}C and ^1H FT-NMR Spectra, 300 MHz,* Aldrich Chemical Company, Milwaukee, WI, 1993.

Pretsch, E., J. P. Buhlmann, and H. Badertscher, *Structure Determination of Organic Compounds. Tables of Spectral Data,* 4th ed., Springer-Verlag, Berlin, 2009. Translated from the German by K. Biemann.

NUCLEAR MAGNETIC RESONANCE SPECTROSCOPY

Part Two: Carbon-13 Spectra, Including Heteronuclear Coupling with Other Nuclei

The study of carbon nuclei through nuclear magnetic resonance (NMR) spectroscopy is an important technique for determining the structures of organic molecules. Using it together with proton NMR and infrared spectroscopy, organic chemists can often determine the complete structure of an unknown compound without "getting their hands dirty" in the laboratory! Fourier transform–NMR (FT-NMR) instrumentation makes it possible to obtain routine carbon spectra easily.

Carbon spectra can be used to determine the number of nonequivalent carbons and to identify the types of carbon atoms (methyl, methylene, aromatic, carbonyl, and so on) that may be present in a compound. Thus, carbon NMR provides direct information about the carbon skeleton of a molecule. Some of the principles of proton NMR apply to the study of carbon NMR; however, structural determination is generally easier with carbon-13 NMR spectra than with proton NMR. Typically, both techniques are used together to determine the structure of an unknown compound.

6.1 THE CARBON-13 NUCLEUS

Carbon-12, the most abundant isotope of carbon, is NMR inactive since it has a spin of zero (see Section 5.1). Carbon-13, or ^{13}C, however, has odd mass and does have nuclear spin, with $I = \frac{1}{2}$. Unfortunately, the resonances of ^{13}C nuclei are more difficult to observe than those of protons (^{1}H). They are about 6000 times weaker than proton resonances, for two major reasons.

First, the natural abundance of carbon-13 is very low; only 1.08% of all carbon atoms in nature are ^{13}C atoms. If the total number of carbons in a molecule is low, it is very likely that a majority of the molecules in a sample will have no ^{13}C nuclei at all. In molecules containing a ^{13}C isotope, it is unlikely that a second atom in the same molecule will be a ^{13}C atom. Therefore, when we observe a ^{13}C spectrum, we are observing a spectrum built up from a collection of molecules, in which each molecule supplies no more than a single ^{13}C resonance. No *single* molecule supplies a complete spectrum.

Second, since the magnetogyric ratio of a ^{13}C nucleus is smaller than that of hydrogen (Table 5.2), ^{13}C nuclei always have resonance at a frequency lower than protons. Recall that at lower frequencies, the excess spin population of nuclei is reduced (Table 5.3); this, in turn, reduces the sensitivity of NMR detection procedures.

For a given magnetic field strength, the resonance frequency of a ^{13}C nucleus is about one-fourth the frequency required to observe proton resonances (see Table 5.2). For example, in a 7.05-Tesla applied magnetic field, protons are observed at 300 MHz, while ^{13}C nuclei are observed at about 75 MHz. With modern instrumentation, it is a simple matter to switch the transmitter frequency from the value required to observe proton resonances to the value required for ^{13}C resonances.

Through the use of modern Fourier transform instrumentation (Section 5.7B), it is possible to obtain ^{13}C NMR spectra of organic compounds even though detection of carbon signals is difficult compared to detection of proton spectra. To compensate for the low natural abundance of carbon, a greater number of individual scans must be accumulated than is common for a proton spectrum.

6.2 CARBON-13 CHEMICAL SHIFTS

A. Correlation Charts

An important parameter derived from carbon-13 spectra is the chemical shift. The correlation chart in Figure 6.1 shows typical ^{13}C chemical shifts, listed in parts per million (ppm) from tetramethylsilane (TMS); the carbons of the methyl groups of TMS (not the hydrogens) are used for reference. Approximate ^{13}C chemical shift ranges for selected types of carbon are also given in Table 6.1. Notice that the chemical shifts appear over a range (0 to 220 ppm) much larger than that observed for protons (0 to 12 ppm). Because of the very large range of values, nearly every nonequivalent carbon atom in an organic molecule gives rise to a peak with a different chemical shift. Peaks rarely overlap as they often do in proton NMR.

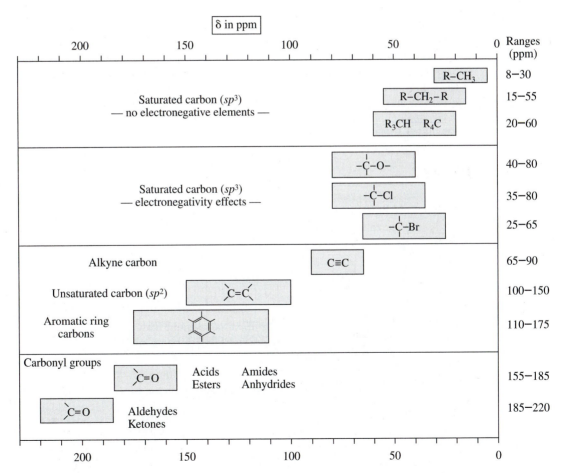

FIGURE 6.1 A correlation chart for ^{13}C chemical shifts (chemical shifts are listed in parts per million from TMS).

TABLE 6.1
APPROXIMATE ^{13}C CHEMICAL SHIFT RANGES (ppm) FOR SELECTED TYPES OF CARBON

$R-CH_3$	8–30	$C \equiv C$	65–90
R_2CH_2	15–55	$C = C$	100–150
R_3CH	20–60	$C \equiv N$	110–140
$C-I$	0–40		110–175
$C-Br$	25–65		
$C-N$	30–65	$R-\overset{O}{\overset{\|}{C}}-OR, R-\overset{O}{\overset{\|}{C}}-OH$	155–185
$C-Cl$	35–80	$R-\overset{O}{\overset{\|}{C}}-NH_2$	155–185
$C-O$	40–80	$R-\overset{O}{\overset{\|}{C}}-R, R-\overset{O}{\overset{\|}{C}}-H$	185–220

The correlation chart is divided into four sections. Saturated carbon atoms appear at highest field, nearest to TMS (8 to 60 ppm). The next section of the correlation chart demonstrates the effect of electronegative atoms (40 to 80 ppm). The third section of the chart includes alkene and aromatic ring carbon atoms (100 to 175 ppm). Finally, the fourth section of the chart contains carbonyl carbons, which appear at the lowest field values (155 to 220 ppm).

Electronegativity, hybridization, and anisotropy all affect ^{13}C chemical shifts in nearly the same fashion as they affect ^{1}H chemical shifts; however, ^{13}C chemical shifts are about 20 times larger.[1] Electronegativity (Section 5.11A) produces the same deshielding effect in carbon NMR as in proton NMR—the electronegative element produces a large downfield shift. The shift is greater for a ^{13}C atom than for a proton since the electronegative atom is directly attached to the ^{13}C atom, and the effect occurs through only a single bond, $C-X$. With protons, the electronegative atoms are attached to carbon, not hydrogen; the effect occurs through two bonds, $H-C-X$, rather than one.

In ^{1}H NMR, the effect of an electronegative element on chemical shift diminishes with distance, but it is always in the same direction (deshielding and downfield). In ^{13}C NMR, an electronegative element also causes a downfield shift in the α and β carbons, but it usually leads to a small *upfield* shift for the γ carbon. This effect is clearly seen in the carbons of 1-hexanol:

$$14.2 \quad 22.8 \quad 32.0 \quad 25.8 \quad 32.8 \quad 61.9 \text{ ppm}$$
$$CH_3-CH_2-CH_2-CH_2-CH_2-CH_2-OH$$
$$\omega \quad \varepsilon \quad \delta \quad \gamma \quad \beta \quad \alpha$$

The shift for C3, the γ carbon, seems quite at odds with the expected effect of an electronegative substituent. This anomaly points up the need to consult detailed correlation tables for ^{13}C chemical shifts. Such tables appear in Appendix 7 and are discussed in the next section.

---

[1]This is sometimes called the *20x Rule*. See Macomber, R., "Proton–Carbon Chemical Shift Correlations," *Journal of Chemical Education, 68(a),* 284–285, 1991.

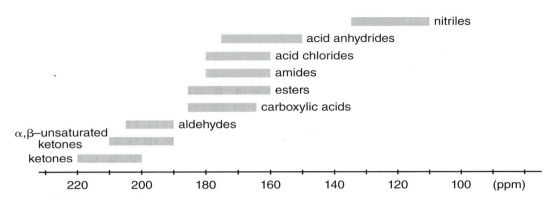

FIGURE 6.2 A ^{13}C correlation chart for carbonyl and nitrile functional groups.

Analogous with ^{1}H shifts, changes in hybridization (Section 5.11B) also produce larger shifts for the carbon-13 that is *directly involved* (no bonds) than they do for the hydrogens attached to that carbon (one bond). In ^{13}C NMR, the carbons of carbonyl groups have the largest chemical shifts, due to both sp^2 hybridization and to the fact that an electronegative oxygen is directly attached to the carbonyl carbon, deshielding it even further. Anisotropy (Section 5.12) is responsible for the large chemical shifts of the carbons in aromatic rings and alkenes.

Notice that the range of chemical shifts is larger for carbon atoms than for hydrogen atoms. Because the factors affecting carbon shifts operate either through one bond or directly on carbon, they are greater than those for hydrogen, which operate through more bonds. As a result, the entire range of chemical shifts becomes larger for ^{13}C (0 to 220 ppm) than for ^{1}H (0 to 12 ppm).

Many of the important functional groups of organic chemistry contain a carbonyl group. In determining the structure of a compound containing a carbonyl group, it is frequently helpful to have some idea of the type of carbonyl group in the unknown. Figure 6.2 illustrates the typical ranges of ^{13}C chemical shifts for some carbonyl-containing functional groups. Although there is some overlap in the ranges, ketones and aldehydes are easy to distinguish from the other types. Chemical shift data for carbonyl carbons are particularly powerful when combined with data from an infrared spectrum.

B. Calculation of ^{13}C Chemical Shifts

Nuclear magnetic resonance spectroscopists have accumulated, organized, and tabulated a great deal of data for ^{13}C chemical shifts. It is possible to predict the chemical shift of almost any ^{13}C atom from these tables, starting with a base value for the molecular skeleton and then adding increments that correct the value for each substituent. Corrections for the substituents depend on both the type of substituent and its position relative to the carbon atom being considered. Corrections for rings are different from those for chains, and they frequently depend on stereochemistry.

Consider *m*-xylene (1,3-dimethylbenzene) as an example. Consulting the tables, you will find that the base value for the carbons in a benzene ring is 128.5 ppm. Next, look in the substituent tables that relate to benzene rings for the methyl substituent corrections (Table A8.7 in Appendix 8). These values are

	ipso	*ortho*	*meta*	*para*
CH$_3$:	9.3	0.7	−0.1	−2.9 ppm

The *ipso* carbon is the one to which the substituent is directly attached. The calculations for *m*-xylene start with the base value and add these increments as follows:

$$C1 = base + ipso + meta \quad = 128.5 + 9.3 + (-0.1) = 137.7 \text{ ppm}$$

$$C2 = base + ortho + ortho = 128.5 + 0.7 + 0.7 \quad = 129.9 \text{ ppm}$$

$$C3 = C1$$

$$C4 = base + ortho + para \ = 128.5 + 0.7 + (-2.9) = 126.3 \text{ ppm}$$

$$C5 = base + meta + meta \ = 128.5 + 2(-0.1) \quad = 128.3 \text{ ppm}$$

$$C6 = C4$$

The observed values for C1, C2, C4, and C5 of *m*-xylene are 137.6, 130.0, 126.2, and 128.2 ppm, respectively, and the calculated values agree well with those actually measured.

Appendix 8 presents some ^{13}C chemical shift correlation tables with instructions. Complete ^{13}C chemical shift correlation tables are too numerous to include in this book. If you are interested, consult the textbooks by Friebolin, Levy, Macomber, Pretsch and Silverstein, which are listed in the references at the end of this chapter. Even more convenient than tables are computer programs that calculate ^{13}C chemical shifts. In the more advanced programs, the operator need only sketch the molecule on the screen, using a mouse, and the program will calculate both the chemical shifts and the rough appearance of the spectrum. Some of these programs are also listed in the references.

6.3 PROTON-COUPLED ^{13}C SPECTRA—SPIN–SPIN SPLITTING OF CARBON-13 SIGNALS

Unless a molecule is artificially enriched by synthesis, the probability of finding two ^{13}C atoms in the same molecule is low. The probability of finding two ^{13}C atoms adjacent to each other in the same molecule is even lower. Therefore, we rarely observe **homonuclear** (carbon–carbon) spin–spin splitting patterns where the interaction occurs between two ^{13}C atoms. However, the spins of protons attached directly to ^{13}C atoms do interact with the spin of carbon and cause the carbon signal to be split according to the *n* + 1 Rule. This is **heteronuclear** (carbon–hydrogen) coupling involving two different types of atoms. With ^{13}C NMR, we generally examine splitting that arises from the protons *directly attached* to the carbon atom being studied. This is a one-bond coupling. Remember that in proton NMR, the most common splittings are *homonuclear* (hydrogen–hydrogen) and occur between protons attached to *adjacent* carbon atoms. In these cases, the interaction is a three-bond coupling, H−C−C−H.

Figure 6.3 illustrates the effect of protons directly attached to a ^{13}C atom. The *n* + 1 Rule predicts the degree of splitting in each case. The resonance of a ^{13}C atom with three attached protons, for instance, is split into a quartet (*n* + 1 = 3 + 1 = 4). The possible spin combinations for the three protons are the same as those illustrated in Figure 5.31, and each spin combination interacts with carbon to give a different peak of the multiplet. Since the hydrogens are directly attached to the ^{13}C (one-bond couplings), the coupling constants for this interaction are quite large, with *J* values of about 100 to 250 Hz. Compare the typical three-bond H−C−C−H couplings that are common in NMR spectra, which have *J* values of about 1 to 20 Hz.

It is important to note while examining Figure 6.3 that you are not "seeing" protons directly when looking at a ^{13}C spectrum (proton resonances occur at frequencies outside the range used to obtain ^{13}C spectra); you are observing only the effect of the protons on ^{13}C atoms. Also, remember that we cannot observe ^{12}C because it is NMR inactive.

Spectra that show the spin–spin splitting, or coupling, between carbon-13 and the protons directly attached to it are called **proton-coupled spectra** or sometimes **nondecoupled spectra** (see the next section). Figure 6.4a is the proton-coupled ^{13}C NMR spectrum of ethyl phenylacetate. In this spectrum, the first quartet downfield from TMS (14.2 ppm) corresponds to the carbon of the methyl group. It is split into a quartet (*J* = 127 Hz) by the three attached hydrogen atoms

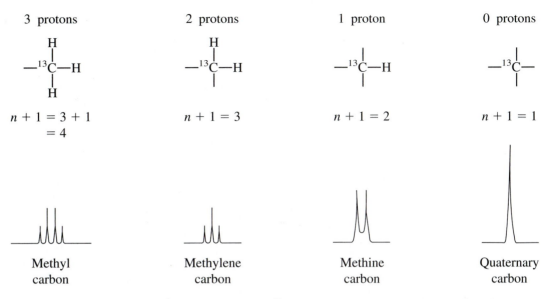

FIGURE 6.3 The effect of attached protons on ^{13}C resonances.

FIGURE 6.4 Ethyl phenylacetate. (a) The proton-coupled ^{13}C spectrum (20 MHz). (b) The proton-decoupled ^{13}C spectrum (20 MHz). (From Moore, J. A., and D. L. Dalrymple, *Experimental Methods in Organic Chemistry*, W. B. Saunders, Philadelphia, 1976.)

(^{13}C—H, one-bond couplings). In addition, although it cannot be seen on the scale of this spectrum (an expansion must be used), each of the quartet lines is split into a closely spaced triplet ($J = $ ca. 1 Hz). This additional fine splitting is caused by the two protons on the adjacent —CH$_2$— group. These are two-bond couplings (H—C—^{13}C) of a type that occurs commonly in ^{13}C spectra, with coupling constants that are generally quite small ($J = 0$–2 Hz) for systems with carbon atoms in an aliphatic chain. Because of their small size, these couplings are frequently ignored in the routine analysis of spectra, with greater attention being given to the larger one-bond splittings seen in the quartet itself.

There are two —CH_2— groups in ethyl phenylacetate. The one corresponding to the ethyl —CH_2— group is found farther downfield (60.6 ppm) as this carbon is deshielded by the attached oxygen. It is a triplet because of the two attached hydrogens (one-bond couplings). Again, although it is not seen in this unexpanded spectrum, the three hydrogens on the adjacent methyl group finely split each of the triplet peaks into a quartet. The benzyl —CH_2— carbon is the intermediate triplet (41.4 ppm). The farthest downfield is the carbonyl-group carbon (171.1 ppm). On the scale of this presentation, it is a singlet (no directly attached hydrogens), but because of the adjacent benzyl —CH_2— group, it is actually split finely into a triplet. The aromatic-ring carbons also appear in the spectrum, and they have resonances in the range from 127 to 136 ppm. Section 6.13 will discuss aromatic-ring ^{13}C resonances.

Proton-coupled spectra for large molecules are often difficult to interpret. The multiplets from different carbons commonly overlap because the ^{13}C—H coupling constants are frequently larger than the chemical shift differences of the carbons in the spectrum. Sometimes, even simple molecules such as ethyl phenylacetate (Fig. 6.4a) are difficult to interpret. Proton decoupling, which is discussed in the next section, avoids this problem.

6.4 PROTON-DECOUPLED ^{13}C SPECTRA

By far the great majority of ^{13}C NMR spectra are obtained as **proton-decoupled spectra.** The decoupling technique obliterates all interactions between protons and ^{13}C nuclei; therefore, only **singlets** are observed in a decoupled ^{13}C NMR spectrum. Although this technique simplifies the spectrum and avoids overlapping multiplets, it has the disadvantage that the information on attached hydrogens is lost.

Proton **decoupling** is accomplished in the process of determining a ^{13}C NMR spectrum by simultaneously irradiating all of the protons in the molecule with a broad spectrum of frequencies in the proper range. Most modern NMR spectrometers provide a second, tunable radiofrequency generator, the **decoupler,** for this purpose. Irradiation causes the protons to become saturated, and they undergo rapid upward and downward transitions, among all their possible spin states. These rapid transitions decouple any spin–spin interactions between the hydrogens and the ^{13}C nuclei being observed. In effect, all spin interactions are averaged to zero by the rapid changes. The carbon nucleus "senses" only one average spin state for the attached hydrogens rather than two or more distinct spin states. In some reference sources proton-decoupled ^{13}C spectra are designated by using the notation ^{13}C{^{1}H}, which means that the ^{13}C spectrum is being observed while the ^{1}H protons are being irradiated with a broadband decoupling signal. In this book, all ^{13}C spectra can be assumed to be proton-decoupled unless otherwise stated.

Figure 6.4b is a proton-decoupled spectrum of ethyl phenylacetate. The proton-coupled spectrum (Fig. 6.4a) was discussed in Section 6.3. It is interesting to compare the two spectra to see how the proton decoupling technique simplifies the spectrum. Every chemically and magnetically distinct carbon gives only a single peak. Notice, however, that the two *ortho* ring carbons (carbons 2 and 6) and the two *meta* ring carbons (carbons 3 and 5) are equivalent by symmetry, and that each gives only a single peak.

Figure 6.5 is a second example of a proton-decoupled spectrum. Notice that the spectrum shows three peaks, corresponding to the exact number of carbon atoms in 1-propanol. If there are no equivalent carbon atoms in a molecule, a ^{13}C peak will be observed for *each* carbon. Notice also that the assignments given in Figure 6.5 are consistent with the values in the chemical shift table (Fig. 6.1). The carbon atom closest to the electronegative oxygen is the farthest downfield, and the methyl carbon is at highest field.

FIGURE 6.5 The proton-decoupled ^{13}C spectrum of 1-propanol (22.5 MHz).

The three-peak pattern centered at $\delta = 77$ ppm is due to the solvent $CDCl_3$. This pattern results from the coupling of a deuterium (^2H) nucleus to the ^{13}C nucleus (see Section 6.13). Often the $CDCl_3$ pattern is used as an internal reference, in place of TMS.

6.5 NUCLEAR OVERHAUSER ENHANCEMENT (NOE)

When we obtain a proton-decoupled ^{13}C spectrum, the intensities of many of the carbon resonances increase significantly above those observed in a proton-coupled experiment. Carbon atoms with hydrogen atoms directly attached are enhanced the most, and the enhancement increases (but not always linearly) as more hydrogens are attached. This effect is known as the nuclear Overhauser effect, and the degree of increase in the signal is called the **nuclear Overhauser enhancement (NOE).** The NOE effect is *heteronuclear* in this case, operating between two dissimilar atoms (carbon and hydrogen). Both atoms exhibit spins and are NMR active. The nuclear Overhauser effect is general, showing up when one of two different types of atoms is irradiated while the NMR spectrum of the other type is determined. If the absorption intensities of the observed (i.e., nonirradiated) atom change, enhancement has occurred. The effect can be either positive or negative, depending on which atom types are involved. In the case of ^{13}C interacting with ^1H, the effect is positive; irradiating the hydrogens increases the intensities of the carbon signals. The maximum enhancement that can be observed is given by the relationship

$$NOE_{max} = \frac{1}{2}\left(\frac{\gamma_{irr}}{\gamma_{obs}}\right)$$

Equation 6.1

where γ_{irr} is the magnetogyric ratio of the nucleus being irradiated, and γ_{obs} is that of the nucleus being observed. Remember that NOE_{max} is the *enhancement* of the signal, and it must be added to the original signal strength:

$$\text{total predicted intensity (maximum)} = 1 + NOE_{max} \qquad \textbf{Equation 6.2}$$

For a proton-decoupled ^{13}C spectrum, we would calculate, using the values in Table 5.2,

$$NOE_{max} = \frac{1}{2}\left(\frac{267.5}{67.28}\right) = 1.988 \qquad \textbf{Equation 6.3}$$

indicating that the ^{13}C signals can be enhanced up to 200% by irradiation of the hydrogens. This value, however, is a theoretical maximum, and most actual cases exhibit less-than-ideal enhancement.

The heteronuclear NOE effect in fact operates in both directions; either atom can be irradiated. If one were to irradiate carbon-13 while determining the NMR spectrum of the hydrogens—the reverse of the usual procedure—the hydrogen signals would increase by a very small amount. However, because there are few ^{13}C atoms in a given molecule, the result would not be very dramatic. In contrast, NOE is a definite *bonus* received in the determination of proton-decoupled ^{13}C spectra. The hydrogens are numerous, and carbon-13, with its low abundance, generally produces weak signals. Because NOE increases the intensity of the carbon signals, it substantially increases the sensitivity (signal-to-noise ratio) in the ^{13}C spectrum.

Signal enhancement due to NOE is an example of **cross-polarization,** in which a polarization of the spin states in one type of nucleus causes a polarization of the spin states in another nucleus. Cross-polarization will be explained in Section 6.6. In the current example (proton-decoupled ^{13}C spectra), when the hydrogens in the molecule are irradiated, they become saturated and attain a distribution of spins very different from their equilibrium (Boltzmann) state. There are more spins than normal in the *excited* state. Due to the interaction of spin dipoles, the spins of the carbon nuclei "sense" the spin imbalance of the hydrogen nuclei and begin to adjust themselves to a new equilibrium state that has more spins in the *lower* state. This increase of population in the lower spin state of carbon increases the intensity of the NMR signal.

In a proton-decoupled ^{13}C spectrum, the total NOE for a given carbon increases as the number of nearby hydrogens increases. Thus, we usually find that the intensities of the signals in a ^{13}C spectrum (assuming a single carbon of each type) assume the order

$$CH_3 > CH_2 > CH \gg C$$

Although the hydrogens producing the NOE effect influence carbon atoms more distant than the ones to which they are attached, their effectiveness drops off rapidly with distance. The interaction of the spin–spin dipoles operates through space, not through bonds, and its magnitude decreases as a function of the inverse of r^3, where r is the radial distance from the hydrogen of origin.

$$C \xrightarrow{r} H \qquad NOE = f\left(\frac{1}{r^3}\right)$$

Thus, nuclei must be rather close together in the molecule in order to exhibit the NOE effect. The effect is greatest for hydrogens that are directly attached to carbon.

In advanced work, NOEs are sometimes used to verify peak assignments. Irradiation of a selected hydrogen or group of hydrogens leads to a greater enhancement in the signal of the closer of the two carbon atoms being considered. In dimethylformamide, for instance, the two methyl groups are nonequivalent, showing two peaks at 31.1 and 36.2 ppm, because free rotation is restricted about the C—N bond due to resonance interaction between the lone pair on nitrogen and the π bond of the carbonyl group.

anti, 31.1 ppm

syn, 36.2 ppm

Dimethylformamide

Irradiation of the aldehyde hydrogen leads to a larger NOE for the carbon of the *syn* methyl group (36.2 ppm) than for that of the *anti* methyl group (31.1 ppm), allowing the peaks to be assigned. The *syn* methyl group is closer to the aldehyde hydrogen.

It is possible to retain the benefits of NOE even when determining a proton-coupled ^{13}C NMR spectrum that shows the attached hydrogen multiplets. The favorable perturbation of spin-state populations builds up slowly during irradiation of the hydrogens by the decoupler, and it persists for some time after the decoupler is turned off. In contrast, decoupling is available only while the decoupler is in operation and stops immediately when the decoupler is switched off. One can build up the NOE effect by irradiating with the decoupler during a period before the pulse and then turning off the decoupler during the pulse and free-induction decay (FID) collection periods. This technique gives an **NOE-enhanced proton-coupled spectrum,** with the advantage that peak intensities have been increased due to the NOE effect. See Section 9.1 for details.

6.6 CROSS-POLARIZATION: ORIGIN OF THE NUCLEAR OVERHAUSER EFFECT

To see how cross-polarization operates to give nuclear Overhauser enhancement, consider the energy diagram shown in Figure 6.6. Consider a two-spin system between atoms ^{1}H and ^{13}C. These two atoms may be spin coupled, but the following explanation is easier to follow if we simply ignore any spin–spin splitting. The following explanation is applied to the case of ^{13}C NMR spectroscopy, although the explanation is equally applicable to other possible combinations of atoms. Figure 6.6 shows four separate energy levels (N_1, N_2, N_3, and N_4), each with a different combination of spins of atoms ^{1}H and ^{13}C. The spins of the atoms are shown at each energy level.

The selection rules, derived from quantum mechanics, require that the only allowed transitions involve a change of only one spin at a time (these are called **single-quantum transitions**). The allowed transitions, proton excitations (labeled ^{1}H) and carbon excitations (labeled ^{13}C), are shown. Notice that both proton transitions and both carbon transitions have the same energy (remember that we are ignoring splitting due to *J* interactions).

Because the four spin states have different energies, they also have different *populations*. Because the spin states N_3 and N_2 have very similar energies, we can assume that their populations are approximately equal. Now use the symbol B to represent the equilibrium populations of these two spin states. The population of spin state N_1, however, will be higher (by an amount δ), and the population of spin state N_4 will be reduced (also by an amount δ). The intensities of the NMR lines will be proportional to the difference in populations between the energy levels where transitions are occurring. If we compare the populations of each energy level, we can see that the intensities of the two carbon lines (X) will be equal.

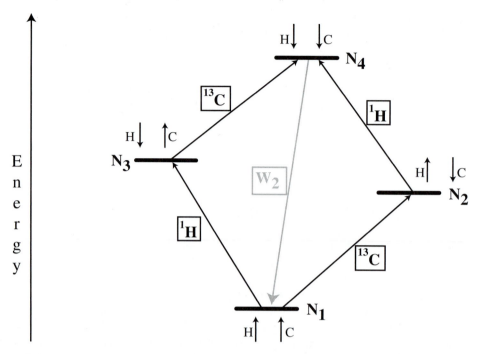

FIGURE 6.6 Spin energy level diagram for an AX System.

Level	Equilibrium Populations
N_1	$B + \delta$
N_2	B
N_3	B
N_4	$B - \delta$

Assuming that the populations of the ^{13}C energy levels are at equilibrium, the carbon signals will have intensities:

^{13}C Energy Levels at Equilibrium

$$N_3 - N_4 = B - B + \delta = \delta$$
$$N_1 - N_2 = B + \delta - B = \delta$$

Consider now what happens when we irradiate the proton transitions during the broad-band decoupling procedure. The irradiation of the protons causes the proton transitions to become **saturated**. In other words, the probability of an upward and a downward transition for these nuclei (the proton transitions shown in Fig. 6.6) now becomes *equal*. The population of level N_4 becomes equal to the population of level N_2, and the population of level N_3 is now equal to the population of level N_1. The populations of the spin states can now be represented by the following expressions:

PROTON DECOUPLED	
Level	Populations
N_1	$B + \frac{1}{2}\delta$
N_2	$B - \frac{1}{2}\delta$
N_3	$B + \frac{1}{2}\delta$
N_4	$B - \frac{1}{2}\delta$

Using these expressions, the intensities of the carbon lines can be represented:

^{13}C Energy Levels with Broad-Band Decoupling

$$N_3 - N_4 = B + \frac{1}{2}\delta - B + \frac{1}{2}\delta = \delta$$

$$N_1 - N_2 = B + \frac{1}{2}\delta - B + \frac{1}{2}\delta = \delta$$

So far, there has been no change in the intensity of the carbon transition.

At this point, we need to consider that there is another process operating in this system. When the populations of the spin states have been disturbed from their equilibrium values, as in this case by irradiation of the proton signal, **relaxation processes** will tend to restore the populations to their equilibrium values. Unlike excitation of a spin from a lower to a higher spin state, relaxation processes are not subject to the same quantum mechanical selection rules. Relaxation involving changes of both spins simultaneously (called **double-quantum transitions**) are allowed; in fact, they are relatively important in magnitude. The relaxation pathway labeled W_2 in Fig. 6.6 tends to restore equilibrium populations by relaxing spins from state N_4 to N_1. We shall represent the number of spins that are relaxed by this pathway by the symbol d. The populations of the spin states thus become as follows:

Level	Populations
N_1	$B + \frac{1}{2}\delta + d$
N_2	$B - \frac{1}{2}\delta$
N_3	$B + \frac{1}{2}\delta$
N_4	$B - \frac{1}{2}\delta - d$

The intensities of the carbon lines can now be represented:

^{13}C Energy Levels with Broad-Band Decoupling and with Relaxation

$$N_3 - N_4 = B + \frac{1}{2}\delta - B + \frac{1}{2}\delta + d = \delta + d$$

$$N_1 - N_2 = B + \frac{1}{2}\delta + d - B + \frac{1}{2}\delta = \delta + d$$

Thus, the intensity of each of the carbon lines has been enhanced by an amount d because of this relaxation.

The theoretical maximum value of d is 2.988 (see Eqs. 6.2 and 6.3). The amount of nuclear Overhauser enhancement that may be observed, however, is often less than this amount. The preceding discussion has ignored possible relaxation from state N_3 to N_2. This relaxation pathway would involve no net change in the total number of spins (a **zero-quantum transition**). This relaxation would tend to *decrease* the nuclear Overhauser enhancement. With relatively small molecules, this second relaxation pathway is much less important than W_2; therefore, we generally see a substantial enhancement.

6.7 PROBLEMS WITH INTEGRATION IN ^{13}C Spectra

Avoid attaching too much significance to peak sizes and integrals in proton-decoupled ^{13}C spectra. In fact, carbon spectra are usually not integrated in the same routine fashion as is accepted for proton spectra. Integral information derived from ^{13}C spectra is usually not reliable unless special techniques are used to ensure its validity. It is true that a peak derived from two carbons is larger than one derived from a single carbon. However, as we saw in Section 6.5, if decoupling is used, the intensity of a carbon peak is NOE enhanced by any hydrogens that are either attached to that carbon or found close by. Nuclear Overhauser enhancement is not the same for every carbon. Recall that as a very rough approximation (with some exceptions), a CH_3 peak generally has a greater intensity than a CH_2 peak, which in turn has a greater intensity than a CH peak, and quaternary carbons, those without any attached hydrogens, are normally the weakest peaks in the spectrum.

A second problem arises in the measurement of integrals in ^{13}C FT-NMR. Figure 6.7 shows the typical pulse sequence for an FT-NMR experiment. Repetitive pulse sequences are spaced at intervals of about 1 to 3 sec. Following the pulse, the time allotted to collect the data (the FID) is called the **acquisition time.** A short **delay** usually follows the acquisition of data. When hydrogen spectra are determined, it is common for the FID to have decayed to zero before the end of the acquisition time. Most hydrogen atoms relax back to their original Boltzmann condition very quickly—within less than a second. For ^{13}C atoms, however, the time required for relaxation is quite variable, depending on the molecular environment of the particular atom (see Section 6.8). Some ^{13}C atoms relax very quickly (in seconds), but others require quite long periods (minutes) compared to hydrogen. If carbon atoms with long relaxation times are present in a molecule, collection of the FID signal may have already ceased before all of the ^{13}C atoms have relaxed. The result of this discrepancy is that some atoms have strong signals, as their contribution to the FID is complete, while others, those that have not relaxed completely, have weaker signals. When this happens, the resulting peak areas do not integrate to give the correct number of carbons.

It is possible to extend the data collection period (and the delay period) to allow all of the carbons in a molecule to relax; however, this is usually done only in special cases. Since repetitive scans are used in ^{13}C spectra, the increased acquisition time means that it would simply take too long to measure a complete spectrum with a reasonable signal-to-noise ratio.

F I G U R E   6 . 7 A typical FT-NMR pulse sequence.

6.8 MOLECULAR RELAXATION PROCESSES

In the absence of an applied field, there is a nearly 50/50 distribution of the two spin states for a nucleus of spin $= \frac{1}{2}$. A short time after a magnetic field is applied, a slight excess of nuclei builds up in the lower-energy (aligned) spin state due to thermal equilibration. We call the relative numbers of nuclei in the upper and lower states the **Boltzmann equilibrium.** In Section 5.5, we used the Boltzmann equations to calculate the expected number of excess nuclei for NMR spectrometers that operate at various frequencies (Table 5.3). We rely on these excess nuclei to generate NMR signals. When we pulse the system at the resonance frequency, we disturb the Boltzmann equilibrium (alter the spin population ratios). Excess nuclei are excited to the upper spin state and, as they **relax,** or return to the lower spin state and equilibrium, they generate the FID signal, which is processed to give the spectrum.

If all of the excess nuclei absorb energy, **saturation,** a condition in which the populations of both spin states are once again equal, is reached, and the population of the upper spin state cannot be increased further. This limitation exists because further irradiation induces just as many downward transitions as there are upward transitions when the populations of both states are equal. Net signals are observed only when the populations are unequal. If irradiation is stopped, either at or before saturation, the excited excess nuclei relax, and the Boltzmann equilibrium is reestablished.

The methods by which excited nuclei return to their ground state and by which the Boltzmann equilibrium is reestablished are called **relaxation processes.** In NMR systems, there are two principal types of relaxation processes: spin–lattice relaxation and spin–spin relaxation. Each occurs as a first-order rate process and is characterized by a **relaxation time,** which governs the rate of decay.

Spin–lattice, or *longitudinal,* **relaxation processes** are those that occur in the direction of the field. The spins lose their energy by transferring it to the surroundings—the *lattice*—as thermal energy. Ultimately, the lost energy heats the surroundings. The **spin–lattice relaxation time T_1** governs the rate of this process. The inverse of the spin–lattice relaxation time $1/T_1$ is the rate constant for the decay process.

Several processes, both within the molecule (intramolecular) and between molecules (intermolecular), contribute to spin–lattice relaxation. The principal contributor is magnetic dipole–dipole interaction. The spin of an excited nucleus interacts with the spins of other magnetic nuclei that are in the same molecule or in nearby molecules. These interactions can induce nuclear spin transitions and exchanges. Eventually, the system relaxes back to the Boltzmann equilibrium. This mechanism is especially effective if there are hydrogen atoms nearby. For carbon nuclei, relaxation is fastest if hydrogen atoms are directly bonded, as in CH, CH_2, and CH_3 groups. Spin–lattice relaxation is also most effective in larger molecules, which tumble (rotate) slowly, and it is very inefficient in small molecules, which tumble faster.

Spin–spin, or *transverse,* **relaxation processes** are those that occur in a plane perpendicular to the direction of the field—the same plane in which the signal is detected. Spin–spin relaxation does not change the energy of the spin system. It is often described as an entropy process. When nuclei are induced to change their spin by the absorption of radiation, they all end up precessing in phase after resonance. This is called **phase coherence.** The nuclei lose the phase coherence by exchanging spins. The phases of the precessing spins randomize (increase entropy). This process occurs only between nuclei of the same type—those that are studied in the NMR experiment. The **spin–spin relaxation time T_2** governs the rate of this process.

Our interest is in spin–lattice relaxation times T_1 (rather than spin–spin relaxation times) as they relate to the intensity of NMR signals and have other implications relevant to structure determination. T_1 relaxation times are relatively easy to measure by the **inversion recovery method.**[2] Spin–spin relaxation times T_2 are more difficult to measure and do not provide useful structural

---

[2]Consult the references listed at the end of the chapter for the details of this method.

information. Spin–spin relaxation (phase randomization) always occurs more quickly than the rate at which spin–lattice relaxation returns the system to Boltzmann equilibrium ($T_2 \leq T_1$). However, for nuclei with spin $= \frac{1}{2}$ and a solvent of low viscosity, T_1 and T_2 are usually very similar.

Spin–lattice relaxation times, T_1 values, are not of much use in proton NMR since protons have very short relaxation times. However, T_1 values are quite important to ^{13}C NMR spectra because they are much longer for carbon nuclei and can dramatically influence signal intensities. One can always expect quaternary carbons (including most carbonyl carbons) to have long relaxation times because they have no attached hydrogens. A common instance of long relaxation times is the carbons in an aromatic ring with a substituent group different from hydrogen. The ^{13}C T_1 values for isooctane (2,2,4-trimethylpentane) and toluene follow.

C	T_1
1, 6, 7	9.3 sec
2	68
3	13
4	23
5, 8	9.8

2,2,4-Trimethylpentane

C	T_1	NOE
α	16 sec	0.61
1	89	0.56
2	24	1.6
3	24	1.7
4	17	1.6

Toluene

Notice that in isooctane the quaternary carbon 2, which has no attached hydrogens, has the longest relaxation time (68 sec). Carbon 4, which has one hydrogen, has the next longest (23 sec) and is followed by carbon 3, which has two hydrogens (13 sec). The methyl groups (carbons 1, 5, 6, 7, and 8) have the shortest relaxation times in this molecule. The NOE factors for toluene have been listed along with the T_1 values. As expected, the *ipso* carbon 1, which has no hydrogens, has the longest relaxation time and the smallest NOE. In the ^{13}C NMR of toluene, the *ipso* carbon has the lowest intensity.

Remember also that T_1 values are greater when a molecule is small and tumbles rapidly in the solvent. The carbons in cyclopropane have a T_1 of 37 sec. Cyclohexane has a smaller value, 20 sec. In a larger molecule such as the steroid cholesterol, all of the carbons except those that are quaternary would be expected to have T_1 values less than 1 to 2 sec. The quaternary carbons would have T_1 values of about 4 to 6 sec due to the lack of attached hydrogens. In solid polymers, such as polystyrene, the T_1 values for the various carbons are around 10^{-2} sec.

To interpret ^{13}C NMR spectra, you should know what effects of NOE and spin–lattice relaxation to expect. We cannot fully cover the subject here, and there are many additional factors besides those that we have discussed. If you are interested, consult more advanced textbooks, such as the ones listed in the references.

The example of 2,3-dimethylbenzofuran will close this section. In this molecule, the quaternary (*ipso*) carbons have relaxation times that exceed 1 min. As discussed in Section 6.7, to obtain a decent spectrum of this compound, it would be necessary to extend the data acquisition and delay periods to determine the entire spectrum of the molecule and see the carbons with high T_1 values.

C	T_1	NOE
2	83 sec	1.4
3	92	1.6
3a	114	1.5
7a	117	1.3
Others	<10	1.6–2

2,3-Dimethylbenzofuran

6.9 OFF-RESONANCE DECOUPLING

The decoupling technique that is used to obtain typical proton-decoupled spectra has the advantage that all peaks become singlets. For carbon atoms bearing attached hydrogens, an added benefit is that peak intensities increase, owing to the nuclear Overhauser effect, and signal-to-noise ratios improve. Unfortunately, much useful information is also lost when carbon spectra are decoupled. We no longer have information about the number of hydrogens that are attached to a particular carbon atom.

In many cases, it would be helpful to have the information about the attached hydrogens that a proton-coupled spectrum provides, but frequently the spectrum becomes too complex, with overlapping multiplets that are difficult to resolve or assign correctly. A compromise technique called **off-resonance decoupling** can often provide multiplet information while keeping the spectrum relatively simple in appearance.

In an off-resonance-decoupled ^{13}C spectrum, the coupling between each carbon atom and each hydrogen attached directly to it is observed. The $n + 1$ Rule can be used to determine whether a given carbon atom has three, two, one, or no hydrogens attached. However, when off-resonance decoupling is used, the *apparent magnitude* of the coupling constants is reduced, and overlap of the resulting multiplets is a less-frequent problem. The off-resonance-decoupled spectrum retains the couplings between the carbon atom and directly attached protons (the one-bond couplings) but effectively removes the couplings between the carbon and more remote protons.

In this technique, the frequency of a second radiofrequency transmitter (the **decoupler**) is set either upfield or downfield from the usual sweep width of a normal proton spectrum (i.e., *off resonance*). In contrast, the frequency of the decoupler is set to *coincide exactly* with the range of proton resonances in a true decoupling experiment. Furthermore, in off-resonance decoupling, the power of the decoupling oscillator is held *low* to avoid complete decoupling.

Off-resonance decoupling can be a great help in assigning spectral peaks. The off-resonance-decoupled spectrum is usually obtained separately, along with the proton-decoupled spectrum. Figure 6.8 shows the off-resonance-decoupled spectrum of 1-propanol, in which the methyl carbon atom is split into a quartet, and each of the methylene carbons appears as a triplet. Notice that the observed multiplet patterns are consistent with the $n + 1$ Rule and with the patterns shown in Figure 6.3. If TMS had been added, its methyl carbons would have appeared as a quartet centered at $\delta = 0$ ppm.

6.10 A QUICK DIP INTO DEPT

Despite its utility, off-resonance decoupling is now considered an old-fashioned technique. It has been replaced by more modern methods, the most important of which is **D**istortionless **E**nhancement by **P**olarization **T**ransfer, better known as **DEPT.** The DEPT technique requires an FT-pulsed spectrometer. It is more complicated than off-resonance decoupling, and it requires a computer, but it gives the same information more reliably and more clearly. Chapter 9 will discuss the DEPT method in detail; only a brief introduction to the method and the results it provides will be provided here.

F I G U R E   6 . 8 The off-resonance-decoupled ^{13}C spectrum of 1-propanol (22.5 MHz).

In the DEPT technique, the sample is irradiated with a complex sequence of pulses in both the ^{13}C and ^{1}H channels. The result of these pulse sequences[3] is that the ^{13}C signals for the carbon atoms in the molecule will exhibit different **phases,** depending on the number of hydrogens attached to each carbon. Each type of carbon will behave slightly differently, depending on the **duration** of the complex pulses. These differences can be detected, and spectra produced in each experiment can be plotted.

One common method of presenting the results of a DEPT experiment is to plot four different **subspectra**. Each subspectrum provides different information. A sample DEPT plot for **isopentyl acetate** is shown in Figure 6.9.

$$\overset{5}{CH_3}-\overset{\overset{\displaystyle O}{\displaystyle \|}}{\underset{6}{C}}-O-\overset{4}{CH_2}-\overset{3}{CH_2}-\overset{2}{\underset{\underset{1}{\overset{|}{CH_3}}}{CH}}-\overset{1}{CH_3}$$

The lowest trace in the figure is the usual broad-band-decoupled ^{13}C spectrum. The second trace from the bottom is the result of a pulse sequence (called a **DEPT-45**) in which the only signals detected are those that arise from protonated carbons. You will notice that the carbonyl carbon (labeled **6**), at 172 ppm, is not seen. The solvent peaks arising from $CDCl_3$ (77 ppm) are also not seen. Deuterium (D or ^{2}H) behaves differently from ^{1}H, and as a result the carbon of $CDCl_3$ behaves as if it were not protonated. The third trace is the result of a slightly different pulse sequence (called a **DEPT-90**). In this trace, only those carbons that bear a single hydrogen are seen. Only the carbon at position **2** (25 ppm) is observed.

---

[3]Pulse sequences were introduced in Section 6.7.

FIGURE 6.9 DEPT spectra of isopentyl acetate.

The uppermost trace is more complicated than the previous subspectra. The pulse sequence that gives rise to this subspectrum is called **DEPT-135**. Here, all carbons that have an attached proton provide a signal, but the **phase** of the signal will be different, depending on whether the number of attached hydrogens is an odd or an even number. Signals arising from CH or CH_3 groups will give positive peaks, while signals arising from CH_2 groups will form negative (inverse) peaks. When we examine the upper trace in Figure 6.9, we can identify all of the carbon peaks in the spectrum of isopentyl acetate. The positive peaks at 21 and 22 ppm must represent CH_3 groups as those peaks are not represented in the DEPT-90 subspectrum. When we look at the original ^{13}C spectrum, we see that the peak at 21 ppm is not as strong as the peak at 22 ppm. We conclude, therefore, that the peak at 21 ppm must come from the CH_3 carbon at position **5**, while the stronger peak at 22 ppm comes from the pair of equivalent CH_3 carbons at position **1**. We have already determined that the positive peak at 25 ppm is due to the CH carbon at position **2** as it appears in both the DEPT-135 and the DEPT-90 subspectra. The inverse peak at 37 ppm is due to a CH_2 group, and we can identify it as coming from the carbon at position **3**. The inverse peak at 53 ppm is clearly caused by the CH_2 carbon at position **4**, deshielded by the attached oxygen atom. Finally, the downfield peak at 172 ppm has already been labeled as arising from the carbonyl carbon at **6**. This peak appears only in the original ^{13}C spectrum; therefore, it must not have any attached hydrogens.

Through the mathematical manipulation of the results of each of the different DEPT pulse sequences, it is also possible to present the results as a series of subspectra in which only the CH carbons appear in one trace, only the CH_2 carbons appear in the second trace, and only the CH_3 carbons appear in the third trace. Another common means of displaying DEPT results is to show only the result of the DEPT-135 experiment. The spectroscopist generally can interpret the results of this spectrum by applying knowledge of likely chemical shift differences to distinguish between CH and CH_3 carbons.

The results of DEPT experiments may be used from time to time in this textbook to help you solve assigned exercises. In an effort to save space, most often only the results of the DEPT experiment, rather than the complete spectrum, will be provided.

6.11 SOME SAMPLE SPECTRA—EQUIVALENT CARBONS

Equivalent ^{13}C atoms appear at the same chemical shift value. Figure 6.10 shows the proton-decoupled carbon spectrum for 2,2-dimethylbutane. The three methyl groups at the left side of the molecule are equivalent by symmetry.

$$CH_3-\underset{\underset{CH_3}{|}}{\overset{\overset{CH_3}{|}}{C}}-CH_2-CH_3$$

Although this compound has a total of six carbons, there are only four peaks in the ^{13}C NMR spectrum. The ^{13}C atoms that are equivalent appear at the same chemical shift. The single methyl carbon **a** appears at highest field (9 ppm), while the three equivalent methyl carbons **b** appear at 29 ppm. The quaternary carbon **c** gives rise to the small peak at 30 ppm, and the methylene carbon **d** appears at 37 ppm. The relative sizes of the peaks are related, in part, to the number of each type of carbon atom present in the molecule. For example, notice in Figure 6.10 that the peak at 29 ppm (**b**) is much larger than the others. This peak is generated by three carbons. The quaternary carbon at 30 ppm (**c**) is very weak. Since no hydrogens are attached to this carbon, there is very little NOE enhancement. Without attached hydrogen atoms, relaxation times are also longer than for other carbon atoms. Quaternary carbons, those with no hydrogens attached, frequently appear as weak peaks in proton-decoupled ^{13}C NMR spectra (see Sections 6.5 and 6.7).

Figure 6.11 is a proton-decoupled ^{13}C spectrum of cyclohexanol. This compound has a plane of symmetry passing through its hydroxyl group, and it shows only four carbon resonances. Carbons **a** and **c** are doubled due to symmetry and give rise to larger peaks than carbons **b** and **d**. Carbon **d**, bearing the hydroxyl group, is deshielded by oxygen and has its peak at 70.0 ppm. Notice that this peak has the lowest intensity of all of the peaks. Its intensity is lower than that of carbon **b** in part

F I G U R E   6 . 1 0 The proton-decoupled ^{13}C NMR spectrum of 2,2-dimethylbutane.

FIGURE 6.11 The proton-decoupled ^{13}C NMR spectrum of cyclohexanol.

because the carbon **d** peak receives the least amount of NOE; there is only one hydrogen attached to the hydroxyl carbon, whereas each of the other carbons has two hydrogens.

A carbon attached to a double bond is deshielded due to its sp^2 hybridization and some diamagnetic anisotropy. This effect can be seen in the ^{13}C NMR spectrum of cyclohexene (Fig. 6.12). Cyclohexene has a plane of symmetry that runs perpendicular to the double bond. As a result, we observe only three absorption peaks. There are two of each type of carbon. Each of the double-bond carbons **c** has only one hydrogen, whereas each of the remaining carbons has two. As a result of a reduced NOE, the double-bond carbons have a lower-intensity peak in the spectrum.

In Figure 6.13, the spectrum of cyclohexanone, the carbonyl carbon has the lowest intensity. This is due not only to reduced NOE (no hydrogen attached) but also to the long relaxation time of the carbonyl carbon. As you have already seen, quaternary carbons tend to have long relaxation times. Also notice that Figure 6.1 predicts the large chemical shift for this carbonyl carbon.

FIGURE 6.12 The proton-decoupled ^{13}C NMR spectrum of cyclohexene.

F I G U R E 6 . 1 3 The proton-decoupled ^{13}C spectrum of cyclohexanone.

6.12 NONEQUIVALENT CARBON ATOMS

The spectrum of 2,2-dimethylbutane shown in Figure 6.10 has three methyl groups (b) attached to carbon c. They are all equivalent and appear as a singlet in the proton-decoupled ^{13}C spectrum. Often, methyl groups are nonequivalent and will appear at different chemical shifts in the spectrum. Several cases where nonequivalent methyl groups may appear include groups located near a stereocenter and for compounds that are conformationally rigid. The two nonequivalent methyl groups adjacent to the stereocenter are diastereotopic. This concept is discussed in greater detail in Sections 7.3 and 7.4. Two examples of nonequivalent carbon atoms are shown in Figure 6.14.

2-Methyl-3-pentanol
* indicates stereocenter

6 carbon atoms appear in the spectrum. Unique carbon atoms indicated with a dot.

equatorial methyl groups appear at the same place

axial methyl groups appear at the same place

3,3,5,5-Tetramethylcyclohexanol

The two equatorial methyl groups are equivalent and appear as single peak at a different place than the two axial methyl groups. There are a total of 6 peaks in the spectrum: 4 unique carbon atoms on the ring, and two for the non-equivalent methyl groups.

F I G U R E 6 . 1 4 Examples of compounds with non-equivalent methyl groups.

6.13 COMPOUNDS WITH AROMATIC RINGS

Compounds with carbon–carbon double bonds or aromatic rings give rise to chemical shifts in the range from 100 to 175 ppm. Since relatively few other peaks appear in this range, a great deal of useful information is available when peaks appear here.

A **monosubstituted** benzene ring shows *four* peaks in the aromatic carbon area of a proton-decoupled ^{13}C spectrum since the *ortho* and *meta* carbons are doubled by symmetry. Often, the carbon with no protons attached, the *ipso* carbon, has a very weak peak due to a long relaxation time and a weak NOE. In addition, there are two larger peaks for the doubled *ortho* and *meta* carbons and a medium-sized peak for the *para* carbon. In many cases, it is not important to be able to assign all of the peaks precisely. In the example of toluene, shown in Figure 6.15, notice that carbons **c** and **d** are not easy to assign by inspection of the spectrum. However, use of chemical shift correlation tables (see Section 6.2B and Appendix 8) would enable us to assign these signals.

Toluene c,d c,d

Difficult to assign without using
chemical shift correlation tables

In an off-resonance-decoupled or proton-coupled ^{13}C spectrum, a monosubstituted benzene ring shows three doublets and one singlet. The singlet arises from the *ipso* carbon, which has no attached hydrogen. Each of the other carbons in the ring (*ortho, meta,* and *para*) has one attached hydrogen and yields a doublet.

Figure 6.4b is the proton-decoupled spectrum of ethyl phenylacetate, with the assignments noted next to the peaks. Notice that the aromatic ring region shows four peaks between 125 and 135 ppm, consistent with a monosubstituted ring. There is one peak for the methyl carbon (13 ppm) and two peaks for the methylene carbons. One of the methylene carbons is directly attached to an electronegative oxygen atom and appears at 61 ppm, while the other is more shielded (41 ppm). The carbonyl

FIGURE 6.15 The proton-decoupled ^{13}C NMR spectrum of toluene.

carbon (an ester) has resonance at 171 ppm. All of the carbon chemical shifts agree with the values in the correlation chart (Fig. 6.1).

Depending on the mode of substitution, a symmetrically **disubstituted** benzene ring can show two, three, or four peaks in the proton-decoupled ^{13}C spectrum. The following drawings illustrate this for the isomers of dichlorobenzene:

Three unique carbon atoms   Four unique carbon atoms   Two unique carbon atoms

Figure 6.16 shows the spectra of all three dichlorobenzenes, each of which has the number of peaks consistent with the analysis just given. You can see that ^{13}C NMR spectroscopy is very useful in the identification of isomers.

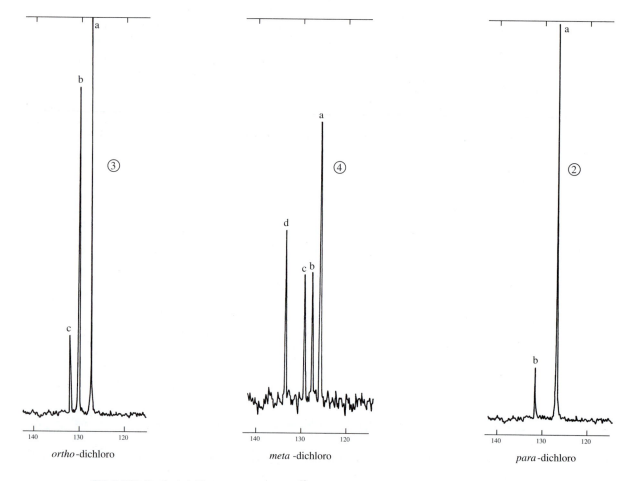

FIGURE 6.16 The proton-decoupled ^{13}C NMR spectra of the three isomers of dichlorobenzene (25 MHz).

Most other polysubstitution patterns on a benzene ring yield six different peaks in the proton-decoupled ^{13}C NMR spectrum, one for each carbon. However, when identical substituents are present, watch carefully for planes of symmetry that may reduce the number of peaks.

☐ 6.14 CARBON-13 NMR SOLVENTS—HETERONUCLEAR COUPLING OF CARBON TO DEUTERIUM

Most FT-NMR spectrometers require the use of deuterated solvents because the instruments use the deuterium resonance signal as a "lock signal," or reference signal, to keep the magnet and the electronics adjusted correctly. Deuterium is the ^{2}H isotope of hydrogen and can easily substitute for it in organic compounds. Deuterated solvents present few difficulties in hydrogen spectra as the deuterium nuclei are largely invisible when a proton spectrum is determined. Deuterium has resonance at a different frequency from hydrogen. In ^{13}C NMR, however, these solvents are frequently seen as part of the spectrum as they all have carbon atoms. In this section, we explain the spectra of some of the common solvents and, in the process, examine heteronuclear coupling of carbon and deuterium. Figure 6.17 shows the ^{13}C NMR peaks due to the solvents chloroform-d and dimethylsulfoxide-d$_6$.

Chloroform-d, CDCl$_3$, is the most commonly used solvent for NMR. It is also called deuteriochloroform or deuterated chloroform. Its use gives rise to a three-peak multiplet in the spectrum, with the center peak having a chemical shift of about 77 ppm. Figure 6.17 shows an example. Notice that this "triplet" is different from the triplets in a proton spectrum (from two neighbors) or in a proton-coupled ^{13}C spectrum (from two attached hydrogens); the intensities are different. In this triplet, all three peaks have approximately the same intensity (1:1:1), whereas the other types of triplets have intensities that follow the entries in Pascal's triangle, with ratios of 1:2:1.

In contrast with hydrogen (spin = $\frac{1}{2}$), deuterium has spin = 1. A single deuterium nucleus can adopt three different spins ($2I + 1 = 3$), where the spins have quantum numbers of -1, 0, and $+1$. In a solution of CDCl$_3$, molecules can have a deuterium with any one of these spins, and as they are equally probable, we see three different chemical shifts for the carbon atom in chloroform-d. The ^{13}C—D one-bond coupling constant for this interaction is about 45 Hz. At 75 MHz, these three peaks are about 0.6 ppm apart (45 Hz/75 MHz = 0.60 ppm).

(a)   (b)

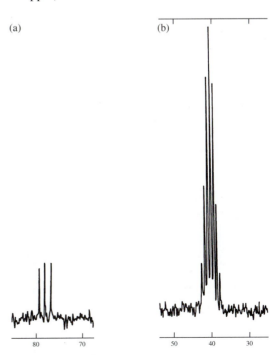

FIGURE 6.17 The ^{13}C NMR peaks of two common solvents. (a) Chloroform-d. (b) Dimethylsulfoxide-d$_6$.

Because deuterium is not a spin $= \frac{1}{2}$ nucleus, the $n + 1$ Rule does not correctly predict the multiplicity of the carbon resonance. The $n + 1$ Rule works only for spin $= \frac{1}{2}$ nuclei and is a specialized case of a more general prediction formula:

$$\text{multiplicity} = 2nI + 1 \hspace{3cm} \textbf{Equation 6.4}$$

where n is the number of nuclei, and I is the spin of that type of nucleus. If we use this formula, the correct multiplicity of the carbon peak with *one deuterium* attached is predicted by

$$2 \cdot 1 \cdot 1 + 1 = 3$$

If there are *three hydrogens*, the formula correctly predicts a quartet for the proton-coupled carbon peak:

$$2 \cdot 3 \cdot \frac{1}{2} + 1 = 4$$

Dimethylsulfoxide-d$_6$, CD_3—SO—CD_3, is frequently used as a solvent for carboxylic acids and other compounds that are difficult to dissolve in $CDCl_3$. Equation 6.4 predicts a septet for the multiplicity of the carbon with three deuterium atoms attached:

$$2 \cdot 3 \cdot 1 + 1 = 7$$

This is exactly the pattern observed in Figure 6.17. The pattern has a chemical shift of 39.5 ppm, and the coupling constant is about 40 Hz.

Because deuterium has spin $= 1$ instead of spin $= \frac{1}{2}$ like hydrogen, the Pascal triangle (Fig. 5.33 in Section 5.16) does not correctly predict the intensities in this seven-line pattern. Instead, a different intensity triangle must be used for splittings caused by deuterium atoms. Figure 6.18 is this intensity triangle, and Figure 6.19 is an analysis of the intensities for three-line and five-line multiplets. In the latter figure, an upward arrow represents spin $= 1$, a downward arrow represents spin $= -1$, and a large dot represents spin $= 0$. Analysis of the seven-line multiplet is left for the reader to complete.

Acetone-d$_6$, CD_3—CO—CD_3, shows the same ^{13}C septet splitting pattern as dimethylsulfoxide-d$_6$, but the multiplet is centered at 29.8 ppm with the carbonyl peak at 206 ppm. The carbonyl carbon is a singlet; three-bond coupling does not appear.

Acetone-d$_5$ frequently appears as an impurity in spectra determined in acetone-d$_6$. It leads to interesting results in both the hydrogen and the carbon-13 spectra. Although this chapter is predominantly about carbon-13 spectra, we will examine both cases.

n	$2nI + 1$ Lines	Relative Intensities												
0	1	1												
1	3	1	1	1										
2	5	1	2	3	2	1								
3	7	1	3	6	7	6	3	1						
4	9	1	4	10	16	19	16	10	4	1				
5	11	1	5	15	30	45	51	45	30	15	5	1		
6	13	1	6	21	50	90	126	141	126	90	50	21	6	1

FIGURE 6.18 An intensity triangle for deuterium multiplets (n = number of deuterium atoms).

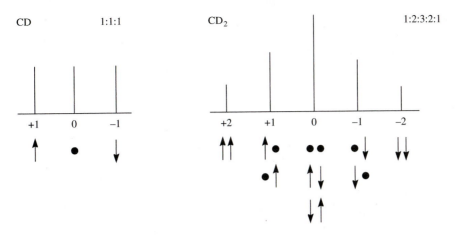

FIGURE 6.19 An intensity analysis of three- and five-line deuterium multiplets.

Hydrogen Spectrum

In proton (^{1}H) NMR spectra, a commonly encountered multiplet arises from a small amount of acetone-d_5 impurity in acetone-d_6 solvent. Figure 6.20 shows the multiplet, which is generated by the hydrogen in the $-CHD_2$ group of the $CD_3-CO-CHD_2$ molecule. Equation 6.4 correctly predicts that there should be a quintet in the proton spectrum of acetone-d_5:

$$2 \cdot 2 \cdot 1 + 1 = 5$$

and this is observed.

Carbon Spectrum

The proton-coupled ^{13}C spectrum of the $-CHD_2$ group is more complicated as both hydrogen (spin $= \frac{1}{2}$) and deuterium (spin $= 1$) interact with carbon. In this case, we use the following formula, which is extended from Equation 6.4:

$$\text{total multiplicity} = \Pi_i \,(2n_iI_i + 1) \qquad \textbf{Equation 6.5}$$

$$\text{Condition: } I \geq \frac{1}{2}$$

The large Π_i indicates a product of terms for each different type of atom i that couples to the atom being observed. These atoms must have spin $\geq \frac{1}{2}$; atoms of spin $= 0$ do not cause splitting. In the present case ($-CHD_2$), there are two terms, one for hydrogen and one for deuterium.

$$\text{total multiplicity} = (2 \cdot 1 \cdot \frac{1}{2} + 1)(2 \cdot 2 \cdot 1 + 1) = 10$$

The ^{13}C$-$H and ^{13}C$-$D coupling constants would most likely be different, resulting in 10 lines that would not all be equally spaced. In addition, acetone has a second "methyl" group on the other side of the carbonyl group. The $-CD_3$ group (seven peaks) would overlap the 10 peaks from $-CHD_2$ and make a pattern that would be quite difficult to decipher! Fortunately, since deuterated acetone typically has an isotopic purity of 99.9% (only 0.1% of the solvent molecules contain a ^{1}H) the complicated pattern from a CHD_2 carbon is not observed under normal acquisition conditions. The ^{1}H and ^{13}C chemical shifts for common NMR solvents are provided in Appendix 10.

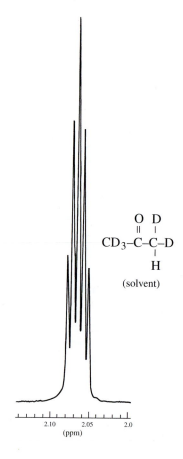

FIGURE 6.20 The 300-MHz ^{1}H spectrum of acetone-d$_5$ (CD$_3$-CO-CHD$_2$).

6.15 HETERONUCLEAR COUPLING OF CARBON-13 TO FLUORINE-19

Organic compounds that contain C, H, O, Cl, and Br will show only singlets when the proton decoupler is turned on. The oxygen, chlorine, and bromine atoms will not couple to a carbon-13 atom under normal conditions. However, when the organic compound has a fluorine atom attached to a carbon-13 atom, you will observe heteronuclear ^{13}C–^{19}F coupling even though the proton decoupler is turned on (proton but not fluorine are decoupled). Figures 6.21 and 6.22 are two spectra that exhibit this effect. The $n + 1$ Rule can be used to determine what the pattern will look like. Fluorine has a nuclear spin that is the same as a proton and a phosphorus. Thus, with one attached fluorine atom, you would expect the carbon-13 atom to be split into a doublet. Two attached fluorine atoms will give rise to a triplet for the carbon-13 atom.

one bond connecting
C to F = ^1J

two bonds connecting C to F = ^2J
The carbon-13 atom is connected
through the common isotope,
carbon-12, to fluorine-19.

FIGURE 6.21 The ^{13}C proton-decoupled spectrum of CFBr$_3$ (75 MHz).

FIGURE 6.22 The ^{13}C proton-decoupled spectrum of CF$_3$CH$_2$OH (75 MHz).

The spectrum of CFBr$_3$ shown in Figure 6.21 has Hertz (Hz) values recorded on top of each peak in the doublet rather than parts-per-million (ppm) values, which is more typical. The chemical shift values for each of the peaks can be calculated by dividing the Hertz values by the field strength of the NMR spectrometer (75 MHz), giving 43.85 and 48.79 ppm. The actual chemical shift for the carbon atom would be in the center of the doublet: 46.32 ppm. The ^{13}C—^{19}F coupling constant in

Hertz is easily determined by subtracting the two Hertz values, yielding 370 Hz. This huge coupling constant is typical for direct one-bond coupling of the fluorine nucleus to a carbon-13 atom (^1J).

The second example for fluorine coupling to ^{13}C is shown in Figure 6.22. This spectrum shows both one-bond and two-bond coupling of ^{13}C to ^{19}F. The large quartet centering on about 125 ppm for C-2 results from the one-bond coupling of three attached fluorine atoms (^1J) to a ^{13}C atom ($n + 1 = 4$). Again, Hertz values are included on each peak in the large quartet. Subtracting the Hertz values on the center two peaks in the quartet yields 278 Hz. Also notice that there is another quartet centering on about 62 ppm for C-1. This quartet results from the three fluorine atoms that are further away from the ^{13}C. Notice that the spacings in that quartet are about 35 Hz. This is described as a two-bond coupling (^2J). Notice that the coupling falls off with distance (see Appendix 9 for typical ^{13}C to ^{19}F coupling constants).

6.16 HETERONUCLEAR COUPLING OF CARBON-13 TO PHOSPHORUS-31

The two spectra in Figures 6.23 and 6.24 demonstrate coupling between ^{13}C and ^{31}P. In the first compound, shown in Figure 6.23, the carbon atom of the methyl group at about 12 ppm is split by one adjacent phosphorus atom into a doublet with a coupling constant equal to 56.1 Hz (919.3 − 863.2 = 56.1 Hz). Notice that the $n + 1$ Rule predicts how the pattern will appear (doublet). The nuclear spin number for phosphorus is the same as for a proton and for a fluorine atom ($\frac{1}{2}$). This interaction is an example of one-bond coupling (^1J).

The second compound, shown in Figure 6.24, shows both one-bond and two-bond coupling of ^{13}C to ^{31}P. The one-bond coupling occurs between the phosphorus atom and the ^{13}C atom of the directly attached methyl group, $^{31}P-^{13}CH_3$, has a value of 144 Hz (819.2 − 675.2). This doublet is found at about 10 ppm. The other CH_3 group, $^{31}P-O-^{13}CH_3$, is two bonds away from the phosphorus atom, and it appears as a doublet about 52 ppm. This two-bond coupling constant equals about 6 Hz (3949.6 − 3943.5). One-bond coupling constants can vary because of the differences in hybridization of the phosphorus atom.

F I G U R E   6 . 2 3 The ^{13}C proton-decoupled spectrum of tetramethylphosphonium chloride, $(CH_3)_4P^+Cl^-$ (75 MHz).

FIGURE 6.24 The ^{13}C proton-decoupled spectrum of $CH_3PO(OCH_3)_2$ (75 MHz).

6.17 CARBON AND PROTON NMR: HOW TO SOLVE A STRUCTURE PROBLEM

How do you approach determining a structure of an unknown compound utilizing carbon and proton NMR spectra? Let's look at the proton NMR spectrum shown in Figure 6.25. The spectrum is for a compound with formula $C_6H_{10}O_2$. The index of hydrogen deficiency for this compound is calculated to be 2.

Proton chemical shift. The first thing you should do is to look at the chemical shift values for the peaks that appear in the spectrum. You will find Figure 5.20 to be very helpful as an overview of where protons would be expected to appear.

0.8 to 1.8 ppm: Protons in this region are generally associated with sp^3 carbon atoms such as CH, CH_2, and CH_3 groups at some distance from electronegative atoms. Groups with more attached protons are more shielded and will appear upfield (closer to TMS). Thus, CH_3 is more shielded than a CH_2 group and will appear at a lower parts-per-million (ppm) value.

1.8 to 3.0 ppm: This region is generally associated with protons on a sp^3 carbon atom next to C=O, C=C, and aromatic groups. Examples include CH_2—C=O, C=C—CH_2—, and CH_2—Ar. One exception to this is a proton directly attached to a triple bond, C≡C—H, that also appears in this range.

FIGURE 6.25 The proton NMR spectrum for an unknown compound.

3.0 to 4.5 ppm: This region is generally associated with protons on a sp^3 carbon atom that are directly attached to an electronegative atom, generally oxygen or a halogen atom. Examples include $-CH_2-Cl$, $-CH_2-Br$, and $-CH_2O-$. The most common oxygen-containing groups are associated with alcohols, ethers, and esters. A value of 3.5 ppm is a good number to remember for $-O-CH_2-$ or $-O-CH_3$.

4.5 to 7.0 ppm: This region is generally associated with protons **directly** attached to C=C sp^2 carbon atoms in alkenes (vinyl protons). Example: C=C$-$H. However, it should be remembered that multiple electronegative atoms attached to a carbon can shift protons downfield into this region. Examples include $-O-CH_2-O-$ and $Cl-CH_2-Cl$.

6.5 to 8.5 ppm: This region is generally associated with protons **directly** attached to C=C sp^2 carbon atoms in a benzene ring or other aromatic compounds.

9.0 to 10 ppm: This region is always associated with aldehyde protons, protons directly attached to a C=O group.

11.0 to 13.0 ppm: Carboxylic acid protons usually appear in this region. Carboxylic acid protons give rise to very broad peaks. In some cases, the peaks are so broad that the peak is not observed and disappears into the baseline.

Using the chemical shift information and the index of hydrogen deficiency, you should be able to determine that the unknown compound contains a C=C$-$H and a COOH group by observing the peaks at 6.8 and 12.5 ppm. Since there is only one peak in the alkene region, you might suggest that the double bond is trisubstituted.

Proton integration. The number of protons on a given carbon atom can be determined from the numbers printed just below the peaks. Referring to Section 5.9 you can easily round off the numbers shown in Figure 6.25 to whole numbers without doing the math. Remember that the numbers are approximate. From right to left, you can determine by inspection that the triplet at 1 ppm represents (3 H), the singlet at 1.7 ppm (3 H), the quintet at 2.3 ppm (2 H), and the triplet at 6.8 ppm (1 H). The remaining proton for the carboxyl group at 12.5 ppm is shown in the inset, and it integrates for about (1 H). Notice that the number of protons you determined equals the number of protons in the formula $C_6H_{10}O_2$. Life is good!

Proton spin–spin splitting. The next thing to look at in Figure 6.25 is the multiplicity of the proton peaks. Here, you look for singlets, doublets, and triplet patterns in the proton spectrum. The $n + 1$ Rule is helpful to determine the number of adjacent protons (^3J, see Sections 5.13–5.18). Typical ^3J coupling constants usually have a value around 7.5 Hz. You will need to remember that most spectra obtained on high-field NMR spectrometers at 300 to 500 MHz will need to be expanded to see the splitting patterns. In this textbook, all spectra obtained on high-field NMR spectrometers will be expanded so that you will be able to observe the splitting patterns. Notice that the NMR spectrum shown in Figure 6.25 did not include the full typical range of 0 to 10 ppm. In some cases, an inset spectrum for part of the compound may appear above the baseline that is out of the typical range. The carboxylic acid proton shown in the inset on Figure 6.25 illustrates this. In other cases, you may find an inset spectrum for protons that need to be expanded to fully see the pattern. An example of this might be a septet pattern (seven peaks) or nonet pattern (nine peaks) which can be expanded in both the x and y directions so that you can observe all of the peaks in the pattern. See the proton NMR spectrum in problem 5d at the end of the chapter as an example.

For our unknown compound shown in Figure 6.25, we expected the triplet at about 1 ppm to result from two adjacent protons. The singlet at about 1.7 ppm results from no adjacent protons. The quintet at 2.3 ppm would indicate four adjacent protons on two different carbon atoms. Finally, the single vinyl proton appearing as a triplet at 6.8 ppm results from two adjacent protons.

At this point, the structure is likely to be the following:

An isomer of the above structure, shown below, would not fit the observed multiplicities and can be ruled out as a possible structure.

We can seek confirmation for the structure by looking at the proton decoupled carbon-13 spectrum shown in Figure 6.26. Notice that the spectrum has six singlet peaks along with the group of three peaks for the solvent, $CDCl_3$, at about 77 ppm (see Fig 6.17).

^{13}**C chemical shift.** The most useful correlation charts are shown in Figure 6.1 and Table 6.1.

10 to 50 ppm: The most common solvent used in NMR spectroscopy is $CDCl_3$, which appears as a three-peak pattern centering on about 77 ppm. Typically, the sp^3 carbon-13 atoms appear to the right of the solvent. CH_3 groups are more shielded than CH_2 groups and usually appear at lower parts-per-million (ppm) values than for CH_2.

F I G U R E 6 . 2 6 The carbon-13 spectrum for an unknown compound.

35 to 80 ppm: As expected, attached electronegative atoms cause a downfield shift similar to that observed in proton NMR spectroscopy. The carbon atoms in this group include $-CH_2-Br$, $-CH_2-Cl$, $-CH_2-O-$. The $C\equiv C$ appears in this range from 65 to 80 ppm.

110 to 175 ppm: The $C=C$ group in alkenes and aromatic compounds appear to the left of the $CDCl_3$ peaks. Generally, aromatic carbon-13 atoms appear further downfield than alkenes, but there are numerous exceptions, and you should expect carbon peaks for both alkenes and aromatic compounds to overlap and appear in the same range.

160 to 220 ppm: The carbonyl group appears to the extreme left-hand part of the carbon-13 spectrum (downfield). Esters and carboxylic acid $C=O$ groups appear at the lower end of the range (160 to 185 ppm), while ketones and aldehydes appear near the higher end of the range (185 to 220 ppm). These $C=O$ peaks can be very weak and can sometimes be missed when looking at the carbon-13 spectrum. Correlation charts that include $C=O$ peaks are shown in Figure 6.1 and 6.2.

Carbon-13 to proton spin–spin splitting. Carbon-13 spectra are generally determined with the proton decoupler turned on. This leads to spectra that consist of singlet peaks (Section 6.4). It is useful to know, however, which carbon atoms have three attached protons (a CH_3 group), or two attached protons (a CH_2 group), or one attached proton (a CH group), and which carbon has no attached protons (a quaternary or *ipso* carbon atom). The most modern way of determining the multiplicity of the carbon-13 atoms is to run a DEPT experiment. Section 6.10, describes how this experiment can determine the multiplicities for each carbon-13 atom. Figure 6.9 shows a typical result for isopentyl acetate. The most useful of these routines are the DEPT-135, which shows CH_3 and CH groups as positive peaks and CH_2 groups as negative peaks. The DEPT-90 experiment shows only CH groups (positive peaks). Carbon atoms with no attached protons (quaternary and *ipso* carbon atoms) do not show up in either experiment. The DEPT experimental results for the unknown compound are shown here. The DEPT experimental results are consistent with our structure shown here.

Normal Carbon	DEPT-135	DEPT-90	Conclusion
11.79 ppm	Positive	No peak	CH_3
12.89	Positive	No peak	CH_3
22.24	Negative	No peak	CH_2
126.62	No peak	No peak	C
146.71	Positive	Positive	CH
174.19	No peak	No peak	C=O

PROBLEMS

*1. A compound with the formula $C_3H_6O_2$ gives the following proton-decoupled and off-resonance-decoupled spectra. Determine the structure of the compound.

Off-resonance-decoupled

quartet

quartet

singlet

CDCI$_3$
(solvent)

200   150   100   50   0

*2. Predict the number of peaks that you would expect in the proton-decoupled ^{13}C spectrum of each of the following compounds. Problems 2a and 2b are provided as examples. Dots are used to show the nonequivalent carbon atoms in these two examples.

(a)

$$CH_3-C(=O)-O-CH_2-CH_3$$

Four peaks

(b)

Five peaks

(c)

Five peaks

(d)

Four peaks

(e)

CH₃
|
CH   CH₂
/   \   /   \
H₃C   C   CH₃
/ \
H₃C   OH

(f)

CH₃
‖
O
CH₃   CH₂

(g)

CH₃
|
CH₂   CH₂
\   /
HO   C
/ \
HO—CH₂   CH₂
|
CH₃

(h)

O
‖

O

(i)

O
‖
CH₃

O

O

(j)

CH₂—CH₃

Br

(k)

CH₂—CH₃

Br

*3. Following are proton-decoupled ¹³C spectra for three isomeric alcohols with the formula C₄H₁₀O. A DEPT or an off-resonance analysis yields the multiplicities shown; **s** = singlet, **d** = doublet, **t** = triplet, and **q** = quartet. Identify the alcohol responsible for each spectrum and assign each peak to an appropriate carbon atom or atoms.

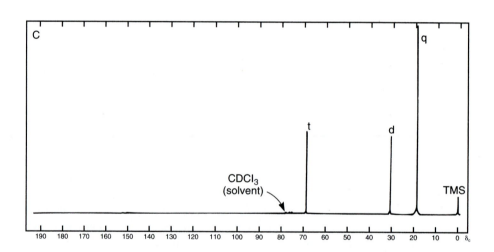

*4. The following spectrum is of an ester with formula $C_5H_8O_2$. Multiplicities are indicated. Draw the structure of the compound and assign each peak.

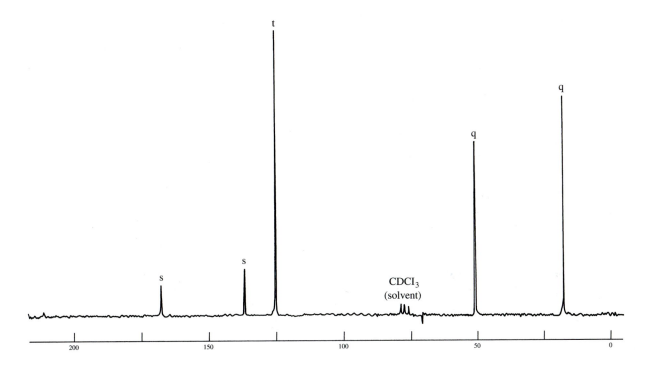

*5. Following are the ^1H and ^{13}C spectra for each of four isomeric bromoalkanes with formula C_4H_9Br. Assign a structure to each pair of spectra.

Carbon spectrum
A

C_4H_9Br

CDCl$_3$
(solvent)

Proton spectrum
A

C₄H₉Br

Carbon spectrum
B

C₄H₉Br

CDCl₃

Proton spectrum
B

C₄H₉Br

0.89   1.78 2.52   2.59

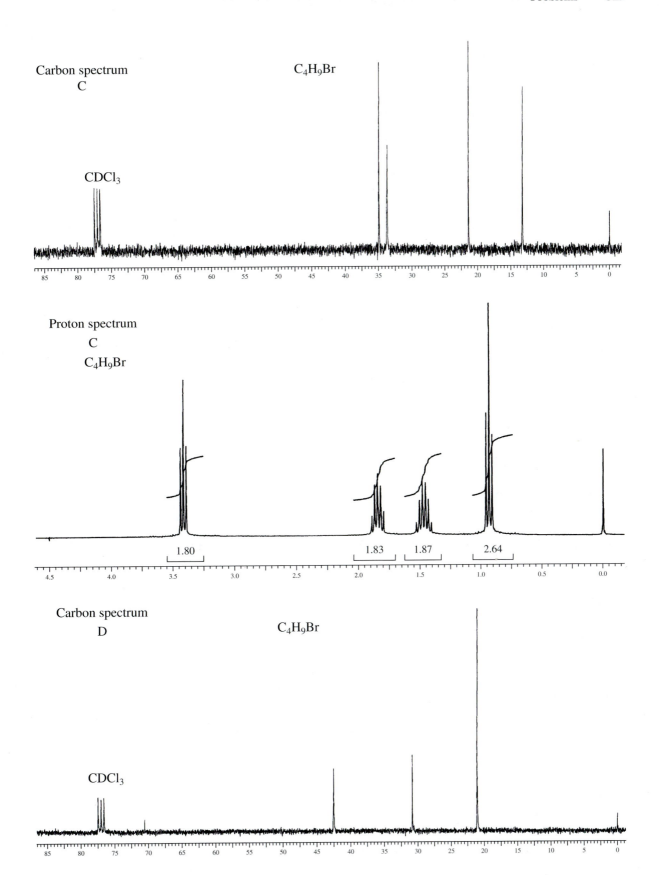

Carbon spectrum
C

C_4H_9Br

CDCl$_3$

Proton spectrum
C
C_4H_9Br

1.80

1.83 1.87

2.64

Carbon spectrum
D

C_4H_9Br

CDCl$_3$

Proton spectrum
D

C_4H_9Br

*6. Following are the ^1H and ^{13}C spectra for each of three isomeric ketones with formula $C_7H_{14}O$. Assign a structure to each pair of spectra.

Carbon spectrum
A

$C_7H_{14}O$

$CDCl_3$

Proton spectrum
A

$C_7H_{14}O$

1.96   2.00   2.91

Carbon spectrum
B

$C_7H_{14}O$

18.55

38.85

218.40

CDCl$_3$

Proton spectrum
B

$C_7H_{14}O$

Carbon spectrum
C

$C_7H_{14}O$

CDCl$_3$
(solvent)

Proton spectrum
C

$C_7H_{14}O$

1.95   2.98   8.91

3.0   2.5   2.0   1.5   1.0   0.5   0.0

7. The proton NMR spectrum for a compound with formula C_8H_{18} shows only one peak at 0.86 ppm. The carbon-13 NMR spectrum has two peaks, a large one at 26 ppm and a small one at 35 ppm. Draw the structure of this compound.

8. The proton NMR spectrum for a compound with formula $C_5H_{12}O_2$ is shown below. The normal carbon-13 NMR spectrum has three peaks. The DEPT-135 and DEPT-90 spectral results are tabulated. Draw the structure of this compound.

Normal Carbon	DEPT-135	DEPT-90
15 ppm	Positive	No peak
63	Negative	No peak
95	Negative	No peak

Proton spectrum

$C_5H_{12}O_2$

2.07   4.03   5.93

5.0   4.5   4.0   3.5   3.0   2.5   2.0   1.5   1.0   0.5   0.0

9. The proton NMR spectrum for a compound with formula $C_5H_{10}O$ is shown below. The normal carbon-13 NMR spectrum has three peaks. The DEPT-135 and DEPT-90 spectral results are tabulated. Draw the structure of this compound.

Normal Carbon	DEPT-135	DEPT-90
26 ppm	Positive	No peak
36	No peak	No peak
84	Negative	No peak

Proton spectrum

$C_5H_{10}O$

4.02

5.95

5.0   4.5   4.0   3.5   3.0   2.5   2.0   1.5   1.0   0.5   0.0

10. The proton NMR spectrum for a compound with formula $C_5H_{10}O_3$ is shown below. The normal carbon-13 NMR spectrum has four peaks. The infrared spectrum has a strong band at 1728 cm^{-1}. The DEPT-135 and DEPT-90 spectral results are tabulated. Draw the structure of this compound.

Normal Carbon	DEPT-135	DEPT-90
25 ppm	Positive	No peak
55	Positive	No peak
104	Positive	Positive
204	No peak	No peak

Proton spectrum

$C_5H_{10}O_3$

1.01   6.14   3.02

5.0   4.5   4.0   3.5   3.0   2.5   2.0   1.5   1.0   0.5   0.0

11. The proton NMR spectrum for a compound with formula C_9H_8O is shown below. The normal carbon-13 NMR spectrum has five peaks. The infrared spectrum has a strong band at 1746 cm^{-1}. The DEPT-135 and DEPT-90 spectral results are tabulated. Draw the structure of this compound.

Normal Carbon	DEPT-135	DEPT-90
44 ppm	Negative	No peak
125	Positive	Positive
127	Positive	Positive
138	No peak	No peak
215	No peak	No peak

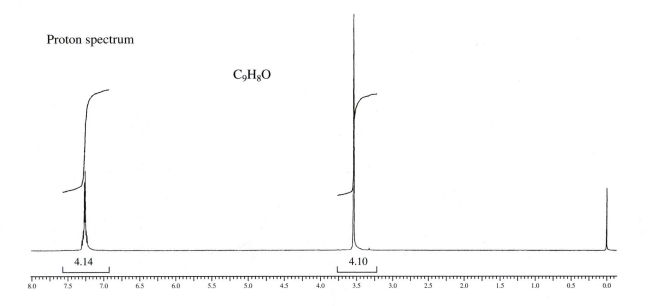

Proton spectrum

C_9H_8O

4.14   4.10

8.0   7.5   7.0   6.5   6.0   5.5   5.0   4.5   4.0   3.5   3.0   2.5   2.0   1.5   1.0   0.5   0.0

12. The proton NMR spectrum for a compound with formula $C_{10}H_{12}O_2$ is shown below. The infrared spectrum has a strong band at 1711 cm^{-1}. The normal carbon-13 NMR spectral results are tabulated along with the DEPT-135 and DEPT-90 information. Draw the structure of this compound.

Normal Carbon	DEPT-135	DEPT-90
29 ppm	Positive	No peak
50	Negative	No peak
55	Positive	No peak
114	Positive	Positive
126	No peak	No peak
130	Positive	Positive
159	No peak	No peak
207	No peak	No peak

Proton $C_{10}H_{12}O_2$

1.91 2.01   3.13 2.15   3.01

7.0   6.5   6.0   5.5   5.0   4.5   4.0   3.5   3.0   2.5   2.0   1.5   1.0   0.5   0.0

13. The proton NMR spectrum of a compound with formula $C_7H_{12}O_2$ is shown. The infrared spectrum displays a strong band at 1738 cm^{-1} and a weak band at 1689 cm^{-1}. The normal carbon-13 and the DEPT experimental results are tabulated. Draw the structure of this compound.

Normal Carbon	DEPT-135	DEPT-90
18 ppm	Positive	No peak
21	Positive	No peak
26	Positive	No peak
61	Negative	No peak
119	Positive	Positive
139	No peak	No peak
171	No peak	No peak

Proton spectrum

$C_7H_{12}O_2$

0.96   1.95   2.92   5.70

14. The proton NMR spectrum of a compound with formula $C_7H_{12}O_3$ is shown. The coupling constant for the triplet at 1.25 ppm is of the same magnitude as the one for the quartet at 4.15 ppm. The pair of distorted triplets at 2.56 and 2.75 ppm are coupled to each other. The infrared spectrum displays strong bands at 1720 and 1738 cm^{-1}. The normal carbon-13 and the DEPT experimental results are tabulated. Draw the structure of this compound.

Normal Carbon	DEPT-135	DEPT-90
14 ppm	Positive	No peak
28	Negative	No peak
30	Positive	No peak
38	Negative	No peak
61	Negative	No peak
173	No peak	No peak
207	No peak	No peak

Proton spectrum

$C_7H_{12}O_3$

2.06   2.07   2.07   2.97   2.97

4.0   3.5   3.0   2.5   2.0   1.5   1.0

15. The proton NMR spectrum of a compound with formula $C_5H_{10}O$ is shown. The normal carbon-13 and the DEPT experimental results are tabulated. The infrared spectrum shows a broad peak at about 3340 cm^{-1} and a medium-sized peak at about 1651 cm^{-1}. Draw the structure of this compound.

Normal Carbon	DEPT-135	DEPT-90
22.2 ppm	Positive	No peak
40.9	Negative	No peak
60.2	Negative	No peak
112.5	Negative	No peak
142.3	No peak	No peak

Proton spectrum $C_5H_{10}O$

2.11   2.13   1.97   1.22   2.99

5.5   5.0   4.5   4.0   3.5   3.0   2.5   2.0   1.5

16. The proton NMR spectrum is shown for a compound with formula $C_5H_9NO_4$. The infrared spectrum displays strong bands at 1750 and 1562 cm^{-1} and a medium-intensity band at 1320 cm^{-1}. The normal carbon-13 and the DEPT experimental results are tabulated. Draw the structure of this compound.

Normal Carbon	DEPT-135	DEPT-90
14 ppm	Positive	No peak
16	Positive	No peak
63	Negative	No peak
83	Positive	Positive
165	No peak	No peak

Proton spectrum

$C_5H_9NO_4$

0.92   2.01   3.00   3.00

5.5   5.0   4.5   4.0   3.5   3.0   2.5   2.0   1.5   1.0   0.5   0.0

17. The proton NMR spectrum of a compound with formula $C_6H_5NCl_2$ is shown. The normal carbon-13 and the DEPT experimental results are tabulated. The infrared spectrum shows peaks at 3432 and 3313 cm^{-1} and a series of medium-sized peaks between 1618 and 1466 cm^{-1}. Draw the structure of this compound.

Normal Carbon	DEPT-135	DEPT-90
118.0 ppm	Positive	Positive
119.5	No peak	No peak
128.0	Positive	Positive
140.0	No peak	No peak

Proton spectrum

$C_6H_5NCl_2$

1.96   0.99   2.13

*18. The following alcohol undergoes elimination in the presence of concentrated sulfuric acid, but the product shown is not its chief product. Instead, another isomeric six-carbon alkene forms. This product shows a large peak at 20.4 ppm and a smaller one at 123.4 ppm in its proton-decoupled ^{13}C NMR spectrum. Draw the structure of the product and interpret the spectrum. Outline a mechanism for the formation of the product that possesses this spectrum.

$$CH_3-CH-CH-CH_2-OH \xrightarrow{H_2SO_4} CH_3-CH-C=CH_2 + H_2O$$

(with CH_3 substituents on the CH carbons and on the C as drawn)

*19. Predict the appearances of the proton-decoupled ^{13}C spectra for the following compounds:

(a)

Cl—C—Cl   Cl—C—Cl
(with H, D substituents and D, D substituents respectively)

$I = 1$

$J_{CD} \cong 20\text{–}30$ Hz (one bond)

(b)

F—C—H   F—C—H   F—C—CH$_2$—Cl   F—C—CH$_2$—Cl
(with H,H; F,F; H,F; F,F substituents respectively)

$I = \frac{1}{2}$

$J_{CF} > 180$ Hz (one bond)

$J_{CF} \cong 40$ Hz (two bonds)

20. The proton-decoupled ^{13}C spectrum (75 MHz) for 1-fluoropentane is shown in this problem. Assign each of the carbon atoms after reviewing Section 6.15. Calculate the coupling constants for the doublets at 27, 30, and 84 ppm, respectively. Indicate the coupling constant notation (J values) for each of these patterns (^{1}J, ^{2}J, and ^{3}J). Can you explain why you do not observe ^{4}J and ^{5}J with this compound? The three-peak pattern centering on 77 ppm is the coupling of ^{13}C to deuterium. Remember that $CDCl_3$ is the solvent (see Section 6.14).

21. The proton-decoupled ^{13}C spectrum (75 MHz) for 1,1,1-trifluoroacetone (1,1,1-trifluoro-2-propanone) is shown in this problem. The quartet appearing at 188.6 ppm is expanded and shown as an inset. Assign each of the carbon atoms. Calculate the coupling constants for the quartets at 188.6 and 115.0 ppm. Indicate the coupling constant notation (J values) for each of these patterns (^{1}J and ^{2}J). Can you explain why you do not observe ^{3}J with this compound? The three-peak pattern centering on 77 ppm is the coupling of ^{13}C to deuterium. Remember that CDCl$_3$ is the solvent (see Section 6.14).

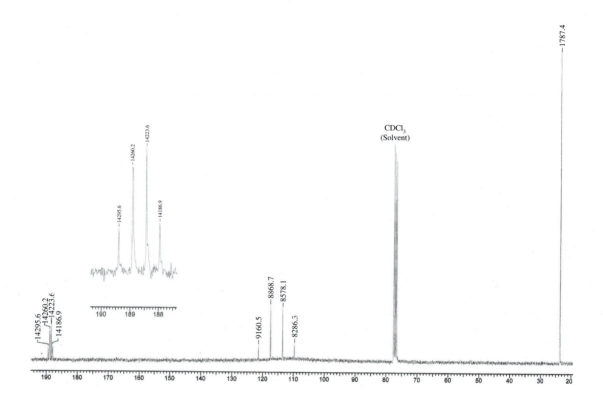

22. The proton-decoupled ^{13}C spectrum (75 MHz) for tetraethylphosphonium chloride is shown in this problem. Assign each of the carbon atoms after reviewing Section 6.16. Calculate the coupling constants for the doublets at 11.7 and 6.0 ppm. Indicate the coupling constant notations (J values) for each of these patterns (^1J and ^2J).

*23. Figure 6.15 is the ^{13}C NMR spectrum of toluene. We indicated in Section 6.13 that it was difficult to assign the **c** and **d** carbons to peaks in this spectrum. Using Table 7 in Appendix 8, calculate the expected chemical shifts of all the carbons in toluene and assign all of the peaks.

*24. Using the tables in Appendix 8, calculate the expected carbon-13 chemical shifts for the indicated carbon atoms in the following compounds:

(a)

(b)

(c)

(d)

All   All CH_3 All

(e)

(f)

(g) $C_6H_5-CH=CH-CH_3$

(h)

All

(i)

(j)

$CH_3CH_2CH_2CH=CH-CH_2CH_2CH_3$

(k)

NH_2 Ring carbons

(l) $CH_3CH_2CH_2-C\equiv C-H$

(m)

(n)

(o)

All

(p)

(q)

NO₂   All

All

(r)

(s)

(t)

REFERENCES

Textbooks

Berger, S., and & S. Braun, *200 and More NMR Experiments,* Wiley-VCH, Weinheim, 2004.

Crews, P., J. Rodriguez, and M. Jaspars, *Organic Structure Analysis*; Oxford University Press, New York, 2009.

Friebolin, H., *Basic One- and Two-Dimensional NMR Spectroscopy*, 5th ed., Wiley-VCH Publishers, New York, 2010.

Gunther, H., *NMR Spectroscopy*, 2nd ed., John Wiley and Sons, New York, 1995.

Lambert, J. B., S. Groenert, H. F. Shurvell, D. A. Lightner, and R. G. Cooks, *Organic Structural Spectroscopy*, Prentice Hall, Upper Saddle River, NJ, 2010.

Levy, G. C., *Topics in Carbon-13 Spectroscopy*, John Wiley and Sons, New York, 1984.

Levy, G. C., R. L. Lichter, and G. L. Nelson, *Carbon-13 Nuclear Magnetic Resonance Spectroscopy*, 2nd ed., John Wiley and Sons, New York, 1980.

Levy, G. C., and G. L. Nelson, *Carbon-13 Nuclear Magnetic Resonance for Organic Chemists*, John Wiley and Sons, New York, 1979.

Macomber, R. S., *NMR Spectroscopy—Essential Theory and Practice*, College Outline Series, Harcourt, Brace Jovanovich, New York, 1988.

Macomber, R. S., *A Complete Introduction to Modern NMR Spectroscopy*, John Wiley and Sons, New York, 1997.

Pretsch, E., P. Buhlmann, and H. Badertscher, *Structure Determination of Organic Compounds. Tables of Spectral Data*, 4th ed., Springer-Verlag, Berlin, 2009.

Sanders, J. K. M., and B. K. Hunter, *Modern NMR Spectroscopy—A Guide for Chemists*, 2nd ed., Oxford University Press, Oxford, England, 1993.

Silverstein, R. M., F. X. Webster, and D. Kiemle, *Spectrometric Identification of Organic Compounds*, 7th ed., John Wiley and Sons, New York, 2005.

Compilations of Spectra

Ault, A., and M. R. Ault, *A Handy and Systematic Catalog of NMR Spectra*, 60 MHz with some 270 MHz, University Science Books, Mill Valley, CA, 1980.

Fuchs, P. L., *Carbon-13 NMR Based Organic Spectral Problems*, 25 MHz, John Wiley and Sons, New York, 1979.

Johnson, L. F., and W. C. Jankowski, *Carbon-13 NMR Spectra: A Collection of Assigned, Coded, and Indexed Spectra*, 25 MHz, Wiley–Interscience, New York, 1972.

Pouchert, C. J., and J. Behnke, *The Aldrich Library of ^{13}C and ^{1}H FT-NMR Spectra,* 75 and 300 MHz, Aldrich Chemical Company, Milwaukee, WI, 1993.

Computer Estimation of Carbon-13 Chemical Shifts

"^{13}C NMR Estimation," CS ChemDraw Ultra, Perkin-Elmer- Cambridge Soft Corp., 100 Cambridge Park Drive, Cambridge, MA 02140.

"Carbon 13 NMR Shift Prediction Module" requires ChemWindow (IBM PC) or ChemIntosh (Macintosh), SoftShell International, Ltd., 715 Horizon Drive, Grand Junction, CO 81506.

"ChemDraw Ultra," Perkin-Elmer - Cambridge Soft Corp., 100 Cambridge Park Drive, Cambridge, MA 02140, www.cambridgesoft .com

"HyperNMR," IBM PC/Windows, Hypercube, Inc., 419 Phillip Street, Waterloo, Ontario, Canada N2L 3X2.

Selected Websites

http://sdbs.riodb.aist.go.jp

This excellent free website presents an integrated Spectral DataBase System (SDBS) for Organic Compounds, National Institute of Advanced Industrial Science and Technology (AIST), Tsukuba, Ibaraki 305-8565, Japan. This database includes infrared, mass spectra, and NMR data (proton and carbon-13) for a number of compounds.

http://www.chem.ucla.edu/~webspectra

UCLA Department of Chemistry and Biochemistry, in connection with Cambridge University Isotope Laboratories, presents WebSpectra an outstanding collection of unknowns for students. The website provides NMR and some IR spectra. No mass spectral data is included. Solutions are provided. The website also provides links to other sites with problems for students to solve.

http://www.nd.edu/~smithgrp/structure/workbook.html

Combined structure problems provided by the Smith group at Notre Dame University.

NUCLEAR MAGNETIC RESONANCE SPECTROSCOPY

Part Three: Spin–Spin Coupling

C hapters 5 and 6 covered only the most essential elements of nuclear magnetic resonance (NMR) theory. Now we will consider applications of the basic concepts to more complicated situations. In this chapter, the emphasis is on the origin of coupling constants and what information can be deduced from them. Enantiotopic and diastereotopic systems will be covered as well as more advanced instances of spin–spin coupling, such as second-order spectra.

7.1 COUPLING CONSTANTS: SYMBOLS

Chapter 5, Sections 5.17 and 5.18, introduced coupling constants. For simple multiplets, the coupling constant J is easily determined by measuring the spacing (in Hertz) between the individual peaks of the multiplet. This coupling constant has the same value regardless of the field strength or operating frequency of the NMR spectrometer. J is a *constant*.[1]

Coupling between two nuclei of the same type is called **homonuclear coupling.** Chapter 5 examined the homonuclear three-bond couplings between hydrogens on adjacent carbon atoms (vicinal coupling, Section 7.2C), which gave multiplets governed by the $n + 1$ Rule. Coupling between two different types of nuclei is called **heteronuclear coupling**. The couplings between ^{13}C and attached hydrogens are one-bond heteronuclear couplings (Section 7.2A).

The magnitude of the coupling constant depends to a large extent on the number of bonds intervening between the two atoms or groups of atoms that interact. Other factors also influence the strength of interaction between two nuclei, but in general, one-bond couplings are larger than two-bond couplings, which in turn are larger than three-bond couplings, and so forth. Consequently, the symbols used to represent coupling are often extended to include additional information about the type of atoms involved and the number of bonds through which the coupling constant operates.

We frequently add a superscript to the symbol J to indicate the number of bonds through which the interaction occurs. If the identity of the two nuclei involved is not obvious, we add this information in parentheses. Thus, the symbol

$$^1J\,(^{13}C-{}^1H) = 156 \text{ Hz}$$

indicates a one-bond coupling between a carbon-13 atom and a hydrogen atom (C—H) with a value of 156 Hz. The symbol

$$^3J\,(^1H-{}^1H) = 8 \text{ Hz}$$

---

[1]We will see, however, the magnitude of J is dependent on the bond angles between the interacting nuclei and can therefore vary with temperature or solvent, to the extent these influence the conformation of the compound.

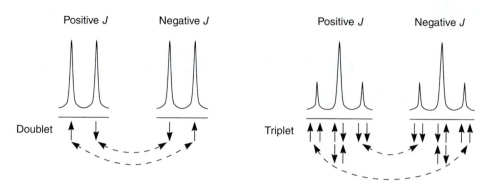

FIGURE 7.1 The dependence of multiplet assignments on the sign of J, the coupling constant.

indicates a three-bond coupling between two hydrogen atoms, as in H−C−C−H. Subscripts may also be used to give additional information. $J_{1,3}$, for instance, indicates a coupling between atoms 1 and 3 in a structure or between protons attached to carbons 1 and 3 in a structure. J_{CH} or J_{HH} clearly indicates the types of atoms involved in the coupling interaction. The different coupling constants in a molecule might be designated simply as J_1, J_2, J_3, and so forth. Expect to see many variants in the usage of J symbols.

Although it makes no difference to the gross appearance of a spectrum, some coupling constants are positive, and others are negative. With a negative J, the meanings of the individual lines in a multiplet are reversed—the upfield and downfield peaks exchange places—as shown in Figure 7.1. In the simple measurement of a coupling constant from a spectrum, it is impossible to tell whether the constant is positive or negative. Therefore, a measured value should always be regarded as the *absolute value* of J ($|J|$).

7.2 COUPLING CONSTANTS: THE MECHANISM OF COUPLING

A physical picture of spin–spin coupling, the way in which the spin of one nucleus influences the spin of another, is not easy to develop. Several theoretical models are available. The best theories we have are based on the Dirac vector model. This model has limitations, but it is fairly easy for the novice to understand, and its predictions are substantially correct. According to the Dirac model, the electrons in the intervening bonds between the two nuclei transfer spin information from one nucleus to another by means of interaction between the nuclear and electronic spins. An electron near the nucleus is assumed to have the lowest energy of interaction with the nucleus when the spin of the electron (small arrow) has its spin direction opposed to (or "paired" with) that of the nucleus (heavy arrow).

Spins of nucleus and electron paired
or opposed (lower energy)

Spins of nucleus and electron parallel
(higher energy)

This picture makes it easy to understand why the size of the coupling constant diminishes as the number of intervening bonds increases. As we will see, it also explains why some coupling constants are negative while others are positive. Theory shows that couplings involving an odd number of intervening bonds (^1J, ^3J, . . .) are expected to be positive, while those involving an even number of intervening bonds (^2J, ^4J, . . .) are expected to be negative.

A. One-Bond Couplings (^1J)

A one-bond coupling occurs when a single bond links two spin-active nuclei. The bonding electrons in a single bond are assumed to avoid each other such that when one electron is near nucleus A, the other is near nucleus B. According to the Pauli Principle, pairs of electrons in the same orbital have opposed spins; therefore, the Dirac model predicts that the most stable condition in a bond is when both nuclei have opposed spins. Following is an illustration of a ^{13}C–^{1}H bond; the nucleus of the ^{13}C atom (heavy solid arrow) has a spin opposite to that of the hydrogen nucleus (heavy open arrow). The alignments shown would be typical for a ^{13}C–^{1}H bond or for any other type of bond in which both nuclei have spin (for instance, ^{1}H–^{1}H or ^{31}P–H).

$$^{13}\text{C} \!-\! \text{H}$$

Notice that in this arrangement the two nuclei prefer to have opposite spins. When two spin-active nuclei prefer an opposed alignment (have opposite spins), the coupling constant J is usually positive. If the nuclei are parallel or aligned (have the same spin), J is usually negative. Thus, most one-bond couplings have positive J values. Keep in mind, however, that there are some prominent exceptions, such as ^{13}C–^{19}F, for which the coupling constants are negative (see Table 7.1).

It is not unusual for coupling constants to depend on the hybridization of the atoms involved. ^1J values for ^{13}C–^{1}H coupling constants vary with the amount of s character in the carbon hybrid, according to the following relationship:

$$^1J_{\text{CH}} = (500 \text{ Hz}) \left(\frac{1}{n+1} \right) \text{ for hybridization type } sp^n \qquad \textbf{Equation 7.1}$$

Notice the specific values given for the ^{13}C–^{1}H couplings of ethane, ethene, and ethyne in Table 7.1.

Using the Dirac nuclear–electronic spin model, we can also develop an explanation for the origin of the multiplets that are the results of spin–spin coupling. As a simple example, consider a ^{13}C–^{1}H bond. Recall that a ^{13}C atom that has one hydrogen attached appears as a doublet (two peaks) in a proton-coupled ^{13}C NMR spectrum (Section 6.3 and Fig. 6.3). There are two lines (peaks) in the ^{13}C NMR spectrum because the hydrogen nucleus can have two spins ($+1/2$ or $-1/2$), leading to two different energy transitions for the ^{13}C nucleus. Figure 7.2 illustrates these two situations.

TABLE 7.1
SOME ONE-BOND COUPLING CONSTANTS (^1J)

^{13}C–^{1}H	110 –270 Hz
	sp^3 115–125 Hz (ethane = 125 Hz)
	sp^2 150–170 Hz (ethene = 156 Hz)
	sp 240–270 Hz (ethyne = 249 Hz)
^{13}C–^{19}F	−165 to −370 Hz
^{13}C–^{31}P	48–56 Hz
^{13}C–D	20 –30 Hz
^{31}P–^{1}H	190 –700 Hz

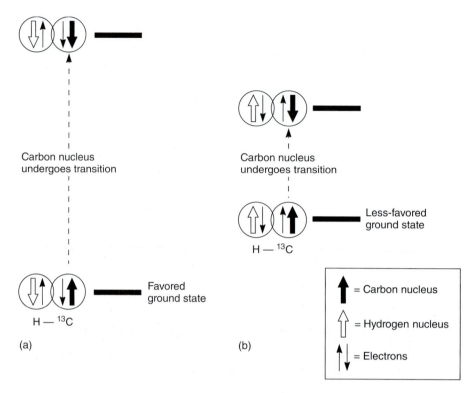

FIGURE 7.2 The two different energy transitions for a ^{13}C nucleus in a C—H bond. (a) The favored ground state (all spins paired); (b) the less-favored ground state (impossible to pair all spins).

At the bottom of Figure 7.2a is the favored ground state for the $^{13}C–^1H$ bond. In this arrangement, the carbon nucleus is in its lowest energy state [spin (^1H) = +1/2], and all of the spins, both nuclear and electronic, are paired, resulting in the lowest energy for the system. The spin of the nucleus of the hydrogen atom is opposed to the spin of the ^{13}C nucleus. A higher energy results for the system if the spin of the hydrogen is reversed [spin (^1H) = –1/2]. This less-favored ground state is shown at the bottom of Figure 7.2b.

Now, assume that the carbon nucleus undergoes transition and inverts its spin. The excited state that results from the less-favored ground state (seen at the top of Fig. 7.2b) turns out to have a lower energy than the one resulting from the favored ground state (top of Fig. 7.2a) because all of its nuclear and electronic spins are paired. Thus, we see two different transitions for the ^{13}C nucleus [spin(^{13}C) = +1/2], depending on the spin of the attached hydrogen. As a result, in a proton-coupled NMR spectrum a doublet is observed for a methine carbon ($^{13}C–^1H$).

B. Two-Bond Couplings (^2J)

Two-bond couplings are quite common in NMR spectra. They are usually called **geminal couplings** because the two nuclei that interact are attached to the same central atom (Latin *gemini* = "twins"). Two-bond coupling constants are abbreviated ^2J. They occur in carbon compounds whenever two or more spin-active atoms are attached to the same carbon atom. Table 7.2 lists some two-bond coupling constants that involve carbon as the central atom. Two-bond coupling constants are typically, although not always, smaller in magnitude than one-bond couplings (Table 7.2). Notice that the most common type of two-bond coupling, HCH, is frequently (but not always) negative.

TABLE 7.2
SOME TWO-BOND COUPLING CONSTANTS (^2J)

C(H)(H)	−9 to −15 Hz	C(H)(^{19}F)	~50 Hz[a]
=C(H)(H)	0 to 2 Hz	C(H)(^{13}C)	~5 Hz[a]
C(H)(D)	~2 Hz[a]	C(H)(^{31}P)	7 – 14 Hz[a]
C(^{19}F)(^{19}F)	~160 Hz[a]		

[a]*Absolute values.*

The mechanistic picture for geminal coupling (^2J) invokes nuclear–electronic spin coupling as a means of transmitting spin information from one nucleus to the other. It is consistent with the Dirac model that we discussed at the beginning of Section 7.2 and in Section 7.2A. Figure 7.3 shows this mechanism. In this case, another atom (without spin) intervenes between two interacting orbitals. When this happens, theory predicts that the interacting electrons, and hence the nuclei, prefer to have parallel spins, resulting in a negative coupling constant. The preferred alignment is shown on the left side of Figure 7.3.

The amount of geminal coupling depends on the HCH angle α. The graph in Figure 7.4 shows this dependence, where the amount of *electronic* interaction between the two C—H orbitals determines the magnitude of the coupling constant ^2J. In general, ^2J geminal coupling constants increase as the angle α decreases. As the angle α decreases, the two orbitals shown in Figure 7.3 move closer, and the electron spin correlations become greater. Note, however, that the graph in Figure 7.4 is very approximate, showing only the general trend; actual values vary quite widely.

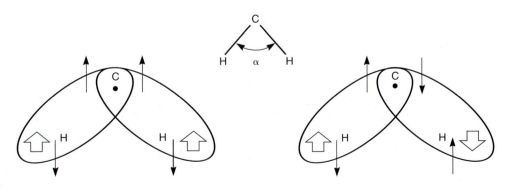

FIGURE 7.3   The mechanism of geminal coupling.

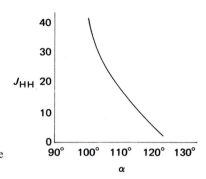

FIGURE 7.4 The dependence of the magnitude of $^2J_{HCH}$, the geminal coupling constant, on the HCH bond angle α.

Following are some systems that show geminal coupling, along with their approximate HCH bond angles. Notice that the coupling constants become smaller, as predicted, when the HCH angle becomes larger. Note also that even small changes in bond angles resulting from stereochemical changes influence the geminal coupling constant.

$\alpha \cong 109°$
$^2J_{HH} \cong 12-18$ Hz

$\alpha \cong 118°$
$^2J_{HH} \cong 5$ Hz

$\alpha \cong 120°$
$^2J_{HH} \cong 0-3$ Hz

$\alpha = 107°$
$^2J_{HH} = 17.5$ Hz

$\alpha = 108°$
$^2J_{HH} = 15.5$ Hz

Table 7.3 shows a larger range of variation, with approximate values for a selected series of cyclic compounds and alkenes. Notice that as ring size decreases, the absolute value of the coupling constant ^2J also decreases. Compare, for instance, cyclohexane, where ^2J is -13, and cyclopropane, where ^2J is -4.

TABLE 7.3
VARIATIONS IN $^2J_{HH}$ WITH HYBRIDIZATION AND RING SIZE

+2	−2	−4	−9	−11	−13	−9 to −15 Hz

As the angle CCC in the ring becomes smaller (as p character increases), the complementary HCH angle grows larger (s character increases), and consequently the geminal coupling constant decreases. Note that hybridization is important, and that the sign of the coupling constant for alkenes changes to positive, except where they have an electronegative element attached.

Plane of symmetry—
no splitting

Free rotation—
no splitting

Geminal coupling between nonequivalent protons is readily observed in the ^{1}H NMR spectrum, and the magnitude of the coupling constant ^2J is easily measured from the line spacings when the resonances are first order (see Sections 7.6 and 7.7). In second-order spectra, the value of ^2J cannot be directly measured from the spectrum but may be determined by computational methods (spectral simulation). In many cases, however, no geminal HCH coupling (no spin–spin splitting) is observed because the geminal protons are magnetically equivalent (see Section 7.3). You have already seen in our discussions of the $n + 1$ Rule that in a hydrocarbon chain the protons attached to the same carbon may be treated as a group and do not split one another. How, then, can it be said that coupling exists in such cases if no spin–spin splitting is observed in the spectrum? The answer comes from deuterium substitution experiments. If one of the hydrogens in a compound that shows no spin–spin splitting is replaced by a deuterium, geminal splitting with deuterium ($I = 1$) is observed. Since deuterium and hydrogen are electronically the same atom (they differ only by a neutron), it can be assumed that if there is interaction for HCD there is also interaction for HCH. The HCH and HCD coupling constants are related by the magnetogyric ratios of hydrogen and deuterium:

$$^2J_{HH} = \gamma H/\gamma D\ (^2J_{HD}) = 6.51(^2J_{HD}) \qquad \text{Equation 7.2}$$

In the following sections of this chapter, whenever coupling constant values are given for seemingly equivalent protons (excluding cases of magnetic inequivalence, see Section 7.3), the coupling values were derived from spectra of deuterium-labeled isomers.

C. Three-Bond Couplings (^3J)

In a typical hydrocarbon, the spin of a hydrogen nucleus in one C—H bond is coupled to the spins of those hydrogens in adjacent C—H bonds. These H—C—C—H couplings are usually called **vicinal couplings** because the hydrogens are on neighboring carbon atoms (Latin *vicinus* = "neighbor"). Vicinal couplings are three-bond couplings and have a coupling constant abbreviated ^3J. In Sections 5.13 through 5.17, you saw that these couplings produce spin–spin splitting patterns that follow the $n + 1$ Rule in simple aliphatic hydrocarbon chains.

Three-bond vicinal coupling

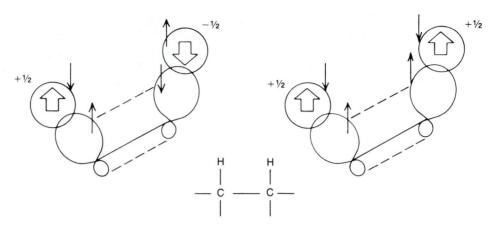

FIGURE 7.5 The method of transferring spin information between two adjacent C—H bonds.

Once again, nuclear and electronic spin interactions carry the spin information from one hydrogen to its neighbor. Since the σ C—C bond is nearly orthogonal (perpendicular) to the σ C—H bonds, there is no overlap between the orbitals, and the electrons cannot interact strongly through the sigma bond system. According to theory, they transfer the nuclear spin information via the small amount of parallel orbital *overlap* that exists between adjacent C—H bond orbitals. The spin interaction between the electrons in the two adjacent C—H bonds is the major factor determining the size of the coupling constant.

Figure 7.5 illustrates the two possible arrangements of nuclear and electronic spins for two coupled protons that are on adjacent carbon atoms. Recall that the carbon nuclei (^{12}C) have zero spin. The drawing on the left side of the figure, where the spins of the hydrogen nuclei are paired and where the spins of the electrons that are interacting through orbital overlap are also paired, is expected to represent the lowest energy and have the favored interactions. Because the interacting nuclei are spin paired in the favored arrangement, three-bond H—C—C—H couplings are expected to be positive. In fact, most three-bond couplings, regardless of atom types, are found to be positive.

That our current picture of three-bond vicinal coupling is substantially correct can be seen best in the effect of the dihedral angle between adjacent C—H bonds on the magnitude of the spin inter-action. Recall that two nonequivalent adjacent protons give rise to a pair of doublets, each proton splitting the other.

The parameter $^3J_{HH}$, the vicinal coupling constant, measures the magnitude of the splitting and is equal to the separation in Hertz between the multiplet peaks. The actual magnitude of the coupling constant between two adjacent C—H bonds can be shown to depend directly on the dihedral angle α between these two bonds. Figure 7.6 defines the dihedral angle α as a perspective drawing and a Newman diagram.

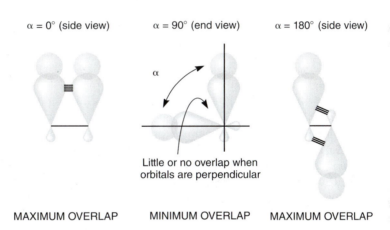

FIGURE 7.6 The definition of a dihedral angle α.

Side View

End View

The magnitude of the splitting between H_A and H_B is greatest when $\alpha = 0°$ or $180°$ and is smallest when $\alpha = 90°$. The side-to-side overlap of the two C—H bond orbitals is at a maximum at $0°$, where the C—H bond orbitals are parallel, and at a minimum at $90°$, where they are perpendicular. At $\alpha = 180°$, overlap with the back lobes of the sp^3 orbitals occurs.

$\alpha = 0°$ (side view) $\alpha = 90°$ (end view) $\alpha = 180°$ (side view)

α

Little or no overlap when
orbitals are perpendicular

MAXIMUM OVERLAP   MINIMUM OVERLAP   MAXIMUM OVERLAP

Martin Karplus was the first to study the variation of the coupling constant $^3J_{HH}$ with the dihedral angle α and developed an equation (Eq. 7.3) that gave a good fit to the experimental data shown in the graph in Figure 7.7. The **Karplus relationship** takes the form

$$^3J_{HH} = A + B \cos \alpha + C \cos 2\alpha$$

$$A = 7 \qquad B = -1 \qquad C = 5 \qquad \qquad \textbf{Equation 7.3}$$

Many subsequent workers have modified this equation—particularly its set of constants, A, B, and C—and several different forms of it are found in the literature. The constants shown are accepted as those that give the best general predictions. Note, however, that actual experimental data exhibit a wide range of variation, as shown by the shaded area of the curve (sometimes called the **Karplus curve**) in Figure 7.7.

The Karplus relationship makes perfect sense according to the Dirac model. When the two adjacent C—H σ bonds are orthogonal ($\alpha = 90°$, perpendicular), there should be minimal orbital overlap, with little or no spin interaction between the electrons in these orbitals. As a result, nuclear spin information is not transmitted, and $^3J_{HH} \cong 0$. Conversely, when these two bonds are parallel ($\alpha = 0°$) or antiparallel ($\alpha = 180°$), the coupling constant should have its greatest magnitude ($^3J_{HH} = $ max).

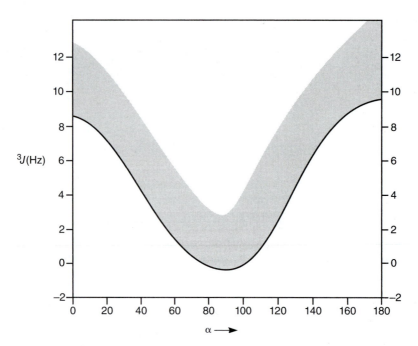

FIGURE 7.7 The Karplus relationship—the approximate variation of the coupling constant ^3J with the dihedral angle α.

The variation of $^3J_{HH}$ indicated by the shaded area in Figure 7.7 is a result of factors other than the dihedral angle α. These factors (Fig. 7.8) include the bond length R_{CC}, the valence angles θ_1 and θ_2, and the electronegativity of any substituents X attached to the carbon atoms.

In any hydrocarbon, the magnitude of interaction between any two adjacent C—H bonds is always close to the values given in Figure 7.7. Cyclohexane derivatives that are conformationally biased are the best illustrative examples of this principle. In the following molecule, the ring is substantially biased to favor the conformation with the bulky *tert*-butyl group in an equatorial position. The coupling constant between two axial hydrogens J_{aa} is normally 10 to 14 Hz ($\alpha = 180°$), whereas the magnitude of interaction between an axial hydrogen and an equatorial hydrogen J_{ae} is generally 2 to 6 Hz ($\alpha = 60°$). A diequatorial interaction also has $J_{ee} = 2$ to 5 Hz ($\alpha = 60°$), but the equatorial-equatorial vicinal coupling constant (J_{ee}) is usually about 1 Hz smaller than the axial-equatorial vicinal coupling constant (J_{ae}) in the same ring system. For cyclohexane derivatives that have more than one solution conformation at room temperature, the observed

dihedral angle   bond length   valence angles   electronegative substituents

FIGURE 7.8 Factors influencing the magnitude of $^3J_{HH}$.

FIGURE 7.9 Vicinal couplings in cyclohexane derivatives.

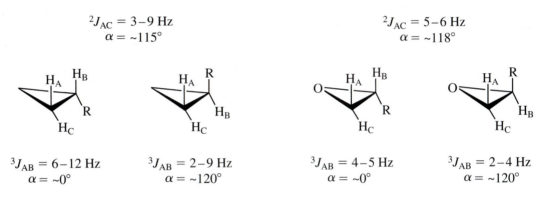

For three-membered rings, $J_{cis} > J_{trans}$

FIGURE 7.10 Vicinal couplings in three-membered ring derivatives.

coupling constants will be the weighted average of the coupling constants for each individual conformation (Fig. 7.9). Cyclopropane derivatives and epoxides are examples of conformationally rigid systems. Notice that J_{cis} ($\alpha = 0°$) is larger than J_{trans} ($\alpha = 120°$) in three-membered rings (Fig. 7.10).

Table 7.4 lists some representative three-bond coupling constants. Notice that in the alkenes the *trans* coupling constant is always larger than the *cis* coupling constant. Spin–spin coupling in

TABLE 7.4
SOME THREE-BOND COUPLING CONSTANTS ($^3J_{XY}$)

H–C–C–H	6–8 Hz	H–C=C–H	*cis*	6–15 Hz
			trans	11–18 Hz
^{13}C–C–C–H	5 Hz	H–C=C–^{19}F	*cis*	18 Hz
			trans	40 Hz
^{19}F–C–C–H	5–20 Hz	^{19}F–C=C–^{19}F	*cis*	30–40 Hz
			trans	–120 Hz
^{19}F–C–C–^{19}F	–3 to –20			
^{31}P–C–C–H	13 Hz			
^{31}P–O–C–H	5–15 Hz			

TABLE 7.5
VARIATION OF $^3J_{HH}$ WITH VALENCE ANGLES IN CYCLIC ALKENES (Hz)

0–2	2–4	5–7	8–11	6–15

alkenes will be discussed in further detail in Sections 7.8 and 7.9. In Table 7.5, an interesting variation is seen with ring size in cyclic alkenes. Larger HCH valence angles in the smaller ring sizes result in smaller coupling constants ($^3J_{HH}$).

D. Long-Range Couplings (^4J–^nJ)

As discussed above, proton–proton coupling is normally observed between protons on *adjacent* atoms (vicinal coupling) and is sometimes observed between protons on the *same* atom (geminal coupling), provided the protons in question are nonequivalent. Only under special circumstances does coupling occur between protons that are separated by four or more covalent bonds, and these are collectively referred to as **long-range couplings.** Long-range couplings are common in allylic systems, aromatic rings, and rigid bicyclic systems. Long-range coupling in aromatic systems will be covered in Section 7.10.

Long-range couplings are communicated through specific overlap of a series of orbitals and as a result have a stereochemical requirement. In alkenes, small couplings between the alkenyl hydrogens and protons on the carbon(s) α to the opposite end of the double bond are observed:

$$|^4J_{ad}| = 0\text{ - }3\text{ Hz}$$
$$|^4J_{bd}| = 0\text{ - }3\text{ Hz}$$

This four-bond coupling (^4J) is called **allylic coupling.** The π electrons of the double bond help transmit the spin information from one nucleus to the other, as shown in Figure 7.11. When the

^4J maximum

C—H σ orbital
orthogonal to π orbital

^4J minimum

C—H σ orbital
parallel to π orbital

FIGURE 7.11 Geometric arrangements that maximize and minimize allylic coupling.

$^4J = 2$ to 4 Hz

FIGURE 7.12 Propargylic coupling.

allylic C—H bond is aligned with the plane of the C—C π bond, there is maximum overlap between the allylic C—H σ orbital and the π orbital, and the allylic coupling interaction assumes the maximum value ($^4J = 3$–4 Hz). When the allylic C—H bond is perpendicular to the C—C π bond, there is *minimum* overlap between the allylic C—H σ orbital and the π orbital, and the allylic coupling is very small ($^4J = \sim 0$ Hz). At intermediate conformations, there is partial overlap of the allylic C—H bond with the π orbital, and intermediate values for ^4J are observed.

In alkenes, the magnitude of allylic coupling (^4J) depends on the extent of overlap of the carbon–hydrogen σ bond with the π bond. A similar type of interaction occurs in alkynes, but with an important difference. In the case of **propargylic coupling** (Fig. 7.12), a C—H σ orbital on the carbon α to the triple bond *always* has partial overlap with the alkyne π system because the triple bond consists of two perpendicular π bonds, effectively creating a cylinder of electron density surrounding the C—C internuclear axis.

In some alkenes, coupling can occur between the C—H σ bonds on either side of the double bond. This **homoallylic coupling** occurs over five bonds (^5J) but is naturally weaker than allylic coupling (^4J) because it occurs over a greater distance. Homoallylic coupling is generally not observed except when *both* C—H σ bonds on either side of the double bond are parallel to the π orbital of the double bond *simultaneously* (Fig. 7.13). This is common when two allylic methyl groups are interacting because of the inherent threefold symmetry of the CH$_3$ group—one of the C—H σ bonds will be partially overlapped with the alkene π bond at all times. For larger or branched alkene substituents, however, the conformations that allow such overlap suffer from significant steric strain (A^{1,3} strain)

FIGURE 7.13 Homoallylic coupling in alkenes and allenes.

and are unlikely to be a significant contribution to the solution structure of such compounds, unless other, more significant, constraints are present, such as rings or steric congestion elsewhere in the molecule. For example, 1,4-cyclohexadiene and 6-methyl-3,4-dihydro-2*H*-pyran both have fairly large homoallylic couplings (^5J, Fig. 7.13). Allenes are also effective at communicating spin–spin splitting over long distances in a type of homoallylic coupling. An example is 1,1-dimethylallene, where $^5J = 3$ Hz (Fig. 7.13).

Unlike the situation for homoallylic coupling in most acyclic alkenes, **homopropargylic coupling** is almost always observed in the ^{1}H NMR spectra of internal alkynes. As we saw above, essentially all conformations of the C—H σ bond on the carbon α to the triple bond allow for partial overlap with the π system of the alkyne, resulting in coupling constants significantly larger than those observed for homoallylic coupling (Fig. 7.14). In conjugated enyne compounds, ^6J is often observed, a result of combination homoallylic/propargylic coupling.

Long-range couplings in compounds without π systems are less common but do occur in special cases. One case of long-range coupling in saturated systems occurs through a rigid arrangement of bonds in the form of a W (^4J), with hydrogens occupying the end positions. Two possible types of orbital overlap have been suggested to explain this type of coupling (Fig. 7.15). The magnitude of ^4J for **W coupling** is usually small except in highly strained ring systems in which the rigid structures reinforce the favorable geometry for the overlaps involved (Fig. 7.16).

In other systems, the magnitude of ^4J is often less than 1 Hz and is not resolved even on high-field spectrometers. Peaks that have spacings less than the resolving capabilities of the spectrometer are usually broadened; that is, two lines very close to each other appear as a single "fat," or broad, peak. Many W couplings are of this type, and small allylic couplings ($^4J < 1$ Hz) can also give rise to peak broadening rather than discrete splitting. Angular methyl groups in steroids and those at the ring junctions in *trans*-decalin systems often exhibit peak broadening due to W coupling with several hydrogens of the ring (Fig. 7.17). Because these systems are relatively unstrained, $^4J_{\rm w}$ is usually quite small.

$$|^5J| = 2 \text{ to } 3 \text{ Hz} \qquad |^5J| = 2.5 \text{ Hz}$$

$$^4J_{ab} = 2.0 \text{ Hz}$$
$$^5J_{ac} = 1.0 \text{ Hz}$$
$$^6J_{ad} = 0.6 \text{ Hz}$$
$$^4J_{bd} = 1.6 \text{ Hz}$$

FIGURE 7.14 Homopropargylic coupling in alkynes.

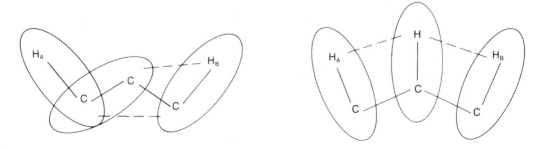

FIGURE 7.15 Possible orbital overlap mechanisms to explain ^4J W coupling.

$^4J = 1.0$ Hz

$^4J = 7.0$ Hz

$^4J = 1.4$ Hz

$^4J_{1,4} = 1.2$ Hz

$^4J_{3n,7a} = 4.2$ Hz

$^4J_{3x,5x} = 2.3$ Hz

$^4J_{5n,7s} = 2.1$ Hz

$^4J_{6n,7s} = 2.3$ Hz

a = *anti*;   s = *syn*

n = *endo*;   x = *exo*

FIGURE 7.16 Examples of ^4J W coupling in rigid bicyclic compounds.

FIGURE 7.17 A steroid ring skeleton showing several possible W couplings (^4J).

7.3 MAGNETIC EQUIVALENCE

In Chapter 5, Section 5.8, we discussed the idea of chemical equivalence. If a plane of symmetry or an axis of symmetry renders two or more nuclei equivalent by symmetry, they are said to be **chemically equivalent.**

In acetone, a plane of symmetry (and a C_2 axis) renders the two methyl groups chemically equivalent. The two methyl carbon atoms yield a single peak in the ^{13}C NMR spectrum. In addition, free rotation of the methyl group around the C—C bond ensures that all six hydrogen atoms are equivalent and resonate at the same frequency, producing a singlet in the ^1H NMR spectrum. In 1,2-dichloroethane, there is also a plane of symmetry, rendering the two methylene (CH_2) groups equivalent. Even though the hydrogens on these two carbon atoms are close enough for vicinal (three-bond) coupling ^3J, all four hydrogens appear as a single peak in the ^1H NMR spectrum, and no spin–spin splitting is seen. In fumaric acid, there is a twofold axis of symmetry that renders the carbons and hydrogens chemically equivalent. Because of symmetry, the adjacent *trans* vinyl hydrogens in fumaric acid do not show spin–spin splitting, and they appear as a singlet (both hydrogens having the same resonance frequency). The two ring hydrogens and methyl groups in *trans*-2,3-dimethylcyclopropanone (axis of symmetry) are also chemically equivalent, as are the two ring hydrogens and methyl groups in *cis*-2,3-dimethylcyclopropanone (plane of symmetry).

plane of symmetry

plane of symmetry

axis of symmetry

fumaric acid

axis of symmetry

plane of symmetry

In most cases, chemically equivalent nuclei have the same resonance frequency (chemical shift), do not split each other, and give a single NMR signal. When this happens, the nuclei are said to be **magnetically equivalent** as well as chemically equivalent. However, it is possible for nuclei to be chemically equivalent but magnetically *inequivalent.* Magnetic equivalence has requirements that are more stringent than those for chemical equivalence. For a group of nuclei to be magnetically equivalent, their magnetic environments, including *all coupling interactions*, must be of identical types. Magnetic equivalence has two strict requirements:

1. Magnetically equivalent nuclei must be **isochronous;** that is, they must have identical chemical shifts.

   and

2. Magnetically equivalent nuclei must have equal coupling (same *J* values) to all other nuclei in the molecule.

A corollary that follows from magnetic equivalence is that magnetically equivalent nuclei, even if they are close enough to be coupled, do not split one another, and they give only one signal (for both nuclei) in the NMR spectrum. This corollary does not imply that no coupling occurs between magnetically equivalent nuclei; it means only that no observable spin–spin splitting results from the coupling.

Some simple examples will help you understand these requirements. In chloromethane, all of the hydrogens of the methyl group are chemically and magnetically equivalent because of the threefold axis of symmetry (coincident with the C—Cl bond axis) and three planes of symmetry (each containing one hydrogen and the C—Cl bond) in this molecule. In addition, the methyl group rotates freely about the C—Cl axis. Taken alone, this rotation would ensure that all the three hydrogens experience the same *average* magnetic environment. The three hydrogens in chloromethane give a single resonance in the NMR (they are isochronous). Because there are no adjacent hydrogens in this one-carbon compound, by default all three hydrogens are equally coupled to all adjacent nuclei (a null set) and equally coupled to each other.

When a molecule has a plane of symmetry that divides it into equivalent halves, the observed spectrum is that for "half " of the molecule. The ^{1}H NMR spectrum of 3-pentanone shows only one quartet (CH_2 with three neighbors) and one triplet (CH_3 with two neighbors). A plane of symmetry renders the two ethyl groups equivalent; that is, the two methyl groups are chemically equivalent, and the two methylene groups are chemically equivalent. The coupling of any of the hydrogens in the methyl group to any of the hydrogens in the methylene group (^3J) is also equivalent (due to free rotation), and the coupling is the same on one "half" of the molecule as on the other. Each type of hydrogen is chemically equivalent.

$$\text{3-Pentanone} \quad CH_3CH_2 - \overset{\overset{\displaystyle O}{\|}}{C} - CH_2CH_3$$

Now, consider a *para*-disubstituted benzene ring, in which the *para* substituents X and Y are different. This molecule has a plane of symmetry that renders the hydrogens on opposite sides of the ring chemically equivalent. You might expect the ^1H spectrum to be that of one-half of the molecule—two doublets. It is not, however, since the corresponding hydrogens in this molecule are not *magnetically equivalent*. Let us label the chemically equivalent hydrogens H_a and H_a' (and H_b and H_b'). We would expect both H_a and H_a' or H_b and H_b' to have the same chemical shift (be isochronous), but *their coupling constants to the other nuclei are not the same*. H_a, for instance, does not have the same coupling constant to H_b (three bonds, ^3J) as H_a' has to H_b (five bonds, ^5J). Because H_a and H_a' do not have the same coupling constant to H_b, they cannot be magnetically equivalent, even though they are chemically equivalent. This analysis also applies to H_a', H_b, and H_b', none of which has equivalent couplings to the other hydrogens in the molecule.

Why is this subtle difference between the two kinds of equivalence important? Often, protons that are chemically equivalent are also magnetically equivalent; however, when chemically equivalent protons are *not* magnetically equivalent, there are usually consequences in the appearance of the NMR spectrum. Nuclei that are magnetically equivalent will give "first-order spectra" that can be analyzed using the $n + 1$ Rule or a simple "tree diagram" (Section 7.5). Nuclei that are not magnetically equivalent sometimes give second-order spectra, in which unexpected peaks may appear in multiplets (Section 7.7).

A simpler case than benzene, which has chemical equivalence (due to symmetry) but not magnetic equivalence, is 1,1-difluoroethene. Both hydrogens couple to the fluorines (^{19}F, $I = \frac{1}{2}$); however, the two hydrogens are not magnetically equivalent because H_a and H_b do not couple to F_a with the same coupling constant ($^3J_{HF}$). One of these couplings is *cis* ($^3J_{cis}$), and the other is *trans* ($^3J_{trans}$). In Table 7.4, it was shown that *cis* and *trans* coupling constants in alkenes were different in magnitude, with $^3J_{trans}$ having the larger value. Because these hydrogens have *different coupling constants to the same atom*, they cannot be magnetically equivalent. A similar argument applies to the two fluorine atoms, which also are magnetically inequivalent.

para-Disubstituted benzene   1, 1-Difluoroethene

Now consider 1-chloropropane. The hydrogens within a group (those on C1, C2, and C3) are isochronous, but each group is on a different carbon, and as a result, each *group* of hydrogens has a different chemical shift. The hydrogens in each group experience identical *average* magnetic environments, mainly because of free rotation, and are magnetically equivalent. Furthermore, also because of rotation, the hydrogens in each group are equally coupled to the hydrogens in the other groups. If we consider the two hydrogens on C2, H_b and H_b' and pick any other hydrogen on either C1 or C3, both H_b and H_b' will have the same coupling constant to the designated hydrogen. Without free rotation (see the preceding illustration) there would be no magnetic equivalence. Because of the fixed unequal dihedral angles (H_a—C—C—C_b versus H_a—C—C—H_b'), J_{ab} and J_{ab}' would not be the same. Free rotation can be slowed or stopped by lowering the temperature, in which case H_b and H_b' would become magnetically inequivalent. This type of magnetic inequivalence is often seen in 1,2-disubstituted ethane groups in which the substituents have sufficient steric bulk to hinder free rotation around the C—C axis enough that it becomes slow on the NMR time-scale.

1-Chloropropane CH_3—CH_2—CH_2—Cl

                              c   **b**   **a**

If conformation is locked (no rotation)

As one can see, it is a frequent occurrence that one needs to determine whether two groups attached to the same carbon (geminal groups) are equivalent or nonequivalent. Methylene groups (geminal protons) and isopropyl groups (geminal methyl groups) are frequently the subjects of interest. There are three possible relationships for such geminal groups: they can be homotopic, enantiotopic, or diastereotopic.

Methylene group: Geminal dimethyl group:

Homotopic groups are always equivalent, and in the absence of couplings from another group of nuclei, they are isochronous and give a single NMR absorption. Homotopic groups are interchangeable by rotational symmetry. The simplest way to recognize homotopic groups is by means of a "hypothetical" or "mental" substitution test. In this test, first one member of the group is substituted for a different group, then the other is substituted in the same fashion. The results of the substitution are examined to see the relationship of the resulting new structures. If the new structures are *identical*, the two original groups are homotopic. Figure 7.18a shows the substitution procedure for a molecule with two homotopic methylene hydrogens. In this molecule, the structures resulting from the replacement of first H_A and then H_B are identical. Notice that for this homotopic molecule, the substituents X are the same. The starting compound is completely symmetric because it has both a plane and a twofold axis of symmetry.

FIGURE 7.18 Replacement tests for homotopic, enantiotopic, and diastereotopic groups.

Enantiotopic groups appear to be equivalent, and they are typically isochronous and give a single NMR absorption—except when they are placed in a chiral environment or acted on by a chiral reagent. Enantiotopic groups can also be recognized by the hypothetical substitution test. Figure 7.18b shows the substitution procedure for a molecule with two enantiotopic methylene hydrogens. In this molecule, the resultant structures from the replacement of first H_A and then H_B are *enantiomers*. Although these two hydrogens appear to be equivalent and are isochronous in a typical NMR spectrum, they are not equivalent on replacement, each hydrogen giving a different enantiomer. Notice that the structure of this enantiotopic molecule is not chiral, but that substituents X and Y are different groups. There is a *plane* of symmetry, but no rotational axis of symmetry. Enantiotopic groups are sometimes called **prochiral** groups. When one or the other of these groups is replaced by a different one, a *chiral* molecule results. The reaction of prochiral molecules with a chiral reagent, such as an enzyme in a biological system, produces a chiral result. If these molecules are placed in a chiral environment, the two groups are no longer equivalent. We will examine a chiral environment induced by chiral shift reagents in Chapter 8 (Section 8.9).

Diastereotopic groups are not equivalent and are not isochronous; they have different chemical shifts in the NMR spectrum. When the diastereotopic groups are hydrogens, they frequently split each other with a geminal coupling constant ^2J. Figure 7.18c shows the substitution procedure for a molecule with two diastereotopic hydrogens. In this molecule, the replacement of first H_A and then H_B yields a pair of *diastereomers*. Diastereomers are produced when substituent Y* already contains an adjacent stereocenter. Diastereotopic groups are also found in prochiral compounds in which the substitution test simultaneously creates two stereogenic centers (Figure 7.18d). Section 7.4 covers both types of diastereotopic situations in detail.

7.4 SPECTRA OF DIASTEREOTOPIC SYSTEMS

In this section, we examine some molecules that have diastereotopic groups (discussed in Section 7.3). Diastereotopic groups are not equivalent, and two different NMR signals are observed. The most common instance of diastereotopic groups is when two similar groups, G and G', are substituents on a carbon *adjacent to a stereocenter*. If first group G and then group G' are replaced by another group, a pair of diastereomers forms (see Fig. 7.18c).[2]

A. Diastereotopic Hydrogens: Ethyl 3-Hydroxybutanoate

A good example of a simple compound with diastereotopic methylene protons is ethyl 3-hydroxybutanoate (Fig. 7.19). The stereocenter at C-3 of the compound makes protons H_c and H_d magnetically inequivalent. When H_c and H_d in a methylene group are in different environments, they will show a ^2J (two-bond, or geminal) coupling (Section 7.2B). For an sp^3 hybridized carbon atom (~109° for a H—C—H angle), values of ^2J are larger (9–18 Hz, typically 15 Hz) than usual ^3J coupling constants (6–8 Hz). Values of ^2J in methylene groups vary considerably with the degree of hybridization of carbon atoms (Table 7.3).

---

[2]Note that groups farther down the chain are also diastereotopic, but the effect becomes smaller as the distance from the stereocenter increases and eventually becomes unobservable. Also note that it is not essential that the stereocenter be a carbon atom.

Ethyl (*R*)-3-Hydroxybutanoate   Ethyl (*S*)-3-Hydroxybutanoate

Stereocenter indiacted with *

FIGURE 7.19 The structures of the enantiomers in a racemic mixture of ethyl 3-hydroxybutanoate.

The 500 MHz proton NMR spectrum of racemic ethyl 3-hydroxybutanoate is shown in Figure 7.20. The high-field NMR spectrum makes it possible to do a detailed analysis of the compound.[3] The expansions of the regions in Figure 7.20 are shown in Figure 7.21. The Hertz values on the top of the peaks can be used to calculate the coupling constants (see Section 7.6). The methyl protons (H_a) appear as a doublet at 1.23 ppm. The ^3J coupling constant is calculated as 6.3 Hz (618.92–612.62 Hz), and the methyl protons (H_b) appear as a triplet at 1.28 ppm with a ^3J coupling constant of about 7.1 Hz (646.63–639.45).[4] The hydroxyl group appears at about 3.15 ppm. The

FIGURE 7.20 ^{1}H NMR spectrum of racemic ethyl 3-hydroxybutanoate (500 MHz, CDCl$_3$).

[3]Often lower-field NMR spectra are more difficult to analyze because lower chemical shift dispersion leads to more closely spaced peaks and second-order resonances (Section 7.7).

[4]The method by which coupling constants are measured is discussed in detail in Section 7.6.

Ethyl 3-Hydroxybutanoate
Stereocenter indicated with *

FIGURE 7.21 Expansions of the ^{1}H NMR spectrum of ethyl 3-hydroxybutanoate (500 MHz, CDCl$_3$).

quartet at 4.18 ppm results from the methylene protons (H$_e$) split by the methyl protons (H$_b$), with ^3J = 7.1 Hz (2090.44–2083.38 Hz). The methine proton (H$_f$) is buried under the quartet from methylene protons H$_e$ at about 4.2 ppm. The pattern between 2.39 and 2.52 ppm is individually analyzed in Figure 7.22.

The methylene protons (H$_c$ and H$_d$) shown in Figure 7.22 are diastereotopic (nonequivalent) and appear at 2.42 and 2.49 ppm. Notice that the protons H$_c$ and H$_d$ have different chemical shifts because they are in different environments: H$_c$ is nearer the OH group (~60° dihedral angle) and H$_d$ is farther away from the OH group (180° dihedral) in the lowest energy conformation of the compound. The resonance for H$_c$ at 2.42 ppm is split first into a doublet by H$_d$ ($^2J_{cd}$), and each leg of the doublet is split again into doublets by the neighboring proton H$_f$ ($^3J_{cf}$). The observed pattern is referred to as a doublet of doublets. Likewise, the resonance for H$_d$ at 2.49 ppm is also a doublet of doublets. The calculations of the coupling constants for the diastereotopic protons H$_c$ and H$_d$ from Figure 7.22 are shown in Table 7.6. The value of $^2J_{cd}$ = 16.5 Hz. The ^3J values vary because of the differences in the dihedral angles between the protons: $^3J_{cf}$ = 8.7 Hz and $^3J_{df}$ = 3.5 Hz.

The reason for the differences between $^3J_{cf}$ and $^3J_{df}$ is that the dihedral angle for H$_c$ to H$_f$ (in the lowest energy conformation) is 180 degrees, whereas the dihedral angle for H$_d$ to H$_f$ in the same

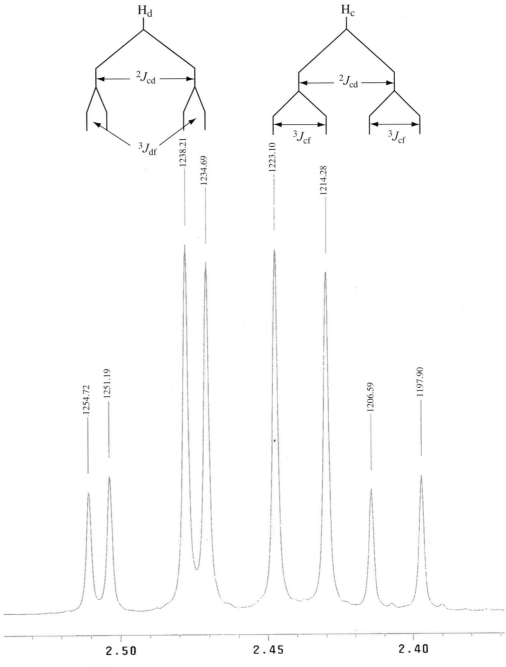

FIGURE 7.22 Expansion of the diasterotopic methylene protons in the ^{1}H NMR spectrum of ethyl 3-hydroxybutanoate (500 MHz, CDCl$_3$).

conformation is 60 degrees. The dihedral angles largely determine the value of the ^3J coupling constants (see Figures 7.6 and 7.7 in Section 7.2C). In systems where rotation around single bonds is fast relative to the NMR measurement, the observed ^3J is the weighted average of the couplings for each conformer that is a local energy minimum. In systems where the potential energy difference between conformers is large, it is usually sufficient to consider only the conformation that is the global energy minimum when assigning J values.

TABLE 7.6
CALCULATION OF ^2J AND ^3J COUPLING CONSTANTS FROM FIGURE 7.22

H_c = 2.42 ppm

$^2J_{cd}$ = 1223.10 – 1206.59 Hz = 16.51 Hz and 1214.28 – 1197.90 = 16.38 Hz.

$^2J_{cd} \approx$ **16.5 Hz**

$^3J_{cf}$ = 1223.10 – 1214.28 = 8.82 Hz and 1206.59 – 1197.90 = 8.69 Hz

$^3J_{cf} \approx$ **8.7 Hz**

H_d = 2.48 ppm

$^2J_{cd}$ = 1254.72 – 1238.21 = 16.51 Hz and 1251.19 – 1234.69 = 16.50 Hz

$^2J_{cd} \approx$ **16.5 Hz**

$^3J_{df}$ = 1254.72 – 1251.19 = 3.53 Hz and 1238.21 – 1234.69 = 3.52 Hz

$^3J_{df} \approx$ **3.5 Hz**

B. Diastereotopic Hydrogens: The Diels-Alder Adduct of Anthracene-9-methanol and *N*-Methylmaleimide

Another example of a compound with diastereotopic methylene protons is the product resulting from the Diels-Alder reaction of anthracene-9-methanol with *N*-methylmaleimide as shown in Figure 7.23. One observes that the molecule is not symmetrical, leading to the methylene group having diastereotopic protons. Notice that the hydrogen atoms on the methylene group, H_f and H_g, are in different environments. Proton H_f is on the same side as the nitrogen-containing ring, while H_g is on the opposite side of the nitrogen-containing ring. If the nitrogen-containing ring were not present, the molecule would have been symmetrical, and the methylene protons would have been equivalent and thus not diastereotopic.

The 500 MHz NMR spectrum of this product is shown in Figure 7.24, with insets showing expansions of the regions from 3.25 to 3.40 ppm and 4.7 to 5.2 ppm. The frequency values that

FIGURE 7.23 The Diels-Alder product of anthracene-9-methanol and *N*-methylmaleimide. This compound possesses diastereotopic methylene protons, H_f and H_g.

FIGURE 7.24 The 500 MHz NMR spectrum (CDCl₃) of the Diels-Alder product of anthracene-9-methanol and *N*-methylmaleimide. The insets show expansions of the regions from 3.25 to 3.40 ppm and 4.7 to 5.2 ppm. The eight aromatic ring protons are not shown. Pavia, D. L., G. M. Lampman, G. S. Kriz, R. G. Engel, *A Microscale Approach to Organic Laboratory Techniques*, 5th ed., Brooks/Cole, Belmont, CA, 2011. p. 926.

appear above the peaks in the expansions may be used to calculate the coupling constants. The diastereotopic protons H_f and H_g appear at 4.96 ppm and 5.13 ppm (we don't know the assignments, precisely). Assuming the correct assignment, proton H_g appears at 4.96 ppm as a doublet ($^2J_{fg}$ = 2490.12 – 2478.40 Hz = about 11.7 Hz). However, each leg of the doublet is split again into doublets ($^3J_{bg}$ = 2490.12 – 2485.36 Hz = about 4.8 Hz). The 4.8 Hz coupling constant is for the ^3J coupling of H_g to the proton on the OH group, H_b. The resulting pattern at 4.96 ppm is a doublet of doublets. Another doublet of doublets appears at 5.13 ppm for the diastereotopic proton H_f.

The other coupling constants may be extracted from the pattern for H_c at 3.27 ppm, a doublet of doublets, $^3J_{cd}$ = 8.8 Hz and $^3J_{ce}$ = 3.3 Hz. Proton H_d appears at 3.36 ppm as doublet, $^3J_{cd}$ = 8.8 Hz. Finally, H_e appears as a doublet at 4.75 ppm with $^3J_{ce}$ = 3.3 Hz. The coupling constants may be summarized, as follows:

$^2J_{fg}$ = 11.7 Hz (diastereotopic protons at 4.96 ppm and 5.13 ppm)

$^3J_{bf}$ = $^3J_{bg}$ = 4.8 Hz (coupling of OH to each of the diastereotopic protons)

$^3J_{ce}$ = 3.3 Hz (coupling of H_c to H_e)

$^3J_{cd}$ = 8.8 Hz (coupling of H_c to H_d)

C.   Diastereotopic Hydrogens: 4-Methyl-2-pentanol

Now let's examine a system with a third coupling, where diastereotopic protons have a geminal and two vicinal couplings and appear as doublets of doublets of doublets (ddd). An example of this type of system is exemplified by the diastereotopic methylene protons of 4-methyl-2-pentanol. The full ^{1}H NMR spectrum of 4-methyl-2-pentanol is shown in Figure 7.25, with an expansion showing the diastereotopic methylene protons H_a and H_b along with the hydroxyl proton and methine H_d in Figure 7.26. H_a and H_b split each other, and each is coupled to H_c and H_d, accounting for the three couplings in a ddd. An analysis of the couplings is shown in Figure 7.27.

The geminal coupling constant $^2J_{ab} = 13.7$ Hz, which is a typical value for diastereotopic geminal coupling in acyclic aliphatic systems (Section 7.2B). The coupling constant $^3J_{bc}$ (8.3 Hz) is somewhat larger than $^3J_{ac}$ (5.9 Hz), which is in agreement with the average dihedral angles predicted from the relevant conformations and the Karplus relationship (Section 7.2C). The hydrogen on C-2, H_c, is coupled not only to H_a and H_b but also to the C-1 methyl group, with ^3J (H_cC-CH_3) = 5.9 Hz. Because of the more complex splitting of H_c, a splitting analysis tree is not shown for this proton. Similarly, the hydrogen on C-4 (seen at 1.74 ppm in Fig. 7.26) has a complex splitting pattern due to coupling to both H_a and H_b as well as the two sets of diastereotopic methyl protons on C-5 and C-5'. Measurement of coupling constants from complex first-order resonances like these is discussed in detail in Sections 7.5 and 7.6.

FIGURE 7.25 ^{1}H spectrum of 4-methyl-2-pentanol showing diastereotopic methyl and methylene groups (500 MHz, CDCl$_3$).

FIGURE 7.26 Expansion of the ^{1}H spectrum of 4-methyl-2-pentanol showing diastereotopic methylene protons.

FIGURE 7.27 Splitting diagrams for the diastereotopic methylene protons in 4-methyl-2-pentanol.

D.   Diasteveotopic Methyl Groups: 4-Methyl-2-pentanol

Now let's look at the diastereotopic methyl groups in the ^{1}H and ^{13}C NMR spectra of 4-methyl-2-pentanol are   in Figures 7.25 and 7.28, respectively. This molecule has diastereotopic methyl groups (labeled 5 and 5') on carbon 4. If this compound did not have diastereotopic groups, we would expect only two different peaks for methyl carbons as there are only two chemically distinct types of methyl groups. However, the spectrum shows three methyl peaks. A very closely spaced pair of resonances is observed at 23.18 and 22.37 ppm, representing the diastereotopic methyl groups, and a third resonance at 23.99 ppm from the C-1 methyl group. There are two peaks for the geminal dimethyl groups! Carbon 4, to which the methyl groups are attached, is seen at 24.8 ppm, carbon 3 is at 48.7 ppm, and carbon 2, which has the deshielding hydroxyl attached, is observed downfield at 66.1 ppm.

The two methyl groups have slightly different chemical shifts because of the nearby stereocenter at C-2. The two methyl groups are always nonequivalent in this molecule, even in the presence of free rotation. You can confirm this fact by examining the various fixed, staggered rotational conformations using Newman projections. There are no planes of symmetry in any of these conformations; neither of the methyl groups is ever enantiomeric.

FIGURE 7.28 ^{13}C spectrum of 4-methyl-2-pentanol showing diastereotopic methyl groups.

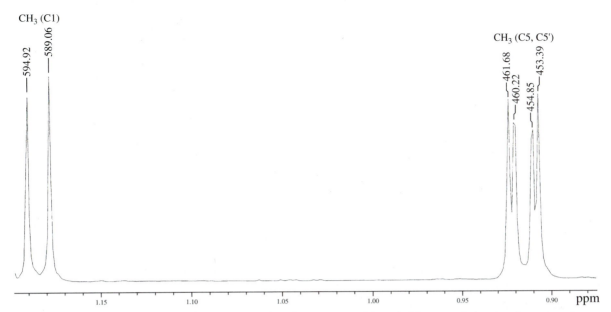

FIGURE 7.29   Upfield region of the ¹H spectrum of 4-methyl-2-pentanol showing diastereotopic methyl groups.

The ¹H NMR spectrum (Figure 7.25 with an expansion of the methyl region in Figure 7.29) is a bit more complicated, but just as the two diastereotopic methyl carbons have different chemical shifts, so do the diastereotopic methyl hydrogens. The hydrogen atom attached to C-4 splits each methyl group into a doublet. The chemical shift difference between the methyl protons is very small, however, and the two doublets are partially overlapped. One of the methyl doublets is observed at 0.92 ppm (J = 6.8 Hz), and the other diastereotopic methyl doublet is seen at 0.91 ppm (J = 6.8 Hz). The C-1 methyl group is also a doublet at 1.18 ppm, split by the hydrogen on C-2 (J = 5.9 Hz).

7.5 NONEQUIVALENCE WITHIN A GROUP—THE USE OF TREE DIAGRAMS WHEN THE *n* + 1 RULE FAILS

When the protons attached to a single carbon are chemically equivalent (have the same chemical shift), the *n* + 1 Rule successfully predicts the splitting patterns. In contrast, when the protons attached to a single carbon are chemically nonequivalent (have different chemical shifts), the *n* + 1 Rule no longer applies. We shall examine two cases, one in which the *n* +1 Rule applies (1,1,2-trichloroethane) and one in which it fails (styrene oxide).

(a) Free rotation
The $n + 1$ rule applies

(b) Biased conformation
A tree diagram is required

$$\delta_A = \delta_B$$
$$J_{AC} = J_{BC}$$
$$J_{AB} = 0$$

$$\delta_A \neq \delta_B$$
$$J_{AC} \neq J_{BC}$$
$$J_{AB} \neq 0$$

FIGURE 7.30   Two cases of splitting.

Chapter 5, Section 5.13, and Figure 5.25, addressed the spectrum of 1,1,2-trichloroethane. This symmetric molecule has a three-proton system, $-CH_2-CH-$ in which the methylene protons are equivalent. Due to free rotation around the C—C bond, the methylene protons each experience the same averaged environment, are isochronous (have the same chemical shift), and do not split each other. In addition, the rotation ensures that they both have the same averaged coupling constant J to the methine (CH) hydrogen. As a result, they behave as a group, and geminal coupling between them does not lead to any splitting. The $n + 1$ Rule correctly predicts a doublet for the CH_2 protons (one neighbor) and a triplet for the CH proton (two neighbors). Figure 7.30a illustrates the parameters for this molecule.

Figure 7.31, the ^1H spectrum of styrene oxide, shows how chemical nonequivalence complicates the spectrum. The three-membered ring prevents rotation, causing protons H_A and H_B to have different chemical shift values; they are chemically and magnetically inequivalent. Hydrogen H_A is on the same side of the ring as the phenyl group; hydrogen H_B is on the opposite side of the ring. These hydrogens have different chemical shift values, $H_A = 2.75$ ppm and $H_B = 3.09$ ppm, and they show geminal splitting with respect to each other. The third proton, H_C, appears at 3.81 ppm and is coupled *differently* to H_A (which is *trans*) than to H_B (which is *cis*). Because H_A and H_B are nonequivalent and because H_C is coupled differently to H_A than to H_B ($^3J_{AC} \neq {}^3J_{BC}$), the $n + 1$ Rule fails, and the spectrum of styrene oxide becomes more complicated. To explain the spectrum, one must examine each hydrogen individually and take into account its coupling with every other hydrogen independent of the others. Figure 7.30b shows the parameters for this situation.

An analysis of the splitting pattern in styrene oxide can be carried out splitting-by-splitting with **graphical analyses, or tree diagrams** (Fig. 7.32). Begin with an examination of hydrogen H_C. First, the two possible spins of H_B split H_C ($^3J_{BC}$) into a doublet; second, H_A splits each of the doublet peaks ($^3J_{AC}$) into another doublet. The resulting pattern of two doublets is called a **doublet of doublets.** You may also look at the same splitting from H_A first and from H_B second. It is customary to *show the largest splitting first*, but it is not necessary to follow this convention to obtain the correct result. If the actual coupling constants are known, it is very convenient to perform this analysis (*to scale*) on graph paper with 1-mm squares.

Note that $^3J_{BC}$ (*cis*) is larger than $^3J_{AC}$ (*trans*). This is typical for small ring compounds in which there is more interaction between protons that are *cis* to each other than between protons that are *trans* to each other (see Section 7.2C and Fig. 7.10). Thus, we see that H_C gives rise to a set of *four* peaks (another doublet of doublets) centered at 3.81 ppm. Similarly, the resonances for H_A and H_B are each a doublet of doublets at 2.75 ppm and 3.09 ppm, respectively. Figure 7.32 also shows these splittings. Notice that the magnetically nonequivalent protons H_A and H_B give rise to geminal splitting ($^2J_{AB}$) that is quite significant.

FIGURE 7.31   The ^{1}H NMR spectrum of styrene oxide.

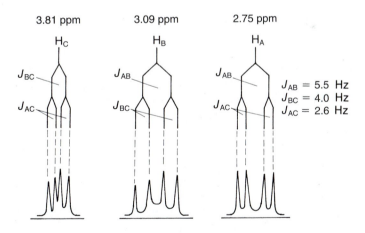

FIGURE 7.32   An analysis of the splitting pattern in styrene oxide.

As you see, the splitting situation becomes quite complicated for molecules that contain non-equivalent groups of hydrogens. In fact, you may ask, how can one be sure that the graphic analysis just given is the correct one? First, this analysis explains the entire pattern; second, it is internally consistent. Notice that the coupling constants have the same magnitude wherever they are used. Thus, in the analysis, $^3J_{BC}$ (*cis*) is given the same magnitude when it is used in splitting H_C as when it is used in splitting H_B. Similarly, $^3J_{AC}$ (*trans*) has the same magnitude in splitting H_C as in splitting H_A. The coupling constant $^2J_{AB}$ (geminal) has the same magnitude for H_A as for H_B. If this kind of self-consistency were not apparent in the analysis, the splitting analysis would have been incorrect. To complete the analysis, note that the NMR peak at 7.28 ppm is due to the protons of the phenyl ring. It integrates for five protons, while the other three multiplets integrate for one proton each.

We must sound one note of caution at this point. In some molecules, the splitting situation becomes so complicated that it is virtually impossible for the beginning student to derive it. Section 7.6 describes the process by which to determine coupling constants in greater detail to assist you. There are also situations involving apparently simple molecules for which a graphical analysis of the type we have just completed does not suffice (second-order spectra). Section 7.7 will describe a few of these cases.

We have now discussed three situations in which the $n + 1$ Rule fails: (1) when the coupling involves nuclei other than hydrogen that do not have spin = 1/2 (e.g., deuterium, Section 6.13), (2) when there is nonequivalence in a set of protons attached to the same carbon; and (3) when the chemical shift difference between two sets of protons is small compared to the coupling constant linking them (see Sections 7.7 and 7.8).

▢ 7.6 MEASURING COUPLING CONSTANTS FROM FIRST-ORDER SPECTRA

When one endeavors to measure the coupling constants from an actual spectrum, there is always some question of how to go about the task correctly. In this section, we will provide guidelines that will help you to approach this problem. The methods given here apply to first-order spectra; analysis of second-order spectra is discussed in Section 7.7. What does "first-order" mean, as applied to NMR spectra? For a spectrum to be first-order, the frequency difference (Δv, in Hz) between any two coupled resonances must be significantly larger than the coupling constant that relates them. A first-order spectrum has $\Delta v/J > \sim 6$.[5]

First-order resonances have a number of helpful characteristics, some of which are related to the number of individual couplings, n:

1. symmetry about the midpoint (chemical shift) of the multiplet. Note that a number of second-order patterns are also centrosymmetric, however (Section 7.7);

2. the maximum number of lines in the multiplet = 2^n; the actual number of lines is often less than the maximum number, though, due to overlap of lines arising from coincidental mathematical relationships among the individual J values;

3. the sum of the line intensities in the multiplet = 2^n;

4. the line intensities of the multiplet correspond to Pascal's triangle (Section 5.16);

5. the J values can be determined directly by measuring the appropriate line spacings in the multiplet;

6. the distance between the outermost lines in the multiplet is the sum of all the individual couplings, ΣJ.

A. Simple Multiplets—One Value of J (One Coupling)

For simple multiplets, where only one value of J is involved (one coupling), there is little difficulty in measuring the coupling constant. In this case it is a simple matter of determining the spacing (in Hertz) between the successive peaks in the multiplet. This was discussed in Chapter 5, Section 5.17.

---

[5]The choice of $\Delta v/J > 6$ for a first-order spectrum is not a hard-and-fast rule. Some texts suggest a $\Delta v/J$ value of >10 for first-order spectra. In some cases, multiplets appear essentially first-order with $\Delta v/J$ values slightly less than 6.

TABLE 7.7
THE HERTZ EQUIVALENT OF A ppm UNIT AT
VARIOUS SPECTROMETER OPERATING FREQUENCIES

Spectrometer Frequency	Hertz Equivalent of 1 ppm
60 MHz	60 Hz
100 MHz	100 Hz
300 MHz	300 Hz
500 MHz	500 Hz

Also discussed in that section was the method of converting differences in parts per million (ppm) to Hertz (Hz). The relationship

$$1 \text{ ppm (in Hertz)} = \text{Spectrometer Frequency in Hertz} \div 1{,}000{,}000$$

gives the simple correspondence values given in Table 7.7, which shows that if the spectrometer frequency is *n* **MHz,** one ppm of the resulting spectrum will be *n* **Hz.** This relationship allows an easy determination of the coupling constant linking two peaks when their chemical shifts are known only in ppm; just find the chemical shift difference in ppm and multiply by the Hertz equivalent.

The current processing software for most modern FT-NMR instruments allows the operator to display peak locations in both Hertz and ppm. Figure 7.33 is an example of the printed output from a modern 300-MHz FT-NMR. In this septet, the chemical shift values of the peaks (ppm) are obtained from the scale printed at the bottom of the spectrum, and the values of the peaks in Hertz are printed vertically above each peak. To obtain the coupling constant it is necessary only to subtract the Hertz values for the successive peaks. In doing this, however, you will note that not all of the differences are

FIGURE 7.33 A septet determined at 300 MHz showing peak positions in ppm and Hz values.

TABLE 7.8
ANALYSIS OF FIRST-ORDER MULTIPLETS AS SERIES OF DOUBLETS

Number of Identical Couplings	Multiplet Appearance	Equivalent Series of Doublets	Sum of Line Intensities
1	d	d	2
2	t	dd	4
3	q	ddd	8
4	quintet (pentet)	dddd	16
5	sextet	ddddd	32
6	septet	dddddd	64
7	octet	ddddddd	128
8	nonet	dddddddd	256

identical. In this case (starting from the downfield side of the resonance) they are 6.889, 6.858, 6.852, 6.895, 6.871, and 6.820 Hz. There are two reasons for the inconsistencies. First, these values are given to more places than the appropriate number of significant figures would warrant. The inherent linewidth of the spectrum makes differences less than 0.1 Hz insignificant. When the above values are rounded off to the nearest 0.1 Hz, the line spacings are 6.9, 6.9, 6.9, 6.9, 6.9, and 6.8 Hz—excellent agreement. Second, the values given for the peaks are not always precise depending on the number of data points in the spectrum. If an insufficient number of points are recorded during the acquisition of the FID (large value of Hz/pt), the apex of a peak may not correspond exactly with a recorded data point and this situation results in a small chemical shift error.

When conflicting J values are determined for a multiplet it is usually appropriate to round them off to two significant figures, or to take an average of the similar values and round that average to two significant figures. For most purposes, it is sufficient if all the measured J values agree to <0.3 Hz difference. In the septet shown in Figure 7.33, the average of all the differences is 6.864 Hz, and an appropriate value for the coupling constant would be 6.9 Hz.

Before we consider multiplets with more than one distinct coupling relationship, it is helpful to review simple multiplets, those adequately described by the $n + 1$ Rule, and begin to consider them as series of doublets by considering each individual coupling relationship separately. For example, a triplet (t) can be considered a doublet of doublets (dd) where two identical couplings ($n = 2$) are present ($J_1 = J_2$). The sum of the triplet's line intensities (1:2:1) is equal to 2^n where $n = 2$ ($1 + 2 + 1$) $= 2^2 = 4$). Similarly, a quartet can be considered a doublet of doublet of doublets where three identical couplings ($n = 3$) are present ($J_1 = J_2 = J_3$) and the sum of the quartet's line intensities (1:3:3:1) equals 2^n where $n = 3$ ($1 + 3 + 3 + 1 = 2^3 = 8$). This analysis is continued in Table 7.8.

B.   Is the $n + 1$ Rule Ever *Really* Obeyed?

In a linear chain, the $n + 1$ Rule is strictly obeyed only if the vicinal proton coupling constants (^3J) are *exactly the same* for every successive pair of carbons.

$$\text{(A)}\quad\text{(B)}\quad\text{(C)}\quad\text{(D)}$$

$$-CH_2-CH_2-CH_2-CH_2-$$

$$J_{AB} = J_{BC} = J_{CD}, \text{ and so on}$$

Condition under which the $n + 1$ Rule is strictly obeyed

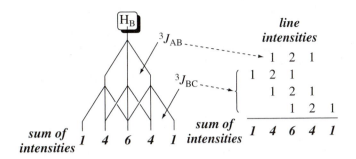

FIGURE 7.34 Construction of a quintet for a methylene group with four neighbors all with identical coupling values.

Consider a three-carbon chain as an example. The protons on carbons A and C split those on carbon B. If there is a total of four protons on carbons A and C, the $n + 1$ Rule predicts a pentet. This occurs only if $^3J_{AB} = {}^3J_{BC}$. Figure 7.34 represents the situation graphically.

One way to describe the situation is as a triplet of triplets, since the methylene protons labeled B above should be split into a triplet by the neighboring methylene protons labeled A and into a triplet by the neighboring methylene protons labeled C. First, the protons on carbon A split those on carbon B ($^3J_{AB}$), yielding a triplet (intensities 1:2:1). The protons on carbon C then split *each component* of the triplet ($^3J_{BC}$) into another triplet (1:2:1). At this stage, many of the lines from the second splitting interaction *overlap* those from the first splitting interaction because they have the same spacings (J value). Because of this coincidence, only five lines are observed. But we can easily confirm that they arise in the fashion indicated by adding the intensities of the splittings to predict the intensities of the final five-line pattern (see Fig. 7.34). These intensities agree with those predicted by the use of Pascal's triangle (Section 5.16). Thus, the $n + 1$ Rule depends on a special condition— that all of the vicinal coupling constants are identical.

Another way to describe the situation above is to consider the H_B methylene protons as a doublet of doublet of doublet of doublets (dddd) where $^3J_{AB} = {}^3J_{A'B} = {}^3J_{BC} = {}^3J_{AC'}$. With four distinct couplings, the sum of the line intensities for the H_B multiplet will be $2^4 = 16$. By constructing a splitting tree for H_B and distributing the intensities for each doublet, one arrives at the same conclusion: H_B is an apparent quintet with line intensities 1:4:6:4:1 = 16 (Figure 7.35).

In many molecules, however, J_{AB} is slightly different from J_{BC}. This leads to peak broadening in the multiplet, since the lines do not perfectly overlap. (Broadening occurs because the peak separation in Hertz is too small in magnitude for the NMR instrument to be able to distinguish the separate peak components.)

Sometimes the perturbation of the quintet is only slight, and then either a shoulder is seen on the side of a peak or a dip is obvious in the middle of a peak. At other times, when there is a large difference between $^3J_{AB}$ and $^3J_{BC}$, distinct peaks, more than five in number, can be seen. Deviations of this type are most common in a chain of the type $X-CH_2CH_2CH_2-Y$, where X and Y are widely different in character. Figure 7.36 illustrates the origin of some of these deviations.

Chains of any length can exhibit this phenomenon, whether or not they consist solely of methylene groups. For instance, the spectrum of the protons in the second methylene group of propylbenzene is simulated as follows. The splitting pattern gives a crude sextet, but the second line has a shoulder on the left, and the fourth line shows an unresolved splitting. The other peaks are somewhat broadened.

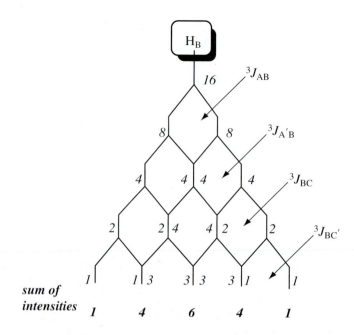

FIGURE 7.35 Construction of a quintet for a methylene group with four neighbors by considering as a dddd.

Broadening due to partial overlap

FIGURE 7.36 Loss of a simple quintet when $^3J_{AB} \neq {}^3J_{BC}$.

C. More Complex Multiplets—More Than One Value of *J*

When analyzing more complicated resonances with more than one distinct coupling, measuring all of the coupling constants presents a challenge. Many chemists take the lazy way out and simply call a complex resonance a "multiplet." This presents problems on multiple levels. First, coupling constants give valuable information about both the two-dimensional (2-D) structure (connectivity) and three-dimensional (3-D) structure (stereochemistry) of compounds. With the availability of high-field instruments with pulsed field gradients (PFGs), chemists often turn immediately to 2-D NMR techniques such as COSY and NOESY (Chapter 9) to determine connectivity within spin systems and three-dimensional structure, respectively. Often the same information (provided the resonances are not too severely overlapping or second-order) may be extracted from the simple 1-D ^{1}H NMR spectrum if one knows how. Thus, *it is always worth the effort to determine all coupling constants from a first-order resonance.*

When measuring coupling constants in a system with more than one coupling you will often notice that none of the multiplet peaks is located at the appropriate chemical shift values to directly determine a value for an intermediate *J* value. This is illustrated in Figure 7.37a where a doublet of

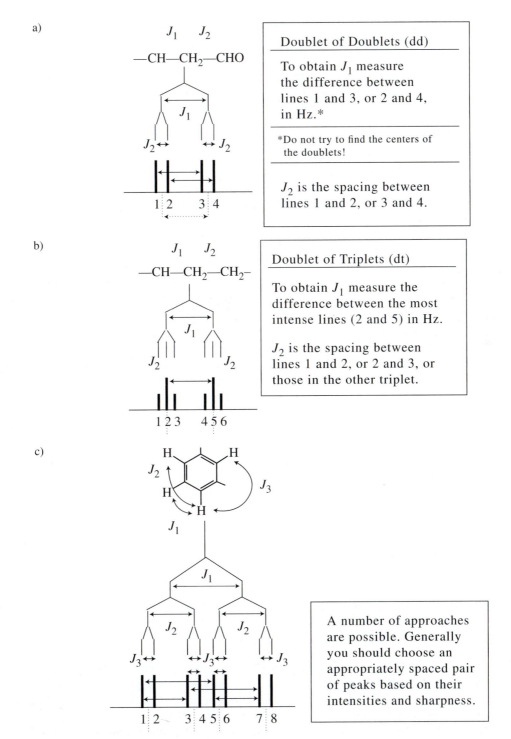

F I G U R E   7 . 3 7   Determining coupling constants for a) doublet of doublets (dd), b) doublet of triplets (dt), and c) doublet of doublet of doublets (ddd) patterns.

doublets is illustrated. In this case, none of the peaks is located at the chemical shift values that would result from the first coupling J_1. To a beginning student, it may be tempting to average the chemical shift values for peaks 1 and 2, and for peaks 3 and 4, and then take the difference (dotted lines). This is not necessary. With a little thought you will see that the distance between peaks 1 and 3, and also the distance between peaks 2 and 4 (solid arrows), can yield the desired value with much less work. This type of situation will occur whenever there is an even number of sub-peaks (doublets, quartets, etc.) in the separated multiplets. In these systems you should look for an appropriately spaced pair of offset subpeaks that will yield the value you need. It will usually be necessary to construct a splitting diagram (tree) in order to decide which of the peaks are the appropriate ones.

When the separated multiplets have an odd number of subpeaks, one of the subpeaks will inevitably fall directly on the desired chemical shift value without a need to look for appropriate offset peaks. Figure 7.37b shows a doublet of triplets. Note that peaks 2 and 5 are ideally located for a determination of J_1.

Figure 7.37c shows a pattern that may be called a doublet of doublets of doublets. After constructing a tree diagram, it is relatively easy to select appropriate peaks to use for the determination of the three coupling constants (solid arrows).

A number of approaches to measuring coupling constants are possible. Generally you should choose an appropriately spaced pair of peaks based on their intensities and sharpness. With experience, most practicing synthetic chemists have the skills to measure coupling constants for all manner of resonances containing two or three unequal J values, i.e., doublet of doublet of doublet (ddd) resonances, including the doublet of triplets (dt) and triplet of doublet (td) permutations using the methods described above and in Fig. 7.37.

Even experienced chemists, however, often struggle to extract all of the coupling constants from resonances that have four couplings in them (doublet of doublet of doublet of doublets, or dddd) and even more complex multiplets. A straightforward systematic method exists, however, that allows for complete analysis of any (even the most complex) first-order multiplet. Practicing this method on the more easily analyzed ddd multiplets discussed above will allow the student to gain confidence in its usefulness. This systematic multiplet analysis was most succinctly presented by Hoye and Zhao, and is presented below.

Analysis of a first-order multiplet begins with numbering each line in the resonance from left-to-right.[6] The outermost line will be relative intensity = 1. Lines of relative intensity > 1 get more than one component number. A line of relative intensity 2 gets two component numbers, one with relative intensity 3 gets three component numbers, and so on. The line component numbers and the relative line intensities must sum to a 2^n number. This is illustrated in Figure 7.38. In Figure 7.38a, there are eight lines of equal intensity ($2^3 = 8$), and each line has one component number. In Figure 7.38b, there is some coincidence of lines; the middle line has double intensity and therefore gets two component numbers. Figure 7.38c and 7.38d show line numbering for multiplets with lines having relative intensity 3 and 6, respectively. The assignment of line components sometimes requires a bit of trial and error as partial overlap of lines and "leaning" of the multiplet may make determining the relative intensities more difficult. Remember, though, that a first-order multiplet is always symmetric about its midpoint.

Once the relative intensities of the lines of the multiplet are determined and the component numbers assigned to arrive at 2^n components, the measurement of coupling constants is, in fact, fairly easy. We will go through the analysis of a dddd pattern step-by-step (Figure 7.39). The distance between the first component and the second component (referred to as {1 to 2} by Hoye) is the smallest coupling constant J_1 (Figure 7.39, step i). The distance between component 1 and component 3 of the multiplet ({1 to 3}) is the next largest coupling constant J_2 (Figure 7.39, step ii).

---

[6]Since first-order resonances are symmetric, one could number the lines of a resonance from right-to-left just as easily. This is useful when part of the multiplet is obscured due to overlap of another resonance. One should also check for internal consistency within a resonance, as on occasion one "half" of the multiplet may be sharper than the other due to the digitization of the spectrum, as discussed previously in Section 7.6A.

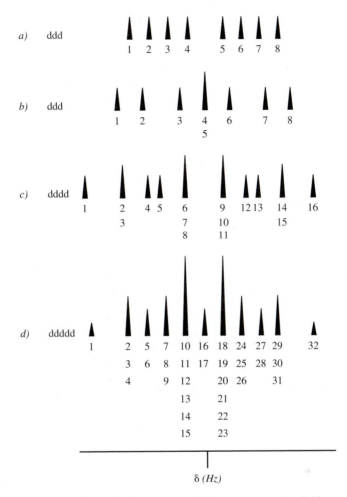

FIGURE 7.38 Numbering lines of a first-order multiplet to account for all 2^n components of the resonance. (From Hoye, T. R. and H. Zhao, *Journal of Organic Chemistry* 2002, 67, 4014–4016.) Reprinted by permission.

Note that if the second line of the resonance has more than one component number, there will be more than one identical J value. If the second line of a resonance has three components, for example, there will be three identical J values, etc. After measuring J_1 and J_2, the next step in the analysis is to "remove" the component of the multiplet corresponding to $(J_1 + J_2)$ (Figure 7.39 step iii, component 5 is crossed out). The reason for removing one of the components is to eliminate from consideration lines that are not due to a unique coupling interaction, but rather from coincidence of lines due to the sum of two smaller couplings. In other words, it shows whether or not the two "halves" of the resonance have "crossed" due to J_3 being smaller than the sum of $J_1 + J_2$. Now, J_3 is the distance between component 1 and the *next highest remaining* component (component 4 or 5, depending on which component was removed in step iii, in this example $J_3 = \{1 \text{ to } 4\}$) (Figure 7.39, step iv). This process now becomes iterative. The next step is to remove the component(s) that correspond to the remaining combinations of the first three J values: $(J_1 + J_3)$, $(J_2 + J_3)$, and $(J_1 + J_2 + J_3)$ (Figure 7.39, step v, components 6, 7, and 9 are crossed out). The next coupling constant, J_4, will be the distance between the first component and the next highest remaining component. In the example case shown in Figure 7.39, J_4 corresponds to $\{1 \text{ to } 8\}$. This iterative process repeats until all the coupling constants are found. Remember that the total number of coupling interactions and the total number of line

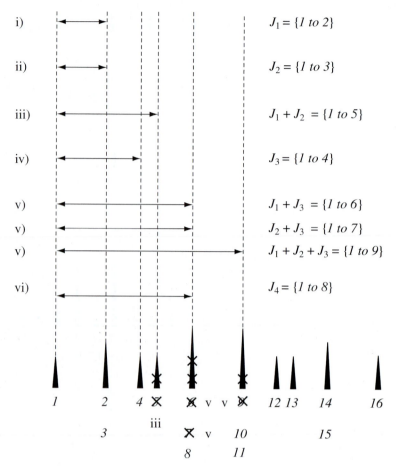

FIGURE 7.39   Assignment of $J_1 - J_4$ of a dddd by systematic analysis. (From Hoye, T. R. and H. Zhao, *Journal of Organic Chemistry* 2002, *67*, 4014–4016.) Reprinted by permission.

components must equal 2^n, and the overall width of the multiplet *must* equal the sum of all the individual coupling constants! This is a convenient check of your work.

☐ 7.7   SECOND-ORDER SPECTRA—STRONG COUPLING

A.   First-Order and Second-Order Spectra

In earlier sections, we have discussed **first-order spectra,** spectra that can be interpreted by using the $n + 1$ Rule or a simple graphical analysis (splitting trees). In certain cases, however, neither the $n + 1$ Rule nor graphical analysis suffices to explain the splitting patterns, intensities, and numbers of peaks observed. In these last cases, a mathematical analysis must be carried out, usually by computer, to explain the spectrum. Spectra that require such advanced analysis are said to be **second-order spectra.**

Second-order spectra are most commonly observed when the difference in chemical shift between two groups of protons is similar in magnitude (in Hertz) to the coupling constant J (also in Hertz), which links them. That is, second-order spectra are observed for couplings between nuclei that have *nearly equivalent chemical shifts* but are not exactly identical. In contrast, if two sets of nuclei are separated by a large chemical shift difference, they show first-order coupling.

Strong coupling, second-order spectra
($\Delta v/J$ small)

Weak coupling, first-order spectra
($\Delta v/J$ large)

Another way of expressing this generalization is by means of the ratio $\Delta v/J$, where Δv is the chemical shift difference, and J is the coupling constant that links the two groups. Both values are expressed in Hertz, and their absolute values are used for the calculation. When $\Delta v/J$ is large (> ~6), the splitting pattern typically approximates first-order splitting. However, when the chemical shifts of the two groups of nuclei move closer together and $\Delta v/J$ approaches unity, we see second-order changes in the splitting pattern. When $\Delta v/J$ is large and we see first-order splitting, the system is said to be **weakly coupled;** if $\Delta v/J$ is small and we see second-order coupling, the system is said to be **strongly coupled.**

We have established that even complex looking first-order spectra may be analyzed in a straight-forward fashion to determine all of the relevant coupling constants, which provide valuable information about connectivity and stereochemistry. Second-order spectra can be deceptive in their appearance and often tempt the novice into trying to extract coupling constant values, which ultimately proves an exercise in futility. How, then, does one determine if a resonance is first order or second order? How can one determine $\Delta v/J$ if one does not know the relevant coupling values in the first place? Herein lays the importance of being familiar with typical coupling constant values for commonly encountered structural features. One should first *estimate* $\Delta v/J$ by finding the chemical shift difference between resonances that are likely to be coupled (based on one's knowledge of the structure or in some cases the 2-D COSY spectra (Chapter 9, Section 9.6) and divide that value by a typical or *average* coupling constant for the relevant structural type. The estimated $\Delta v/J$ value allows one to make a judgment about whether detailed analysis of the resonance is likely to be useful ($\Delta v/J$ > ~ 6) or not ($\Delta v/J$ < ~ 6).

B. Spin System Notation

Nuclear Magnetic Resonance (NMR) spectroscopists have developed a convenient shorthand notation, sometimes called **Pople notation,** to designate the type of spin system. Each chemically different type of proton is given a capital letter: A, B, C, and so forth. If a group has two or more protons of one type, they are distinguished by subscripts, as in A_2 or B_3. Protons of similar chemical shift values are assigned letters that are close to one another in the alphabet, such as A, B, and C. Protons of widely different chemical shift are assigned letters far apart in the alphabet: X, Y, Z versus A, B, C. A two-proton system where H_A and H_X are widely separated, and that exhibits first-order splitting, is called an AX system. A system in which the two protons have similar chemical shifts, and that exhibits second-order splitting, is called an AB system. When the two protons have identical chemical shifts, are magnetically equivalent, and give rise to a singlet, the system is designated A_2. Two protons that have the same chemical shift but are not magnetically equivalent are designated as AA'. If three protons are involved and they all have very different chemical shifts, a letter from the middle of the alphabet is used, usually M, as in AMX. The ^{1}H NMR spectrum of styrene oxide in Figure 7.31 is an example of an AMX pattern. In contrast, ABC would be used for the strongly coupled situation in which all three protons have similar chemical shifts. We will use designations similar to these throughout this section.

FIGURE 7.40   A first-order AX system: Δv large, and $n + 1$ Rule applies.

C.   The A_2, AB, and AX Spin Systems

Start by examining the system with two protons, H_A and H_B, on adjacent carbon atoms. Using the $n + 1$ Rule, we expect to see each proton resonance as a doublet with components of equal intensity in the ^1H NMR spectrum. In actuality, we see two doublets of equal intensity in this situation only if the difference in chemical shift (Δv) between H_A and H_B is large compared to the magnitude of the coupling constant ($^3J_{AB}$) that links them. Figure 7.40 illustrates this case.

Figure 7.41 shows how the splitting pattern for the two-proton system H_AH_B changes as the chemical shifts of H_A and H_B come closer together and the ratio $\Delta v/J$ becomes smaller. The figure is drawn to scale, with $^3J_{AB} = 7$ Hz. When $\delta H_A = \delta H_B$ (that is, when the protons H_A and H_B have the same chemical shift), then $\Delta v = 0$, and no splitting is observed; both protons give rise to a single absorption peak. Between one extreme, where there is no splitting due to chemical shift equivalence ($\Delta v/J = 0$), and the other extreme, the simple first-order spectrum ($\Delta v/J = 15$) that follows the $n + 1$ Rule, subtle and continuous changes in the splitting pattern take place. Most obvious is the decrease in intensity of the outer peaks of the doublets, with a corresponding increase in the intensity of the inner peaks. Other changes that are not as obvious also occur.

Mathematical analysis by theoreticians has shown that although the chemical shifts of H_A and H_B in the simple first-order AX spectrum correspond to the center point of each doublet, a more complex situation holds in the second-order cases: The chemical shifts of H_A and H_B are closer to the inner peaks than to the outer peaks. The actual positions of δ_A and δ_B must be calculated. The difference in chemical shift must be determined from the line positions (in Hertz) of the individual peak components of the group, using the equation

$$(\delta_A - \delta_B) = \sqrt{(\delta_1 - \delta_4)(\delta_2 - \delta_3)}$$

where δ_1 is the position (in Hertz downfield from TMS) of the first line of the group, and δ_2, δ_3, and δ_4 are the second, third, and fourth lines, respectively (Fig. 7.42). The chemical shifts of H_A and H_B are then displaced $\frac{1}{2}(\delta_A - \delta_B)$ to each side of the center of the group, as shown in Figure 7.42.

D.   The AB_2 . . . AX_2 and A_2B_2 . . . A_2X_2 Spin Systems

To provide some idea of the magnitude of second-order variations from simple behavior, Figures 7.43 and 7.44 illustrate the calculated ^1H NMR spectra of two additional systems ($-CH-CH_2-$ and $-CH_2-CH_2-$). The first-order spectra appear at the top ($\Delta v/J > 10$), while increasing amounts of second-order complexity are encountered as we move toward the bottom ($\Delta v/J$ approaches zero).

The two systems shown in Figures 7.43 and 7.44 are, then, AB_2 ($\Delta v/J < 10$) and AX_2 ($\Delta v/J > 10$) in one case and A_2B_2 ($\Delta v/J < 10$) and A_2X_2 ($\Delta v/J > 10$) in the other. We will leave discussion of these types of spin systems to more advanced texts, such as those in the reference list at the end of this chapter.

Figures 7.45 through 7.48 show actual 60-MHz ^1H NMR spectra of some molecules of the A_2B_2 type. It is convenient to examine these spectra and compare them with the expected patterns in Figure 7.44; which were calculated from theory using a computer.

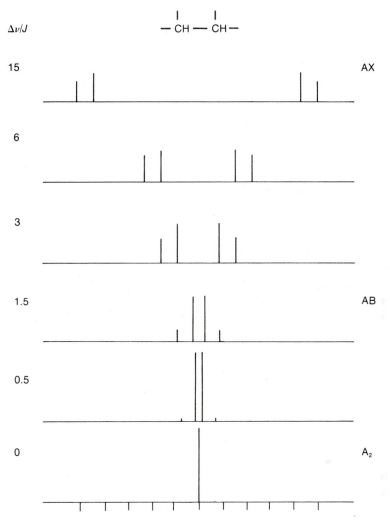

FIGURE 7.41 Splitting patterns of a two-proton system for various values of $\Delta\nu/J$. Transition from an AB to an AX pattern.

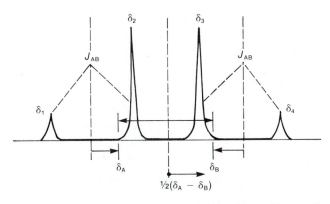

FIGURE 7.42 The relationships among the chemical shifts, line positions, and coupling constant in a two-proton AB system that exhibits second-order effects.

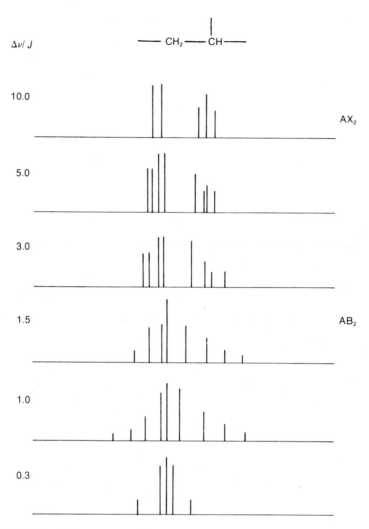

FIGURE 7.43   The splitting patterns of a three-proton system $-CH-CH_2-$ for various $\Delta\nu/J$ values.

E.   Simulation of Spectra

We will not consider all the possible second-order spin systems in this text. Splitting patterns can often be more complicated than expected, especially when the chemical shifts of the interacting groups of protons are very similar. In many cases, only an experienced NMR spectroscopist using a computer can interpret spectra of this type. Today, there are many computer programs, for both PC and UNIX workstations, that can simulate the appearances of NMR spectra (at any operating frequency) if the user provides a chemical shift and a coupling constant for each of the peaks in the interacting spin system. In addition, there are programs that will attempt to match a calculated spectrum to an actual NMR spectrum. In these programs, the user initially provides a best guess at the parameters (chemical shifts and coupling constants), and the program varies these parameters until it finds the best fit. Some of these programs are included in the reference list at the end of this chapter.

F.   The Absence of Second-Order Effects at Higher Field

With routine access to NMR spectrometers with ^1H operating frequencies >300 MHz, chemists today encounter fewer second-order spectra than in years past. In Sections 5.17 and 5.18, you saw

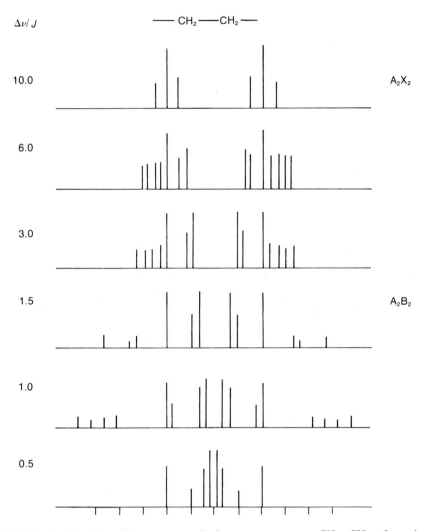

FIGURE 7.44 The splitting patterns of a four-proton system $-CH_2-CH_2-$ for various $\Delta v/J$ values.

that the chemical shift increases when a spectrum is determined at higher field, but that the coupling constants do not change in magnitude (see Fig. 5.38). In other words, Δv (the chemical shift difference in Hertz) increases, but J (the coupling constant) does not. This causes the $\Delta v/J$ ratio to increase, and second-order effects begin to disappear. At high field, many spectra are first order and are therefore easier to interpret than spectra determined at lower field strengths.

As an example. Figure 7.48a is the 60-MHz ^1H NMR spectrum of 2-chloroethanol. This is an A_2B_2 spectrum showing substantial second-order effects ($\Delta v/J$ is between 1 and 3). In Figure 7.48b, which shows the ^1H spectrum taken at 300 MHz, the formerly complicated and second-order patterns have *almost* reverted to two triplets just as the $n + 1$ Rule would predict ($\Delta v/J$ is between 6 and 8). At 500 MHz (Figure 7.48c), the predicted A_2X_2 pattern ($\Delta v/J \sim 12$) is observed.

G. Deceptively Simple Spectra

It is not always obvious when a spectrum has become completely first order. Consider the A_2B_2 to A_2X_2 progression shown in Figure 7.44. At which value of $\Delta v/J$ does this spectrum become truly first order? Somewhere between $\Delta v/J = 6$ and $\Delta v/J = 10$ the spectrum seems to become A_2X_2. The

FIGURE 7.45   The 60-MHz ¹H NMR spectrum of diethyl succinate.

FIGURE 7.46   The 60-MHz ¹H NMR spectrum of phenylethyl acetate.

number of observed lines decreases from 14 lines to only 6 lines. However, if spectra are simulated, incrementally changing $\Delta v/J$ slowly from 6 to 10, we find that the change is not abrupt but gradual. Some of the lines disappear by decreasing in intensity, and some merge together, increasing their intensities. It is possible for weak lines to be lost in the noise of the baseline or for merging lines to approach so closely that the spectrometer cannot resolve them any longer. In these cases, the spectrum would appear to be first order, but in fact it would not quite be so. A common deceptively

(a)

FIGURE 7.47 ¹H NMR spectrum of β-chlorophenetole: (a) 60 MHz, (b) 300 MHz (7.22 peak CHCl₃), (c) 500 MHz (7.24 peak CHCl₃).

FIGURE 7.48 ^{1}H NMR spectrum of 2-chloroethanol: (a) 60 MHz, (b) 300 MHz (OH not shown), (c) 500 MHz (OH not shown).

simple pattern is that encountered with *para*-disubstituted aromatics, an AA'BB' spectrum (see Section 7.10B).

Also notice in Figure 7.41 that the AB spectra with $\Delta v/J$ equal to 3, 6, and 15 all appear roughly first order, but the doublets observed in the range $\Delta v/J = 3$ to 6 have chemical shifts that do not correspond to the center of the doublet (see Fig. 7.42). Unless the worker recognizes the possibility of second-order effects and does a *mathematical* extraction of the chemical shifts, the chemical shift values will be in error. Spectra that appear to be first order, but actually are not, are called **deceptively simple spectra.** The pattern appears to the casual observer to be first order and capable of being explained by the $n + 1$ Rule. However, there may be second-order lines that are either too weak or too closely spaced to observe, and there may be other subtle changes.

Is it important to determine whether a system is deceptively simple? In many cases, the system is so close to first order that it does not matter. However, there is always the possibility that if we assume the spectrum is first order and measure the chemical shifts and coupling constants, we will get incorrect values. Only a complete mathematical analysis tells the truth. For the organic chemist trying to identify an unknown compound, it rarely matters whether the system is deceptively simple. However, if you are trying to use the chemical shift values or coupling constants to prove an important or troublesome structural point, take the time to be more careful. Unless they are obvious cases, we will treat deceptively simple spectra as though they follow the $n + 1$ Rule or as though they can be analyzed by simple tree diagrams. In doing your own work, always realize that there is a considerable margin for error.

7.8 ALKENES

Just as the protons attached to double bonds have characteristic chemical shifts due to a change in hybridization (sp^2 vs. sp^3) and deshielding due to the diamagnetic anisotropy generated by the π electrons of the double bond, alkenyl protons have characteristic splitting patterns and coupling constants. For monosubstituted alkenes, three distinct types of spin interaction are observed:

$^3J_{AB} = 6\text{--}15$ Hz (typically 9–12 Hz)

$^3J_{AC} = 14\text{--}19$ Hz (typically 15–18 Hz)

$^2J_{BC} = 0\text{--}5$ Hz (typically 1–3 Hz)

Protons substituted *trans* on a double bond couple most strongly, with a typical value for ^3J of about 16 Hz. The *cis* coupling constant is slightly more than half this value, about 10 Hz. Coupling between terminal methylene protons (geminal coupling) is smaller yet, less than 5 Hz. These coupling constant values decrease with electronegative substituents in an additive fashion, but $^3J_{trans}$ is always larger than $^3J_{cis}$ for a given system.

As an example of the ^{1}H NMR spectrum of a simple *trans*-alkene, Figure 7.49 shows the spectrum of *trans*-cinnamic acid. The phenyl protons appear as a large group between 7.4 and 7.6 ppm, and the acid proton is a singlet that appears off scale at 13.2 ppm. The two vinyl protons H_A and H_C split each other into two doublets, one centered at 7.83 ppm downfield of the phenyl resonances and the other at 6.46 ppm upfield of the phenyl resonances. The proton H_C, attached to the carbon bearing the phenyl ring, is assigned the larger chemical shift as it resides on the more electron-poor β-carbon of the α,β-unsaturated carbonyl system in addition to its position in a deshielding area of the anisotropic field generated by the π electrons of the aromatic ring. The coupling constant $^3J_{AC}$ can be determined quite easily from the 300-MHz spectrum shown in Figure 7.49. The *trans* coupling constant in this

FIGURE 7.49   The ¹H NMR spectrum of *trans*-cinnamic acid.

case is 15.8 Hz—a common value for *trans* proton–proton coupling across a double bond. The *cis* isomer would exhibit a smaller splitting.

A molecule that has a symmetry element (a plane or axis of symmetry) passing through the C=C double bond does not show any *cis* or *trans* splitting since the vinyl protons are chemically and magnetically equivalent. An example of each type can be seen in *cis*- and *trans*-stilbene, respectively. In each compound, the vinyl protons H$_A$ and H$_B$ give rise to only a single *unsplit* resonance peak.

Plane of symmetry
cis-Stilbene

Twofold axis of symmetry
trans-Stilbene

FIGURE 7.50 The ^{1}H NMR spectrum of vinyl acetate (AMX).

Vinyl acetate gives an NMR spectrum typical of a compound with a terminal alkene. Each alkenyl proton has a chemical shift and a coupling constant different from those of each of the other protons. This spectrum, shown in Figure 7.50 is not unlike that of styrene oxide (Fig. 7.31). Each hydrogen is split into a doublet of doublets (four peaks). Figure 7.51 is a graphical analysis of the vinyl portion. Notice that $^3J_{BC}$ (*trans,* 14 Hz) is larger than $^3J_{AC}$ (*cis,* 6.3 Hz), and that $^2J_{AB}$ (geminal, 1.5 Hz) is very small—the usual situation for vinyl compounds.

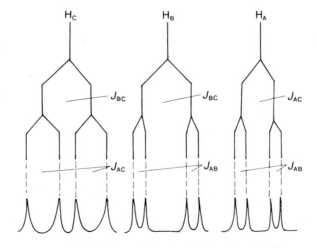

FIGURE 7.51 Graphical analysis of the splittings in vinyl acetate (AMX).

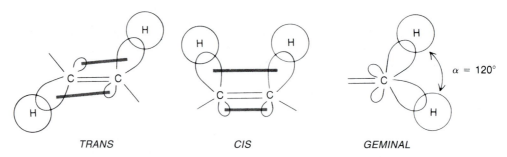

F I G U R E 7.52 Coupling mechanisms in alkenes.

The mechanism of *cis* and *trans* coupling in alkenes is no different from that of any other three-bond vicinal coupling, and that of the terminal methylene protons is just a case of two-bond geminal coupling. All three types have been discussed already and are illustrated in Figure 7.52. To obtain an explanation of the relative magnitudes of the ^3J coupling constants, notice that the two C—H bonds are parallel in *trans* coupling, while they are angled away from each other in *cis* coupling. Also note that the H—C—H angle for geminal coupling is nearly 120°, a virtual minimum in the graph of Figure 7.4. In addition to these three types of coupling, alkenes often show small long-range (allylic) couplings (Section 7.2D).

Figure 7.53 is a spectrum of crotonic acid. See if you can assign the peaks and explain the couplings in this compound (draw a tree diagram). The acid peak is not shown on the full-scale

F I G U R E 7.53 The 300-MHz ^{1}H NMR spectrum of crotonic acid (AMX$_3$).

spectrum, but is shown in the expansions at 12.2 ppm. Also remember that $^3J_{trans}$ is quite large in an alkene while the allylic couplings will be small. The multiplets may be described as a doublet of doublets (1.92 ppm), a doublet of quartets (5.86 ppm), and a doublet of quartets (7.10 ppm) with the peaks of the two quartets overlapping.

7.9 MEASURING COUPLING CONSTANTS—ANALYSIS OF AN ALLYLIC SYSTEM

In this section, we will work through the analysis of the 300-MHz FT-NMR spectrum of 4-allyloxyanisole. The complete spectrum is shown in Figure 7.54. The hydrogens of the allylic system are labeled a–d. Also shown are the methoxy group hydrogens (three-proton singlet at 3.78 ppm) and the *para*-disubstituted benzene ring resonances (second-order multiplet at 6.84 ppm). The origin of the *para*-disubstitution pattern will be discussed in Section 7.10B. The main concern here will be to explain the allylic splitting patterns and to determine the various coupling constants. The exact assignment of the multiplets in the allylic group depends not only on their chemical shift values, but also on the splitting patterns observed. Some initial analysis must be done before any assignments can be definitely made.

Initial Analysis

The allylic OCH_2 group (4.48 ppm) is labeled **a** on the spectrum and is the easiest multiplet to identify because it has an integral of 2H. It is also in the chemical shift range expected for a group

FIGURE 7.54 The 300-MHz ^1H NMR spectrum of 4-allyloxyanisole.

of protons on a carbon atom that is attached to an oxygen atom. It has a larger chemical shift than the upfield methoxy group (3.77 ppm) because it is attached to the carbon–carbon double bond as well as the oxygen atom.

The hydrogen attached to the same carbon of the double bond as the OCH_2 group will be expected to have the broadest and most complicated pattern and is labeled **d** on the spectrum. This pattern should be spread out because the first splitting it will experience is a large splitting $^3J_{cd}$ from the *trans*-H_c, followed by another large coupling $^3J_{bd}$ from the *cis*-H_b. The adjacent OCH_2 group will yield additional (and smaller) splitting into triplets $^3J_{ad}$. Finally, this entire pattern integrates for only 1H.

Assigning the two terminal vinyl hydrogens relies on the difference in the magnitude of a *cis* and a *trans* coupling. H_c will have a *wider* pattern than H_b because it will have a *trans* coupling $^3J_{cd}$ to H_d, while H_b will experience a smaller $^3J_{bd}$ *cis* coupling. Therefore, the multiplet with wider spacing is assigned to H_c, and the narrower multiplet is assigned to H_b. Notice also that each of these multiplets integrates for 1H.

The preliminary assignments just given are tentative, and they must pass the test of a full tree analysis with coupling constants. This will require expansion of all the multiplets so that the exact value (in Hertz) of each subpeak can be measured. Within reasonable error limits, all coupling constants must agree in magnitude wherever they appear.

Tree-Based Analysis and Determination of Coupling Constants

The best way to start the analysis of a complicated system is to start with the **simplest** of the splitting patterns. In this case, we will start with the OCH_2 protons in multiplet **a**. The expansion of this multiplet is shown in Figure 7.55a. It appears to be a doublet of triplets (dt). However, examination of the molecular structure (see Fig. 7.54) would lead us to believe that this multiplet should be a doublet of doublets of doublets (ddd), the OCH_2 group being split first by H_d ($^3J_{ad}$), then by H_b ($^4J_{ab}$), and then by H_c ($^4J_{ac}$), each of which is a single proton. A doublet of triplets could result only if (by coincidence) $^4J_{ab} = {}^4J_{ac}$. We can find out if this is the case by extracting the coupling constants and constructing a tree diagram. Figure 7.55b gives the positions of the peaks in the multiplet. By taking appropriate differences (see Section 7.6), we can extract two coupling constants with magnitudes of 1.5 Hz and 5.2 Hz. The larger value is in the correct range for a vicinal coupling ($^3J_{ad}$), and the smaller value must be identical for both the *cis* and *trans* allylic couplings ($^4J_{ab}$ and $^4J_{ac}$). This

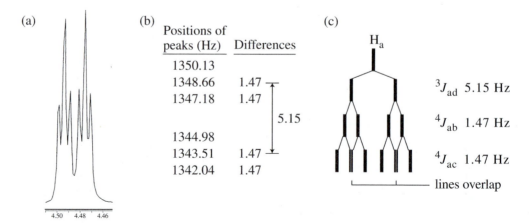

FIGURE 7.55 Allyloxyanisole. (a) Expansion of H_a. (b) Peak positions (Hz) and selected frequency differences. (c) Splitting tree diagram showing the origin of the splitting pattern.

would lead to the tree diagram shown in Figure 7.55c. Notice that when the two smaller couplings are equivalent (or nearly equivalent) the central lines in the final doublet coincide, or overlap, and effectively give triplets instead of pairs of doublets. We will begin by assuming that this is correct. If we are in error, there will be a problem in trying to make the rest of the patterns consistent with these values.

Next consider H_b. The expansion of this multiplet (Fig. 7.56a) shows it to be an apparent doublet of quartets. The largest coupling should be the *cis* coupling $^3J_{bd}$, which should yield a doublet. The geminal coupling $^2J_{bc}$ should produce another pair of doublets (dd), and the allylic geminal coupling $^4J_{ab}$ should produce triplets (two H_a protons). The expected final pattern would be a doublet of doublet of triplets (ddt) with six peaks in each half of the splitting pattern. Since only four peaks are observed, there must be overlap such as was discussed for H_a. Figure 7.56c shows that could happen if $^2J_{bc}$ and $^4J_{ab}$ are both small and have nearly the same magnitude. In fact, the two *J* values appear to be coincidentally the same (or similar), and this is not unexpected (see the typical geminal and allylic values in Section 7.2D). Figure 7.56b also shows that only two different *J* values can be extracted from the positions of the peaks (1.5 and 10.3 Hz). Examine the tree diagram in Figure 7.56c to see the final solution, a doublet of doublet of triplets (ddt) pattern, which appears to be a doublet of quartets due to the coincidental overlap.

H_c is also expected to be a doublet of doublet of triplets (ddt) but shows a doublet of quartets for reasons similar to those explained for H_b. Examination of Figure 7.57 explains how this occurs. Notice that the first coupling ($^3J_{cd}$) is larger than $^3J_{bd}$.

At this point, we have extracted all six of the coupling constants for the system

$$^3J_{cd}\text{-}trans = 17.3 \text{ Hz} \qquad ^2J_{bc}\text{-}gem = 1.5 \text{ Hz}$$

$$^3J_{bd}\text{-}cis = 10.3 \text{ Hz} \qquad ^4J_{ab}\text{-}allylic = 1.5 \text{ Hz}$$

$$^3J_{ad} = 5.2 \text{ Hz} \qquad ^4J_{ac}\text{-}allylic = 1.5 \text{ Hz}$$

H_d has not been analyzed, but we will do this *by prediction* in the next paragraph. Note that three of the coupling constants (all of which are expected to be small ones) are equivalent or nearly equivalent. This is either pure coincidence or could have to do with an inability of the NMR spectrometer to resolve the very small differences between them more clearly. In any event, note that one small inconsistency is seen in Figure 7.56b; one of the differences is 1.1 Hz instead of the expected 1.5 Hz.

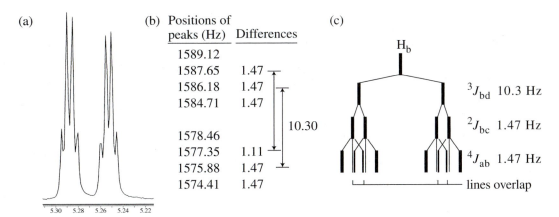

FIGURE 7.56 Allyloxyanisole. (a) Expansion of H_b. (b) Peak positions (Hz) and selected frequency differences. (c) Splitting tree diagram showing the origin of the splitting pattern.

FIGURE 7.57   Allyloxyanisole. (a) Expansion of H_c. (b) Peak positions (Hz) and selected frequency differences. (c) Splitting tree diagram showing the origin of the splitting pattern.

Proton d—A Prediction Based on the *J* Values Already Determined

An expansion of the splitting pattern for H_d is shown in Figure 7.58a, and the peak values in Hz are given in Figure 7.58b. The observed pattern will be predicted using the *J* values just determined as a way of checking our results. If we have extracted the constants correctly, we should be able to correctly predict the splitting pattern. This is done in Figure 7.58c, in which

FIGURE 7.58   Allyloxyanisole. (a) Expansion of H_d. (b) Peak positions (Hz) and selected frequency differences. (c) Splitting tree diagram showing the origin of the splitting pattern.

the tree is constructed to scale using the J values already determined. The predicted pattern is a doublet of doublet of triplets (ddt), which should have six peaks on each half of the symmetrical multiplet. However, due to overlaps, we see what appear to be two overlapping quintets. This agrees well with the observed spectrum, thereby validating our analysis. Another small inconsistency is seen here. The *cis* coupling ($^3J_{bd}$) measured in Figure 7.56 was 10.3 Hz. The same coupling measured from the H_d multiplet gives $^3J_{bd} = 10.7$ Hz. What is the true value of $^3J_{bd}$? The lines in the H_d resonance are sharper than those in the H_b resonance because H_d does not experience the small long-range allylic couplings that are approximately identical in magnitude. In general, J values measured from sharp, uncomplicated resonances are more reliable than those measured from broadened peaks. The true coupling magnitude for $^3J_{bd}$ is likely closer to 10.7 Hz than to 10.3 Hz.

The Method

Notice that we started with the *simplest* pattern, determined its splitting tree, and extracted the relevant coupling constants. Then, we moved to the next most complicated pattern, doing essentially the same procedure, making sure that the values of any coupling constants shared by the two patterns agreed (within experimental error). If they do not agree, something is in error, and you must go back and start again. With the analysis of the third pattern, all of the coupling constants were obtained. Finally, rather than extracting constants from the last pattern, the pattern was predicted using the constants already determined. It is always a good idea to use prediction on the final pattern as a method of validation. If the predicted pattern matches the experimentally determined pattern, then it is almost certainly correct.

7.10 AROMATIC COMPOUNDS—SUBSTITUTED BENZENE RINGS

Phenyl rings are so common in organic compounds that it is important to know a few facts about NMR absorptions in compounds that contain them. In general, the ring protons of a benzenoid system appear around 7 ppm; however, electron-withdrawing ring substituents (e.g., nitro, cyano, carboxyl, or carbonyl) move the resonance of these protons downfield, and electron-donating ring substituents (e.g., methoxy or amino) move the resonance of these protons upfield. Table 7.9 shows these trends for a series of symmetrically *para*-disubstituted benzene compounds. The *p*-disubstituted compounds were chosen because their two planes of symmetry render all of the hydrogens equivalent. Each compound gives only one aromatic peak (a singlet) in the proton NMR spectrum. Later you will see that some positions are affected more strongly than others in systems with substitution patterns different from this one. Table A6.3 in Appendix 6 enables us to make rough estimates of some of these chemical shifts.

In the sections that follow, we will attempt to cover some of the most important types of benzene ring substitution. In many cases, it will be necessary to examine sample spectra taken at both 60 and 300 MHz. Many benzenoid rings show second-order splittings at 60 MHz but are essentially first order at 300 MHz or higher field.

A. Monosubstituted Rings

Alkylbenzenes

In monosubstituted benzenes in which the substituent is neither a strongly electron-withdrawing nor a strongly electron-donating group, all the ring protons give rise to what appears to be a *single resonance* when the spectrum is determined at 60 MHz. This is a particularly common occurrence in alkyl-substituted benzenes. Although the protons *ortho, meta,* and *para* to the substituent are not chemically equivalent, they generally give rise to a single unresolved absorption peak. All of the protons are nearly equivalent under these conditions. The NMR spectra of the aromatic portions of

TABLE 7.9
^{1}H CHEMICAL SHIFTS IN *p*-DISUBSTITUTED BENZENE COMPOUNDS

Substituent X	δ (ppm)	
—OCH$_3$	6.80	Electron donating
—OH	6.60	
—NH$_2$	6.36	
—CH$_3$	7.05	
—H	7.32	
—COOH	8.20	Electron withdrawing
—NO$_2$	8.48	

alkylbenzene compounds are good examples of this type of circumstance. Figure 7.59a is the 60-MHz ^{1}H spectrum of ethylbenzene.

The 300-MHz spectrum of ethylbenzene, shown in Figure 7.59b, presents quite a different picture. With the increased frequency shifts at higher field (see Fig. 5.35), the aromatic protons (that were nearly equivalent at 60 MHz) are neatly separated into two groups. The *ortho* and *para* protons appear upfield from the *meta* protons. The splitting pattern is clearly second order.

FIGURE 7.59 The aromatic ring portions of the ^{1}H NMR spectrum of ethylbenzene at (a) 60 MHz and (b) 300 MHz.

Electron-Donating Groups

When electron-donating groups are attached to the aromatic ring, the ring protons are not equivalent, even at 60 MHz. A highly activating substituent such as methoxy clearly increases the electron density at the *ortho* and *para* positions of the ring (by resonance) and helps give these protons greater shielding than those in the *meta* positions and thus a substantially different chemical shift.

At 60 MHz, this chemical shift difference results in a complicated second-order splitting pattern for anisole (methoxybenzene), but the protons do fall clearly into two groups, the *ortho/para* protons and the *meta* protons. The 60-MHz NMR spectrum of the aromatic portion of anisole (Fig. 7.60) has a complex multiplet for the *o, p* protons (integrating for three protons) that is upfield from the *meta* protons (integrating for two protons), with a clear separation between the two types. Aniline (aminobenzene) provides a similar spectrum, also with a 3:2 split, owing to the electron-releasing effect of the amino group.

FIGURE 7.60 The aromatic ring portions of the ^{1}H NMR spectrum of anisole at (a) 60 MHz and (b) 300 MHz.

FIGURE 7.61 Anisotropic deshielding of the *ortho* protons of benzaldehyde.

The 300-MHz spectrum of anisole shows the same separation between the *ortho/para* hydrogens (upfield) and the *meta* hydrogens (downfield). However, because the actual shift Δv (in Hertz) between the two types of hydrogens is greater, there is less second-order interaction, and the lines in the pattern are sharper at 300 MHz. In fact, it might be tempting to try to interpret the observed pattern as if it were first order, but remember that the protons on opposite sides of the ring are not magnetically equivalent even though there is a plane of symmetry (see Section 7.3). Anisole is an AA'BB'C spin system.

Anisotropy—Electron-Withdrawing Groups

A carbonyl or a nitro group would be expected to show (aside from anisotropy effects) a reverse effect since these groups are electron withdrawing. One would expect that the group would act to decrease the electron density around the *ortho* and *para* positions, thus deshielding the *ortho* and *para* hydrogens and providing a pattern exactly the reverse of the one shown for anisole (3:2 ratio, downfield:upfield). Convince yourself of this by drawing resonance structures. Nevertheless, the actual NMR spectra of nitrobenzene and benzaldehyde do not have the appearances that would be predicted on the basis of resonance structures. Instead, the *ortho* protons are much more deshielded than the *meta* and *para* protons due to the magnetic anisotropy of the π bonds in these groups.

Anisotropy is observed when a substituent group bonds a carbonyl group directly to the benzene ring (Fig. 7.61). Once again, the ring protons fall into two groups, with the *ortho* protons downfield from the *meta/para* protons. Benzaldehyde (Fig. 7.62) and acetophenone both show this effect in their NMR spectra. A similar effect is sometimes observed when a carbon–carbon double bond is attached to the ring. The 300-MHz spectrum of benzaldehyde (Fig. 7.62b) is a nearly first-order spectrum (probably a deceptively simple AA'BB'C spectrum) and shows a doublet (H_C, 2 H), a triplet (H_B, 1 H), and a triplet (H_A, 2 H).

B.   *para*-Disubstituted Rings

Of the possible substitution patterns of a benzene ring, only a few are easily recognized. One of these is the *para*-disubstituted benzene ring. Examine anethole (Fig. 7.63a) as a first example. Because this compound has a plane of symmetry (passing through the methoxy and propenyl groups), the protons H_a and H_a' (both *ortho* to OCH$_3$) would be expected to have the same chemical shift. Similarly, the protons H_b and H_b' should have the same chemical shift. This is found to be the case. You might think that both sides of the ring should then have identical splitting patterns. With this assumption, one is tempted to look at each side of the ring separately, expecting a pattern in which proton H_b splits proton H_a into a doublet, and proton H_a splits proton H_b into a second doublet.

FIGURE 7.62 The aromatic ring portions of the ^{1}H NMR spectrum of benzaldehyde at (a) 60 MHz and (b) 300 MHz.

FIGURE 7.63 The planes of symmetry present in (a) a *para*-disubstituted benzene ring (anethole) and (b) a symmetric *ortho*-disubstituted benzene ring.

Examination of the NMR spectrum of anethole (Fig. 7.64a) shows (crudely) just such a four-line pattern for the ring protons. In fact, a *para*-disubstituted ring is easily recognized by this four-line pattern. However, the four lines do not correspond to a simple first-order splitting pattern. That is because the two protons H_a and H_a' are *not magnetically equivalent* (Section 7.3). Protons H_a and H_a' interact with each other and have finite coupling constant J_{aa}'. Similarly, H_b

FIGURE 7.64 The aromatic ring portions of the 300-MHz ^{1}H NMR spectra of (a) anethole and (b) 4-allyloxyanisole.

and $H_b{}'$ interact with each other and have coupling constant $J_{bb}{}'$. More importantly, H_a does not interact equally with H_b (*ortho* to H_a) and with $H_b{}'$ (*para* to H_a); that is, $J_{ab} \neq J_{ab}{}'$. If H_b and $H_b{}'$ are coupled differently to H_a, they cannot be magnetically equivalent. Turning the argument around, H_a and $H_a{}'$ also cannot be magnetically equivalent because they are coupled differently to H_b and to $H_b{}'$. This fact suggests that the situation is more complicated than it might at first appear. A closer look at the pattern in Figure 7.64a shows that this is indeed the case. With an expansion of the parts-per-million scale, this pattern in fact resembles four distorted triplets, as shown in Figure 7.65. The pattern is an AA'BB' spectrum.

We will leave the analysis of this second-order pattern to more advanced texts. Note, however, that a crude four-line spectrum is characteristic of a *para*-disubstituted ring. It is also characteristic of an *ortho*-disubstituted ring of the type shown in Figure 7.63b, in which the two *ortho* substituents are identical, leading to a plane of symmetry.

As the chemical shifts of H_a and H_b approach each other, the *para*-disubstituted pattern becomes similar to that of 4-allyloxyanisole (Fig. 7.64b). The inner peaks move closer together, and the outer ones become smaller or even disappear. Ultimately, when H_a and H_b approach each other closely enough in chemical shift, the outer peaks disappear, and the two inner peaks merge into a *singlet*; *p*-xylene, for instance, gives a singlet at 7.05 ppm (Table 7.9). Hence, a single aromatic resonance integrating for four protons could easily represent a *para*-disubstituted ring, but the substituents would obviously be identical.

C. Other Substitution

Other modes of ring substitution can often lead to splitting patterns more complicated than those of the aforementioned cases. In aromatic rings, coupling usually extends beyond the adjacent carbon atoms. In fact, *ortho, meta,* and *para* protons can all interact, although the last interaction (*para*) is not usually observed. Following are typical J values for these interactions:

ortho
$^3J = 7-10$ Hz

meta
$^4J = 1-3$ Hz

para
$^5J = 0-1$ Hz

(ppm)

FIGURE 7.65   The expanded *para*-disubstituted benzene AA'BB' pattern.

The trisubstituted compound 2,4-dinitroanisole shows all of the types of interactions mentioned. Figure 7.66 shows the aromatic-ring portion of the ^{1}H NMR spectrum of 2,4-dinitroanisole, and Figure 7.67 is its analysis. In this example, as is typical, the coupling between the *para* protons is essentially zero. Also notice the effects of the nitro groups on the chemical shifts of the adjacent protons. Proton H_D, which lies between two nitro groups, has the largest chemical shift (8.72 ppm). Proton H_C, which is affected only by the anisotropy of a single nitro group, is not shifted as far downfield.

Figure 7.68 gives the 300-MHz ^{1}H spectra of the aromatic-ring portions of 2-, 3-, and 4-nitroaniline (the *ortho, meta,* and *para* isomers). The characteristic pattern of a *para*-disubstituted ring makes it easy to recognize 4-nitroaniline. Here, the protons on opposite sides of the ring are not magnetically equivalent, and the observed splitting is a second order. In contrast, the splitting patterns for 2- and

FIGURE 7.66 The aromatic ring portion of the 60-MHz ¹H NMR spectrum of 2,4-dinitroanisole.

FIGURE 7.67 An analysis of the splitting pattern in the ¹H NMR spectrum of 2,4-dinitroanisole.

3-nitroaniline are simpler, and at 300 MHz a first-order analysis will suffice to explain the spectra. As an exercise, see if you can analyze these patterns, assigning the multiplets to specific protons on the ring. Use the indicated multiplicities (s, d, t, etc.) and expected chemical shifts to help your assignments. You may ignore any *meta* or *para* interactions, remembering that ^4J and ^5J couplings will be too small in magnitude to be observed on the scale that these figures are presented.

In Figures 7.69 and 7.70 the expanded ring-proton spectra of 2-nitrophenol and 3-nitrobenzoic acid are presented. The phenol and acid resonances, respectively, are not shown. In these spectra, the position of each subpeak is given in Hertz. For these spectra, it should be possible not only to assign peaks to specific hydrogens but also to derive tree diagrams with discrete coupling constants for each interaction (see Problem 1 at the end of this chapter).

FIGURE 7.68 The 300-MHz ^{1}H NMR spectra of the aromatic ring portions of 2-, 3-, and 4-nitroaniline.

FIGURE 7.69 Expansions of the aromatic ring proton multiplets from the 300-MHz ^{1}H NMR spectrum of 2-nitrophenol. The hydroxyl resonance is not shown.

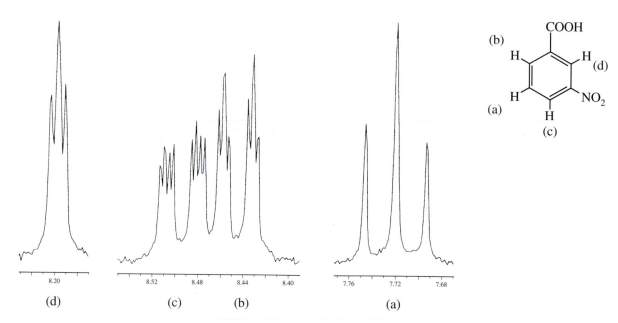

FIGURE 7.70 Expansions of the aromatic ring proton multiplets from the 300-MHz ^{1}H NMR spectrum of 3-nitrobenzoic acid. The acid resonance is not shown.

▨ 7.11 COUPLING IN HETEROAROMATIC SYSTEMS

Heteroaromatic systems (furans, pyrroles, thiophenes, pyridines, etc.) show couplings analogous to those in benzenoid systems. In furan, for instance, couplings occur between all of the ring protons. Typical values of coupling constants for furanoid rings follow. The analogous couplings in pyrrole systems are similar in magnitude.

$$^3J_{\alpha\beta} = 1.6 - 2.0 \text{ Hz}$$

$$^4J_{\alpha\beta'} = 0.3 - 0.8 \text{ Hz}$$

$$^4J_{\alpha\alpha'} = 1.3 - 1.8 \text{ Hz}$$

$$^3J_{\beta\beta'} = 3.2 - 3.8 \text{ Hz}$$

The structure and spectrum for furfuryl alcohol are shown in Figure 7.71. Only the ring hydrogens are shown—the resonances of the hydroxymethyl side chain (—CH$_2$OH) are not included. Determine a tree diagram for the splittings shown in this molecule and determine the magnitude of the coupling constants (see Problem 1 at the end of this chapter). Notice that proton H$_a$ not only

FIGURE 7.71 Expansions of the ring proton resonances from the 300-MHz ^{1}H NMR spectrum of furfuryl alcohol. The resonances from the hydroxymethyl side chain are not shown.

shows coupling to the other two ring hydrogens (H$_b$ and H$_c$) but also appears to have small unresolved *cis*-allylic interaction with the methylene (CH$_2$) group.

Figure 7.72 shows the ring-proton resonances of 2-picoline (2-methylpyridine)—the methyl resonance is not included. Determine a tree diagram that explains the observed splittings and extract the values of the coupling constants (see Problem 1 at the end of this chapter). Typical values of coupling constants for a pyridine ring are different from the analogous couplings in benzene:

$^3J_{ab} = 4–6$ Hz $^3J_{bc} = 7–9$ Hz

$^4J_{ac} = 0–2.5$ Hz $^4J_{bd} = 0.5–2$ Hz

$^5J_{ad} = 0–2.5$ Hz

$^4J_{ae} = < 1$ Hz

Notice that the peaks originating from proton H$_d$ are quite broad, suggesting that some long-range splitting interactions may not be completely resolved. There may also be some coupling of this hydrogen to the adjacent nitrogen ($I = 1$) or a quadrupole-broadening effect operating (Section 8.5). Coupling constant values for other heterocycles may be found in Appendix 5.

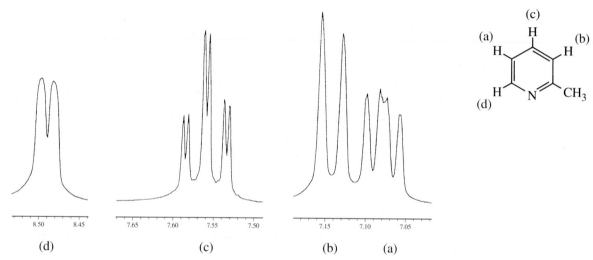

FIGURE 7.72   Expansions of the ring proton resonances from the 300-MHz ^{1}H NMR spectrum of 2-picoline (2-methylpyridine). The methyl resonance is not shown.

7.12 HETERONUCLEAR COUPLING OF ^{1}H TO ^{19}F AND ^{31}P

Heteronuclear coupling of ^{13}C to ^{19}F and ^{13}C to ^{31}P was discussed in Chapter 6, Sections 6.15 and 6.16. Fluorine-19 is the only natural isotope for this element, and while phosphorus has more than one isotope, phosphorus-31 is the most abundant. This section will cover ^{1}H couplings to ^{19}F and ^{31}P. Since ^{19}F and ^{31}P each have a nuclear spin of 1/2 , the principles discussed earlier in this chapter apply directly to interpreting ^{1}H resonances coupled to these elements.

A. ^{1}H to ^{19}F Couplings

Coupling constants for ^{19}F to ^{1}H can be very large, and one may typically expect ^2J and ^3J couplings values to be much larger than typical ^{1}H to ^{1}H coupling constants. The following examples show the effects of ^2J and ^3J couplings of ^{1}H to ^{19}F as the proton moves away from the electronegative fluorine atom.

$^2J_{HF} \sim 50$ Hz $^3J_{HF} \sim 20$ Hz

Remember that protons in the spectrum will couple with the fluorine atoms as well as adjacent protons. Consider as a first example the ^{1}H spectra for 2-fluoroethanol shown in Figure 7.73.

Notice that the protons H_b on C2 appear downfield at 4.52 ppm in Figure 7.73. Fluorine is very electronegative and deshields the adjacent protons. The pattern for this hydrogen atom can be described as a doublet of triplets. The H_b to ^{19}F coupling constant is easily obtained from the Hertz values on the middle peaks in each triplet, as $^2J_{HbF} = 47.8$ Hz (1379.2 – 1331.5). The H_{ab} coupling constant is obtained from the triplet patterns as $^3J = 4.1$ Hz by observing the spacing of the individual peaks in the two triplets (for example, 1379.3 – 1375.2 = 4.1 Hz).

Now look at the doublet of triplets for H_a centered on 3.83 ppm. Because the fluorine atom is farther away from the protons on C1, the value of the coupling constant of H_b to ^{19}F (^3J) is smaller, 29.8 Hz. The value is calculated by subtracting the Hertz values on the two middle peaks in each triplet pattern from each other (1164.5 – 1134.7). We expect, and confirm, that the H_{ab} values obtained from C1 to be nearly identical to those obtained from C2 (for example, 1168.6 – 1164.5 = 4.1 Hz). The other spacing of the peaks in the triplets would yield similar values.

F I G U R E   7 . 7 3 ^{1}H NMR Spectrum of 2-fluoroethanol (300 MHz, CDCl$_3$)

FIGURE 7.74 ^{1}H NMR Spectrum of 3,3,3-trifluoroethanol (300 MHz, CDCl$_3$)

A spectrum for a simpler fluorine-containing compound, 2,2,2-trifluoroethanol, is shown in Figure 7.74. The two protons on C1 see three fluorine atoms on C2, so the $n + 1$ rule predicts a quartet. Thus, we predict that the protons attached to C1 at 3.93 ppm appear at a quartet with ^3J for ^{1}H to ^{19}F = 8.8 Hz (1192.1–1183.3). There are no other protons in the compound to complicate the analysis.

B. ^{1}H to ^{31}P Couplings

Proton-phosphorus coupling constants vary considerably according to the hybridization of the phosphorus atom in addition to the distance between the proton and the phosphorous. For example, tetravalent phosphorus atoms, such as in tetraethylphosphonium chloride, have coupling constants that are larger than ^{1}H–^{1}H coupling constants. Often, coupling constants decrease with distance, but in this case, they increase with distance.

$$^2J_{HP} \sim 14 \text{ Hz} \qquad\qquad ^3J_{HP} \sim 20 \text{ Hz}$$

In Figure 7.75, the resonance centering on 1.33 ppm from the methyl protons shows a doublet of triplets. The larger coupling is a three-bond coupling between the phosphorus and the methyl

FIGURE 7.75 ¹H NMR Spectrum of tetraethylphosphonium chloride (300 MHz, CDCl₃). The pattern centering on 1.33 ppm is a doublet of triplets. The pattern centering on 2.55 ppm can be described as a doublet of quartets. One of the quartets is indicated with dots. The other quartet is shown with x.

hydrogens and $^3J_{HP}$ = 18.1 Hz (407.1–389.0). The smaller coupling is between the methyl and methylene groups with $^3J_{HH}$ = 7.7 Hz (414.8–407.1). The resonance centering on 2.55 ppm in Figure 7.75 can be described as a doublet of quartets. Again, the larger coupling involves the phosphorous nucleus and $^2J_{HP}$ = 13.2 Hz (775.1–761.9 or 767.4–754.2). The $^3J_{HH}$ coupling constant is, of course, 7.7 Hz (782.8–775.1) in agreement with the $^3J_{HH}$ coupling measured on the other signal.

In contrast to tetravalent phosphorus compounds, trivalent phosphorus atoms, such as in trimethylphosphine, have very small coupling constants, ^2J = 1.8 Hz. Pentavalent phosphorus compounds such as those shown, have ^1J, ^2J, and ^3J coupling constants that vary. Usually, $^4J_{HP}$ coupling is ~0 Hz.

diethyl methylphosphonate

H_a	1.35 ppm	triplet	$^3J_{ac} = 7.1$ Hz
H_b	1.48 ppm	doublet	$^2J_{HP} = 17.6$ Hz
H_c	4.10 ppm	multiplet	$^3J_{ac} = 7.1$ Hz

tris(2-chloroethyl) phosphate

H_a	3.74 ppm	triplet	$^3J_{ab} = 5.5$ Hz
H_b	4.34 ppm	doublet of	$^3J_{HP} = 7.7$ Hz
		triplets	$^3J_{ab} = 5.5$ Hz

ethyl methylphosphonate

H_a	1.35 ppm	triplet	$^3J_{ac} = 7$ Hz
H_b	1.49 ppm	doublet	$^2J_{HP} = 17.6$ Hz
H_c	4.10 ppm	app quintet	$^3J_{ac} = 7$ Hz
			$^3J_{HP} = 8$ Hz

dimethyl phosphonate

H_a	3.77 ppm	doublet	$^3J_{HP} = 11.8$ Hz
H_b	6.77 ppm	doublet	$^1J_{HP} = 697.5$ Hz

7.13 HOW TO SOLVE PROBLEMS INVOLVING COUPLING CONSTANT ANALYSIS

In this section, we will use coupling constant analysis to assist us in solving structure determination problems. As with any problem-solving exercise, having a systematic approach simplifies the task and helps ensure we do not overlook any information. We will begin with a brief survey of the NMR data, looking at chemical shifts, integration, and a superficial description of the splitting of each 1H resonance. Usually from this information we can determine the two-dimensional structure of the compound. From there, more detailed analysis of the magnitudes of the couplings can be used to determine three-dimensional information such as olefin geometry or relative configuration of stereogenic centers.

■ SOLVED EXAMPLE 1

The proton NMR spectral information (300 MHz, $CDCl_3$) shown in this problem is for a compound with formula $C_{10}H_{10}O_3$. Expansions are shown for the downfield section of the spectrum. Determine the structure of the compound. Note the offset resonance that appears at 12.3 ppm.

(ppm)

■ SOLUTION

From the molecular formula, $C_{10}H_{10}O_3$ we can determine that this compound has six units of unsaturation. In a brief survey of the NMR data, we can reasonable conclude the compound is a carboxylic acid based on the broad resonance at 12.3 ppm. The carboxylic acid accounts for two oxygen atoms and one unit of unsaturation. There are six protons in the region from 8 to 6 ppm, and based on the chemical shifts are attached to sp^2 carbons. The singlet at 3.8 ppm integrates to three hydrogens, and based on the chemical shift and molecular formula must belong to a methoxy group. At this point it seems likely we have a carboxylic acid (one unit of unsaturation), a benzene ring (four units of unsaturation), and alkene (the last unit of unsaturation) and a methoxy group. How can we put these pieces together?

Let's look at the splitting patterns and coupling constants of the benzene/alkene protons for more clues. Measuring the coupling constants for each resonance provides the values listed in the following summary table.

Signal	δ (ppm)	int.	Multiplicity, *J* (Hz)	Assignment
a	12.3	1	broad singlet	carboxylic acid
b	7.77	1	doublet, *J* = 16	*trans*-alkene deshielded
c	7.33	1	apparent triplet, *J* = 8	aryl H with 2 *ortho* neighbors
d	7.16	1	doublet, *J* = 7.7	aryl H with one *ortho* neighbor
e	7.08	1	br triplet, very small *J*	isolated aryl H (?)
f	6.98	1	doublet of doublets, *J* = 8, 2.5	aryl H with *ortho* and *meta*
g	6.45	1	doublet, *J* = 16	*trans*-alkene
h	3.81	3	singlet	methoxy

The coupling constants of the aryl and alkene protons give critical information. The signals at 7.77 and 6.45 ppm are part of a *trans*-1,2-disubstituted alkene based on their large coupling constant of 16 Hz (Section 7.8). The splitting pattern and coupling constants for the aromatic protons (Section 7.10C) are consisted with a 1,3-disubstituted ring. The most logical structure based on this information is 3-methoxycinnamic acid:

■ **SOLVED EXAMPLE 2**

A portion of the ¹H NMR spectrum (300 MHz, CDCl₃) of commercially available 2-tert-butylcyclohexyl acetate (a mixture of cis/trans isomers) is reproduced here. The resonances shown in the expansion correspond to the proton indicated in bold on the structure. Which diastereomer is the major component, and which is the minor component?

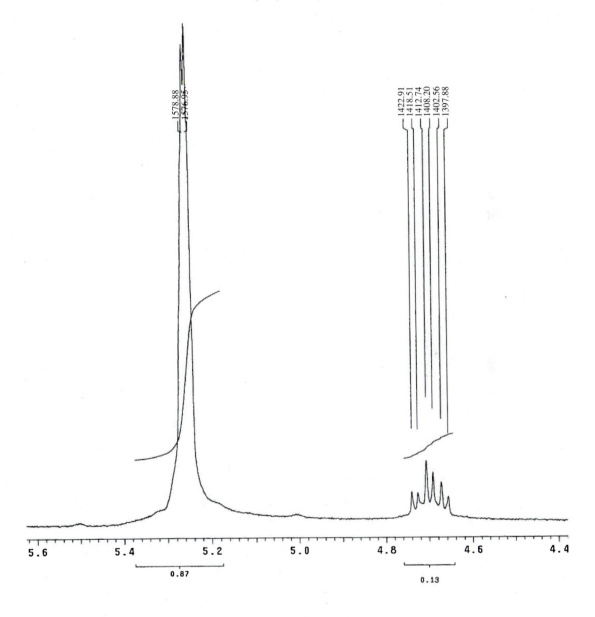

2-*tert*-butyl-cyclohexyl acetate

■ SOLUTION

The downfield proton of the compound in question should appear as a doublet of doublet of doublets (ddd) as a result of having three neighboring protons, two of which are part of a diastereotopic methylene group. The resonance at 4.7 ppm is relatively easier to interpret, so let's start there. The multiplicity of this resonance could be described as a triplet of doublets (td) or a doublet of doublet of doublets (ddd) where two of the individual couplings are identical (Section 7.6). The coupling constants are 10.2, 10.2, and 4.4 Hz. The resonance at 5.23 ppm is a little more difficulty to decipher. There is only one resolved coupling of ~2 Hz. This hardly makes sense for a proton with three neighbors. If all of the couplings for this proton were sufficiently small and similar in magnitude, however, the overlap could result in a broadened resonance like the one in this example. Remember that the total width of a first-order multiplet is the sum of the individual coupling constants (Section 7.6), which in this case means that the sum of the three couplings must be ~10 Hz.

To determine which diastereomers is present in the larger amount, we will need to consider the conformation of the stereoisomers in solution and use our knowledge of coupling constants. The dominant conformation for each diastereomers will have the bulky *tert*-butyl group in the equatorial position:

lowest energy conformer for
trans diastereomer

lowest energy conformer for
cis diastereomer

In the *trans* isomer, the **H** is axial and therefore will experience two large axial-axial couplings and one smaller axial-equatorial coupling (Section 7.2C). This proton appears at 4.70 ppm (ddd, *J* = 10.2, 10.2, 4.4 Hz). In the *cis* isomer, the **H** is equatorial and therefore experiences three small couplings: 2 axial-equatorial and 1 equatorial-equatorial. This proton appears at 5.25 ppm and is technically a ddd, but the small *J* values result in significant overlap of the lines as just discussed. Therefore, the *cis* diastereomer is the major component and the *trans* diastereomer is the minor component.

PROBLEMS

*1. Determine the coupling constants for the following compounds from their NMR spectra shown in this chapter. Draw tree diagrams for each of the protons.

(a) Vinyl acetate (Fig. 7.50).

(b) Crotonic acid (Fig. 7.53).

(c) 2-Nitrophenol (Fig. 7.69).

(d) 3-Nitrobenzoic acid (Fig. 7.70).

(e) Furfuryl alcohol (Fig. 7.71).

(f) 2-Picoline (2-methylpyridine) (Fig. 7.72).

*2. Estimate the expected splitting (J in Hertz) for the lettered protons in the following compounds; i.e., give J_{ab}, J_{ac}, J_{bc}, and so on. You may want to refer to the tables in Appendix 5.

(a)   (b)   (c)

(d)   (e)   (f)

(g)   (h)   (i)

*3. Determine the coupling constants for methyl vinyl sulfone. Draw tree diagrams for each of the three protons shown in the expansions, using Figures 7.55–7.58 as examples. Assign the protons to the structure shown using the letters a, b, c, and d. Hertz values are shown above each of the peaks in the expansions.

*4. The proton NMR spectrum shown in this problem is of *trans*-4-hexen-3-one. Expansions are shown for each of the five unique types of protons in this compound. Determine the coupling constants. Draw tree diagrams for each of the protons shown in the expansions and label them with the appropriate coupling constants. Also determine which of the coupling constants are ^3J and which are ^4J. Assign the protons to the structure using the letters a, b, c, d, and e. Hertz values are shown above each of the peaks in the expansions.

e

2016.57
2009.75
2002.94
2000.86
1996.12
1994.05
1987.24
1980.42

d

1787.54
1785.91
1784.28
1782.50
1771.83
1770.06
1768.43
1766.80

c

719.42
712.02
704.76
697.35

(ppm) 6.72 6.68 6.64 6.60

(ppm) 5.96 5.92 5.88

(ppm) 2.40 2.36 2.32

b

513.06
511.43
506.24
504.61

a

271.14
263.88
256.47

(ppm) 1.72 1.68

(ppm) 0.92 0.88 0.84

*5. The proton NMR spectrum shown in this problem is of *trans*-2-pentenal. Expansions are shown for each of the five unique types of protons in this compound. Determine the coupling constants. Draw tree diagrams for each of the protons shown in the expansions and label them with the appropriate coupling constants. Also determine which of the coupling constants are ^3J and which are ^4J. Assign the protons to the structure using the letters a, b, c, d, and e. Hertz values are shown above each of the peaks in the expansions.

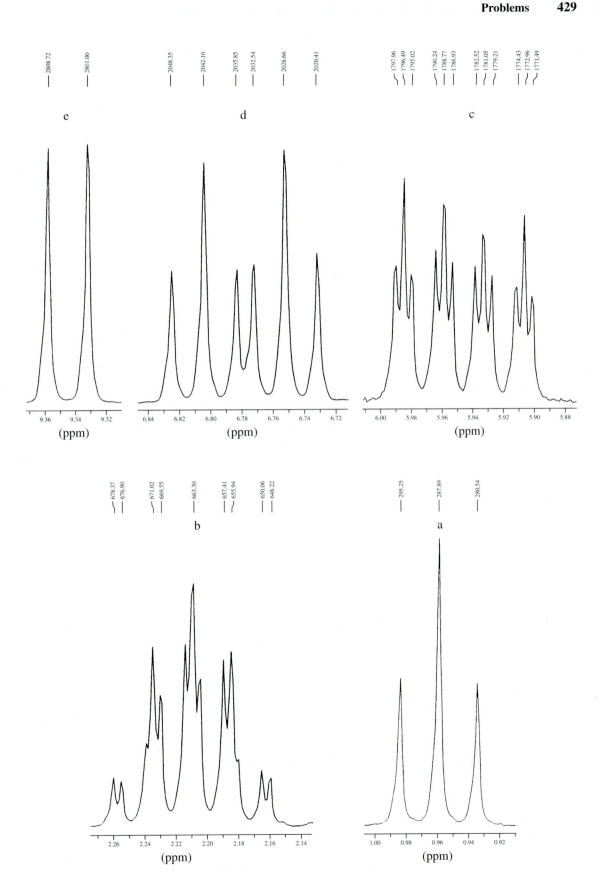

*6. In which of the following two compounds are you likely to see allylic (^4J) coupling?

A   B

7. The reaction of dimethyl malonate with acetaldehyde (ethanal) under basic conditions yields a compound with formula $C_7H_{10}O_4$. The proton NMR is shown here. The normal carbon-13 and the DEPT experimental results are tabulated:

Normal Carbon	DEPT-135	DEPT-90
16 ppm	Positive	No peak
52.2	Positive	No peak
52.3	Positive	No peak
129	No peak	No peak
146	Positive	Positive
164	No peak	No peak
166	No peak	No peak

Determine the structure and assign the peaks in the proton NMR spectrum to the structure.

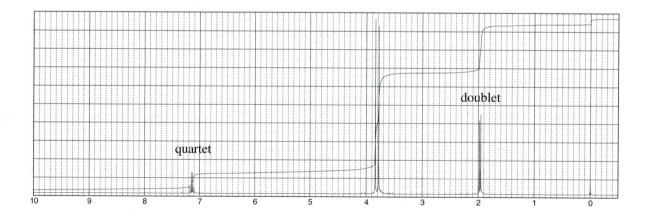

8. Diethyl malonate can be monoalkylated and dialkylated with bromoethane. The proton NMR spectra are provided for each of these alkylated products. Interpret each spectrum and assign an appropriate structure to each spectrum.

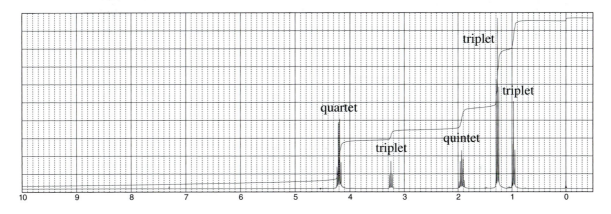

9. The proton NMR spectral information shown in this problem is for a compound with formula $C_8H_8O_3$. An expansion is shown for the region between 8.2 and 7.0 ppm. Analyze this region to determine the structure of this compound. A broad peak (1H) appearing near 12.0 ppm is not shown in the spectrum. Draw the structure of this compound and assign each of the peaks in the spectrum.

10. The proton NMR spectral information shown in this problem is for a compound with formula $C_{12}H_8N_2O_4$. An expansion is shown for the region between 8.3 and 7.2 ppm. No other peaks appear in the spectrum. Analyze this region to determine the structure of this compound. Strong bands appear at 1352 and 1522 cm^{-1} in the infrared spectrum. Draw the structure of this compound.

(ppm)

11. The proton NMR spectral information shown in this problem is for a compound with formula $C_9H_{11}NO$. Expansions of the protons appearing in the range 9.8 and 3.0 ppm are shown. No other peaks appear in the full spectrum. The usual aromatic and aliphatic C–H stretching bands appear in the infrared spectrum. In addition to the usual C–H bands, two weak bands also appear at 2720 and 2842 cm^{-1}. A strong band appears at 1661 cm^{-1} in the infrared spectrum. Draw the structure of this compound.

12. The fragrant natural product anethole ($C_{10}H_{12}O$) is obtained from anise by steam distillation. The proton NMR spectrum of the purified material follows. Expansions of each of the peaks are also shown, except for the singlet at 3.75 ppm. Deduce the structure of anethole, including stereochemistry, and interpret the spectrum.

*13. Determine the structure of the following aromatic compound with formula C_8H_7BrO:

*14. The following spectrum of a compound with formula $C_5H_{10}O$ shows interesting patterns at about 2.4 and 9.8 ppm. Expansions of these two sets of peaks are shown. Expansions of the other patterns (not shown) in the spectrum show the following patterns: 0.92 ppm (triplet), 1.45 ppm (sextet), and 1.61 ppm (quintet). Draw a structure of the compound. Draw tree diagrams of the peaks at 2.4 and 9.8 ppm, including coupling constants.

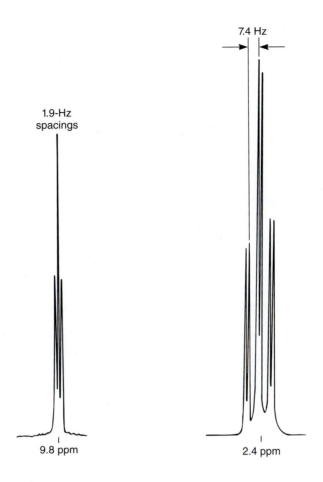

*15. The proton NMR spectral information shown in this problem is for a compound with formula $C_{10}H_{12}O_3$. A broad peak appearing at 12.5 ppm is not shown in the proton NMR reproduced here. The normal carbon-13 spectral results, including DEPT-135 and DEPT-90 results, are tabulated:

Normal Carbon	DEPT-135	DEPT-90
15 ppm	Positive	No peak
40	Negative	No peak
63	Negative	No peak
115	Positive	Positive
125	No peak	No peak
130	Positive	Positive
158	No peak	No peak
179	No peak	No peak

Draw the structure of this compound.

16. The proton NMR spectral information shown in this problem is for a compound with formula $C_{10}H_9N$. Expansions are shown for the region from 8.7 to 7.0 ppm. The normal carbon-13 spectral results, including DEPT-135 and DEPT-90 results, are tabulated:

Normal Carbon	DEPT-135	DEPT-90
19 ppm	Positive	No peak
122	Positive	Positive
124	Positive	Positive
126	Positive	Positive
128	No peak	No peak
129	Positive	Positive
130	Positive	Positive
144	No peak	No peak
148	No peak	No peak
150	Positive	Positive

Draw the structure of this compound and assign each of the protons in your structure. The coupling constants should help you do this (see Appendix 5).

17. The proton NMR spectral information shown in this problem is for a compound with formula $C_9H_{14}O$. Expansions are shown for all the protons. The normal carbon-13 spectral results, including DEPT-135 and DEPT-90 results, are tabulated:

Normal Carbon	DEPT-135	DEPT-90
14 ppm	Positive	No peak
22	Negative	No peak
27.8	Negative	No peak
28.0	Negative	No peak
32	Negative	No peak
104	Positive	Positive
110	Positive	Positive
141	Positive	Positive
157	No peak	No peak

Draw the structure of this compound and assign each of the protons in your structure. The coupling constants should help you do this (see Appendix 5).

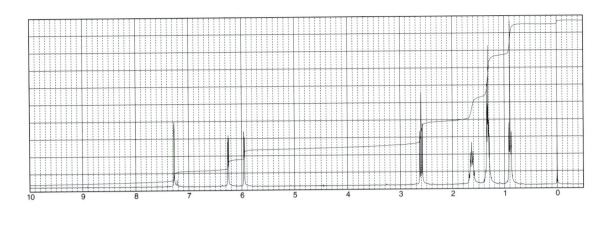

2189.54
2188.81
2187.71
2186.97

1884.74
1882.90
1881.80
1879.96

1792.82
1792.08
1789.87
1789.14

7.30   7.28

(ppm)

6.28   6.26

(ppm)

5.98   5.96   5.94

(ppm)

2.70   2.60   2.50   1.70   1.60   1.40   1.30   1.20   1.00   0.90   0.80

(ppm)

18. The proton NMR spectral information shown in this problem is for a compound with formula $C_{10}H_{12}O_2$. One proton, not shown, is a broad peak that appears at about 12.8 ppm. Expansions are shown for the protons absorbing in the region from 3.5 to 1.0 ppm. The monosubstituted benzene ring is shown at about 7.2 ppm but is not expanded because it is uninteresting. The normal carbon-13 spectral results, including DEPT-135 and DEPT-90 results, are tabulated:

Normal Carbon	DEPT-135	DEPT-90
22 ppm	Positive	No peak
36	Positive	Positive
43	Negative	No peak
126.4	Positive	Positive
126.6	Positive	Positive
128	Positive	Positive
145	No peak	No peak
179	No peak	No peak

Draw the structure of this compound and assign each of the protons in your structure. Explain why the interesting pattern is obtained between 2.50 and 2.75 ppm. Draw tree diagrams as part of your answer.

19. The spectrum shown in this problem is of 1-methoxy-1-buten-3-yne. Expansions are shown for each proton. Determine the coupling constants for each of the protons and draw tree diagrams for each. The interesting part of this problem is the presence of significant long-range coupling constants. There are ^3J, ^4J, and ^5J couplings in this compound. Be sure to include all of them in your tree diagram (graphical analysis).

20. The partial proton NMR spectra (**A** and **B**) are given for the *cis* and *trans* isomers of the compound shown below (the bands for the three phenyl groups are not shown in either NMR). Draw the structures for each of the isomers and use the magnitude of the coupling constants to assign a structure to each spectrum. It may be helpful to use a molecular modeling program to determine the dihedral angles for each compound. The finely spaced doublet at 3.68 ppm in spectrum **A** is the band for the O—H peak. Assign each of the peaks in spectrum **A** to the structure. The O—H peak is not shown in spectrum **B**, but assign the pair of doublets to the structure using chemical shift information.

Spectrum A

Spectrum B

(ppm)

21. The proton NMR spectrum is shown for a compound with formula $C_6H_8Cl_2O_2$. The two chlorine atoms are attached to the same carbon atom. The infrared spectrum displays a strong band 1739 cm^{-1}. The normal carbon-13 and the DEPT experimental results are tabulated. Draw the structure of this compound.

Normal Carbon	DEPT-135	DEPT-90
18 ppm	Positive	No peak
31	Negative	No peak
35	No peak	No peak
53	Positive	No peak
63	No peak	No peak
170	No peak	No peak

22. The proton NMR spectrum of a compound with formula $C_8H_{14}O_2$ is shown. The DEPT experimental results are tabulated. The infrared spectrum shows medium-sized bands at 3055, 2960, 2875, and 1660 cm^{-1} and strong bands at 1725 and 1185 cm^{-1}. Draw the structure of this compound.

Normal Carbon	DEPT-135	DEPT-90
10.53 ppm	Positive	No peak
12.03	Positive	No peak
14.30	Positive	No peak
22.14	Negative	No peak
65.98	Negative	No peak
128.83	No peak	No peak
136.73	Positive	Positive
168.16	No peak	No peak (C=O)

0.97   2.05   7.92   2.96

7.0   6.5   6.0   5.5   5.0   4.5   4.0   3.5   3.0   2.5   2.0   1.5   1.0

23. The proton NMR spectrum of a compound with formula $C_5H_{10}O$ is shown. The DEPT experimental results are tabulated. The infrared spectrum shows medium-sized bands at 2968, 2937, 2880, 2811, and 2711 cm^{-1} and strong bands at 1728 cm^{-1}. Draw the structure of this compound.

Normal Carbon	DEPT-135	DEPT-90
11.35 ppm	Positive	No peak
12.88	Positive	No peak
23.55	Negative	No peak
47.78	Positive	Positive
205.28	Positive	Positive (C=O)

*24. (a) Predict the appearance of the proton NMR spectrum of $F-CH_2-O-CH_3$.

(b) Scientists using modern instruments directly observe many different NMR-active nuclei by changing the frequency of the spectrometer. How would the fluorine NMR spectrum for $F-CH_2-O-CH_3$ appear?

*25. The proton NMR spectral information shown in this problem is for a compound with formula $C_9H_8F_4O$. Expansions are shown for all of the protons. The aromatic ring is disubstituted. In the region from 7.10 to 6.95 ppm, there are two doublets (1H each). One of the doublets is partially overlapped with a singlet (1H). The interesting part of the spectrum is the one proton pattern found in the region from 6.05 to 5.68 ppm. Draw the structure of the compound and draw a tree diagram for the proton at 5.85 ppm. (see Section 7.12 for proton-to-fluorine coupling constants).

26. A compound with the formula C_2H_4BrF has the following NMR spectrum. Draw the structure for this compound. Using the Hertz values on the expansions, calculate the coupling constants. Completely explain the spectrum.

300 MHz

1433.66 1427.91 1422.16   1387.00 1381.25 1375.50   1084.59 1078.73 1072.98 1063.58 1057.83 1052.08

4.80 4.76 4.72 4.68 4.64 4.60 4.56
(ppm)

3.64 3.60 3.56 3.52 3.48
(ppm)

*27. Predict the proton and deuterium NMR spectra of $D-CH_2-O-CH_3$, remembering that the spin quantum number for deuterium = 1. Compare the proton spectrum to that of $F-CH_2-O-CH_3$ (Problem 24a).

*28. Although the nuclei of chlorine ($I = \frac{3}{2}$), bromine ($I = \frac{3}{2}$), and iodine ($I = \frac{5}{2}$) exhibit nuclear spin, the geminal and vicinal coupling constants, J_{HX} (vic) and J_{HX} (gem), are normally zero. These atoms are simply too large and diffuse to transmit spin information via their plethora of electrons. Owing to strong electrical quadrupole moments, these halogens are completely decoupled from directly attached protons or from protons on adjacent carbon atoms. Predict the proton NMR spectrum of $Br-CH_2-O-CH_3$ and compare it to that of $F-CH_2-O-CH_3$ (Problem 24a).

29. The proton NMR spectra for methyltriphenylphosphonium halide and its carbon-13 analogue are shown in this problem. Concentrate your attention on the doublet at 3.25 ppm ($^2J_{HP}$ = 13.2 Hz) in the first spectrum and the doublet of doublets at 3.25 ppm in the 13C-labeled compound. The coupling constants in the second spectrum are 135 and 13.2 Hz. Interpret the two spectra. You may need to refer to Section 7.12 and Appendix 9. Ignore the phenyl groups in your interpretation.

30. All three of the compounds, a, b, and c, have the same mass (300.4 amu). Identify each compound and assign as many peaks as you can, paying special attention to methyl and vinyl hydrogens. There is a small $CHCl_3$ peak near 7.3 ppm in each spectrum that should be ignored when analyzing the spectra.

(a)   (b)   (c)

*31. Calculate the chemical shifts for the indicated protons using Table A6.1 in Appendix 6.

(a) $CH_3-\overset{\overset{\displaystyle O}{\|}}{C}-CH_3$

(b) $CH_3-CH_2-\overset{\overset{\displaystyle O}{\|}}{C}-CH_2-\overset{\overset{\displaystyle O}{\|}}{C}-O-CH_3$

(c) $Cl-CH_2-\overset{\overset{\displaystyle O}{\|}}{C}-O-CH_3$

(d) $CH_3-CH_2-CH_2-C\equiv C-H$

(e) $CH_3-CH_2-\overset{\overset{\displaystyle Cl}{|}}{\underset{\underset{\displaystyle Cl}{|}}{C}}-H$

(f) $CH_2=\overset{\overset{\displaystyle CH_3}{|}}{C}-CH_2-O-H$

*32. Calculate the chemical shifts for the vinyl protons using Table A6.2 in Appendix 6.

(a)
$$\underset{H}{\overset{H}{>}}C=C\underset{CH_3}{\overset{\overset{\displaystyle O}{\|}}{\overset{C-O-CH_3}{<}}}$$

(b)
$$\underset{H}{\overset{CH_3}{>}}C=C\underset{\overset{\displaystyle O}{\|}}{\overset{H}{<}}{\overset{}{C-O-CH_3}}$$

(c)
$$\underset{H}{\overset{H}{>}}C=C\underset{H}{\overset{C_6H_5}{<}}$$

(d)
$$\underset{H}{\overset{C_6H_5}{>}}C=C\underset{\overset{\displaystyle O}{\|}}{\overset{H}{<}}{\overset{}{C-CH_3}}$$

(e)
$$\underset{CH_3}{\overset{H}{>}}C=C\underset{H}{\overset{CH_2-OH}{<}}$$

(f)
$$\underset{CH_3}{\overset{CH_3}{>}}C=C\underset{\overset{\displaystyle O}{\|}}{\overset{H}{<}}{\overset{}{C-CH_3}}$$

33. A trisubstituted benzene derivative possessing one bromine and two methoxy substituents exhibits three aromatic proton resonances: δ 6.40, 6.46, and 7.41. What is the identity of the compound?

34. Partial NMR data (500 MHz, CDCl₃) is given below for the methine proton adjacent to the hydroxyl group of a methylcyclohexanol isomer. Determine whether the data given belongs to the *cis* or *trans* isomer in each case.

a) 2-methylcyclohexanol: 3.78 ppm (ddd, 1H, J = 5.2, 2.7, 2.7 Hz)
b) 3-methylcyclohexanol: 4.05 ppm (dddd, 1H, J = 4.4, 4.4, 3.0, 3.0 Hz)
c) 4-methylcyclohexanol: 3.53 ppm (tt, 1H, J = 10.9, 4.3 Hz)

35. The heliannuols are a family of natural products isolated from the sunflower, *Helianthus annuus*, with unusual sesquiterpenoid structures. Heliannuol G was originally assigned the structure shown [*Journal of Natural Products, 62* (1999): 1636–9]. When a compound with that structure was later synthesized, however, the NMR spectra of the synthetic compound did not match the data from the natural material. Using the following NMR data for the natural product, propose a different reasonable structure for heliannuol G. Note that the OH proton is not visible in the NMR spectrum.

originally proposed
(incorrect) structure of
heliannuol G

Natural heliannuol G: ¹H NMR (400 MHz, CDCl₃): δ (ppm): 6.59 (s, 1H), 6.56 (s, 1H), 5.98 (d, 1H, J = 15.5 Hz), 5.79 (dd, 1H, J = 15.5, 8.4 Hz), 5.09 (dd, 1H, J = 8.4, 7.4 Hz), 3.43 (dq, 1H, J = 7.2, 7.2 Hz), 2.19 (s, 3H), 1.34 (s, 6H), 1.10 (d, 3H, J = 7.2 Hz).

REFERENCES

Books and Monographs

Becker, E. D., *High Resolution NMR: Theory and Chemical Applications,* 3rd ed., Academic Press, New York, 1999.

Breitmaier, E., *Structure Elucidation by NMR in Organic Chemistry: A Practical Guide*, 3rd ed., Wiley, New York, 2002.

Claridge, T. D. W., *High Resolution NMR Techniques in Organic Chemistry*, 2nd ed., Pergamon, Oxford, 2009.

Crews, P., J. Rodriguez, and M. Jaspars, *Organic Structure Analysis*, 2nd ed., Oxford University Press, New York, 2010.

Friebolin, H., *Basic One- and Two-Dimensional NMR Spectroscopy*, 5th ed., Wiley-VCH, Weinheim, 2011.

Günther, H., *NMR Spectroscopy*, 2nd ed., John Wiley and Sons, New York, 1995.

Jackman, L. M., and S. Sternhell, *Applications of Nuclear Magnetic Resonance Spectroscopy in Organic Chemistry*, 2nd ed., Pergamon Press, London, 1969.

Lambert, J. B., S. Gronert, H. F. Shurvell, D. A. Lightner, *Organic Structural Spectroscopy*, 2nd ed., Prentice Hall, Upper Saddle River, NJ, 2011.

Macomber, R. S., *A Complete Introduction to Modern NMR Spectroscopy*, John Wiley and Sons, New York, 1998.

Nelson, J. H., *Nuclear Magnetic Resonance Spectroscopy*, Prentice Hall, Upper Saddle River, NJ, 2003.

Pople, J. A., W. C. Schneider, and H. J. Bernstein, *High Resolution Nuclear Magnetic Resonance*, McGraw–Hill, New York, 1959.

Pretsch, E., T. P. Buhlmann, and C. Affolter, *Structure Determination of Organic Compounds. Tables of Spectral Data*, 4th ed., Springer-Verlag, Berlin and New York, 2009.

Roberts, J. D., *An Introduction to the Analysis of Spin–Spin Splitting in High Resolution Nuclear Magnetic Resonance Spectra*, W. A. Benjamin, New York, 1962.

Roberts, J. D., *ABCs of FT-NMR*, University Science Books, Sausalito, CA, 2000.

Sanders, J. K. M., and B. K. Hunter, *Modern NMR Spectroscopy—A Guide for Chemists*, 2nd ed., Oxford University Press, Oxford, England, 1993.

Silverstein, R. M., F. X. Webster, and D. Kiemle, *Spectrometric Identification of Organic Compounds*, 7th ed., John Wiley and Sons, New York, 2005.

Vyvyan, J. R., Ph.D. Thesis, University of Minnesota, 1995.

Compilations of Spectra

Ault, A., and M. R. Ault, *A Handy and Systematic Catalog of NMR Spectra*, 60 MHz with some 270 MHz, University Science Books, Mill Valley, CA, 1980.

Pouchert, C. J., and J. Behnke, *The Aldrich Library of ^{13}C and ^1H FT-NMR Spectra*, 300 MHz, Aldrich Chemical Company, Milwaukee, WI, 1993.

Computer Programs

Bell, H., Virginia Tech, Blacksburg, VA. (Dr. Bell has a number of NMR programs available from http://www.chemistry.vt.edu/chem-dept/hbell/bellh.htm or e-mail: hmbell@vt.edu.

Reich, H. J., University of Wisconsin, WINDNMR-Pro, a Windows program for simulating high-resolution NMR spectra. http://www.chem.wisc.edu/areas/reich/plt/windnmr.htm

Papers

Helms, E., N. Arpaia, and M. Widener, "Assigning the NMR Spectrum of Glycidol: An Advanced Organic Chemistry Exercise," *Journal of Chemical Education, 84* (2007): 1328–1330.

Hoye, T. R., P. R. Hanson, and J. R. Vyvyan, "A Practical Guide to First-Order Multiplet Analysis in ^1H NMR Spectroscopy," *Journal of Organic Chemistry 59* (1994): 4096.

Hoye, T. R., and H. Zhao, "A Method for Easily Determining Coupling Constant Values: An Addendum to 'A Practical Guide to First-Order Multiplet Analysis in ^1H NMR Spectroscopy,'" *Journal of Organic Chemistry 67* (2002): 4014.

Mann, B. "The Analysis of First-Order Coupling Patterns in NMR Spectra," *Journal of Chemical Education, 72* (1995): 614.

Selected Websites

National Magnetic Resonance Facility at Madison (NMRFAM)
http://www.nmrfam.wisc.edu/
National High Magnetic Field Laboratory
http://www.magnet.fsu.edu/scientificdivisions/nmr/overview.html
National Institute of Materials and Chemical Research, Tsukuba, Ibaraki, Japan
Integrated Spectra Data Base System for Organic Compounds (SDBS):
http://sdbs.riodb.aist.go.jp/sdbs/cgi-bin/cre_index.cgi
WebSpectra, UCLA
NMR and IR spectroscopy problems for students to interpret. They provide links to other sites with problems for students to solve.
http://www.chem.ucla.edu/~webspectra/
Organic Structure Elucidation, Notre Dame
Combined structure problems.
http://www.nd.edu/~smithgrp/structure/workbook.html

NUCLEAR MAGNETIC RESONANCE SPECTROSCOPY

Part Four: Other Topics in One-Dimensional NMR

In this chapter, we shall consider some additional topics in one-dimensional nuclear magnetic resonance (NMR) spectroscopy. Among the topics that will be covered will be the variability in chemical shifts of protons attached to electronegative elements such as oxygen and nitrogen, the special characteristics of protons attached to nitrogen, the effects of solvent on chemical shift, lanthanide shift reagents, methods for determining stereochemical configuration, and spin decoupling experiments.

8.1 PROTONS ON OXYGEN: ALCOHOLS

For most alcohols, no coupling is observed between the hydroxyl hydrogen and vicinal hydrogens on the carbon atom to which the hydroxyl group is attached (^3J for R—CH—OH) under typical conditions of determining the ^{1}H NMR spectrum. Coupling does, in fact, exist between these hydrogens, but the spin–spin splitting is often not observed due to other factors. Whether or not spin–spin splitting involving the hydroxyl hydrogen is observed for a given alcohol depends on several factors, including temperature, purity of the sample, and the solvent used. These variables are all related to the rate at which hydroxyl protons exchange with one another (or the solvent) in solution. Under normal conditions, the rate of exchange of protons between alcohol molecules is faster than the rate at which the NMR spectrometer can respond.

$$R-O-H_a + R'-O-H_b \rightleftharpoons R-O-H_b + R'-O-H_a$$

About 10^{-2} to 10^{-3} sec is required for an NMR transitional event to occur and be recorded. At room temperature, a typical pure liquid alcohol sample undergoes intermolecular proton exchange at a rate of about 10^5 protons/sec. This means that the average residence time of a single proton on the oxygen atom of a given alcohol is only about 10^{-5} sec. This is a much shorter time than is required for the nuclear spin transition that the NMR spectrometer measures. Because the NMR spectrometer cannot respond rapidly to these situations, the spectrometer "sees" the proton as unattached more frequently than it is attached to oxygen, and the spin interaction between the hydroxyl proton and any other proton in the molecule is effectively decoupled. *Rapid chemical exchange decouples spin interactions,* and the NMR spectrometer records only the *time-averaged environment* it detected for the exchanging proton. The hydroxyl proton, for instance, often exchanges between alcohol molecules so rapidly that that proton "sees" all the possible spin orientations of hydrogens attached to the carbon as a single time-averaged spin configuration. Similarly, the α hydrogens see so many different protons on the hydroxyl oxygen (some with spin $+\frac{1}{2}$ and some with spin $-\frac{1}{2}$) that the spin configuration they sense is an average or intermediate value between $+\frac{1}{2}$ and $-\frac{1}{2}$, that is, zero. In effect, the NMR spectrometer is like a camera with a slow shutter speed that is used to photograph a fast event. Events that are faster than the click of the shutter mechanism are blurred or averaged.

If the rate of exchange in an alcohol can be slowed to the point at which it approaches the "time-scale of the NMR" (i.e., <10^2 to 10^3 exchanges per second), then coupling between the hydroxyl proton and vicinal protons on the hydroxyl-bearing carbon can be observed. For instance, the NMR spectrum of methanol at 25°C (ca. 300 K) consists of only two peaks, both singlets, integrating for one proton and three protons, respectively. However, at temperatures below −33°C (<240 K), the spectrum changes dramatically. The one-proton O−H resonance becomes a quartet (^3J = 5 Hz), and the three-proton methyl resonance becomes a doublet (^3J = 5 Hz). Clearly, at or below −33°C (<240 K) chemical exchange has slowed to the point at which it is within time-scale of the NMR spectrometer, and coupling to the hydroxyl proton is observed. At temperatures between 25°C and −33°C (300 K to 240 K), transitional spectra are seen. Figure 8.1 is a stacked plot of NMR spectra of methanol determined at a range of temperatures from 290 K to 200 K.

The room temperature spectrum of an ordinary sample of ethanol (Fig. 8.2) shows no coupling of the hydroxyl proton to the methylene protons. Thus, the hydroxyl proton is seen as a broad singlet, and the methylene protons (split by the methyl group) are seen as a simple quartet. The rate of hydroxyl proton exchange in such a sample is faster than the NMR time-scale, and coupling between the hydroxyl and methylene protons is effectively removed. However, if a sample of ethanol is purified to eliminate all traces of impurity (especially of acids and water, further slowing the O−H proton exchange rate), the hydroxyl–methylene coupling can be observed in the form of increased complexity

FIGURE 8.1 Stacked plot of NMR spectra of methanol determined at a range of temperatures from 290 K to 200 K.

FIGURE 8.2 The NMR spectrum of an ordinary ethanol sample.

of the spin–spin splitting patterns. The hydroxyl absorption becomes a triplet, and the methylene absorptions are seen as an overlapping pair of quartets. The hydroxyl resonance is split (just as the methyl group is, but with a different J value) into a triplet by its two neighbors on the methylene carbon.

$J = 5$ Hz $J = 7$ Hz

Two neighbors ($n + 1 = 3$);
gives a triplet
$J = 5$ Hz

Two different Js; requires
graphical analysis

Two neighbors ($n + 1 = 3$)
gives a triplet
$J = 7$ Hz

The coupling constant for the methylene–hydroxyl interaction is found to be ^3J (CH$_2$, OH) = ~5 Hz. The methyl triplet is found to have a different coupling constant, ^3J (CH$_3$, CH$_2$) = ~7 Hz, for the methylene–methyl coupling. As discussed in Chapter 7, the $n + 1$ Rule does not apply in such an instance; each coupling interaction is independent of the other, but the algorithm described in Section 7.6 makes the analysis straightforward.

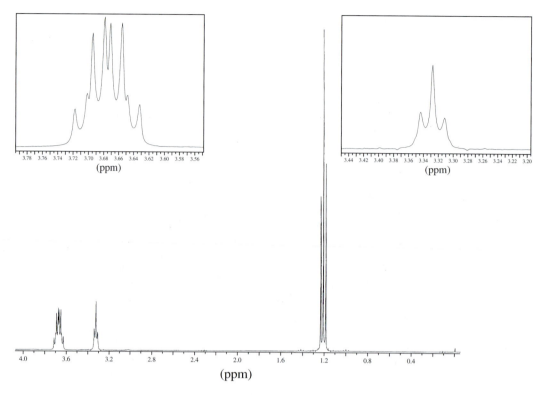

F I G U R E 8.3 The NMR spectrum of an ultrapure sample of ethanol. Expansions of the splitting patterns are included.

Figure 8.3 shows the spectrum of ultrapure ethanol. Notice in the expanded splitting patterns that the methylene protons are split into two overlapping quartets (a doublet of quartets).[1] If even a drop of acid (including water) is added to the ultrapure ethanol sample, proton exchange becomes so fast that the methylene and hydroxyl protons are decoupled, and the simpler spectrum (Fig. 8.2) is obtained.

8.2 EXCHANGE IN WATER AND D₂O

A. Acid/Water and Alcohol/Water Mixtures

When two compounds, each of which contains an O—H group, are mixed, one often observes only a single NMR resonance due to O—H. For instance, consider the spectra of (1) pure acetic acid, (2) pure water, and (3) a 1:1 mixture of acetic acid and water. Figure 8.4 indicates their general appearances. Mixtures of acetic acid and water might be expected to show three peaks since there are two distinct types of hydroxyl groups in the solutions—one on acetic acid and one on water. In addition, the methyl group on acetic acid should give an absorption peak. In actuality, however, mixtures of these two reagents produce only two peaks. The methyl peak occurs at its normal position in the mixture, but there is only a single hydroxyl peak *between* the hydroxyl positions of the pure substances. Apparently, exchange of the type shown in Figure 8.4 occurs so rapidly that the NMR again "sees" the hydroxyl protons only in an averaged environment intermediate between

---

[1]By convention, this pattern would best be referred to as a "quartet of doublets" since the quartet coupling (7 Hz) is larger than the doublet coupling (5 Hz).

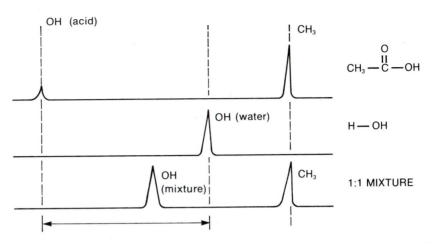

FIGURE 8.4 A schematic comparison of the spectra of acetic acid, water, and a 1:1 mixture of the two.

the two extremes of the pure substances. The exact position of the O—H resonance depends on the relative amounts of acid and water. In general, if there is more acid than water, the resonance appears closer to the pure acid hydroxyl resonance. With the addition of more water, the resonance moves closer to that of pure water. Samples of ethanol and water show a similar type of behavior, except that at low concentration of water in ethanol (1%) both peaks are still often observed. As the amount of water is increased, however, the rate of exchange increases, and the peaks coalesce into the single averaged peak.

$$CH_3-\overset{O}{\underset{O-H_a}{C}} + H-O-H_b \rightleftharpoons CH_3-\overset{O}{\underset{O-H_b}{C}} + H-O-H_a$$

B.   Deuterium Exchange

When compounds with acidic hydrogen atoms are placed in D_2O, the acidic hydrogens exchange with deuterium. Sometimes, a drop of acid or base catalyst is required, but frequently the exchange occurs spontaneously. The catalyst, however, allows a faster approach to equilibrium, a process that can require anywhere from several minutes to an hour or more. Acids, phenols, alcohols, and amines are the functional groups that exchange most readily. Basic catalysis works best for acids and phenols, while acidic catalysis is most effective for alcohols and amines.

Basic catalyst

$$RCOOH + D_2O \rightleftharpoons RCOOD + DOH$$

$$ArOH + D_2O \rightleftharpoons ArOD + DOH$$

Acidic catalyst

$$ROH + D_2O \rightleftharpoons ROD + DOH$$

$$RNH_2 + D_2O \rightleftharpoons RND_2 + DOH$$

The result of each deuterium exchange is that the peaks due to the exchanged hydrogens "disappear" from the ^{1}H NMR spectrum. Since all of the hydrogens end up in HOD molecules, the "lost" hydrogens generate a new peak, that of the hydrogen in HOD. If the NMR spectrum of a particular substance is complicated by the presence of an OH or NH proton, it is possible to simplify the spectrum by removing the peak arising from the exchangeable protons: simply add a few drops of deuterium oxide to the NMR tube containing the solution of the compound being studied. After recapping and shaking the tube vigorously for a few seconds, return the sample tube to the probe and acquire a new spectrum. The added deuterium oxide is immiscible with many NMR solvents, like $CDCl_3$, and forms a layer on top of the solution. The presence of this layer, however, does not usually interfere in the determination of the spectrum. The resonance from the exchangeable proton will either disappear or greatly diminish in intensity, and a new peak, owing to the presence of H$-$O$-$D, will likely be observed, generally between 4.5 and 5.0 ppm. An interesting spectral simplification resulting from D_2O exchange is observed in the case of 2-chloroethanol (Fig. 8.5). The bottom ^{1}H NMR spectrum of 2-chloroethanol clearly shows the OH proton as a broad unsymmetric resonance centered at 2.22 ppm. Also note the complicated appearance of the resonances for the methylene protons at 3.68 and 3.87 ppm resulting from vicinal coupling of the hydroxyl group to the adjacent methylene (**HO**$-$**CH$_2$**$-$**CH$_2$**$-$**Cl**). After addition of D_2O to the sample and thorough mixing, the ^{1}H NMR spectrum was acquired again (Fig. 8.5, top spectrum). Note the nearly complete disappearance of the OH resonance, which is reduced to a very weak, broad signal at

FIGURE 8.5 The 500-MHz ^{1}H NMR spectrum of 2-chloroethanol before (bottom) and after treatment with D_2O (top).

2.38 ppm. Furthermore, the coupling of the hydroxyl proton to the adjacent methylene is removed, and the two methylene groups appear as nearly first-order multiplets.

D_2O can be used as a solvent for NMR, and it is useful for highly polar compounds that do not dissolve well in the standard organic NMR solvents. For instance, some carboxylic acids are difficult to dissolve in $CDCl_3$. A basic solution of NaOD in D_2O is easily produced by adding a small chip of sodium metal to D_2O. This solution readily dissolves carboxylic acids since it converts them to water-soluble (D_2O-soluble) sodium carboxylate salts. The peak due to the hydrogen of the carboxyl group is lost, and a new HOD peak appears.

$$2\,D_2O + 2\,Na \longrightarrow 2\,NaOD + D_2$$

$$RCOOH + NaOD \rightleftharpoons RCOONa + DOH$$

This D_2O/NaOD solvent mixture can also be used to exchange the α hydrogens in some ketones, aldehydes, and esters thereby removing them from the spectrum. Care must be taken, however, that the basic conditions do not promote aldol type condensations in the sample.

$$R-CH_2-\overset{\overset{\displaystyle O}{\|}}{C}-R + 2\,NaOD \rightleftharpoons R-CD_2-\overset{\overset{\displaystyle O}{\|}}{C}-R + 2\,NaOH$$

$$NaOH + D_2O \rightleftharpoons NaOD + DOH$$

Amines dissolve in solutions of D_2O to which the acid DCl has been added. The amino protons end up in the HOD peak.

$$R-NH_2 + 3\,DCl \rightleftharpoons R-ND_3^+\,Cl^- + 2\,HCl$$

$$HCl + D_2O \rightleftharpoons DCl + DOH$$

It is important to note that the presence of deuterium in a compound can, in fact, complicate a proton spectrum in some cases. Since deuterium has $I = 1$, multiplets can end up with more peaks than they originally had. Consider the methine hydrogen in the following case. This hydrogen would be a triplet in the all-hydrogen compound, but it would be a five-line pattern in the deuterated compound. The proton-coupled ^{13}C spectrum would also show an increased complexity due to deuterium (see Section 6.14).

$$-CH_2-\boxed{CH}-\quad\text{versus}\quad-CD_2-\boxed{CH}-$$

Triplet   Five lines

C. Peak Broadening Due to Exchange

Rapid intermolecular proton exchange often (but not always!) leads to *peak broadening*. Rather than having a sharp and narrow line shape, the peak sometimes increases in width at the base and loses height as a result of rapid exchange. Note the hydroxyl peak in Figure 8.2. An O—H peak can often be distinguished from all other singlets on the basis of this shape difference. Peak broadening is caused by factors that are rather complicated, and we will leave their explanation to more advanced texts. We note only that the phenomenon is *time dependent,* and that the intermediate transitional stages of peak coalescence are sometimes seen in NMR spectra when the rate of exchange is neither

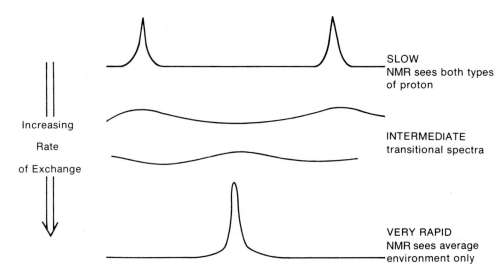

FIGURE 8.6 The effect of the rate of exchange on the NMR spectrum of a hydroxylic compound dissolved in water.

slower nor faster than the NMR time-scale but instead is on approximately the same order of magnitude. Figure 8.6 illustrates these situations.

Also, do not forget that when the spectrum of a pure acid or alcohol is determined in an inert solvent (e.g., $CDCl_3$ or CCl_4), the NMR absorption position is concentration dependent. You will recall that this is due to hydrogen-bonding differences. If you have not, now is a good time to reread Sections 5.11C and 5.19F.

8.3 OTHER TYPES OF EXCHANGE: TAUTOMERISM

The exchange phenomena that have been presented thus far in this chapter have been essentially *intermolecular* in nature. They are examples of **dynamic NMR**, in which the NMR spectrometer is used to study processes that involve the rapid interconversion of molecular species. The rates of these interconversions as a function of temperature can be studied, and they can be compared with the NMR time-scale.

Molecules with structures that differ markedly in the arrangement of atoms but that exist in equilibrium with one another are called **tautomers**. The most common type of tautomerism is **keto–enol tautomerism**, in which the species differ mainly by the position of a hydrogen atom.

In general, the keto form is much more stable than the enol form, and the equilibrium lies strongly in favor of the keto form. Keto–enol tautomerism is generally considered an *intermolecular* process. 1,3-Dicarbonyl compounds are capable of exhibiting keto–enol tautomerism; this is illustrated for the case of **acetylacetone**. For most 1,3-dicarbonyl compounds, the equilibrium lies substantially to the right, favoring the *enol*. The enol form is stabilized through the formation of a strong *intramolecular* hydrogen bond. Note that both methyl groups are equivalent in the enol due to resonance (see arrows).

The proton NMR spectrum of acetylacetone is shown in Figure 8.7. The O—H proton of the enol form (not shown) appears very far downfield, at $\delta = 15.5$ ppm. The vinyl C—H proton is at $\delta = 5.5$ ppm. Note also the strong CH$_3$ peak from the enol form (2.0 ppm) and compare it with the much weaker CH$_3$ peak from the keto form (2.2 ppm). Also note that the CH$_2$ peak at 3.6 ppm is weak. Clearly, the enol form predominates in this equilibrium. The fact that we can see the spectra of both tautomeric forms, superimposed on each other, suggests that the rate of conversion of keto form

FIGURE 8.7 ^{1}H NMR spectrum of acetylacetone. The O—H proton of the enol tautomer is not shown.

to enol form, and vice versa, must be slow on the NMR time-scale. By comparing the integrals of the two different methyl peaks, one can easily calculate the equilibrium distribution of the two tautomers.

Another type of tautomerism, *intramolecular* in nature, is called **valence tautomerism** (or **valence isomerization**). Valence tautomers rapidly interconvert with one another, but the tautomeric forms differ principally by the positions of *covalent bonds* rather than by the positions of protons. There are many examples of valence tautomerism in the literature. An example is the isomerization of **bullvalene**, an interesting compound with threefold symmetry. At low temperatures (below $-85°C$), the proton NMR spectrum of bullvalene consists of four complex multiplets (each of the hydrogens labeled H_a—H_d on the structure below are in unique environments; there are three equivalent H_a positions, three equivalent hydrogens for each of H_b and H_c, and a single hydrogen in environment H_d). As the temperature is raised, however, the multiplets broaden and move closer together. Eventually, at $+120°C$, the entire spectrum consists of *one* sharp singlet—all of the hydrogens are equivalent on the NMR time-scale at that temperature.

To explain the temperature-dependent behavior of the spectrum of bullvalene, chemists have determined that bullvalene rearranges through a series of isomerizations known as **Cope rearrangements**. Notice that repeated Cope rearrangements involve all positions, and as a result all 10 hydrogens in bullvalene become equivalent if the rate of Cope rearrangement is faster than the NMR time-scale. An examination of the temperature at which the different multiplets coalesce into one very broad singlet ($+15°C$) allows the energy of activation, and thus the rate constant, for the isomerization to be determined. This process would be virtually impossible to study by any other technique except NMR spectroscopy.

bullvalene

8.4 PROTONS ON NITROGEN: AMINES

In simple amines, as in alcohols, intermolecular proton exchange is usually fast enough to decouple spin–spin interactions between protons on nitrogen and those on the α carbon atom. Under such conditions, the amino hydrogens usually appear as a broad singlet (unsplit), and in turn the hydrogens on the α carbon are also unsplit by amino hydrogens. The rate of exchange can be made slower by making the solution strongly acidic (pH < 1) and forcing the protonation equilibrium to favor the quaternary ammonium cation rather than the free amine.

FIGURE 8.8 The NMR spectrum of *n*-butylamine.

Under these conditions, the predominant species in solution is the protonated amine, and intermolecular proton exchange is slowed, often allowing us to observe spin–spin coupling interactions that are decoupled and masked by exchange in the free amine. In amides, which are less basic than amines, proton exchange is slow, and coupling is often observed between the protons on nitrogen and those on the α carbon of an alkyl substituent that is substituted on the same nitrogen. The spectra of *n*-butylamine (Fig. 8.8) and 1-phenylethylamine (Fig. 8.9) are examples of uncomplicated spectra (no ^3J HN—CH splitting).

Unfortunately, the spectra of amines are not always this simple. Another factor can complicate the splitting patterns of both amines and amides: Nitrogen itself has a nuclear spin, which is unity ($I = 1$).

FIGURE 8.9 The NMR spectrum of 1-phenylethylamine.

Nitrogen can therefore adopt three spin states: +1, 0, and −1. On the basis of what we know so far of spin–spin coupling, we can predict the following possible types of interaction between H and N:

Direct coupling
$^1J \sim 50$ Hz

Geminal coupling
^2J and $^3J \sim$ negligible
(i.e., almost always zero)

Vicinal coupling

Of these types of coupling, the geminal and vicinal types are very rarely seen, and we can dismiss them. Observation of direct coupling is infrequent but not unknown. Direct coupling is not observed if the hydrogen on the nitrogen is undergoing rapid exchange. The same conditions that decouple NH—CH or HO—CH proton–proton interactions also decouple N—H nitrogen–proton interactions. When direct coupling is observed, the coupling constant is found to be quite large: $^1J \sim 50$ Hz.

One of the cases in which both N—H and CH—NH proton–proton coupling can be observed is the NMR spectrum of methylamine in aqueous hydrochloric acid solution (pH < 1). The species actually being observed in this medium is methylammonium chloride, that is, the hydrochloride salt of methylamine. Figure 8.10 simulates this spectrum. The peak at about 2.2 ppm is due to water (of which there is plenty in aqueous hydrochloric acid solution!). Figures 8.11 and 8.12 analyze the remainder of the spectrum.

FIGURE 8.10 ^{1}H NMR spectrum of $CH_3NH_3^+$ in H_2O (pH < 1).

ANALYSIS OF THE PROTONS ON NITROGEN

FIGURE 8.11 An analysis of the ^{1}H NMR spectrum of methylammonium chloride: protons on nitrogen.

ANALYSIS OF THE METHYL PROTONS

Three neighbors

The ammonium protons
split the methyl
resonance into a
quartet ($n + 1 = 4$)
intensities 1:3:3:1

FIGURE 8.12 An analysis of the ^{1}H NMR spectrum of methylammonium chloride: methyl protons.

8.5 PROTONS ON NITROGEN: QUADRUPOLE BROADENING AND DECOUPLING

Elements that have $I = \frac{1}{2}$ have approximately spherical distributions of charge within their nuclei. Those that have $I > \frac{1}{2}$ have ellipsoidal distributions of charge within their nuclei and as a result have a **quadrupole moment.** Thus, a major factor determining the magnitude of a quadrupole moment is the symmetry about the nucleus. Unsymmetrical nuclei with a large quadrupole moment are very sensitive both to interaction with the magnetic field of the NMR spectrometer and to magnetic and electric perturbations of their valence electrons or their environment. Nuclei with large quadrupole moments undergo nuclear spin transitions at faster rates than nuclei with small moments and easily reach *saturation*—the condition in which nuclear spin transitions (both absorption and emission) occur at a rapid rate. Rapid nuclear transitions lead to an effective decoupling of the nucleus with a quadrupole moment from the adjacent NMR-active nuclei. These adjacent nuclei see a single averaged spin ($I_{\text{effective}} = 0$) for the nucleus with the quadrupole moment, and no splitting occurs. Chlorine, bromine, and iodine have large quadrupole moments and are effectively decoupled from interaction with adjacent protons. Note, however, that fluorine ($I = \frac{1}{2}$) has no quadrupole moment, and it does couple with protons.

Nitrogen has a moderate-size quadrupole moment, and its spin transitions do not occur as rapidly as those in the heavier halogens. Furthermore, the transitional rates and lifetimes of its excited spin states (i.e., its quadrupole moments) vary slightly from one molecule to another. Solvent environment and temperature also seem to affect the quadrupole moment. As a result, three distinct situations are possible with a nitrogen atom:

1. **Small quadrupole moment for nitrogen.** In this case, coupling is seen. An attached hydrogen (as in N−H) is split into three absorption peaks because of the three possible spin states of nitrogen (+1, 0, −1). This first situation is seen in the spectrum of methylammonium chloride (Figs. 8.10 to 8.12). Ammonium and methylammonium salts place the nitrogen nucleus in a very symmetrical environment, and ^{1}H−^{15}N coupling is observed. A similar circumstance occurs in borohydride ion, where ^{1}H−^{11}B and ^{1}H−^{10}B couplings are readily observed.

2. **Large quadrupole moment for nitrogen.** In this case, no coupling is seen. Due to rapid transitions among the three spin states of nitrogen, an attached proton (as in N−H) "sees" an averaged (zero) spin state for nitrogen. A singlet is observed for the hydrogen. This second situation is seen frequently in primary aromatic amines, such as substituted anilines.

FIGURE 8.13 ^{1}H NMR spectrum of pyrrole. The inset shows expansions of the resonances of the ring C—H protons.

3. **Moderate quadrupole moment for nitrogen.** This intermediate case leads to peak broadening, called **quadrupole broadening,** rather than splitting. The attached proton (as in N—H) is "not sure of what it sees." Figure 8.13, the NMR spectrum of pyrrole, shows an extreme example of quadrupole broadening in which the NH absorption extends from 7.5 to 8.5 ppm.

8.6 AMIDES

Quadrupole broadening usually affects only the proton (or protons) attached directly to nitrogen. In the proton NMR spectrum of an amide, we usually expect to see the NH proton appear as a broadened singlet. In some cases, the broadening is due to proton exchange, but recall that the lower acidity of the amide proton slows chemical exchange (Section 8.4). In many instances, one will observe the protons on a carbon atom adjacent to the nitrogen split by the NH proton (^3J H—C—N—H). Nevertheless, the NH peak will still appear as a broad singlet; nuclear quadrupole broadening obscures any coupling to the NH. This is illustrated in the ^{1}H NMR spectrum of *N*-ethylnicotinamide (Fig. 8.14). Note the methylene protons at 3.5 ppm are split by the vicinal methyl protons and the N—H proton and should be a doublet of quartets. In this case, the

FIGURE 8.14 ^{1}H NMR spectrum of *N*-ethylnicotinamide.

resonance is an apparent pentet (apparent quintet) because the two types of vicinal couplings are approximately equal in magnitude. The amide N—H is a broadened singlet at 6.95 ppm.

While considering the NMR spectra of amides, note that groups attached to an amide nitrogen often exhibit different chemical shifts. For instance, the NMR spectrum of *N,N*-dimethylformamide shows two distinct methyl peaks (Fig. 8.15). Normally, one might expect the two identical groups attached to nitrogen to be chemically equivalent because of free rotation around the C—N bond to the carbonyl group. However, the rate of rotation around this bond is slowed by resonance interaction between the unshared pair of electrons on nitrogen and the carbonyl group.

The resonance delocalization requires that the molecule adopt a planar geometry, and it thus interferes with free rotation. If the free rotation is slowed to the point that it takes longer than an NMR transition, the NMR spectrometer sees two different methyl groups, one on the same side of the C=N bond as the carbonyl group and the other on the opposite side. Thus, the groups are in magnetically different environments and have slightly different chemical shifts.

FIGURE 8.15 The ^{1}H NMR spectrum of *N,N*-diethylformamide.

If one successively raises the temperature of the dimethylformamide sample and redetermines the spectrum, the two peaks first broaden (80–100°C), then merge to a single broad peak (~120°C), and finally give a sharp singlet (150°C). The increase of temperature apparently speeds up the rate of rotation to the point at which the NMR spectrometer records an "average" methyl group. That is, the methyl groups exchange environments so rapidly that during the period of time required for the NMR excitation of one of the methyl protons, that proton is simultaneously experiencing all of its possible conformational positions. Figure 8.16 illustrates changes in the appearance of the methyl resonances of *N,N*-dimethylformamide with temperature.

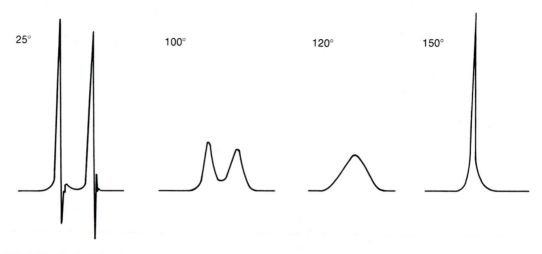

FIGURE 8.16 The appearance of the methyl resonances of *N,N*-dimethylformamide with increasing temperature.

In Figure 8.17, the spectrum of chloroacetamide appears to show quadrupole broadening of the —NH$_2$ resonance. Also, notice that there are *two* N—H peaks. In amides, restricted rotation often occurs about the C—N bond, leading to nonequivalence of the two hydrogens on the nitrogen as was observed for the methyl groups of *N,N'*-dimethylformamide. Even in a substituted amide (RCONHR'), the single hydrogen could have two different chemical shifts.

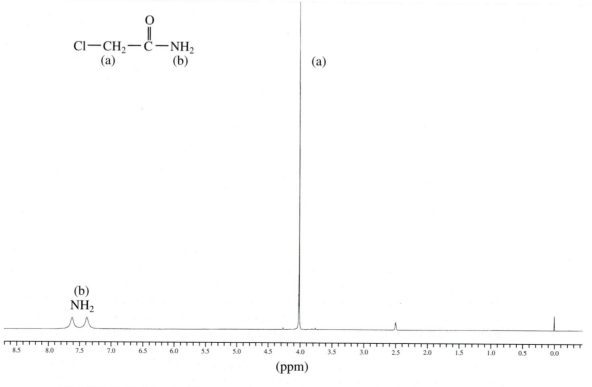

FIGURE 8.17 The ^{1}H NMR spectrum of chloroacetamide.

Depending on the rate of rotation, an averaging of the two NH absorptions could lead to peak broadening (see Sections 8.1, 8.2C, and 8.4). Thus, in amides, three different peak-broadening factors must always be considered:

1. Quadrupole broadening

2. An intermediate rate of hydrogen exchange on nitrogen

3. Nonequivalence of the NH hydrogen(s) due to restricted rotation

The last two effects should disappear at higher temperatures, which increase either the rate of rotation or the rate of proton exchange.

8.7 SOLVENT EFFECTS

Deuteriochloroform (chloroform-d, $CDCl_3$) is the solvent used for the vast majority of NMR measurements of small molecules. It has all the attributes of a desirable solvent for NMR: it dissolves most compounds, is inexpensive, is easily removed from the sample after the spectrum is acquired, and is available in high isotopic purity. The deuterium in the solvent is used for locking and shimming the FT-NMR spectrometer and eliminates proton resonances from the solvent that may obscure the signals from the sample. Chemists often give little consideration to the role of the solvent in determining the spectrum beyond this point.

Intermolecular forces between the solvent and solute, however, sometimes affect the NMR spectrum in significant ways. The local shielding experienced by the nuclei in the sample molecules is a function of both the molecule's electronic structure (Section 5.6) and the magnetic anisotropy of the medium. This is one reason we use a chemical shift reference in NMR spectroscopy. Even so, if the solvent used interacts to a significantly different degree with the solute than with the chemical shift reference compound (TMS), the result is that the observed chemical shift of the sample molecule of interest will be shifted with respect to the observed chemical shift of the reference compound. The magnitude of this **solvent-induced shift** can sometimes be on the order of several tenths of a parts-per-million in a proton spectrum. If one uses a polar solvent (e.g., acetone-d_6, acetonitrile-d_3, dimethylsulfoxide-d_6, methanol-d_3, etc.), the interactions between the polar solvent and a polar solute are likely to be stronger than the interactions between the solvent and nonpolar tetramethylsilane (TMS). Conversely, a non-polar solvent may interact more strongly with TMS than with a more polar solute.

One can get a sense of how common solvent-induced shifts actually are by looking at a series of spectra in a reference work such as *The Aldrich Library of ^{13}C and ^1H FT-NMR Spectra*. All of the spectra in this library were carefully referenced to TMS = 0.00 ppm. Looking through the spectra of nonaromatic esters and lactones in the *Aldrich Library*, for example, one sees the resonance from the residual chloroform peak (the small amount of $CHCl_3$ remaining in the $CDCl_3$) varies from 7.25 to 7.39 ppm. This chemical shift variability is from the small changes in the local shielding environment of the $CHCl_3$ induced by the solute (and vice versa) via intermolecular interactions. A couple tenths of a ppm may not sound like much, but when one is trying to match the spectrum of a painstakingly prepared synthetic material to literature data for a natural product, every Hertz counts. Great care must be taken when comparing one's own experimental data with tabulated spectral data from the literature for chemical shift matches. Many researchers use NMR solvents that do not contain TMS and thus reference their chemical shift to the residual protonated solvent signal, which, as we have just seen, can vary. One should be sure to reference spectra in the same way as described in the literature. When making such comparisons, it is not at all uncommon to have consistent chemical shift mismatches across a spectrum, with all of the resonances having a slightly higher (or lower) resonant frequency than the literature data.

Simply changing the concentration of the solute can sometimes result in significant chemical shift changes, especially for environments near a hydrogen bond donor/acceptor or π-systems with significant diamagnetic anisotropy. For example, the ^1H resonances of acridine in CD_3OD shift

FIGURE 8.18 ^{1}H NMR spectra of acridine in CD_3OD at different concentrations (mol/L). (From Mitra, A., P. J. Seaton, R. A. Assapour, T. Williamson; *Tetrahedron 54* (1998): 15489–15498.)

between 0.1 and 0.5 ppm upfield, depending on the position on the ring, as the concentration of the acridine is increased (Figure 8.18).

If the solvent has a strong diamagnetic anisotropy (e.g., benzene-d_6, pyridine-d_5, or nitromethane-d_3), the interaction between the solute and the anisotropic field of the solvent may give rise to significant chemical shift changes. Solvents such as benzene and pyridine will generally cause the observed resonance of a given proton to be shifted to a higher field (smaller δ), while other solvents, such as acetonitrile, cause a shift in the opposite direction. This difference appears to be dependent, in part, on the shape of the solvent molecules affecting the nature of the solute–solvent complexes that are formed in solution. Aromatic solvents such as benzene and pyridine are flat, of course, while acetonitrile has a rodlike shape.

The chemist can use these solvent-induced chemical shift changes to her advantage to clarify complex spectra that feature overlapping resonances. Often, by adding just a small amount (5–20%) of benzene-d_6 (C_6D_6) or pyridine-d_5 to a $CDCl_3$ solution of an unknown, significant effects on the appearance of the spectrum can often be observed. The use of this technique is an easy way to simplify a crowded spectrum. In a recent example, titration of a sample with C_6D_6 was used to resolve three overlapping methyl resonances, which together with other spectroscopic and computational experiments proved that the natural products phomopsichalasin and diaporthichalasin were in fact identical (Figure 8.19).

For samples that contain basic or acid-sensitive functional groups, trace amounts of acid in NMR solvents can make reproducible NMR measurements difficult. Slow decomposition of $CDCl_3$ creates small amounts of DCl. This acid can be removed by passing the solvent through a small plug of basic alumina immediately before preparing the NMR sample. In other cases, deliberate addition of acid becomes necessary to reproduce spectral data obtained from the literature. In the case of the indole-derived natural product mersicarpine, an acid titration demonstrated that the chemical shifts of all of the aromatic protons and the diastereotopic methylenes of the ring system all changed dramatically depending on the amount of acid present (Figure 8.20). Note that the spectral data

FIGURE 8.19 Insets of methyl regions of ^{1}H NMR spectra of phompsichalasin/diaporthichalasin in methanol-d_4 (500 MHz) containing varying percentages of added benzene-d_6 in order to verify the presence of two methyl doublets with J values of 7.3 and 6.5 Hz. (From Brown, S. G., M. J. Jansma, T. R. Hoye; *Journal of Natural Products, 75* (2012): 1326–1331.) Reprinted by Permission.

FIGURE 8.20 Variation of ^{1}H NMR resonances with added trifluoroacetic acid. Vertical lines indicate literature values for mersicarpine. (From Magolan, J., C. A. Carson, M. A. Kerr, *Organic Letters, 10* (2008): 1437–1440.) Reprinted with permission.

obtained in base-washed $CDCl_3$ do not match the literature data, indicating an unknown amount of an acidic species was present in the sample of natural material.

We have seen that hydrogen bonding of exchangeable groups alters the appearance of NMR spectra (Section 8.2). Hydrogen bonding interactions (or lack thereof) with the solvent can in some cases change the lowest energy conformation of a compound, significantly altering observed chemical shifts and coupling constants. This phenomenon is illustrated by the case of *cis-3*-methoxycyclohexanol. When the H^1 NMR spectrum of *cis-3*-methoxycyclohexanol is acquired in cyclohexane-d_{12}, the vicinal coupling constants indicate the diaxial conformation is, in fact, lower in energy than the diequatorial one (calculations based on the coupling data indicate a 67:33 preference for diaxial). An intramolecular hydrogen bond apparently compensates for the destabilizing effect of the 1,3-diaxial steric repulsion between the hydroxyl and methoxy groups. When DMSO-d_6 is used for the solvent, however, the diequatorial conformation is lower in energy to the tune of a 97:3 ratio. In this case, the strong dipolar interactions between the solute and polar DMSO are stronger that the intramolecular hydrogen bond.[2]

solvent	$J_{H_1\text{-}H_{2a}}$ (Hz)	$J_{H_1\text{-}H_{2e}}$ (Hz)
cyclohexane-d_{12}	6.6	3.4
DMSO-d_6	10.9	4.2

Solvents also play a role in NMR spectroscopy as common impurities in samples, especially in synthetic work, for which trace amounts of solvents that were not removed completely by rotary evaporation often remain in samples. Other common trace impurities in spectra include water (either from the deuterated solvent or from the surface of the sample tube) and stopcock grease. Occasionally, one will see resonances in an NMR spectrum from plasticizer that has leached from laboratory tubing. Being able to spot these trace impurities for what they are and "mentally edit" the spectrum to avoid distraction by the extraneous resonances is a valuable skill. Just as chemical shifts of sample resonances can change in different solvents, the chemical shifts of these trace impurities also appear at different places in the spectrum in different solvents. Tables listing the properties of common NMR solvents will often include an entry for the chemical shift of residual water as well. Trace water, for example, appears at 1.56 ppm in $CDCl_3$, but at 0.40 ppm in benzene-d_6 (C_6D_6) and at 2.13 ppm and 4.78 ppm in acetonitrile-d_3 (CD_3CN) and methanol-d_4 (CD_3OD), respectively. Two of the most widely cited papers in the last 15 years are tabulations of the ^1H and ^{13}C chemical shifts of common laboratory solvents and other impurities in commonly used deuterated solvents.

---

[2]de Oliveira, P. R., R. Rittner, "The relevant effect of an intramolecular hydrogen bond on the conformational equilibrium of *cis-3*-methyoxycyclohexanol compared to *trans-3*-methoxycyclohexanol and *cis-1,3*-dimethoxycyclohexanol," *Spectrochemica Acta A 61* (2005): 1737–1745.

8.8 CHEMICAL SHIFT REAGENTS

Often, the low-field (60- or 90-MHz) spectrum of an organic compound, or a portion of it, is almost undecipherable because the chemical shifts of several groups of protons are all very similar. In such a case, all of the proton resonances occur in the same area of the spectrum, and often peaks overlap so extensively that individual peaks and splittings cannot be extracted. One of the ways in which such a situation can be simplified is by the use of a spectrometer that operates at a frequency higher. Although coupling constants do not depend on the operation frequency or the field strength of the NMR spectrometer, chemical shifts in Hertz *are* dependent on these parameters (as Section 5.17 discussed). This circumstance can often be used to simplify an otherwise-undecipherable spectrum.

Suppose, for instance, that a compound contains three multiplets: a quartet and two triplets derived from groups of protons with very similar chemical shifts. At 60 MHz, these peaks may overlap and simply give an unresolved envelope of absorptions. In redetermining the spectrum at higher field strengths, the coupling constants do not change, but the chemical shifts in Hertz (not parts per million) of the proton groups (H_A, H_B, H_C) responsible for the multiplets do increase. At 300 MHz, the individual multiplets are cleanly separated and resolved (see, for example, Fig. 5.35). Also remember that second-order effects disappear at higher fields, and that many second-order spectra become first order at or above 300 MHz (Sections 7.7A and 7.7F).

Researchers have known for some time that interactions between molecules and solvents, such as those due to hydrogen bonding, can cause large changes in the resonance positions of certain types of protons (e.g., hydroxyl and amino). They have also known that changing from the usual NMR solvents such as $CDCl_3$ to solvents such as benzene, which impose local anisotropic effects on surrounding molecules, can greatly affect the resonance positions of some groups of protons (just discussed in Section 8.7). In many cases, it is possible to resolve partially overlapping multiplets by such a solvent change. However, the use of **chemical shift reagents**, an innovation dating from the late 1960s, allows a rapid and relatively inexpensive means of resolving overlapping multiplets in some spectra. Most of these chemical shift reagents are organic complexes of paramagnetic rare-earth metals from the lanthanide series. When such metal complexes are added to the compound for which the spectrum is being determined, profound shifts in the resonance positions of the various groups of protons are observed. The direction of the shift (upfield or downfield) depends primarily on which metal is used. Complexes of europium, erbium, thulium, and ytterbium shift resonances to lower field (larger δ), while complexes of cerium, praseodymium, neodymium, samarium, terbium, and holmium generally shift resonances to higher field. The advantage of using such reagents is that shifts similar to those observed at higher field can be induced without the purchase of a high-field NMR instrument.

Of the lanthanides, europium is probably the most commonly used metal for shift reagents. Two of its widely used complexes are *tris*-(dipivalomethanato) europium and *tris*-(6,6,7,7,8,8,8-heptafluoro-2,2-dimethyl-3,5-octanedionato) europium, frequently abbreviated $Eu(dpm)_3$ and $Eu(fod)_3$, respectively.

Eu(dpm)₃   Eu(fod)₃

These lanthanide complexes produce spectral simplifications in the NMR spectrum of any compound with a relatively basic pair of electrons (an unshared pair) which can coordinate with Eu^{3+}. Typically, aldehydes, ketones, alcohols, thiols, ethers, and amines all interact:

$$2 \text{ B: } + \text{ Eu(dpm)}_3 \longrightarrow \quad \begin{array}{c} \text{B:} \\ \\ \text{B:} \end{array} \begin{array}{c} \text{dpm} \\ \text{Eu---dpm} \\ \text{dpm} \end{array}$$

The amount of shift a given group of protons experiences depends on (1) the distance separating the metal (Eu^{3+}) and that group of protons and (2) the concentration of the shift reagent in the solution. Because of the latter dependence, it is necessary to include the number of mole equivalents of shift reagent used or its molar concentration when reporting a lanthanide-shifted spectrum.

The spectra of 1-hexanol (Figs. 8.21 and 8.22) beautifully illustrate the distance factor. In the absence of shift reagent, the spectrum shown in Figure 8.21 is obtained. Only the triplet of the terminal methyl group and the triplet of the methylene group next to the hydroxyl are resolved in the spectrum. The other protons (aside from O—H) are found together in a broad, unresolved group. With the shift reagent added (Fig. 8.22), each of the methylene groups is clearly separated and is resolved into the proper multiplet structure. The spectrum is in every sense *first order* and thus simplified; all of the splittings are explained by the $n + 1$ Rule.

Note one final consequence of the use of a shift reagent. Figure 8.22 shows that the multiplets are not as nicely resolved into sharp peaks as one usually expects. The europium cation of the shift reagent causes a small amount of line broadening by decreasing the relaxation time of the protons in the sample. At high shift-reagent concentrations this problem becomes serious, but at most useful concentrations the amount of broadening is tolerable.

F I G U R E   8 . 2 1   The normal 60-MHz ^{1}H NMR spectrum of 1-hexanol.

FIGURE 8.22 The 100-MHz NMR spectrum of 1-hexanol with 0.29-mole equivalent of Eu(dpm)$_3$ added. (From Sanders, J. K. M., and D. H. Williams, *Chemical Communications* (1970): 442. Reprinted by permission.)

Today, most laboratories have access to high-field NMR spectrometers (operating at a ^{1}H frequency of 300 MHz or greater), and simple chemical shift reagents as discussed above are infrequently used. Lanthanide complexes in which the organic ligand on the metal is optically active, however, create a **chiral shift reagent**. One such reagent commonly used for this purpose is tris [3-(heptafluoropropylhydroxymethylene)-*d*-camphorato] europium(III) [Eu(hfc)$_3$]. When Eu(hfc)$_3$ complexes to a chiral molecule, diastereomeric complexes are formed, which gives rise to different chemical shifts for protons that were previously identical.

Tris[3-(heptafluoropropylhydroxymethylene)-*d*-camphorato] europium(III) [Eu(hfc)$_3$]

8.9 CHIRAL RESOLVING AGENTS

A group attached to a stereocenter normally has the same chemical shift whether the stereogenic center has *R* or *S* configuration. However, the group can be made diastereotopic in the NMR (have different chemical shifts) when the racemic parent compound is treated with an optically pure **chiral resolving agent** to produce diastereomers. In this case, the group is no longer present in two enantiomers but in two different *diastereomers*, and its chemical shift is different in each environment.

For instance, if a mixture containing both the *R* and *S* enantiomers of α-phenylethylamine is mixed with an equimolar amount of optically pure (*S*)-(+)-*O*-acetylmandelic acid in an NMR tube containing $CDCl_3$, two diastereomeric salts form:

$$
\overset{(R/S)}{CH_3-\underset{\underset{Ph}{|}}{CH}-NH_2} + \overset{(S)}{Ph-\underset{\underset{OAc}{|}}{CH}-COOH} \longrightarrow \left[\overset{(R)}{CH_3-\underset{\underset{Ph}{|}}{CH}-NH_3^+} + \overset{(S)}{Ph-\underset{\underset{OAc}{|}}{CH}-COO^-} \right]
$$

α-Phenylethylamine   *S*-(+)-*O*-acetyl-
mandelic acid

$$
+
$$

$$
\left[\overset{(S)}{CH_3-\underset{\underset{Ph}{|}}{CH}-NH_3^+} + \overset{(S)}{Ph-\underset{\underset{OAc}{|}}{CH}-COO^-} \right]
$$

Diastereomers

The methyl groups in the amine portion of the salts are attached to a stereocenter, *S* in one case and *R* in the other. As a result, the methyl groups themselves are now diastereotopic, and they have different chemical shifts. In this case, the *R* isomer is downfield, and the *S* isomer is upfield. Since the methyl groups are adjacent to a methine (CH) group, they appear as doublets at approximately 1.1 and 1.2 ppm, respectively, in the NMR spectrum of the mixture (the exact chemical shifts vary slightly with concentration) (Fig. 8.23).

These doublets may be integrated to determine the exact percentages of the *R* and *S* amines in the mixture. In the example shown, the NMR spectrum was determined with a mixture made by dissolving equal quantities of unresolved (±)-α-phenylethylamine and a student's resolved product, which contained predominantly (*S*)-(−)-α-phenylethylamine.

FIGURE 8.23 The 300-MHz ^1H spectrum of a 50–50 mixture of (*S*)-α-phenylethylamine from a resolution and unresolved (racemic) α-phenylethylamine in $CDCl_3$ with the chiral resolving agent (*S*)-(+)-*O*-acetylmandelic acid added.

Similarly, an optically pure amine can be used as a chiral resolving agent to analyze the optical purity of a chiral carboxylic acid. For example, addition of optically pure (S)-(−)-α-phenylethylamine to a CDCl$_3$ solution of O-acetylmandelic acid will form diastereomeric salts as illustrated earlier. In this case, one would look for the two doublets (one for each enantiomer) from the Ph–**CH**–OAc methine between 5 and 6 ppm in the ^{1}H NMR spectrum.

When one needs to determine the optical purity of a compound that is not amenable to salt formation (i.e., not a carboxylic acid or amine), analysis by NMR becomes slightly more difficult. It is frequently necessary to determine the enantiomeric excesses of chiral secondary alcohols, for example. In these cases, derivatization of the alcohol through covalent attachment of an optically pure auxiliary provides the mixture of diastereomers for analysis. This requires reacting a small sample of chiral alcohol with the optically pure derivatizing agent. Sometimes, purification of the products is necessary. In the example shown below, a chiral secondary alcohol is reacted with (S)-2-methoxyphenylacetic acid [(S)-MPA] using dicyclohexylcarbodiimide (DCC) to form diastereomeric esters. After workup, the ^{1}H NMR spectrum of product mixture is acquired, and the resonances from the oxygenated methine (**H**CR$_1$R$_2$– O–Aux, there will be one signal for each diastereomer) are integrated to determine the optical purity (enantiomeric excess) of the original alcohol sample. Because the products are diastereomers, other methods of analysis (for example, gas chromatography) could also be used for this purpose.

This process is illustrated in Figure 8.24 for 2-pentanol and α-methoxyphenylacetic acid (MPA). To simplify the discussion, ^{1}H NMR spectra from two separate samples are shown. The ester formed

FIGURE 8.24 Use of 2-methoxyphenylacetic acid (MPA) as a chiral derivatizing reagent. (From Seco, J. M., E. Quinoa, and R. Riguera, *Chemical Reviews 104* (2004): 17–117.) Reprinted by permission.

from (*R*)-2-pentanol and (*R*)-MPA produced the top spectrum in Figure 8.24, and the ester formed from (*R*)-2-pentanol and (*S*)-MPA produced the bottom spectrum. Most diagnostic are the chemical shifts of the methyl doublets. The lowest energy conformation of the (*R*, *R*) ester places position 3' in the shielding region of the phenyl ring, and the methyl group (position 1') is not significantly perturbed, and its doublet appears at 1.18 ppm. In the lowest energy conformation of the (*R*, *S*) ester, however, the methyl group is shielded by the phenyl ring, and its doublet appears upfield at 1.05 ppm. One can imagine an analogous set of spectra would be produced by esters formed by reaction of just one enantiomer of MPA with a mixture of 2-pentanol enantiomers. Integration of the two different methyl doublets would give the enantiomeric ratio of the alcohol sample.

☐ 8.10 DETERMINING ABSOLUTE AND RELATIVE CONFIGURATION VIA NMR

A.   Determining Absolute Configuration

The methods described in Section 8.9 are very useful for determining optical purities (enantiomeric excesses), but it is usually not possible to determine with certainty the *absolute* configuration of the major enantiomer present unless one has access to authentic samples of each pure enantiomer. This is rarely the case in natural product isolation or synthesis research. In 1973, Mosher described a method to determine the absolute configuration of secondary alcohols by NMR analysis, and since that time his method has been expanded and refined. In Mosher's method, the alcohol is reacted separately with each enantiomer of methoxytrifluoromethylphenylacetic acid (MTPA) or the corresponding acid chloride (MTPA-Cl) (Fig. 8.25). Note that the carboxylic acid and the acid chloride have the same three-dimensional arrangement of substituents on the stereogenic center but have opposite *R*/*S* configurations as a result of a Cahn–Ingold–Prelog priority change in converting the –OH of the acid to the –Cl of the acid chloride. This unfortunate circumstance has resulted in many instances of confusion and incorrect stereochemical assignments.

After the two MTPA esters are prepared, the NMR spectrum (^{19}F, ^{1}H, and/or ^{13}C) of each derivative is acquired, and the chemical shifts of each resonance are compared. The chemical shift of the

FIGURE 8.25 Formation of Mosher ester derivatives (From Seco, J. M., E. Quinoa, and R. Riguera, *Chemical Reviews 104* (2004): 17–117.) Reprinted by permission.

FIGURE 8.26 Analysis of Mosher ester derivatives to determine absolute configuration. (From Seco, J. M., E. Quinoa, and R. Riguera, *Chemical Reviews* 104 (2004): 17–117.)

resonances for the groups directly attached to the stereocenter in the spectrum of the (*R*) ester is subtracted from the corresponding chemical shifts for those resonances in the spectrum of the (*S*) ester [$\delta(S) - \delta(R) = \Delta\delta^{SR}$]. The absolute configuration of the substrate is then deduced by interpreting the signs of the $\Delta\delta$ values using certain empirical models for the most stable conformation of the esters (Fig. 8.26). Based on his experiments, Mosher concluded that the CF_3 group, $C\alpha$, the carboxyl group of the ester, and the oxygenated methine (C1') are all coplanar. This conformation results in differential shielding of L_1 and L_2 by the phenyl group of the MTPA ester (see Section 5.12 for a discussion of shielding effects of aromatic rings). In the (*R*)-MTPA ester, L_2 is shielded by the phenyl group (Fig. 8.26a). The opposite is true in the (*S*)-MTPA ester—L_1 is shielded by the phenyl group (Fig. 8.26b). As a result, all the protons (or carbons) that are relatively shielded in the (*R*)-MTPA ester will have a positive $\Delta\delta^{SR}$ value (L_2 in Fig. 8.26c), and those not shielded by the phenyl will have a negative $\Delta\delta^{SR}$ value (L_1 in Fig. 8.26c). If the alcohol has the opposite configuration, the shielding environments are reversed (Fig. 8.26d). Once the $\Delta\delta^{SR}$ values are determined for the groups flanking the MTPA ester, one can use the structural models in Figure 8.26c and 8.26d to assign L_1 and L_2 and thereby determine the absolute configuration of the original alcohol. In common practice, most researchers use the *modified Mosher method*, which involves examination of the $\Delta\delta^{SR}$ values not just for the groups directly attached to the stereocenter in question, but to *all* the protons (or carbons) in the compound. In this way, a representative sign of $\Delta\delta^{SR}$ for the substituents L_1 and L_2 can be determined, thus helping to prevent confusion that could arise from an anomalous chemical shift.

The Mosher method can also be applied to β-chiral primary alcohols and α-chiral tertiary alcohols. Mosher amides can be prepared from chiral amines and analyzed in a similar fashion. A number of other chiral derivatizing reagents for the determination of absolute configuration of alcohols, amines, carboxylic acids, and sulfoxides have been developed over the years. In general, these chiral auxiliaries all have three features in common: (1) a functional group that allows efficient covalent attachment of the auxiliary to the substrate; (2) a polar or bulky group to fix the compound of interest in a particular conformation; and (3) a group that is able to produce a significant anisotropic effect in the dominant conformation that results in differential shielding in the two species (diastereomers) used in the determination.

Mosher originally used ^{19}F spectroscopy to determine absolute configuration of MTPA derivatives, but today most researchers use ^{1}H NMR for this purpose. ^{19}F has the advantage of an uncrowded spectrum since the only fluorine signals are likely from the MTPA auxiliary itself. ^1H NMR is useful in most circumstances, but overlap of resonances can still be a problem, even with a high-field spectrometer, if $\Delta\delta^{SR}$ is small. ^{13}C NMR spectroscopy has the advantage of a wider chemical shift range and therefore less likelihood of resonance overlap. Furthermore, ^{13}C NMR provides useful information even when one or more of the substituents on the stereocenter have no protons. The low sensitivity of ^{13}C, however, presents a limitation if only minute quantities of the substrates are available.

B.   Determining Relative Configuration

In Chapter 7, we saw many instances when ^1H–^1H coupling constants could be used to assign relative configuration, especially when the conformation of the compound can be inferred. We will not expand on that discussion here. For some classes of compounds, simple ^{13}C NMR spectroscopy can be used very reliably to assign relative stereochemical configuration. One of the most reliable examples is the [^{13}C]acetonide method for determining relative configuration of acyclic 1,3-diols. The conformational preferences for 2,2-dimethyl-l,3-dioxolanes (acetone ketals, acetonides) were already well known by 1990, when Rychnovsky correlated the ^{13}C chemical shifts of acetonide methyl groups to stereochemical configuration. Acetonides of *syn*-1,3-diols adopted a chair conformation in which one methyl group of the acetonide is in an axial position and the other methyl group is in an equatorial position. The methyl group in the more shielded axial position has a chemical shift of ~19 ppm in the ^{13}C NMR spectrum and the less-shielded methyl group in the equatorial position appears at ~30 ppm (Fig. 8.27). Conversely, the acetonide derivatives of *anti*-1,3-diols exist in a twist boat conformation to alleviate steric repulsions in the chair conformations. In the *anti*-1,3-diol acetonides, the two methyl groups both appear at ~25 ppm in the ^{13}C NMR spectrum. The chemical shift of the acetal carbon also correlates well to stereochemical configuration, with the acetal carbon of *syn*-1,3-diol acetonides appearing at 98.5 ppm and that of the *anti*-1,3-diol acetonide appearing at 100.6 ppm in the ^{13}C NMR spectrum.

Analysis of literature ^{13}C NMR data for hundreds of 1,3-diol acetonides have proven this method reliable. Only a few types of substituents (R_1 and/or R_2) are problematic. The chemical shift correlations shown in Figure 8.27 only become unreliable when the substituents in the 4 and/or 6 position of the dioxolane ring are an *sp*-hybridized carbon (alkyne or nitrile). Use of the acetal carbon chemical shift correlation is not quite as reliable, but of the hundreds of acetonides examined, fewer than 10% of *syn*-l,3-diol acetonides and 5% of *anti*-1,3-diol acetonides would be misassigned based on the chemical shift of the acetal carbon alone—and practially none will be misassigned if the acetal chemical shift is considered in conjunction with the acetonide methyl chemical shifts. The only

F I G U R E 8 . 2 7 ^{13}C NMR chemical shift correlations for 1,3-diol acetonides. (From Rychnovsky, S. D., B. N. Rogers, and T. I. Richardson, *Accounts of Chemical Research 31* (1998): 9–17.)

drawbacks to this method is that the acetonide derivatives must be prepared from the diol substrates, but this is easily accomplished with a mixture of acetone, 2,2-dimethoxypropane, and pyridinium/ p-toluenesulfonate (PPTS). When only a small amount of sample is available, ^{13}C-enriched acetone can be used to prepare the acetonides. The [^{13}C] acetonide method is also readily applied to complex natural products containing several different 1,3-diols.

8.11 NUCLEAR OVERHAUSER EFFECT DIFFERENCE SPECTRA

In many cases of interpretation of NMR spectra, it would be helpful to be able to distinguish protons by their *spatial* location within a molecule. For example, for alkenes it would be useful to determine whether two groups are *cis* to each other or whether they represent a *trans* isomer. In bicyclic molecules, the chemist may wish to know whether a substituent is in an *exo* or in an *endo* position. Many of these types of problems cannot be solved by an analysis of chemical shift or by examination of spin–spin splitting effects.

A handy method for solving these types of problems is nuclear Overhauser effect (**NOE**) **difference spectroscopy**. This technique is based on the same phenomenon that gives rise to the nuclear Overhauser effect (Section 6.5), except that it uses *homonuclear*, rather than a heteronuclear, decoupling. In the discussion of the nuclear Overhauser effect, attention was focused on the case in which a hydrogen atom was directly bonded to a ^{13}C atom, and the hydrogen nucleus was saturated by a broadband signal. In fact, however, for two nuclei to interact via the nuclear Overhauser effect, the two nuclei do not need to be directly bonded; it is sufficient that they be *near* each other (generally within about 4 Å). Nuclei that are in close spatial proximity are capable of relaxing one another by a **dipolar** mechanism. (In this context we refer to magnetic dipoles, not electronic dipoles.) If the magnetic moment of one nucleus, as it precesses in the presence of an applied magnetic field, happens to generate an oscillating field that has the same frequency as the resonance frequency of a nearby nucleus, the two affected nuclei will undergo a mutual exchange of energy, and they will relax one another. The two groups of nuclei that interact by this dipolar process must be very near each other; the magnitude of the effect decreases as r^{-6}, where r is the distance between the nuclei.

We can take advantage of this dipolar interaction with an appropriately timed application of a low-power decoupling pulse. If we irradiate one group of protons, any nearby protons that interact with it by a dipolar mechanism will experience an enhancement in signal *intensity*.

The typical NOE difference experiment consists of *two* separate spectra. In the first experiment, the decoupler frequency is tuned to match exactly the group of protons that we wish to irradiate. The second experiment is conducted under conditions identical to the first experiment, except that the frequency of the decoupler is adjusted to a value far away in the spectrum from any peaks. The two spectra are subtracted from each other (this is done by treating digitized data within the computer), and the *difference* spectrum is plotted.

The NOE difference spectrum thus obtained would be expected to show a *negative* signal for the group of protons that had been irradiated. *Positive* signals should be observed *only* for those nuclei that interact with the irradiated protons by means of a dipolar mechanism. In other words, only those nuclei that are located within about 3 to 4 Å of the irradiated protons will give rise to a positive signal. All other nuclei that are not affected by the irradiation will appear as very weak or absent signals.

The spectra presented in Figure 8.28 illustrate an NOE difference analysis of **ethyl methacrylate**.

F I G U R E   8 . 2 8   NOE difference spectrum of ethyl methacrylate. Top spectrum: proton NMR spectrum of ethyl methacrylate without decoupling. Middle spectrum: NOE difference spectrum with irradiation at 1.9 ppm. Bottom spectrum: NOE difference spectrum with irradiation at 6.1 ppm.

The upper spectrum shows the normal proton NMR spectrum of this compound. We see peaks arising from the two vinyl hydrogens at 5.5 to 6.1 ppm. It might be assumed that H_E should be shifted further downfield than H_Z owing to the through-space deshielding effect of the carbonyl group. It is necessary, however, to confirm this prediction through experiment to determine unambiguously which of these peaks corresponds to H_Z and which corresponds to H_E.

The second spectrum was determined with the simultaneous irradiation of the methyl resonance at 1.9 ppm. We immediately see that the 1.9-ppm peak appears as a strongly negative peak. The only peak in the spectrum that appears as a positive peak is the vinyl proton peak at 5.5 ppm. The other vinyl peak at 6.1 ppm has nearly disappeared, as have most of the other peaks in the spectrum. The presence of a positive peak at 5.5 ppm confirms that this peak must come from proton H_Z; proton H_E is too far away from the methyl group to experience any dipolar relaxation effects.

The above result could have been obtained by conducting the experiment in the opposite direction. Irradiation of the vinyl proton at 5.5 ppm would have caused the methyl peak at 1.9 ppm to be positive. The results, however, would not be very dramatic; it is always more effective to irradiate the group with the larger number of equivalent hydrogens and observe the enhancement of the group with the smaller number of hydrogens rather than vice versa.

Finally, the third spectrum was determined with the simultaneous irradiation of the H_E peak at 6.1 ppm. The only peak that appears as a positive peak is the H_Z peak at 5.5 ppm, as expected. The methyl peak at 1.9 ppm does not show any enhancement, confirming that the methyl group is distant from the proton responsible for the peak at 6.1 ppm.

This example is intended to illustrate how NOE difference spectroscopy can be used to solve complex structural problems. This technique is particularly well suited to the solution of problems involving the location of substituents around an aromatic ring and stereochemical differences in alkenes or in bicyclic compounds.

8.12 HOW TO SOLVE PROBLEMS INVOLVING ADVANCED 1-D METHODS

In this section, we will use some advanced 1-D NMR data to solve problems. As with any problem-solving exercise, having a systematic approach simplifies the task and helps ensure we do not overlook any information. We will begin with analysis of what sort of NMR experiments are likely to provide definitive information relevant to the question. It is then prudent to think about what one would observe for each possible answer. Sometimes we may have to think about what would *not* be observed, but this is a potentially risky proposition. It is generally unwise to base conclusions on what might be called absent data.

■ SOLVED EXAMPLE 1

The Horner-Emmons reaction shown here produces two stereoisomeric alkenes (Z and E isomers). The isomers are separable by column chromatography but have nearly identical 1-D ^1H NMR spectra. How can you tell them apart?

■ SOLUTION

It is unusual but occasionally diastereomers such as these have indistinguishable ^{1}H NMR spectra. This is a case where the through-space relationships in the two compounds are different, so an NOE experiment is called for (Section 8.11). Irradiating the alkenyl methine adjacent to the ester in the *E*-alkene isomer should result in an enhancement of an aromatic proton but not the methyl group on the alkene. Conversely, irradiating the alkenyl methine proton in the *Z*-alkene isomer should result in an enhancement of the methyl group on the alkene only. Indeed, this is observed when the NOE spectra of each compound are acquired.

■ SOLVED EXAMPLE 2

You find a vial labeled "optically active α—methylbenzyl amine." After preparing the (*S*)- and (*R*)-Mosher amides of the material, you obtain the following NMR data. Which enantiomer of α—methylbenzyl amine is in the vial?

(*S*)-Mosher amide ^{1}H NMR (400 MHz, CDCl$_3$): δ 7.57-7.25 (m, 10H), 7.01 (d, *J* = 7.2 Hz, 1H), 5.19 (dq, *J* = 7.2, 6.9Hz, 1H), 3.37 (s, 3H), and 1.51 (d, *J* = 6.9 Hz, 1H).

(*R*)- Mosher amide ^{1}H NMR (400 MHz, CDCl$_3$): δ 7.44-7.24 (m, 10H), 7.01 (d, *J* = 7.0 Hz, 1H), 5.18 (dq, *J* = 7.0, 6.9 Hz, 1H), 3.41 (s, 3H), and 1.55 (d, *J* = 6.9 Hz, 1H).

■ SOLUTION

As with the Mosher esters described in Section 8.10, finding the δ^{SR} values by subtracting the chemical shifts of the groups flanking the stereocenter in the *(R)*-Mosher amide from those of the *(S)*-Mosher amide tells which group is shielded by the phenyl group on the MTPA stereocenter. In this case we use the methyl doublet. The δ^{SR} value is 1.51 ppm – 1.55 ppm = –0.04 ppm. Since this value is <0, the methyl group is shielded by the aromatic ring. This leads to the assignment of the *(S)* configuration of the original α—methylbenzyl amine.

PROBLEMS

*1. The spectrum of an ultrapure sample of ethanol is shown in Figure 8.3. Draw a tree diagram for the methylene groups in ethanol that takes into account the coupling to both the hydroxyl and methyl groups.

*2. The following spectrum is for a compound with the formula $C_5H_{10}O$. The peak at about 1.9 ppm is solvent and concentration dependent. Expansions are included, along with an indication of the spacing of the peaks in Hertz. The pairs of peaks at about 5.0 and 5.2 ppm have fine structure. How do you explain this small coupling? Draw the structure of the compound, assign the peaks, and include tree diagrams for the expanded peaks in the spectrum.

*3. Determine the structure of the aromatic compound with formula C_6H_5BrO. The peak at about 5.6 ppm is exchangeable and shifts readily when the sample is diluted. The expansions that are provided show ^4J couplings of about 1.6 Hz.

*4. The compound with the spectrum shown is derived from 2-methylphenol. The formula of the product obtained is $C_9H_{10}O_2$. The infrared spectrum shows prominent peaks at 3136 and 1648 cm^{-1}. The broad peak at 8.16 ppm is exchangeable and solvent dependent. Determine the structure of this compound using the spectrum provided and calculations of the chemical shifts (see Appendix 6). The calculated values will be only approximate but should allow you to determine the correct structure.

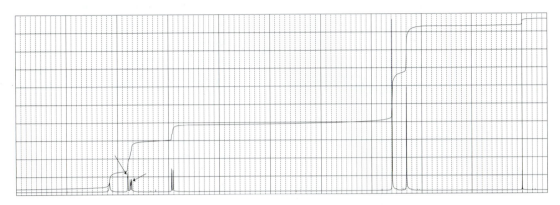

*5. The spectrum and expansions provided in this problem are for one of the compounds shown below. The broad peak at 5.25 ppm is exchangeable and solvent dependent. Using calculations of the *approximate* chemical shifts and the appearance and position of the peaks (singlet and doublets), determine the correct structure. The chemical shifts may be calculated from the information provided in Appendix 6. The calculated values will be only approximate but should allow you to determine the correct structure.

*6. The proton NMR spectrum for a compound with formula $C_5H_{10}O$ is shown. Determine the structure of this compound. The peak at 2.1 ppm is solvent dependent. Expansions are provided for some of the protons. Comment on the fine structure on the peak at 4.78 ppm. The normal carbon-13, DEPT-135, and DEPT-90 spectra data are tabulated.

Normal Carbon	DEPT-135	DEPT-90
22 ppm	Positive	No peak
41	Negative	No peak
60	Negative	No peak
112	Negative	No peak
142	No peak	No peak

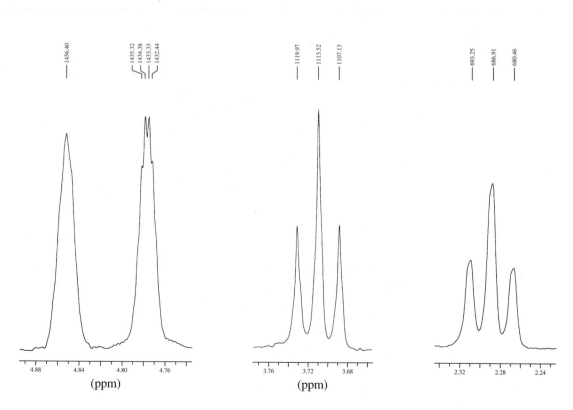

(ppm)   (ppm)

7. The proton NMR spectrum of a compound with formula $C_5H_{10}O$ is shown. The peak at 2.1 ppm is solvent dependent. The infrared spectrum shows a broad and strong peak at 3332 cm^{-1}. The normal carbon-13, DEPT-135, and DEPT-90 spectra data are tabulated.

Normal Carbon	DEPT-135	DEPT-90
11 ppm	Negative	No peak
18	No peak	No peak
21	Positive	No peak
71	Negative	No peak

8. Determine the structure of the aromatic compound with formula $C_9H_9ClO_3$. The infrared spectrum shows a very broad band from 3300 to 2400 cm^{-1} and a strong band at 1714 cm^{-1}. The full proton NMR spectrum and expansions are provided. The compound is prepared by a nucleophilic substitution reaction of the sodium salt of 3-chlorophenol on a halogen-bearing substrate.

Offset: 1.4 ppm.

(ppm)

(ppm)

(ppm)

*9. Determine the structure of a compound with formula $C_{10}H_{15}N$. The proton NMR spectrum is shown. The infrared spectrum has medium bands at 3420 and 3349 cm^{-1} and a strong band at 1624 cm^{-1}. The broad peak at 3.5 ppm in the NMR shifts when DCl is added, while the other peaks stay in the same positions.

*10. Determine the structure of a compound with formula $C_6H_5Br_2N$. The proton NMR spectrum is shown. The infrared spectrum has medium bands at 3420 and 3315 cm^{-1} and a strong band at 1612 cm^{-1}. The normal carbon, DEPT-135, and DEPT-90 spectra data are tabulated.

Normal Carbon	DEPT-135	DEPT-90
109 ppm	No peak	No peak
119	Positive	Positive
132	Positive	Positive
142	No peak	No peak

11. There are three spectra shown in this problem along with three structures of aromatic primary amines. Assign each spectrum to the appropriate structure. You should calculate the *approximate* chemical shifts (Appendix 6) and use these values along with the appearance and position of the peaks (singlet and doublets) to assign the correct structure.

*12. When aniline is chlorinated, a product with the formula $C_6H_5NCl_2$ is obtained. The spectrum of this compound is shown. The expansions are labeled to indicate couplings, in Hertz. Determine the structure and substitution pattern of the compound and assign each set of peaks. Explain the splitting patterns.

*13. A naturally occurring amino acid with the formula $C_3H_7NO_2$ gives the following proton NMR spectrum when determined in deuterium oxide solvent. The amino and carboxyl protons merge into a single peak at 4.9 ppm in the D_2O solvent (not shown); the peaks of each multiplet are separated by 7 Hz. Determine the structure of this amino acid.

(ppm)

14. Determine the structure of a compound with formula C_7H_9N. The proton NMR spectrum is shown, along with expansions of the region from 7.10 to 6.60 ppm. The three-peak pattern for the two protons at about 7 ppm involves overlapping peaks. The broad peak at 3.5 ppm shifts when DCl is added, while the other peaks stay in the same positions. The infrared spectrum shows a pair of peaks near 3400 cm^{-1} and an out-of-plane bending band at 751 cm^{-1}.

(ppm)

15. A naturally occurring amino acid with the formula $C_9H_{11}NO_3$ gives the following proton NMR spectrum when determined in deuterium oxide solvent with DCl added. The amino, carboxyl, and hydroxyl protons merge into a single peak at 5.1 ppm (4 H) in D_2O. Determine the structure of this amino acid and explain the pattern that appears in the range 3.17 to 3.40 ppm, including coupling constants.

16. Determine the structure of a compound with formula $C_6H_{10}O_2$. The proton NMR spectrum with expansions is provided. Comment regarding why the proton appearing at 6.91 ppm is a triplet of quartets, with spacing of 1.47 Hz. Also comment on the "singlet" at 1.83 that shows fine structure. The normal carbon, DEPT-135, and DEPT-90 spectral results are tabulated.

Normal Carbon	DEPT-135	DEPT-90
12 ppm	Positive	No peak
13	Positive	No peak
22	Negative	No peak
127	No peak	No peak
147	Positive	Positive
174	No peak	No peak

Offset: 2.7 ppm.

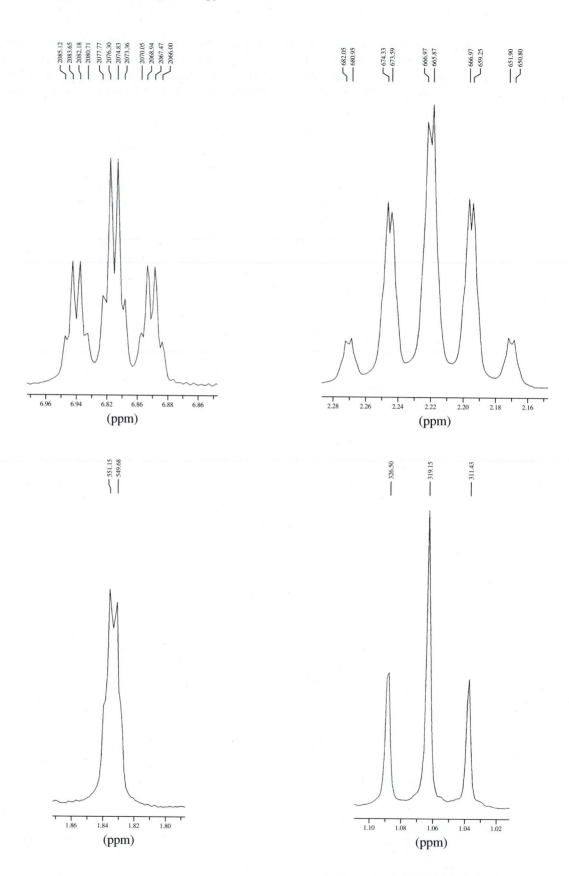

17. The following proton NMR spectrum is of a discontinued analgesic drug, phenacetin ($C_{10}H_{13}NO_2$). Phenacetin is structurally related to the very popular and current analgesic drug acetaminophen. Phenacetin contains an amide functional group. Two tiny impurity peaks appear near 3.4 and 8.1 ppm. Give the structure of this compound and interpret the spectrum.

300 MHz

| 1210.78 1203.79 1196.80 1189.82 | | 426.14 419.16 412.17 |

(ppm)   (ppm)   (ppm)

18. The proton NMR spectrum shown in this problem is for a common insect repellent, *N,N*-diethyl-*m*-toluamide, determined at 360 K. This problem also shows a stacked plot of this compound determined in the temperature range of 290 to 360 K (27–87°C). Explain why the spectrum changes from two pairs of broadened peaks near 1.2 and 3.4 ppm at low temperature to a triplet and quartet at the higher temperatures.

19. The proton NMR spectral information shown in this problem is for a compound with formula C_4H_7Cl. Expansions are shown for each of the unique protons. The original "quintet" pattern centering on 4.52 ppm is simplified to a doublet by irradiating (decoupling) the protons at 1.59 ppm (see Section 8.10). In another experiment, decoupling the proton at 4.52 ppm simplifies the original pattern centering on 5.95 ppm to the four-peak pattern shown. The doublet at 1.59 ppm becomes a singlet when the proton at 4.52 ppm is irradiated (decoupled). Determine the coupling constants and draw the structure of this compound. Notice that there are ^2J, ^3J, and ^4J couplings present in this compound. Draw a tree diagram for the proton at 5.95 ppm (nondecoupled) and explain why irradiation of the proton at 4.52 ppm simplified the pattern. Assign each of the peaks in the spectrum.

(ppm)   (ppm)

irradiation of
proton at
4.52 ppm

irradiation of
proton at
1.59 ppm

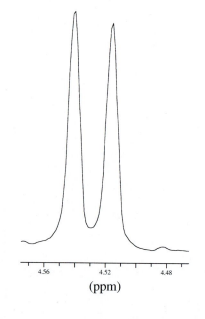

20. In Problem 11, calculations proved to be a good way of assigning structures to the spectra of some aromatic amines. Describe an experimental way of differentiating between the following amines:

*21. At room temperature, the NMR spectrum of cyclohexane shows only a single resonance peak. As the temperature of the sample is lowered, the sharp single peak broadens until at −66.7°C it begins to split into two peaks, both broad. As the temperature is lowered further to −100°C, each of the two broad bands begins to give a splitting pattern of its own. Explain the origin of these two families of bands.

*22. In *cis*-1-bromo-4-*tert*-butylcyclohexane, the proton on carbon-4 is found to give resonance at 4.33 ppm. In the *trans* isomer, the resonance of the C4 hydrogen is at 3.63 ppm. Explain why these compounds should have different chemical shift values for the C4 hydrogen. Can you explain the fact that this difference is not seen in the 4-bromomethylcyclohexanes except at very low temperature?

23. Both enantiomers of menthol are commercially available. Use the following ^{1}H NMR chemical shift data for the Mosher esters of menthol to determine which enantiomer of menthol was used to prepare the samples.

(*S*)-Mosher ester: ^{1}H NMR (500 MHz, CDCl$_3$) δ 7.54 (m, 2H), 7.39 (m, 3H), 4.90 (ddd, 1H, J = 4.5, 11.0, 11.0 Hz), 3.59 (q, 3H, J = 1.0 Hz), 2.13 (dddd, 1H, J = 2.0, 3.5, 4.0, 11.5 Hz), 1.70 (dddd, 1H, J = 3.5, 3.5, 3.5, 12.5 Hz), 1.67 (dddd, 1H, J = 3.5, 3.5, 3.5, 13 Hz), 1.56 (dsept, 1H, J = 3.0, 7.0), 1.54 (ddddq, 1H, J = 3.5, 3.5, 12, 12, 7.0 Hz), 1.42 (dddd, 1H, J = 3.0, 3.0, 11.0, 12.5 Hz), 1.12 (ddd, 1H, J = 12, 12, 12 Hz), 1.04 (dddd, 1H, J = 3.5, 12.5, 12.5, 12.5 Hz), 0.94 (d, 3H, J = 7.0 Hz), 0.87 (dddd, 1H, J = 4.0, 12.5, 13, 13 Hz), 0.74 (d, 3H, J = 7.0 Hz), and 0.67 (d, 3H, J = 7.0 Hz).

(*R*)-Mosher ester: ^{1}H NMR (500 MHz, CDCl$_3$) δ 7.52 (m, 2H), 7.40 (m, 3H), 4.88 (ddd, 1H, J = 4.5, 11.0, 11.0 Hz), 3.53 (q, 3H, J = 1.0 Hz), 2.08 (dddd, 1H, J = 2.0, 3.5, 4.0, 12 Hz), 1.88 (dsept, 1H, J = 3.0, 7.0), 1.70 (dddd, 1H, J = 3.5, 3.5, 3.5, 13 Hz), 1.69 (dddd, 1H, J = 3, 3, 3, 13 Hz), 1.52 (ddddq, 1H, J = 3.5, 3.5, 12, 12, 6.5 Hz), 1.45 (dddd, 1H, J = 3.0, 3.0, 11.0, 12.5 Hz), 1.06 (dddd, 1H, J = 3.5, 13, 13, 13 Hz), 0.98 (ddd, 1H, J = 12, 12, 12 Hz), 0.91 (d, 3H, J = 6.5 Hz), 0.87 (d, 3H, J = 7.0 Hz), 0.86 (dddd, 1H, J = 3.5, 12.5, 12.5, 12.5 Hz), and 0.77 (d, 3H, J = 7.0 Hz).

REFERENCES

Books and Monographs

Claridge, T. D. W., *High Resolution NMR Techniques in Organic Chemistry*, 2nd ed., Pergamon, Oxford, 2009.

Crews, P., J. Rodriguez, and M. Jaspars, *Organic Structure Analysis*, 2nd ed., Oxford University Press, New York, 2010.

Friebolin, H., *Basic One- and Two-Dimensional NMR Spectroscopy*, 5th ed., Wiley-VCH, Weinheim, Germany, 2011.

Gunther, H., *NMR Spectroscopy*, 2nd ed., John Wiley and Sons, New York, 1995.

Lambert, J. B., S. Gronert, H. F. Shurvell, D. A. Lightner, *Organic Structural Spectroscopy*, 2nd ed., Prentice Hall, Upper Saddle River, NJ, 2011.

Macomber, R. S., *A Complete Introduction to Modern NMR Spectroscopy*, John Wiley and Sons, New York, 1997.

Nelson, J. H., *Nuclear Magnetic Resonance Spectroscopy*, Prentice Hall, Upper Saddle River, NJ, 2003.

Pople, J. A., W. C. Schneider, and H. J. Bernstein, *High Resolution Nuclear Magnetic Resonance*, McGraw–Hill, New York, 1969.

Pretsch, E., T. P. Buhlmann, and C. Affolter, *Structure Determination of Organic Compounds. Tables of Spectral Data*, 4th ed., Springer-Verlag, Berlin and New York, 2009.

Reichardt, C., and T. Welton, *Solvents and Solvent Effects in Organic Chemistry*, 4th ed., Wiley-VCH, Weinheim, Germany, 2011.

Sanders, J. K. M., and B. K. Hunter, *Modern NMR Spectroscopy—A Guide for Chemists*, 2nd ed., Oxford University Press, Oxford, England, 1993.

Silverstein, R. M., F. X. Webster, and D. J. Kiemle, Spectrometric Identification of Organic Compounds, 7th ed., John Wiley, New York, 2005.

Compilations of Spectra

Pouchert, C. J., and J. Behnke, *The Aldrich Library of ^{13}C and ^{1}H FT-NMR Spectra*, 300 MHz, Aldrich Chemical Company, Milwaukee, 1993.

Papers

Cavaleiro, J. A. S., "'Sovent Effects' in 1H NMR Spectroscopy," *Journal of Chemical Education, 64*(1987): 549–550.

Gottlieb, H. E., V. Kotlyar, and A. Nudelman. "NMR Chemical Shifts of Common Laboratory Solvents as Trace Impurities," *Journal of Organic Chemistry, 62*, (1997): 7512–7515.

Fulmer, Gregory R., A. J. M. Miller, N. H. Sherden, H. E. Gottlieb, A. Nudelman, B. M. Stoltz, J. E. Bercaw, K. I. Goldberg, "NMR Chemical Shifts of Trace Impurities: Common Laboratory Solvents, Organics, and Gases in Deuterated Solvents Relevant to the Organometallic Chemist," *Organometallics, 29* (2010): 2176–2179.

Rothchild, R., "NMR Methods for Determination of Enantiomeric Excess," *Enantiomer 5* (2000): 457–471.

Rychnovsky, S. D., B. N. Rogers, and G. Yang, "Analysis of Two Carbon-13 NMR Correlations for Determining the Stereochemistry of 1,3-diol Acetonides," *Journal of Organic Chemistry 58* (1993): 3511–3515.

Rychnovsky, S. D., B. N. Rogers, and T. I. Richardson, "Configurational Assignment of Polyene Macrolide Antibiotics Using the [^{13}C] acetonide Analysis," *Accounts of Chemical Research 31* (1998): 9–17.

Seco, J. M., E. Quinoa, and R. Riguera, "The Assignment of Absolute Configuration by NMR," *Chemical Reviews 104* (2004): 17–117 and references therein.

Selected Websites

WebSpectra, UCLA

NMR and IR spectroscopy problems for students to interpret. They provide links to other sites with problems for students to solve.

http://www.chem.ucla.edu/~webspectra/

CHAPTER 9

NUCLEAR MAGNETIC RESONANCE SPECTROSCOPY
Part Five: Advanced NMR Techniques

Since the advent of modern, computer-controlled Fourier transform nuclear magnetic resonance (FT-NMR) instruments, it has been possible to conduct more sophisticated experiments than those described in preceding chapters. Although a great many specialized experiments can be performed, the following discussion examines only a few of the most important ones.

9.1 PULSE SEQUENCES

Chapter 6, Section 6.5, introduced the concept of **pulse sequences.** In an FT-NMR instrument, the computer that operates the instrument can be programmed to control the timing and duration of the excitation pulse—the radiofrequency pulse used to excite the nuclei from the lower spin state to the upper spin state. Chapter 5, Section 5.7B, discussed the nature of this pulse and the reasons why it can excite all the nuclei in the sample simultaneously. Precise timing can also be applied to any decoupling transmitters that operate during the pulse sequence. As a simple illustration, Figure 9.1 shows the pulse sequence for the acquisition of a simple proton NMR spectrum. The pulse sequence is characterized by an excitation pulse from the transmitter; an **acquisition time,** during which the free-induction decay (FID) pattern is collected in digitized form in the computer; and a **relaxation delay,** during which the nuclei are allowed to relax in order to reestablish the equilibrium populations of the two spin states. After the relaxation delay, a second excitation pulse marks the beginning of another cycle in the sequence.

There are many possible variations on this simple pulse sequence. For example, in Chapter 6 we learned that it is possible to transmit *two* signals into the sample. In ^{13}C NMR spectroscopy, a pulse sequence similar to that shown in Figure 9.1 is transmitted at the absorption frequency of the ^{13}C nuclei. At the same time, a second transmitter, tuned to the frequency of the hydrogen (^{1}H) nuclei in the sample, transmits a broad band of frequencies to **decouple** the hydrogen nuclei from the ^{13}C nuclei. Figure 9.2 illustrates this type of pulse sequence.

The discussion in Chapter 6, Section 6.5, of methods for determining ^{13}C spectra described how to obtain proton-*coupled* spectra and still retain the benefits of nuclear Overhauser enhancement. In this method, which is called an **NOE-enhanced proton-coupled spectrum** or a **gated decoupling spectrum,** the decoupler is turned on during the interval *before* the pulsing of the ^{13}C nuclei. At the moment the excitation pulse is transmitted, the decoupler is switched off. The decoupler is switched on again during the relaxation delay period. The effect of this pulse sequence is to allow the nuclear Overhauser effect to develop while the decoupler is on. Because the decoupler is switched off during the excitation pulse, spin decoupling of the ^{13}C atoms is not observed (a proton-coupled spectrum is seen). The nuclear Overhauser enhancement decays over a relatively long time period; therefore, most of the enhancement is retained as the FID is collected. Once the FID information has been accumulated, the decoupler is switched on again to allow the nuclear Overhauser enhancement to develop before the next excitation pulse. Figure 9.3a shows the pulse sequence for gated decoupling.

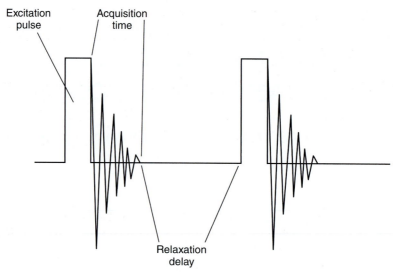

FIGURE 9.1 A simple pulse sequence.

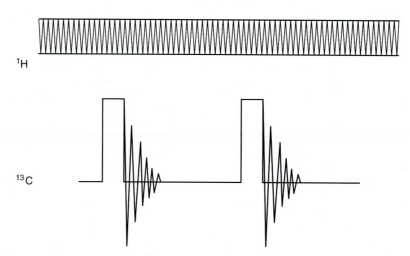

FIGURE 9.2 A proton-decoupled ^{13}C NMR pulse sequence.

The opposite result is obtained if the decoupler is not switched on until *the very instant* the excitation pulse is transmitted. Once the FID data have been collected, the decoupler is switched off until the next excitation pulse. This pulse sequence is called **inverse gated decoupling.** The effect of this pulse sequence is to provide a proton-decoupled spectrum with no NOE enhancement. Because the decoupler is switched off before the excitation pulse, the nuclear Overhauser enhancement is not allowed to develop. Proton decoupling is provided since the decoupler is switched on during the excitation pulse and the acquisition time. Figure 9.3b shows the pulse sequence for inverse gated decoupling. This technique is used when it is necessary to determine integrals in a ^{13}C spectrum.

The computer that is built into modern FT-NMR instruments is very versatile and enables us to develop more complex and more interesting pulse sequences than the ones shown here. For example, we can transmit second and even third pulses, and we can transmit them along any of the Cartesian axes. The pulses can be transmitted with varying durations, and a variety of times can also be programmed into the sequence. As a result of these pulse programs, nuclei may exchange energy, they may affect each other's relaxation times, or they may encode information about spin coupling from one nucleus to another.

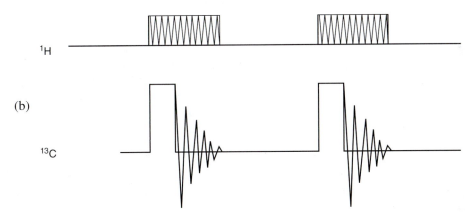

FIGURE 9.3 A simple pulse sequence. (a) A pulse sequence for gated decoupling; (b) a pulse sequence for inverse gated decoupling.

We will not describe these complex pulse sequences further; their description and analysis are beyond the scope of this discussion. Our purpose in describing a few simple pulse sequences in this section is to give you an idea of how a pulse sequence is constructed and how its design may affect the results of an NMR experiment. From this point forward, we shall simply describe the *results* of experiments that utilize some complex sequences and show how the results may be applied to the solution of a molecular structure problem. If you want more detailed information about pulse sequences for the experiments described in the following sections, consult one of the works listed in the references at the end of this chapter.

9.2 PULSE WIDTHS, SPINS, AND MAGNETIZATION VECTORS

To gain some appreciation for the advanced techniques that this chapter describes, you must spend some time learning what happens to a magnetic nucleus when it receives a pulse of radiofrequency energy. The nuclei that concern us in this discussion, ^1H and ^{13}C, are magnetic. They have a finite spin, and a spinning charged particle generates a magnetic field. That means that each individual nucleus behaves as a tiny magnet. The nuclear magnetic moment of each nucleus can be illustrated as a vector (Fig. 9.4a). When the magnetic nuclei are placed in a large, strong magnetic field, they tend to align themselves with the strong field, much as a compass needle aligns itself with the earth's magnetic field. Figure 9.4b shows this alignment. In the following discussion, it would be very inconvenient to continue describing the behavior of each individual nucleus. We can simplify the discussion by considering that the magnetic field vectors for each nucleus give rise to a resultant vector called the **nuclear magnetization vector** or the **bulk-magnetization vector.** Figure 9.4b also shows this vector (**M**). The individual nuclear magnetic vectors precess about the principal magnetic field axis (Z). They have random precessional motions that are not in phase; vector addition produces the resultant, the nuclear (bulk) magnetization vector, which is aligned along the Z axis. We can more easily describe an effect that involves the individual magnetic nuclei by examining the behavior of the nuclear magnetization vector.

The small arrows in Figure 9.4 represent the individual magnetic moments. In this picture, we are viewing the orientations of the magnetic moment vectors from a stationary position, as if we were standing on the laboratory floor watching the nuclei precess inside the magnetic field. This view, or **frame of reference,** is thus known as the **laboratory frame** or the **stationary frame.** We can simplify the study of the magnetic moment vectors by imagining a set of coordinate axes that rotate in the same direction and at the same speed as the average nuclear magnetic moment precesses. This reference frame is called the **rotating frame,** and it rotates about the Z axis. We can visualize these vectors more easily by considering them within the rotating frame, much as we can visualize the complex motions of objects on the earth more easily by first observing their motions from the earth, alone, even though the earth is spinning on its axis, rotating about the sun, and moving through the solar system. We can label the axes of the rotating frames X', Y', and Z' (identical with Z). In this rotating frame, the microscopic magnetic moments are stationary (are not rotating) because the reference frame and the microscopic moments are rotating at the same speed and in the same direction.

Since the small microscopic moments (vectors) from each nucleus add together, what our instrument sees is the *net* or *bulk* magnetization vector for the whole sample. We will refer to this bulk magnetization vector in the discussions that follow.

In a Fourier transform NMR instrument, the radiofrequency is transmitted into the sample in a **pulse** of very short duration—typically on the order of 1 to 10 microseconds (μsec); during this time, the radiofrequency transmitter is suddenly turned on and, after about 10 μsec, suddenly turned off again. The

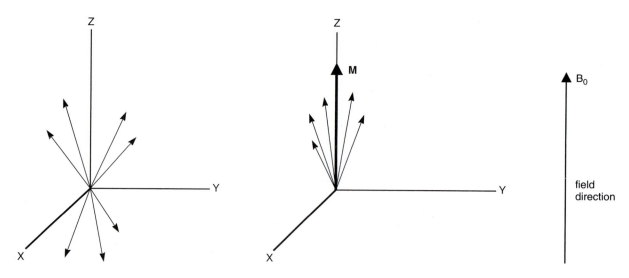

(a) A collection of magnetic nuclei, showing the individual magnetic moments.

(b) Magnetic nuclei aligned in an external magnetic field; M represents the bulk magnetization vector.

F I G U R E   9 . 4   Nuclear magnetization (laboratory frame).

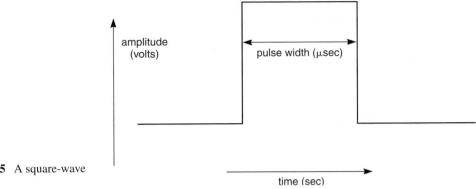

FIGURE 9.5 A square-wave pulse.

pulse can be applied along either the X' or the Y' axis and in either the positive or the negative direction. The shape of the pulse, expressed as a plot of DC voltage versus time, looks like Figure 9.5.

When we apply this pulse to the sample, the magnetization vector of every magnetic nucleus begins to precess about the axis of the new pulse.[1] If the pulse is applied along the X' axis, the magnetization vectors all begin to tip in the same direction at the same time. The vectors tip to greater or smaller extents depending on the duration of the pulse. In a typical experiment, the duration of the pulse is selected to cause a specific **tip angle** of the bulk magnetization vector (the resultant vector of all the individual vectors), and a pulse duration (known as a **pulse width**) is selected to result in a 90° rotation of the bulk magnetization vector. Such a pulse is known as a **90° pulse.** Figure 9.6 shows its effect along the X' axis. At the same time, if the duration of the pulse were twice as long, the bulk magnetization vector would tip to an angle of 180° (it would point straight down in Fig. 9.6). A pulse of this duration is called a **180° pulse.**

What happens to the magnetization vector following a 90° pulse? At the end of the pulse, the B_0 field is still present, and the nuclei continue to precess about it. If we focus for the moment on the

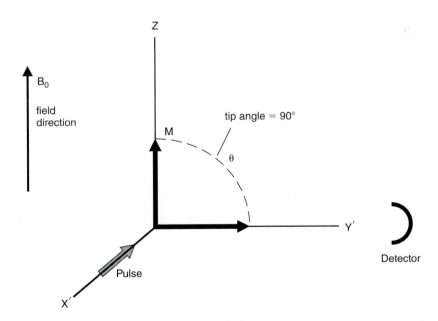

FIGURE 9.6 The effect of a 90° pulse (**M** is the bulk magnetization vector for the sample).

---

[1]Recall from Chapter 5 that if the duration of the pulse is short, the pulse has an uncertain frequency. The range of the uncertainty is sufficiently wide that all of the magnetic nuclei absorb energy from the pulse.

nuclei with precessional frequencies that exactly match the frequency of the rotating frame, we expect the magnetization vector to remain directed along the Y' axis (see Fig. 9.6).

In the **laboratory frame,** the Y' component corresponds to a magnetization vector rotating in the XY plane. The magnetization vector rotates in the XY plane because the individual nuclear magnetization vectors are precessing about Z (the principal field axis). Before the pulse, individual nuclei have random precessional motions and are not in phase. The pulse causes **phase coherence** to develop so that all the vectors precess in phase (see Fig. 9.7). Because all the individual vectors precess about the Z axis, **M**, the resultant of all of these vectors, also rotates in the XY plane.

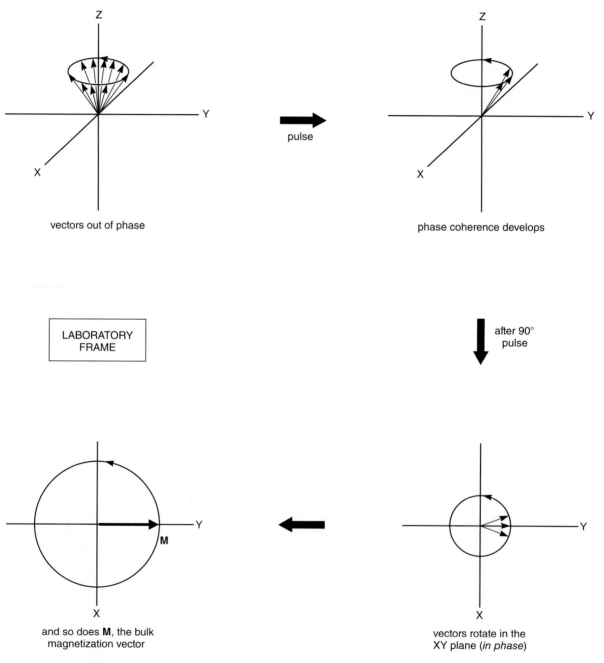

FIGURE 9.7 Precession of magnetization vectors in the XY plane.

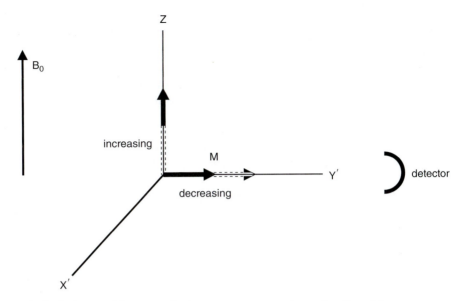

FIGURE 9.8 Decay of the magnetization vector components as a function of time.

Once the pulse stops, however, the excited nuclei begin to relax (lose excitation energy and invert their individual nuclear spins). Over time, such **relaxation processes** diminish the magnitude of the bulk magnetization vector along the Y' axis and increase it along the Z axis, as illustrated in Figure 9.8. These changes in bulk magnetization occur as a result of both spin inversion to restore the Boltzmann distribution (spin–lattice relaxation) *and* loss of phase coherence (spin–spin relaxation). If we wait long enough, eventually the bulk magnetization returns to its equilibrium value, and the bulk magnetization vector points along the Z axis.

A receiver coil is situated in the XY plane, where it senses the rotating magnetization. As the Y' component grows smaller, the oscillating voltage from the receiver coil decays, and it reaches zero when the magnetization has been restored along the Z axis. The record of the receiver voltage as a function of time is called the **free-induction decay** (**FID**) because the nuclei are allowed to precess "freely" in the absence of an X-axis field. Figure 5.15 in Chapter 5 shows an example of a free-induction decay pattern. When such a pattern is analyzed via a Fourier transform, the typical NMR spectrum is obtained.

To understand how some of the advanced experiments work, it is helpful to develop an appreciation of what an excitation pulse does to the nuclei in the sample and how the magnetization of the sample nuclei behaves during the course of a pulsed experiment. At this point, we shall turn our attention to three of the most important advanced experiments.

9.3 PULSED FIELD GRADIENTS

Before determining an NMR spectrum, it is very important that the magnetic field be **shimmed.** The NMR experiment requires that there be a uniform magnetic field over the entire volume of the sample. If the field is not uniform, the result will be broadened peaks, the appearance of spurious side-band peaks, and a general lack of resolution. This means that every time a sample is introduced into the magnetic field, the field must be adjusted slightly to achieve this uniformity of magnetic field (magnetic field homogeneity).

The shimming process allows one to achieve field homogeneity by carefully adjusting a series of controls to vary the amount of current that passes through a set of coils that generate small magnetic fields of their own. These adjustable magnetic fields compensate for inhomogeneity in the overall magnetic field. The result of careful shimming is that the spectral lines will have good line shape, and the resolution will be maximized.

The drawback to this manual shimming process is that it is time consuming, and it does not lend itself well to the determination of spectra in an automated environment. With the advent of **pulsed field gradients,** this shimming process becomes much faster, and it can be applied to the determination of spectra automatically.

In a "normal" NMR experiment, one applies a pulse of magnetic field that is uniform along the length of the sample. Figure 9.9a depicts how this pulse might appear. In a pulsed field gradient experiment, the pulse that is applied varies along the length of the sample tube. Figure 9.9b shows what this might look like.

A field gradient pulse causes the nuclei in molecules in different locations along the length of the sample tube to precess at different frequencies. The result is that the rotating magnetization vectors of each nucleus will rapidly fall out of phase, resulting in destruction of the signal. By the application of a second gradient pulse, directed in opposite directions along the Z axis, peaks that arise from noise and other artifacts will be eliminated. Magnetization vectors that belong to the sample of interest will be "unwound" with this second pulse, and they will appear as clean signals. Thus, the unwanted peaks are destroyed, and only the peaks of interest remain. To selectively refocus the desired signals correctly, the instrument's computer must already have a **field map** in its memory. This field map is determined for each probe head using a sample that gives a strong signal. Water or the deuterium in the solvent is generally used for this purpose. Once the field map has been created for the probe head that is being used, the computer then uses those values to adjust the field gradient to yield the strongest, sharpest signal.

The advantage of field gradient shimming is that it is usually complete within two or three iterations. Manual shimming, by contrast, can be tedious and time consuming requiring many iterations. The automated nature of field gradient shimming lends itself well to the determination of spectra automatically. This is especially useful when an automatic sample changer is attached to the instrument.

The advantages of field gradient shimming may also be applied to a wide variety of two-dimensional spectroscopic techniques. These will be mentioned in succeeding sections of this chapter.

(a) Normal B_0 field gradient   (b) A field gradient lying along the Z direction

F I G U R E   9 . 9 Diagram showing the shape of a magnetic field pulse along the Z axis of an NMR sample tube.

9.4 THE DEPT EXPERIMENT: NUMBER OF PROTONS ATTACHED TO ^{13}C ATOMS

A very useful pulse sequence in ^{13}C spectroscopy is employed in the experiment called **Distortionless Enhancement by Polarization Transfer,** better known as **DEPT.** The DEPT method has become one of the most important techniques available to the NMR spectroscopist for determining the number of hydrogens attached to a given carbon atom. The pulse sequence involves a complex program of pulses and delay times in both the ^{1}H and ^{13}C channels. The result of this pulse sequence is that carbon atoms with one, two, and three attached hydrogens exhibit different **phases** as they are recorded. The phases of these carbon signals will also depend on the duration of the delays that are programmed into the pulse sequence. In one experiment, called a DEPT-45, only carbon atoms that bear any number of attached hydrogens will produce a peak. With a slightly different delay, a second experiment (called a DEPT-90) shows peaks only for those carbon atoms that are part of a **methine** (CH) group. With an even longer delay, a DEPT-135 spectrum is obtained. In a DEPT-135 spectrum, methine and methyl carbons give rise to positive peaks, whereas methylene carbons appear as inverse peaks. Section 9.5 will develop the reasons why carbon atoms with different numbers of attached hydrogens behave differently in this type of experiment. Quaternary carbons, which have no attached hydrogens, give no signal in a DEPT experiment.

There are several variations on the DEPT experiment. In one form, separate spectra are traced on a single sheet of paper. On one spectrum, only the methyl carbons are shown; on the second spectrum, only the methylene carbons are traced; on the third spectrum, only the methine carbons appear; and on the fourth trace, all the carbon atoms that bear hydrogen atoms are shown. In another variation on this experiment, the peaks due to methyl, methylene, and methine carbons are all traced on the same line, with the methyl and methine carbons appearing as positive peaks and the methylene carbons appearing as negative peaks.

In many instances, a DEPT spectrum makes spectral assignments easier than does a proton-decoupled ^{13}C spectrum. Figure 9.10 is the DEPT-135 spectrum of **isopentyl acetate.**

In a DEPT 135 spectrum, methyl and methine groups are positive while methylene groups are negative. The two equivalent methyl carbons (numbered **1**) can be seen as the tallest peak (at 22.3 ppm), whereas the methyl group on the acetyl function (numbered **6**) is a shorter peak at 20.8 ppm. The methine carbon (**2**) is a still smaller peak at 24.9 ppm. The methylene carbons produce the negative peaks: carbon **3** appears at 37.1 ppm, and carbon **4** appears at 63.0 ppm. Carbon **4** is deshielded since it is near the electronegative oxygen atom. The carbonyl carbon (**5**) does not appear in the DEPT spectrum since it has no attached hydrogen atoms. Another way to display a DEPT spectrum for isopentyl acetate is shown on page 307 (Figure 6.9). Both plots support the assignments made.

Clearly, the DEPT technique is a very useful adjunct to ^{13}C NMR spectroscopy. The results of the DEPT experiment can tell us whether a given peak arises from a carbon on a methyl group, a methylene group, or a methine group. By comparing the results of the DEPT spectrum with the original ^{13}C-decoupled ^{13}C NMR spectrum, we can also identify the peaks that must arise from quaternary carbons. Quaternary carbons, which bear no hydrogens, appear in the ^{13}C NMR spectrum but are missing in the DEPT spectrum.

ISOPENTYL ACETATE IN CDCl$_3$—DEPT-135

F I G U R E 9 . 1 0 DEPT-135
spectrum of isopentyl acetate.

Another example that demonstrates some of the power of the DEPT technique is the terpenoid alcohol **citronellol.**

Figure 9.11 is the ^{1}H-decoupled ^{13}C NMR spectrum of citronellol. We can easily assign certain features of the spectrum to particular carbon atoms of the molecule by examining the chemical shifts and intensities. In the region where we expect to find double-bonded sp^2 carbon atoms (120–140 ppm), we see two peaks. The taller peak at 124.6 ppm we assign to carbon 6, while the smaller peak at 131.0 ppm is assigned to carbon 7. Carbon atoms with no attached hydrogen atoms are smaller (no nuclear Overhauser enhancement; see Chapter 6, Section 6.5), whereas carbon 6 has one hydrogen atom (nuclear Overhauser enhanced). The pattern appearing between 15 and 65 ppm, however, is more complex and thus more difficult to interpret.

The stacked full-edited DEPT spectrum of citronellol (Fig. 9.12) makes the specific assignment of individual carbon atoms much easier. Our earlier assignment of the peak at 124.6 ppm to carbon 6 is confirmed because that peak appears in the methine "CH carbon" region. Notice that the peak at 131.0 ppm is missing from the "all protonated carbon" section since carbon 7 has no attached hydrogen atom. The peaks at 60.8, 39.7, 37.1, and 25.33 ppm are all methylene groups. The peak at 29.0 ppm is a methine proton that appears in the "CH region" for single attached protons, and is C3. We can assume that the methylene group appearing at 60.8 ppm can be assigned to the CH$_2$ group next to the oxygen atom (OH group) on C1. We expect this group to be significantly deshielded by the electronegative oxygen

FIGURE 9.11 ^{13}C NMR spectrum of citronellol (125 MHz).

FIGURE 9.12 Stacked full-edited DEPT spectrum of citronellol (125 MHz).

atom. The remaining three methylene groups at 39.7, 37.1, and 25.3 ppm cannot be assigned to C2, C4, or C5 without more information. Likewise, the three methyl group appearing at 25.6, 19.4, and 17.5 cannot be assigned without more information. Section 9.7 and the COSY spectra shown in Figure 9.15a and Figure 9.15b allow us to make the remaining assignments.

These examples should give you an idea of the capabilities of the DEPT technique. It is an excellent means of distinguishing among methyl, methylene, methine, and quaternary carbons in a ^{13}C NMR spectrum.

9.5 DETERMINING THE NUMBER OF ATTACHED HYDROGENS

The DEPT experiment is a modification of a basic NMR experiment called the **attached proton test (APT)** experiment. Although a detailed explanation of the theory underlying the DEPT experiment is beyond the scope of this book, an examination of a much simpler experiment (APT) should give you sufficient insight into the DEPT experiment so that you will understand how its results are determined.

This type of experiment uses two transmitters, one operating at the proton resonance frequency and the other at the ^{13}C resonance frequency. The proton transmitter serves as a proton decoupler; it is switched on and off at appropriate intervals during the pulse sequence. The ^{13}C transmitter provides the usual 90° pulse along the X' axis, but it can also be programmed to provide pulses along the Y' axis.

A.   Methine Carbons (CH)

Consider a ^{13}C atom with one proton attached to it, where J is the coupling constant:

$$-\overset{|}{\underset{|}{C}}-H \qquad ^1J_{CH}$$

After a 90° pulse, the bulk magnetization vector **M** is directed along the Y' axis. The result of this simple experiment should be a single line since there is only one vector that is rotating at exactly the same frequency as the Larmor precessional frequency.

In this case, however, the attached hydrogen splits this resonance into a doublet. The resonance does not occur exactly at the Larmor frequency; rather, coupling to the proton produces *two* vectors. One of the vectors rotates J/2 Hz *faster* than the Larmor frequency, and the other vector rotates J/2 Hz *slower* than the Larmor frequency. One vector results from a coupling to the proton with its magnetic moment aligned with the magnetic field, and the other vector results from a coupling to the proton with its magnetic moment aligned against the magnetic field. The two vectors are separated in the rotating frame.

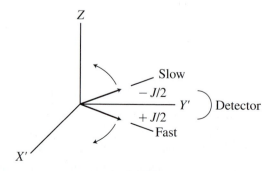

The vectors are moving relative to the rotating frame at a speed of $J/2$ revolutions per second but in opposite directions. The time required for one revolution is therefore the inverse of this speed or $2/J$ seconds per revolution. At time $\frac{1}{4}(2/J)$, $= \frac{1}{2}J$ the vectors have made one-fourth of a revolution and are opposite each other along the X' axis. At this point, no signal is detected by the receiver because there is no component of magnetization along the Y' axis (the resultant of these two vectors is zero).

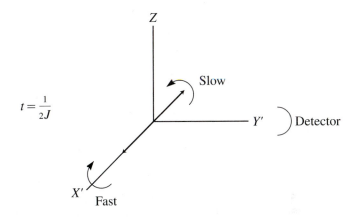

At time $\frac{1}{2}(2/J) = 1/J$, the vectors have realigned along the Y' axis but in the negative direction. An inverted peak would be produced if we collected signal at that time. So, if $t = 1/J$, a methine carbon should show an inverted peak.

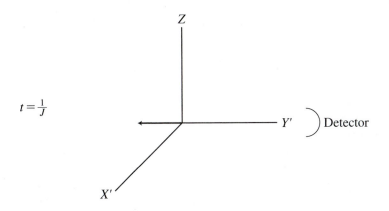

B. Methylene Carbons (CH$_2$)

If we examine the fate of a ^{13}C atom with *two* attached protons, we find different behavior:

$$H - \overset{|}{\underset{|}{C}} - H \qquad {}^1J_{CH}$$

In this case, there are *three* vectors for the ^{13}C nucleus because the two attached protons split the ^{13}C resonance into a triplet. One of these vectors remains stationary in the rotating frame, while the other two move apart with a speed of J revolutions per second.

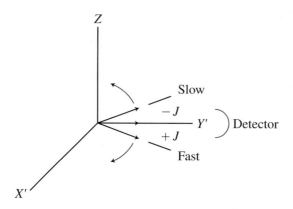

At time $\frac{1}{2}(1/J)$, the two moving vectors have realigned along the negative Y' axis,

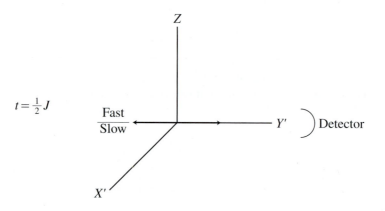

and at time $1/J$, they have realigned along the positive Y axis. The vectors thus produce a normal peak if they are detected at time $t = 1/J$. Therefore, a methylene carbon should show a normal (positive) peak.

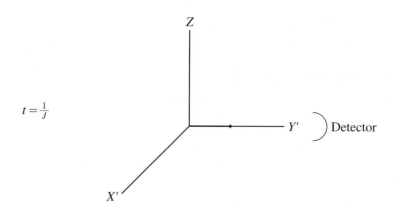

C.   Methyl Carbons (CH₃)

In the case of a methyl carbon,

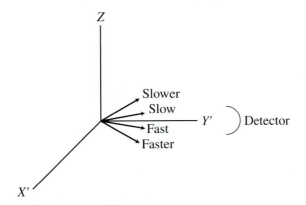

$$^1J_{CH}$$

there should be *four* vectors, corresponding to the four possible spin states of a collection of three hydrogen nuclei.

An analysis of the precessional frequencies of these vectors shows that after time $t = 1/J$, the methyl carbon should also show an inverted peak.

D.   Quaternary Carbons (C)

An unprotonated carbon simply shows one magnetization vector, which precesses at the Larmor frequency (i.e., it always points along the Y' axis). A normal peak is recorded at time $t = 1/J$.

E.   The Final Result

In this type of experiment, we should see a normal peak for every quaternary carbon and methylene carbon and an inverted peak for every methine carbon and methyl carbon. We can thus tell whether the number of hydrogens attached to the carbon is *even* or *odd*.

In the form of this experiment known as the DEPT experiment, the pulse sequence is more complex than those described in the preceding paragraphs. By varying the pulse widths and delay times, it is possible to obtain separate spectra for methyl, methylene, and methine carbons. In the normal manner of presenting DEPT spectra (e.g., a DEPT-135 spectrum), the trace that combines the spectra of all these types of ^{13}C atoms is inverted from the presentation described for the attached proton test (APT). Therefore, in the spectra presented in Figure 9.10, carbon atoms that bear odd numbers of hydrogens appear as positive peaks, carbon atoms that bear even numbers of hydrogens appear as negative peaks, and unprotonated carbon atoms do not appear.

In the DEPT experiment, results similar to those described here for the APT experiment are obtained. A variety of pulse angles and delay times are incorporated into the pulse sequence. The result of the DEPT experiment is that methyl, methylene, methine, and quaternary carbons can be distinguished from one another.

9.6 INTRODUCTION TO TWO-DIMENSIONAL SPECTROSCOPIC METHODS

The methods we have described to this point are examples of one-dimensional experiments. In a **one-dimensional experiment,** the signal is presented as a function of a single parameter, usually the chemical shift. In a **two-dimensional experiment,** there are two coordinate axes. Generally, these axes also represent ranges of chemical shifts. The signal is presented as a function of each of these chemical shift ranges. The data are plotted as a grid; one axis represents one chemical shift range, the second axis represents the second chemical shift range, and the third dimension constitutes the magnitude (intensity) of the observed signal. The result is a form of contour plot in which contour lines correspond to signal intensity.

In a normal pulsed NMR experiment, the 90° **excitation pulse** is followed immediately by a **data acquisition phase** in which the FID is recorded and the data are stored in the computer. In experiments that use complex pulse sequences, such as DEPT, an **evolution phase** is included before data acquisition. During the evolution phase, the nuclear magnetization vectors are allowed to precess, and information may be exchanged between magnetic nuclei. In other words, a given nucleus may become encoded with information about the spin state of another nucleus that may be nearby.

Of the many types of two-dimensional experiments, two find the most frequent application. One of these is **H—H correlation spectroscopy,** better known by its acronym, **COSY.** In a COSY experiment, the chemical shift range of the proton spectrum is plotted on both axes. The second important technique is **heteronuclear correlation spectroscopy,** better known as the **HETCOR** technique. In a HETCOR experiment, the chemical shift range of the proton spectrum is plotted on one axis, while the chemical shift range of the ^{13}C spectrum for the same sample is plotted on the second axis.

9.7 THE COSY TECHNIQUE: ¹H-¹H CORRELATIONS

When we obtain the splitting patterns for a particular proton and interpret it in terms of the numbers of protons located on adjacent carbons, we are using only one of the ways in which NMR spectroscopy can be applied to a structure proof problem. We may also know that a certain proton has two equivalent protons nearby that are coupled with a *J* value of 4 Hz, another nearby proton coupled with a *J* value of 10 Hz, and three others nearby that are coupled by 2 Hz. This gives a very rich pattern for the proton we are observing, but we can interpret it, with a little effort, by using a tree diagram. Selective spin decoupling may be used to collapse or sharpen portions of the spectrum in order to obtain more direct information about the nature of coupling patterns. However, each of these methods can become tedious and very difficult with complex spectra. What is needed is a simple, unbiased, and convenient method for relating coupled nuclei.

A.   An Overview of the COSY Experiment

The pulse sequence for a ^{1}H COSY experiment contains a variable delay time t_1 as well as an acquisition time t_2. The experiment is repeated with different values of t_1, and the data collected during t_2 are stored in the computer. The value of t_1 is increased by regular, small intervals for each experiment, so that the data that are collected consist of a series of FID patterns collected during t_2, each with a different value of t_1.

To identify which protons couple to each other, the coupling interaction is allowed to take place during t_1. During the same period, the individual nuclear magnetization vectors spread as a result of spin-coupling interactions. These interactions modify the signal that is observed during t_2. Unfortunately, the mechanism of the interaction of spins in a COSY experiment is too complex to be described completely in a simple manner. A pictorial description must suffice.

Consider a system in which two protons are coupled to each other:

$$-\overset{|}{\underset{|}{C}}-\overset{|}{\underset{|}{C}}-$$
$$\quad H_a \quad H_x$$

An initial relaxation delay and a pulse **prepare** the spin system by rotating the bulk magnetization vectors of the nuclei by 90°. At this point, the system can be described mathematically as a sum of terms, each containing the spin of only one of the two protons. The spins then **evolve** during the variable delay period (called t_1). In other words, they precess under the influences of both chemical shift and mutual spin–spin coupling. This precession modifies the signal that we finally observe during the acquisition time (t_2). In addition, mutual coupling of the spins has the mathematical effect of converting some of the single-spin terms to **products,** which contain the magnetization components of *both* nuclei. The product terms are the ones we will find most useful in analyzing the COSY spectrum.

Following the evolution period, a second 90° pulse is introduced; this constitutes the next essential part of the sequence, the **mixing period** (which we have not discussed previously). The mixing pulse has the effect of distributing the magnetization among the various spin states of the coupled nuclei. Magnetization that has been encoded by chemical shift during t_1 can be detected at another chemical shift during t_2. The mathematical description of the system is too complex to be treated here. Rather, we can say that two important types of terms arise in the treatment. The first type of term, which does not contain much information that is useful to us, results in the appearance of diagonal peaks in the two-dimensional plot. The more interesting result of the pulse sequences comes from the terms that contain the precessional frequencies of *both* coupled nuclei. The magnetization represented by these terms has been modulated (or "labeled") by the chemical shift of one nucleus during t_1 and, after the mixing pulse, by the precession of the other nucleus during t_2. The resulting off-diagonal peaks (**cross peaks**) show the correlations of pairs of nuclei by means of their spin–spin coupling. When the data are subjected to a Fourier transform, the resulting spectrum plot shows the chemical shift of the first proton plotted along one axis (f_1) and the chemical shift of the second proton plotted along the other axis (f_2). The existence of the off-diagonal peak that corresponds to the chemical shifts of both protons is *proof* of spin coupling between the two protons. If there had been no coupling, their magnetizations would not have given rise to off-diagonal peaks. In the COSY spectrum of a complete molecule, the pulses are transmitted with short duration and high power so that all possible off-diagonal peaks are generated. The result is a complete description of the coupling partners in a molecule.

Since each axis spans the entire chemical shift range, something on the order of a thousand individual FID patterns, each incremented in t_1, must be recorded. With instruments operating at a high spectrometer frequency (high-field instruments), even more FID patterns must be collected. As a result, a typical COSY experiment may require about a half hour to be completed. Furthermore, since each FID pattern must be stored in a separate memory block in the computer, this type of experiment requires a computer with a large available memory. Nevertheless, most modern instruments are capable of performing COSY experiments routinely.

B. How to Read COSY Spectra

2-Nitropropane. To see what type of information a COSY spectrum may provide, we shall consider several examples of increasing complexity. The first is the COSY spectrum of 2-nitropropane. In this simple molecule, we expect to observe coupling between the protons on the two methyl groups and the proton at the methine position.

$$CH_3-CH-CH_3$$
$$|$$
$$NO_2$$

Figure 9.13 is the COSY spectrum of 2-nitropropane. The first thing to note about the spectrum is that the proton NMR spectrum of the compound being studied is plotted along both the horizontal and vertical axes, and each axis is calibrated according to the chemical shift values (in parts per million). The COSY spectrum shows distinct spots on a diagonal, extending from the upper right corner of the spectrum down to the lower left corner. By extending vertical and horizontal lines from each spot on the diagonal, you can easily see that each spot on the diagonal corresponds with the same peak on each coordinate axis. The diagonal peaks serve only as reference points. The important peaks in the spectrum are the **off-diagonal peaks.** In the spectrum of 2-nitropropane, we can extend a horizontal line from the spot at 1.56 ppm (which is labeled **A** and corresponds to the methyl protons). This horizontal line eventually encounters an off-diagonal spot **C** (at the upper left of the COSY spectrum) that corresponds to the methine proton peak at 4.66 ppm (labeled **B**). A vertical line drawn from this off-diagonal spot intersects the spot on the diagonal that corresponds to the methine proton (**B**). The presence of this off-diagonal spot **C**, which correlates the methyl proton spot and the methine proton spot, confirms that the methyl protons are coupled to the methine protons, as we would have expected. A similar result would have been obtained by drawing a vertical line from the 1.56-ppm spot (**A**) and a horizontal line from the 4.66-ppm spot (**B**). The two lines would have intersected at the second off-diagonal spot **D** (at the lower right of the COSY spectrum). The vertical and horizontal lines described in this analysis are drawn on the COSY spectrum in Figure 9.13.

Isopentyl Acetate. In practice, we would not require a COSY spectrum to fully interpret the NMR spectrum of 2-nitropropane. The preceding analysis illustrated how to interpret a COSY spectrum using a simple, easy-to-understand example. A more interesting example is the COSY spectrum of **isopentyl acetate** (Fig. 9.14).

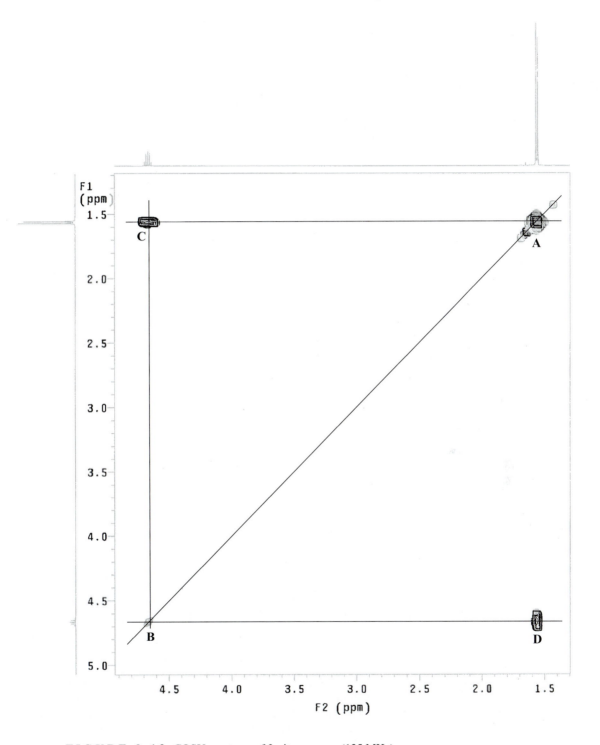

FIGURE 9.13 COSY spectrum of 2-nitropropane (125 MHz).

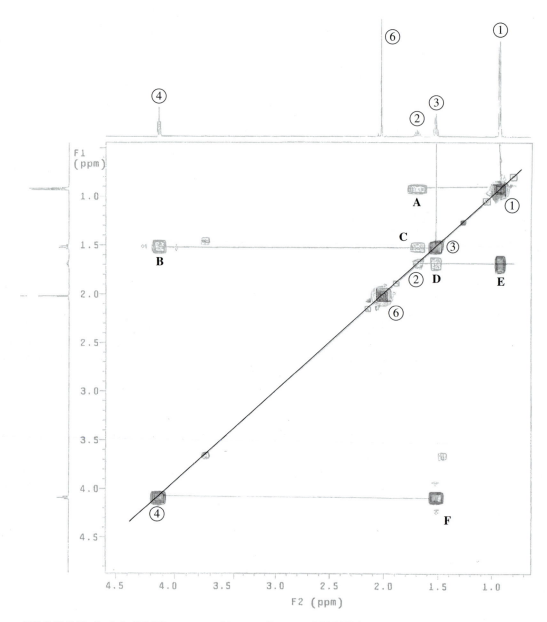

FIGURE 9.14 COSY spectrum of isopentyl acetate (125 MHz).

Again we see coordinate axes; the proton spectrum of isopentyl acetate is plotted along each axis. The COSY spectrum shows a distinct set of spots on a diagonal line, with each spot corresponding to the same peak on each coordinate axis. Lines have been drawn to make the correlations easier to see. The diagonal line shown in Figure 9.14 should be inspected first.

• Start at the upper-right-hand side of the diagonal line for the protons on carbon 1. Follow the horizontal line to the left to the off-diagonal peak at A. This tells us that the protons on carbon 1 correlate with the protons on carbon 2. Thus, we conclude that the protons in the two equivalent methyl groups (1) correlate with the methine proton on carbon **2** at **A**.

- Now look at the diagonal line again, the protons on carbon 3. Following the horizontal line to the left you will observe two off-diagonal peaks at B and C. This tells us that protons on carbon 3 correlate with both the protons on carbon 4 (methylene) and the protons on carbon 2 (methine).

- Again, look closely at the diagonal line at the place labeled 2. The line drawn to the right to the off-diagonal peaks at D and E show that the protons on carbon 2 correlate to the protons on both carbon 3 and carbon 1.

- Now inspect the diagonal line at the protons on carbon 6. This methyl group of the acetyl moiety does not show any off-diagonal peaks. There is no correlation with any other protons because this methyl group labeled as 6 has no adjacent protons. The acetyl methyl protons are not coupled to other protons in the molecule.

- Finally, look at the protons on carbon 4 at the lower end of the diagonal line. The horizontal line drawn to the right shows a correlation to the protons on carbon 3 (F). This shows that the protons on the two methylene groups 3 and 4 correlate with each other.

You may have noticed that each of the COSY spectra shown in this section may show additional spots besides the ones examined in our discussion. Often these "extra" spots indicate the presence of a small amount of an impurity. These "extra" spots have much lower intensities than the principal spots on the plot. However, the COSY method can sometimes detect interactions between nuclei over ranges that extend beyond three bonds (adjacent hydrogen atoms). Besides this long-range coupling, nuclei that are several atoms apart but that are close together spatially also may produce off-diagonal peaks. We learn to ignore these minor peaks in our interpretation of COSY spectra. In some variations of the method, however, spectroscopists make use of such long-range interactions to produce two-dimensional NMR spectra that specifically record this type of information.

Citronellol. The structure of this compound is shown below. The full COSY spectrum is shown in Figure 9.15a. As in the previous example of isopentyl acetate, a diagonal line is drawn. Horizontal lines are drawn from the spots on this diagonal line to off-diagonal spots for protons that are coupled. Because of the symmetry of the plot, vertical lines may also be drawn to off-diagonal spots on the spectrum, but these lines are not shown on the spectrum. Notice that the spectrum is complicated, especially in the region from 2.2 to 0.8 ppm. Because of this, an expansion of this region is shown in Figure 9.15b. In the previous example with isopenyl acetate (Fig. 9.14), we started the analysis on the upper-right part of the diagonal line. This time, we will start at the lower-left part of the spectrum where there are far fewer spots. Concentrate your attention on spots 6, 1, and 5 on the diagonal line. See the horizontal lines drawn in Figure 9.15a.

Citronellol

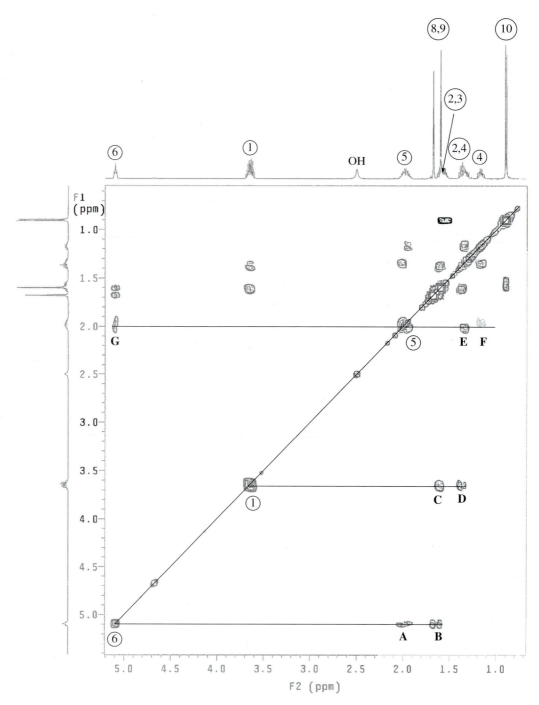

F I G U R E 9 . 1 5 a gCOSY spectrum of citronellol (125 MHz).

Analysis of the Full COSY Spectrum Shown in Figure 9.15a

- The proton on C6 on the diagonal line shown in Figure 9.15a is clearly coupled to the two protons on C5 at **A** (follow the horizontal line). Closer examination of the spectrum also reveals that the proton on C6 is also coupled through allylic (four-bond) coupling to the two methyl groups at C8 and C9 at **B**.

FIGURE 9.15b Expansion of the gCOSY spectrum of citronellol from 0.8 to 2.2 ppm.

- Following the horizontal line connecting position C1 on the diagonal line with off-diagonal spots at **C** (1.6 ppm) and **D** (1.4 ppm) shows that the two protons on C1 are coupled to two nonequivalent protons on C2. They are nonequivalent, owing to the presence of a stereocenter in the molecule at C3, identified with an asterisk on the structure. The protons are examples of diastereotopic hydrogen atoms and were discussed in Chapter 7, Section 7.4. (Note: The HETCOR spectral method described in Section 9.8 and shown in Problem 3 confirms the assignments of the diastereotopic protons.)

- Now focus your attention on the spot on the diagonal line at C5 in Figure 9.15a. First, follow the horizontal line to the right. The two protons on C5 couple to the nonequivalent (diastereotopic) hydrogen atoms on C4 at **E** (1.35 ppm) and **F** (1.17 ppm). These two protons on C4 are nonequivalent since they are next to the stereocenter on C3. (Note: The HETCOR spectral method described in Section 9.8 and shown in Problem 3 confirms the assignments of the diastereotopic protons.) Following the horizontal line to the left, observe that the two hydrogen atoms on C5 also couple with the hydrogen atom on C6, labeled **G**.

Analysis of the Expanded COSY Spectrum Shown in Figure 9.15b

The expansion of the region from 0.8 to 2.2 ppm shown in Figure 9.15b provides the remaining correlations. One of the main problems in interpreting this area is that many of the spots overlap with each other, making the analysis difficult. For example, you may notice that some of the peaks in the ^{1}H spectrum shown at the top of the plot overlap with each other, especially at 1.6 and 1.4 ppm. The pairs of diastereotopic hydrogen atoms on the two methylene groups on C2 and C4 also make the analysis interesting.

- Starting at the upper right in the spectrum on the diagonal line at C10, follow the horizontal line to the left to the spot at **H** showing that the methyl group on C10 correlates to the single hydrogen atom on C3. This methyl group appears as a doublet on the ^{1}H spectrum drawn on the upper axis.

- Now look at the diagonal line, at C4. One of the diastereotopic protons on C4 (1.17 ppm) correlates with the other diastereotopic proton on C4 (1.35 ppm) at spot **I**. It also correlates with one of the two protons on C5 spot **J**. Interestingly, C4 does not correlate with the single proton on C3.

- Notice that there is another C4 indicated on the diagonal line, nearly overlapping with C2. This diastereotopic proton on C4 (1.35 ppm) correlates with the other diastereotopic proton at 1.17 ppm labeled as **M**. It also correlates with the single proton on C3 at **L**. This proton on C4 also correlates with one of the two protons on C5 at spot **K**. The protons on C5 are diastereotopic.

- One of the two diastereotopic protons on C2 (1.37 ppm) correlates with the other diastereotopic proton at about 1.60 ppm, labeled as **L**. It also correlates with the single proton on C3, also at **L**. It also correlates with the two protons on C1 (not shown on the expansion). (Note: The HETCOR spectral method described in Section 9.8 and shown in Problem 3 confirms the assignments of the diastereotopic protons.)

- The other diastereotopic proton at C2 (1.60 ppm) overlaps with C3. It correlates with the other diastereotopic proton on C2 (1.37 ppm) at **N**.

- Finally, C3 correlates with the methyl group located at C10 at **O**. It also correlates with one of the protons on C4 at **N** but not the other proton on C4. C3 also correlates with C2 at **N**.

Pulsed field gradients were introduced in Section 9.3. The COSY method can be combined with the use of pulsed field gradients to produce a result that contains the same information as a COSY spectrum but that has much better resolution and can be obtained in a shorter time. This type of experiment is known as a **gradient-selected** COSY (sometimes known as a **gCOSY**). A gCOSY spectrum can be obtained in as little as 5 min; by contrast, a typical COSY spectrum requires as much as 40 min for data acquisition. This spectrum of citronellol was obtained by this method.

☐ 9.8 THE HETCOR TECHNIQUE: ^{1}H-^{13}C CORRELATIONS

Protons and carbon atoms interact in two very important ways. First, they both have magnetic properties, and they can induce relaxation in one another. Second, the two types of nuclei can be spin-coupled to each other. This latter interaction can be very useful because directly bonded protons and carbons have a *J* value that is at least a power of 10 larger than nuclei related by two-bond or three-bond couplings. This marked distinction between orders of coupling provides us with a sensitive way of identifying carbons and protons that are directly bonded to one another.

To obtain a correlation between carbons and attached protons in a two-dimensional experiment, we must be able to plot the chemical shifts of the ^{13}C atoms along one axis and the chemical shifts of the protons along the other axis. A spot of intensity in this type of two-dimensional spectrum would indicate the existence of a C—H bond. The heteronuclear chemical shift correlation (HETCOR) experiment is designed to provide the desired spectrum.

A.   An Overview of the HETCOR Experiment

As we did in the COSY experiment, we want to allow the magnetization vectors of the protons to precess according to different rates, as dictated by their chemical shifts. Therefore, we apply a 90° pulse to the protons, then include an evolution time (t_1). This pulse tips the bulk magnetization vector into the $X'Y'$ plane. During the evolution period, the proton spins precess at a rate determined by their chemical shifts and the coupling to other nuclei (both protons and carbons). Protons bound to ^{13}C atoms experience not only their own chemical shifts during t_1 but also homonuclear spin coupling and heteronuclear spin coupling to the attached ^{13}C atoms. It is the interaction between ^{1}H nuclei and ^{13}C nuclei that produces the correlation that interests us. After the evolution time, we apply simultaneous 90° pulses to both the protons and the carbons. These pulses transfer magnetization from protons to carbons. Since the carbon magnetization was "labeled" by the proton precession frequencies during t_1, the ^{13}C signals that are detected during t_2 are modulated by the chemical shifts of the coupled protons. The ^{13}C magnetization can then be detected in t_2 to identify a particular carbon carrying each type of proton modulation.

A HETCOR experiment, like all two-dimensional experiments, describes the environment of the nuclei during t_1. Because of the manner in which the HETCOR pulse sequence has been constructed, the only interactions that are responsible for modulating the proton spin states are the proton chemical shifts and homonuclear couplings. Each ^{13}C atom may have one or more peaks appearing on the f_2 axis that correspond to its chemical shift. The proton chemical shift modulation causes the two-dimensional intensity of the proton signal to appear at an f_1 value that corresponds to the proton chemical shift. Further proton modulations of much smaller frequency arise from homonuclear (H—H) couplings. These provide fine structure on the peaks along the f_1 axis. We can interpret the fine structure exactly as we would in a normal proton spectrum, but in this case we understand that the proton chemical shift value belongs to a proton that is attached to a specific ^{13}C nucleus that appears at its own carbon chemical shift value.

We can thus assign carbon atoms on the basis of known proton chemical shifts, or we can assign protons on the basis of known carbon chemical shifts. For example, we might have a crowded proton spectrum but a carbon spectrum that is well resolved (or vice versa). This approach makes the HETCOR experiment particularly useful in the interpretation of the spectra of large, complex molecules. An even more powerful technique is to use results from both the HETCOR and COSY experiments together.

B.   How to Read HETCOR Spectra

2-Nitropropane. Figure 9.16 is an example of a simple HETCOR plot. In this case, the sample substance is 2-nitropropane.

$$CH_3-CH-CH_3$$
$$|$$
$$NO_2$$

It is common practice to plot the proton spectrum of the compound being studied along one axis and the carbon spectrum along the other axis. Each spot of intensity on the two-dimensional plot indicates a carbon atom that bears the corresponding protons. In Figure 9.16, you should be able to see a peak corresponding to the methyl carbons, which appear at 21 ppm in the carbon spectrum (horizontal axis), and a peak at 79 ppm corresponding to the methine carbon. On the vertical axis, you should also be able to find the doublet for the methyl protons at 1.56 ppm (proton spectrum) and a septet for the methine proton at 4.66 ppm. If you drew a vertical line from the methyl peak of the carbon spectrum (21 ppm) and a horizontal line from the methyl peak of the proton spectrum (1.56 ppm), the two lines would intersect at the exact point **A** on the two-dimensional plot where a spot is marked. This spot indicates that the protons at 1.56 ppm and the carbons at 21 ppm represent the same position of the

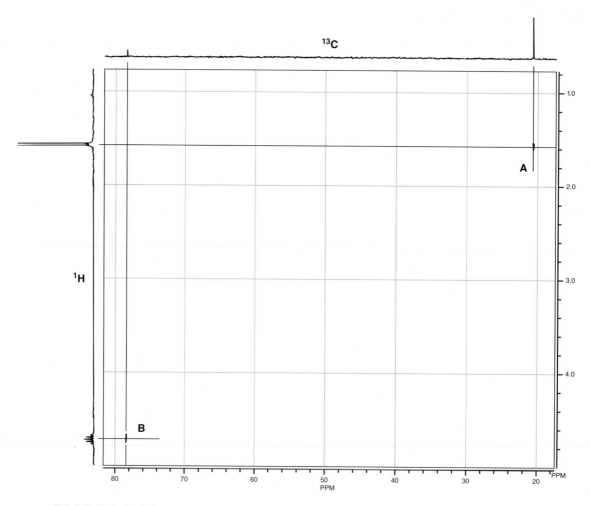

FIGURE 9.16 HETCOR spectrum of 2-nitropropane.

molecule. That is, the hydrogens are attached to the indicated carbon. In the same way, the spot **B** in the lower left corner of the HETCOR plot correlates with the carbon peak at 79 ppm and the proton septet at 4.66 ppm, indicating that these two absorptions represent the same position in the molecule.

Isopentyl Acetate. A second, more complex example is isopentyl acetate. Figure 9.17 is the HETCOR plot for this substance.

$$\underset{6}{CH_3}-\underset{5}{\overset{\overset{\displaystyle O}{\|}}{C}}-O-\underset{4}{CH_2}-\underset{3}{CH_2}-\underset{2}{\underset{\underset{1}{\overset{|}{CH_3}}}{CH}}-\underset{1}{CH_3}$$

Each spot on the HETCOR plot has been labeled with a number, and lines have been drawn to help you see the correlations between proton peaks and carbon peaks. The carbon peak at 23 ppm and the proton doublet at 0.92 ppm correspond to the methyl groups (**1**); the carbon peak at 25 ppm and the proton multiplet at 1.69 ppm correspond to the methine position (**2**); and the carbon peak at

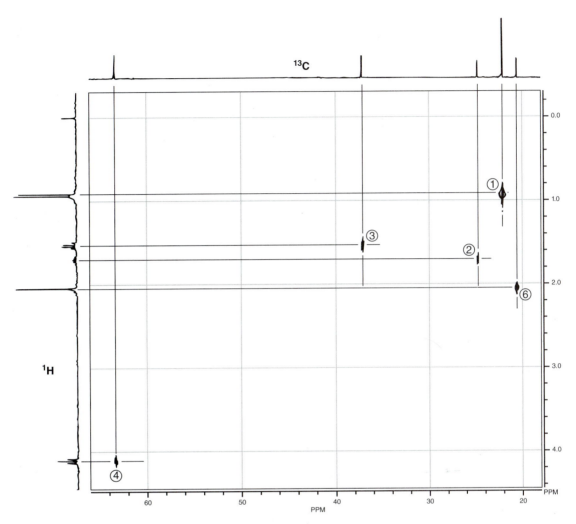

FIGURE 9.17 HETCOR spectrum of isopentyl acetate.

37 ppm and the proton quartet at 1.52 ppm correspond to the methylene group (**3**). The other methylene group (**4**) is deshielded by the nearby oxygen atom. Therefore, a spot on the HETCOR plot for this group appears at 63 ppm on the carbon axis and 4.10 ppm on the proton axis. It is interesting that the methyl group of the acetyl function (**6**) appears downfield of the methyl groups of the isopentyl group (**1**) in the proton spectrum (2.04 ppm). We expect this chemical shift since the methyl protons should be deshielded by the anisotropic nature of the carbonyl group. In the carbon spectrum, however, the carbon peak appears *upfield* of the methyl carbons of the isopentyl group. A spot on the HETCOR plot that correlates these two peaks confirms that assignment.

4-Methyl-2-Pentanol. Figure 9.18 is a final example that illustrates some of the power of the HETCOR technique for 4-methyl-2-pentanol. Lines have been drawn on the spectrum to help you find the correlations.

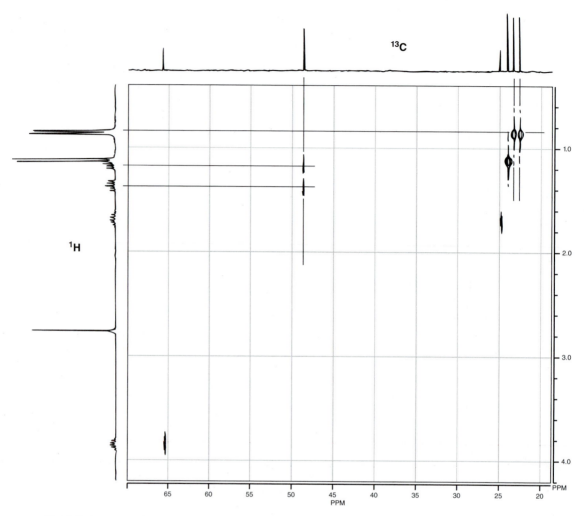

FIGURE 9.18 HETCOR spectrum of 4-methyl-2-pentanol.

This molecule has a stereocenter at carbon 2. An examination of the HETCOR plot for 4-methyl-2-pentanol reveals *two* spots that correspond to the two methylene protons on carbon 3. At 48 ppm on the carbon axis, two contour spots appear, one at about 1.20 ppm on the proton axis and the other at about 1.40 ppm. The HETCOR plot is telling us that there are two nonequivalent protons attached to carbon 3. If we examine a Newman projection of this molecule, we find that the presence of the stereocenter makes the two methylene protons (**a** and **b**) nonequivalent (they are diastereotopic, see Chapter 7, Section 7.4). As a result, they appear at different values of chemical shift.

The *carbon* spectrum also reveals the effect of a stereocenter in the molecule. In the proton spectrum, the apparent doublet (in fact, it is a *pair* of doublets) at 0.91 ppm arises from the six protons on the methyl groups, which are labeled **5** and **6** in the preceding structure. Looking across to the right on the HETCOR plot, you will find two contour spots, one corresponding to 22 ppm and the other corresponding to 23 ppm. These two carbon peaks arise because the two methyl groups are also not quite equivalent; the distance from one methyl group to the oxygen atom is not quite the same as that from the other methyl group when we consider the most stable conformation for the molecule.

A great many advanced techniques can be applied to complex molecules. We have introduced only a few of the most important ones here. As computers become faster and more powerful, as chemists evolve their understanding of what different pulse sequences can achieve, and as scientists write more sophisticated computer programs to control those pulse sequences and treat data, it will become possible to apply NMR spectroscopy to increasingly complex systems.

9.9 INVERSE DETECTION METHODS

The NMR detection probe that is used for most heteronuclear experiments (such as the HETCOR experiment) is designed so that the receiver coils for the less-sensitive nucleus (the "insensitive" nucleus) are located closer to the sample than the receiver coils for the more sensitive (generally the ^1H) nucleus. This design is aimed at maximizing the signal that is detected from the insensitive nucleus. As was described in Chapter 6, owing to a combination of low natural abundance and a low magnetogyric ratio, a ^{13}C nucleus is about 6000 times more difficult to detect than a ^1H nucleus. A ^{15}N nucleus is also similarly more difficult to detect than a ^1H nucleus.

The difficulty with this probe design is that the initial pulse and the detection occur in the insensitive channel, while the evolution period is detected in the ^1H channel. The *resolution* that is possible, however, is much lower in the channel where the evolution of spins is detected. In the case of a carbon–hydrogen correlation (a HETCOR), this means that the greatest resolution will be seen in the ^{13}C spectrum (in which every peak is a singlet), and the lowest resolution will be observed in the ^1H spectrum (in which maximum resolution is required). In effect, the *lowest* resolution is observed along the axis where the *greatest* resolution is required.

Within the past 20 years, the technology of probe design has advanced significantly. Today, an instrument can be fitted with an **inverse detection** probe. In this design, the initial pulse and detection occurs in the ^1H channel, where the resolution is very high. The insensitive nucleus is detected during the evolution time of the pulse sequence, for which high resolution is generally not required. The result is a cleaner two-dimensional spectrum with high resolution. Examples of heteronuclear detection experiments that utilize an inverse detection probe are **heteronuclear multiple-quantum correlation** (HMQC) and **heteronuclear single-quantum correlation** (HSQC). Each of these techniques provides the same information that can be obtained from a HETCOR spectrum but is more suitable when the spectrum contains many peaks that are crowded close together. The improved resolution of the HMQC and HSQC experiments allows the spectroscopist to distinguish between two closely spaced peaks, whereas these peaks might overlap into one broadened peak in a HETCOR spectrum. Furthermore, since the more sensitive proton nucleus is detected in these experiments, the data acquisition takes less time.

9.10 THE NOESY EXPERIMENT

The nuclear Overhauser effect was described in Chapter 6, Sections 6.5 and 6.6. A two-dimensional NMR experiment that takes advantage of the nuclear Overhauser effect is nuclear Overhauser effect spectroscopy, or NOESY. Any ^1H nuclei that may interact with one another through a dipolar relaxation process will appear as cross peaks in a NOESY spectrum. This type of interaction not only includes nuclei that are directly coupled to one another, but it also includes nuclei that are not directly coupled but are located near to one another *through space*. The result is a two-dimensional spectrum that looks very much like a COSY spectrum but includes, besides many of the expected COSY cross peaks, additional cross peaks that arise from interactions of nuclei that interact through space. In practice, the nuclei must be within 5 Å of each other for this spatial interaction to be observed.

NOESY spectroscopy has become especially useful in the study of large molecules, such as proteins and polynucleotides. Very large molecules tend to tumble more slowly in solution, which means that nuclear Overhauser effect interactions have more time to develop. Small molecules tumble more quickly in solution; the nuclei move past one another too quickly to allow a significant development of dipolar interactions. The result is that NOESY cross peaks may be too weak to be observed.

Because the cross peaks in NOESY spectra arise from spatial interactions, this type of spectroscopy is particularly well suited to the study of configurations and conformations of molecules. The example of acetanilide demonstrates the capabilities of the NOESY experiment. The structural formula is shown, with the proton NMR chemical shifts of the relevant protons indicated.

The problem to be solved is to decide which of two possible conformations is the more important for this molecule. The two conformations are shown, with circles drawn around the protons that are close to each other spatially and would be expected to show nuclear Overhauser interactions.

For conformation **A**, the N—H hydrogen is close to the methyl C—H hydrogens. We would expect to see a cross peak in the NOESY spectrum that correlates the N—H peak at 8.8 ppm with the C—H peak at 2.13 ppm. For conformation **B**, the protons that are close to each other are the methyl C—H protons and the *ortho* proton of the aromatic ring. For this conformation, we would expect to see a cross peak that correlates the aromatic proton at 7.49 ppm with the methyl protons at 2.13 ppm. When the actual spectrum is determined, one finds a weak cross peak that links the 8.8-ppm peak with the 2.13-ppm peak. This demonstrates clearly that the preferred conformation for acetanilide is **A.**

Certainly, when one considers solving the three-dimensional structure of a complex molecule such as a polypeptide, the challenge of assigning every peak and every cross peak becomes formidable. Nevertheless, the combination of COSY and NOESY methods finds wide application in the determination of the structures of biomolecules.

9.11 MAGNETIC RESONANCE IMAGING

The principles that govern the NMR experiments described throughout this textbook have applications in the field of medicine. A very important diagnostic tool in medicine is a technique known as **magnetic resonance imaging (MRI).** In the space of only a few years, MRI has found wide use in the diagnosis of injuries and other forms of abnormality. It is quite common for sports fans to hear of an athlete who has sustained a knee injury and had it examined via an MRI scan.

Typical magnetic resonance imaging instruments use a superconducting magnet with a field strength on the order of 1 Tesla. The magnet is constructed with a very large inner cavity so that an entire human body can fit inside. A transmitter–receiver coil (known as a **surface coil**) is positioned outside the body, near the area being examined. In most cases, the ^{1}H nucleus is the one studied since it is found in the water molecules that are present in and around living tissue. In a manner somewhat analogous to that used for an X-ray-based CAT (computerized axial tomography) scan, a series of planar images is collected and stored in the computer. The planar images can be obtained from various angles. When the data have been collected, the computer processes the results and generates a three-dimensional picture of the proton density in the region of the body under study.

The ^{1}H nuclei of water molecules that are not bound within living cells have a relaxation time different from the nuclei of water molecules bound within tissue. Water molecules that appear in a highly ordered state have relaxation times shorter than water molecules that appear in a more random state. The degree of ordering of water molecules within tissues is greater than that of water molecules that are part of the fluid flowing within the body. Furthermore, the degree of ordering of water molecules may be different in different types of tissue, especially in diseased tissue as compared with normal tissue. Specific pulse sequences detect these differences in relaxation time for the protons of water molecules in the tissue being examined. When the results of the scans are processed, the image that is produced shows different densities of signals, depending on the degree to which the water molecules are in an ordered state. As a result, the "picture" that we see shows the various types of tissue clearly. The radiologist can then examine the image to determine whether any abnormality exists.

As a brief illustration of the type of information that can be obtained from MRI, consider the image in Figure 9.19. This is a view of the patient's skull from the spinal column, looking toward

FIGURE 9.19 MRI scan of a skull showing soft tissues of the brain and eyes.

FIGURE 9.20 MRI scan of a skull showing the presence of an infarct.

the top of the patient's head. The light-colored areas represent the locations of soft tissues of the brain. Because bone does not contain a very high concentration of water molecules, the MRI image provides only a somewhat dim view of the bones of the spinal column. The two bulbous features at the top of the image are the person's eyes.

Figure 9.20 is another MRI scan of the same patient shown in Figure 9.19. On the left side of the image is an area that appears brightly white. This patient has suffered an **infarct,** an area of dying tissue resulting from an obstruction of the blood vessels supplying that part of the brain. In other words, the patient has had a stroke, and the MRI scan has clearly shown exactly where this lesion has occurred. The physician can use very specific information of this type to develop a plan for treatment.

The MRI method is not limited to the study of water molecules. Pulse sequences designed to study the distribution of lipids are also used.

The MRI technique has several advantages over conventional X-ray or CAT scan techniques; it is better suited for studies of abnormalities of soft tissue or of metabolic disorders. Furthermore, unlike other diagnostic techniques, MRI is noninvasive and painless, and it does not require the patient to be exposed to large doses of X-rays or radioisotopes.

9.12 SOLVING A STRUCTURAL PROBLEM USING COMBINED 1-D AND 2-D TECHNIQUES

This section shows you how to solve a structural problem using the various spectroscopic techniques. We will make use of ^1H, ^{13}C, HETCOR (gHSQC), COSY, and DEPT NMR techniques. We will also make use of infrared spectroscopy to solve the structure of this example.

FIGURE 9.21 Infrared spectrum of $C_6H_{10}O_2$.

A.   Index of Hydrogen Deficiency and Infrared Spectrum

The compound has a formula $C_6H_{10}O_2$. Your first order of business should be to calculate the index of hydrogen deficiency, which is 2. Let's now look at the infrared spectrum shown in Figure 9.21 to determine the types of functional groups present that would be consistent with an index of 2. The spectrum shows a strong C=O peak at 1716 and a strong peak at 1661 cm^{-1} for a C=C. Even though the C=O peak appears near the expected value for a ketone, the presence of the C=C is more likely to indicate that the compound is a conjugated ester with the C=O stretch shifted from the normal 1735-cm^{-1} value found in unconjugated esters to the lower value due to resonance with the double bond. The strong C—O bands in the region of 1350 to 1100 cm^{-1} would support the idea of an ester. The out-of-plane C—H bending patterns shown in Figure 2.22 on p. 42 may be useful to help decide the type of substitution on the C=C bond. For example, the band at 970 cm^{-1} would indicate a *trans* double bond. Notice that a weak peak appears at 3054 cm^{-1}, indicating the presence of a sp^2 C—H bond. The other C—H stretching bonds below 3000 cm^{-1} indicate sp^3 C—H bonds.

B.   Carbon-13 NMR Spectrum

Next, look at the proton decoupled ^{13}C spectrum shown in Figure 9.22. Notice that there are six peaks in the spectrum matching the six carbons in the formula. Read Section 6.17 starting on p. 319 to determine how to make use of the ^{13}C spectrum. Three of the peaks appear to the right of the solvent peaks (CDCl$_3$) and represent sp^3 carbon atoms. The peak at about 60 ppm suggests a carbon atom attached to an electronegative oxygen atom. Three peaks appear to the left of the solvent peak. Two of them, appearing at about 122 and 144 ppm, are for sp^2 carbon atoms in the C=C bond. The remaining carbon peak at about 166 can be assigned to the C=O carbon atom.

FIGURE 9.22 ^{13}C NMR spectrum for $C_6H_{10}O_2$.

C.   DEPT Spectrum

The DEPT spectrum is shown in Figure 9.23. The beauty of this experiment is that it tells you the number of protons attached to each carbon atom. The type of presentation shown here is the same type of DEPT presentation shown in Figure 9.12. Other ways of displaying the DEPT spectrum are shown in Figure 9.10 and in Chapter 6, Figure 6.9. The plot shown in Figure 9.23 shows the methyl,

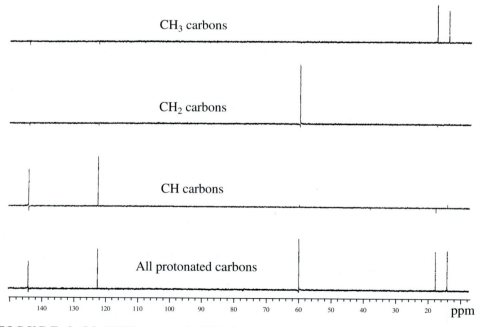

FIGURE 9.23 DEPT spectrum for $C_6H_{10}O_2$.

methylene, and methine carbons on the first three lines as positive peaks. The bottom trace shows all protonated carbon atoms. Carbon atoms without attached protons will not appear in a DEPT spectrum. Thus, the plot does not show the C=O carbon atom because there are no attached protons. However, we know from the normal ^{13}C NMR spectrum that a peak appears at 166.4 ppm and this must be the C=O carbon atom. Notice that the $CDCl_3$ solvent does not appear in the DEPT spectrum but does appear in the normal ^{13}C NMR spectrum as a three-line pattern centering on about 77 ppm. From the DEPT experiment, you can conclude that there are two methyl carbons, appearing at 14.1 and 17.7 ppm. There is one methylene carbon appearing at 59.9 ppm ($-O-CH_2-$) and two methine carbons for the C=C bond that appear downfield at 122.6 and 144.2 ppm. We now know that the compound is a disubstituted alkene, which confirms the IR results. Making use of the IR, ^{13}C NMR, and DEPT experiments yields the following structure:

D. Proton NMR Spectrum

The proton spectrum is shown in Figure 9.24. The integral values need to be determined by using the numbers below the peaks. The 10 protons in the spectrum integrate as follows: 1:1:2:3:3. The

FIGURE 9.24 ^1H (proton) NMR spectrum for $C_6H_{10}O_2$.

protons of special interest are shown as expansions in Figure 9.25. The signal centered on 6.97 ppm is a doublet of quartets. What is most apparent is that there is a pair of overlapping quartets (the right-hand quartet is shaded so that you can more easily see the patterns). The doublet part of the pattern results from a vinyl proton H_e being split by the *trans*-proton H_d into a doublet, $^3J_{trans}$. The peaks are numbered on the expansion in Figure 9.25, counting from left to right. In effect the coupling constant for the doublet can be derived by subtracting the Hertz value for the center of the right quartet from the Hertz value for the center of the left quartet. It turns out that it is easier to simply subtract the Hertz value for line 6 from the Hertz value on line 2 or subtracting the value for line 7 from line 3. The averaged value is $^3J = 15.3$ Hz. Also, one can calculate the coupling constant for the quartet part of the pattern that results from the coupling between the vinyl proton H_e and the methyl protons H_b. This is calculated by subtracting the value on line 2 from line 1, line 3 from 2, and so on, yielding an average value of $^3J = 7.1$ Hz. The overall pattern is described as a doublet of quartets, with a $^3J = 15.3$ and 7.1 Hz.

The other vinyl proton (H_d) centered on 5.84 ppm can also be described as a doublet of quartets. In this case, it is much more obvious that it is a doublet of quartets than for the pattern at 6.97 ppm. The Hertz values for peaks in the quartets yield an average value for $^4J = 1.65$ Hz resulting from the long-range coupling between H_d and H_b. The remaining coupling constant for H_d to H_e can be derived by subtracting 2908.55 Hz from 2924.14 Hz, yielding a value for $^3J_{trans} = 15.5$ Hz. This value agrees within experimental error with the $^3J_{trans}$ obtained from the proton at 6.97 ppm, discussed above.

The methyl group (H_b) appearing at 1.87 ppm is a doublet of doublets. The coupling between proton H_b and H_e is calculated by subtracting 931.99 from 939.08 Hz, $^3J = 7.1$ Hz. Notice that this is the same value that was obtained above for H_e. The average value for the distances in Hertz between the smaller peaks yields $^4J = 1.65$ Hz. This value is identical to the one obtained above for H_d.

FIGURE 9.25 Proton NMR expansions for $C_6H_{10}O_2$.

Finally, the triplet at about 1.3 ppm is assigned to the methyl group (H_a) split by the neighboring methylene group (H_c). In turn, the quartet at about 4.2 results from the coupling to the neighboring methyl group (H_a).

E.   COSY NMR Spectrum

The COSY spectrum is shown in Figure 9.26. A COSY spectrum is an **$^1H-^1H$ correlation** with the proton NMR spectrum plotted on both axes. It helps to confirm that we have made the right assignments for the coupling of neighboring protons in this example: ^3J and ^4J. Following the lines drawn on the spectrum, we see that proton H_a correlates to proton H_c in the ethyl group. We also see that proton H_b correlates to both H_d and H_e. Proton H_d correlates to both H_e and H_b. Finally, proton H_e correlates to both H_d and H_b. Life is very good!

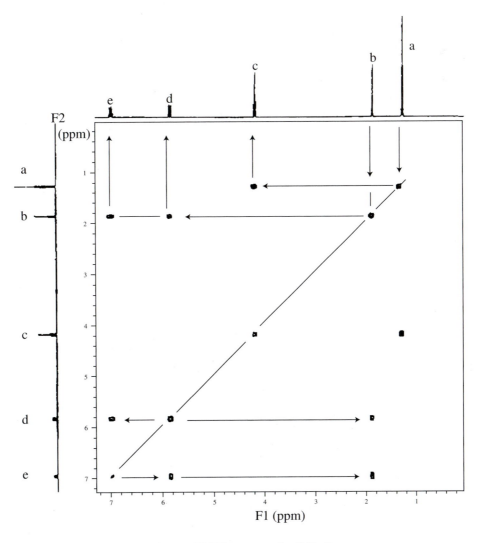

FIGURE 9.26 H—H correlation (COSY) spectrum for $C_6H_{10}O_2$.

F.   HETCOR (HSQC) NMR Spectrum

The HETCOR (HSQC) spectrum is shown in Figure 9.27. This type of spectrum is a $^{13}C-H$ **correlation** with the ^{13}C NMR spectrum and the ^1H NMR spectrum plotted on the two axes. The purpose of this experiment is to assign each of the ^{13}C peaks to the corresponding proton patterns. The results support the conclusions already made about the assignments. No surprises here!

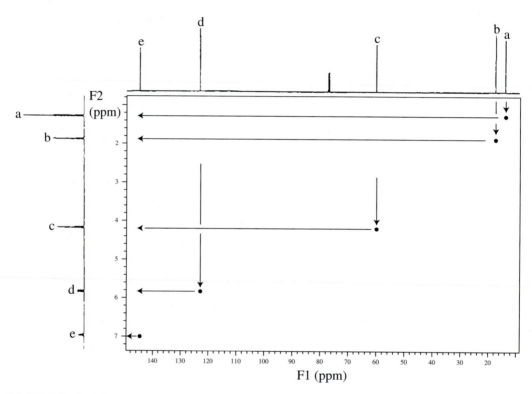

F I G U R E   9 . 2 7   C$-$H correlation (HETCOR/HSQC) spectrum for $C_6H_{10}O_2$.

PROBLEMS

*1. Using the following set of DEPT-135, COSY, and HETCOR spectra, provide a complete assignment of all protons and carbons for C_4H_9Cl.

200   180   160   140   120   100   80   60   40   20   0

PPM

*2. Assign each of the peaks in the following DEPT spectrum of $C_6H_{14}O$. The infrared spectrum shows a strong and broad peak at about 3350 cm^{-1}. (Note: There is more than one possible answer.)

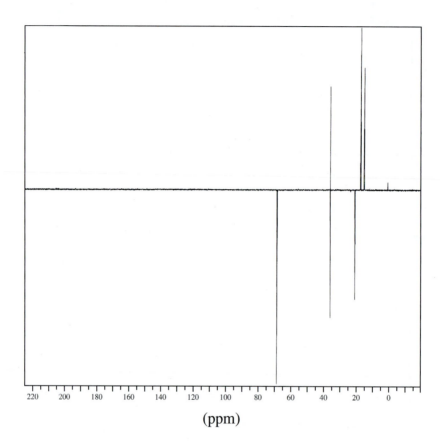

(ppm)

*3. The upper spectrum is a HETCOR (HSQC) spectrum for citronellol in the range of 0.8 to
   5.5 ppm. The lower spectrum is an expansion of the same compound in the region between 0.8
   and 2.2 ppm. Use the structural formula on p. 531, as well as the stacked full-edited DEPT
   spectrum in Figure 9.12 and the gCOSY spectra shown in Figures 9.15a and 9.15b, to provide
   a complete assignment of all carbon and hydrogen atoms in the molecule, especially the
   diastereotopic protons at 25, 37, and 40 ppm. Complete assignments have been provided in the
   Answers to Selected Problems.

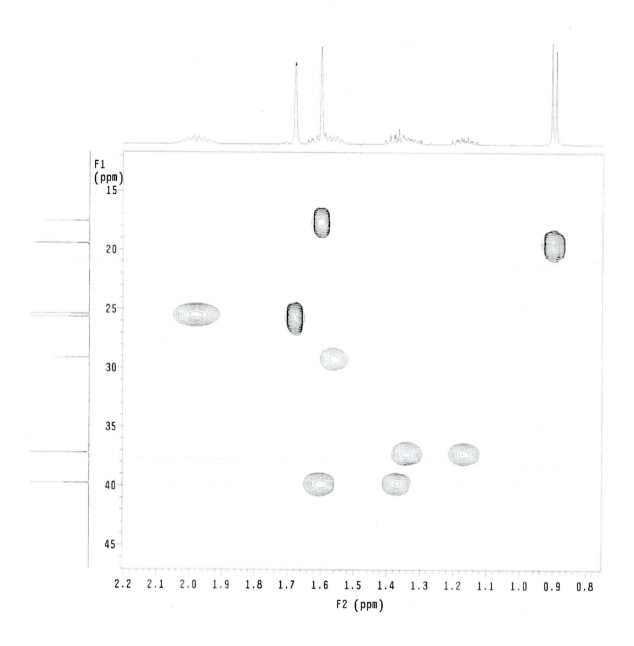

*4. Geraniol has the structure

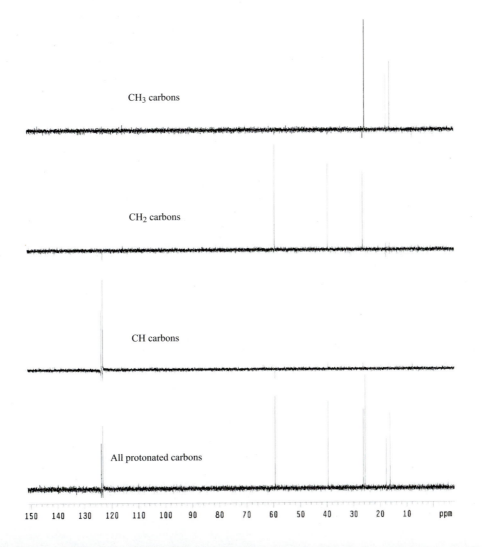

Use the stacked full-edited DEPT spectrum shown on this page, the gCOSY spectrum shown on the next page, and finally the HETCOR (HSQC) spectrum shown on the third page. Carbons 3 and 7 do not show on the DEPT spectrum because the spectrum only shows protonated carbon atoms. The assignments you determined in Problem 3 may help you here. Complete assignments have been provided in the Answers to Selected Problems.

*5. The following set of spectra includes an expansion of the aromatic region of the ^{1}H NMR spectrum of **methyl salicylate** as well as a HETCOR spectrum. Provide a complete assignment of all aromatic protons and unsubstituted ring carbons in methyl salicylate. (*Hint*: Consider the resonance effects of the substituents to determine relative chemical shifts of the aromatic hydrogens. Also try calculating the expected chemical shifts using the data provided in Appendix 6.)

METHYL SALICYLATE EXPANSION

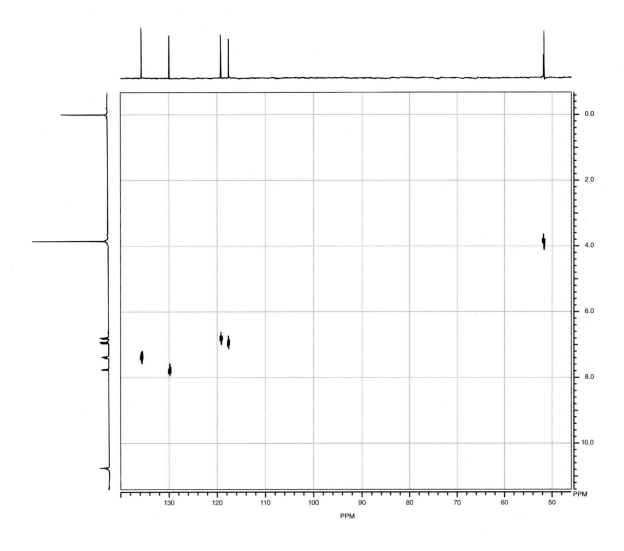

6. Determine the structure for a compound with formula $C_6H_{12}O_2$. The IR spectrum shows a strong and broad band from 3400 to 2400 cm^{-1} and also at 1710 cm^{-1}. The ^{1}H NMR spectrum and expansions are shown, but a peak appearing at 12.0 ppm is not shown in the full spectrum. Fully interpret the ^{1}H NMR spectrum, especially the patterns between 2.1 and 2.4 ppm. A HETCOR spectrum is provided in this problem. Comment on the carbon peaks appearing at 29 and 41 ppm in the HETCOR spectrum. Assign all the protons and carbons for this compound.

(ppm)

HETCOR

7. Determine the structure for a compound with formula $C_{11}H_{16}O$. This compound is isolated from jasmine flowers. The IR spectrum shows strong bands at 1700 and 1648 cm^{-1}. The ^1H NMR spectrum, with expansions, along with the HETCOR, COSY, and DEPT spectra are provided in this problem. The DEPT-90 spectrum is not shown, but it has peaks at 125 and 132 ppm. This compound is synthesized from 2,5-hexanedione by monoalkylation with (Z)-1-chloro-2-pentene, followed by aldol condensation. Assign all the protons and carbons for this compound.

(ppm)

(ppm)

—885.75
—878.76

3.00 2.95 2.90
(ppm)

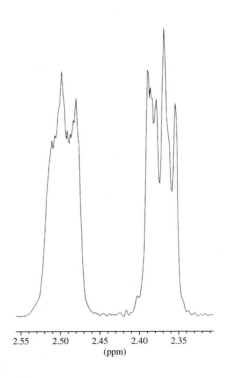

2.55 2.50 2.45 2.40 2.35
(ppm)

HETCOR

(ppm)

(ppm)

Normal Carbon

CDCl₃

DEPT-135

8. Determine the structure for a compound with formula $C_{10}H_8O_3$. The IR spectrum shows strong bands at 1720 and 1620 cm^{-1}. In addition, the IR spectrum has bands at 1580, 1560, 1508, 1464, and 1125 cm^{-1}. The ^{1}H NMR spectrum, with expansions, along with the COSY and DEPT spectra are provided in this problem. Assign all of the protons and carbons for this compound.

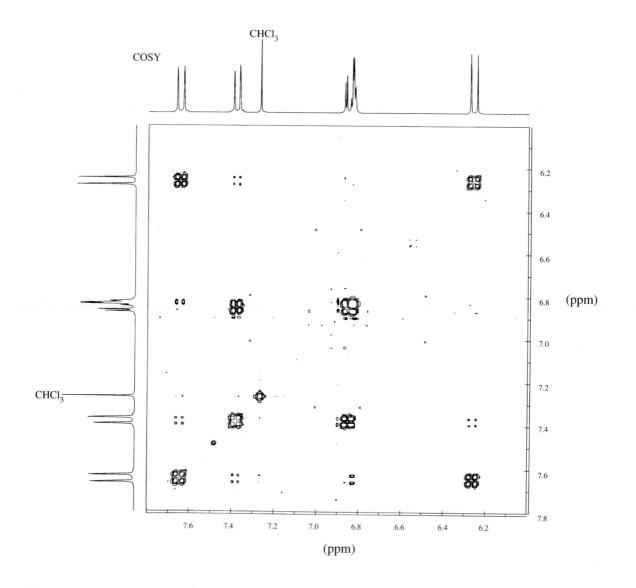

9. Determine the structure for a compound with formula $C_8H_{14}O$. The IR spectrum, ^1H NMR spectrum and expansions, ^{13}C NMR spectrum, DEPT spectrum, COSY spectrum, and HETCOR (HSQC) spectrum are included in this problem.

Infrared spectrum

^1H spectrum and expansions

¹³C spectrum

HETCOR

F2 (ppm)

F1 (ppm)

REFERENCES

Becker, E. D., *High Resolution NMR: Theory and Chemical Applications*, 3rd ed., Academic Press, San Diego, 2000.

Claridge, T. D. W., "High-Resolution NMR Techniques in Organic Chemistry," *Tetrahedron Organic Chemistry*, Vol. 27, 2nd ed., Elsevier, 2008.

Friebolin, H., *Basic One- and Two-Dimensional NMR Spectroscopy,* 5th ed., Wiley-VCH, Weinheim, Germany, 2010.

Jacobsen, N. E., *NMR Spectroscopy Explained: Simplified Theory, Applications and Examples for Organic Chemistry and Structural Biology*, John Wiley & Sons, New York, 2007.

Keeler, J., *Understanding NMR Spectroscopy*, 2nd ed., John Wiley & Sons, New York, 2010.

Sanders, J. K. M., and B. K. Hunter, *Modern NMR Spectroscopy: A Guide for Chemists*, 2nd ed., Oxford University Press, Oxford, England, 1993.

Silverstein, R. M., F. X. Webster, and D. J. Kiemle, *Spectrometric Identification of Organic Compounds*, 7th ed., John Wiley and Sons, New York, 2005.

Simpson, J. H., *Organic Structure Determination Using 2-D NMR Spectroscopy: A Problem-Based Approach*, 2nd ed., Academic Press, San Diego, 2012.

Journal of Chemical Education References

Advanced NMR methods was covered in a series of articles that appeared in the *Journal of Chemical Education* under the general title "The Fourier Transform in Chemistry." The specific volume and page citations are as follows.

Volume 66 (1989), p. A213
Volume 66 (1989), p. A243
Volume 67 (1990), p. A93
Volume 67 (1990), p. A100
Volume 67 (1990), p. A125

Selected Websites

http://www.chem.ucla.edu/~webspectra/
Problems in NMR and IR Spectroscopy, C. A. Merlic, Project Director.

http://www.cis.rit.edu/htbooks/nmr/inside.htm
Basics of NMR, J. P. Hornak, Rochester Institute of Technology

CHAPTER 10

ULTRAVIOLET SPECTROSCOPY

Most organic molecules and functional groups are transparent in the portions of the electromagnetic spectrum that we call the **ultraviolet (UV)** and **visible (VIS)** regions—that is, the regions where wavelengths range from 190 nm to 800 nm. Consequently, absorption spectroscopy is of limited utility in this range of wavelengths. However, in some cases we can derive useful information from these regions of the spectrum. That information, when combined with the details provided by infrared and nuclear magnetic resonance (NMR) spectra, can lead to valuable structural proposals.

10.1 THE NATURE OF ELECTRONIC EXCITATIONS

When continuous radiation passes through a transparent material, a portion of the radiation may be absorbed. If that occurs, the residual radiation, when it is passed through a prism, yields a spectrum with gaps in it, called an **absorption spectrum.** As a result of energy absorption, atoms or molecules pass from a state of low energy (the initial, or **ground state**) to a state of higher energy (the **excited state**). Figure 10.1 depicts this excitation process, which is quantized. The electromagnetic radiation that is absorbed has energy exactly equal to the energy *difference* between the excited and ground states.

In the case of ultraviolet and visible spectroscopy, the transitions that result in the absorption of electromagnetic radiation in this region of the spectrum are transitions between **electronic** energy levels. As a molecule absorbs energy, an electron is promoted from an occupied orbital to an unoccupied orbital of greater potential energy. Generally, the most probable transition is from the **highest occupied molecular orbital (HOMO)** to the **lowest unoccupied molecular orbital (LUMO)**. The energy differences between electronic levels in most molecules vary from 125 to 650 kJ/mole (kilojoules per mole).

For most molecules, the lowest-energy occupied molecular orbitals are the σ orbitals, which correspond to σ bonds. The π orbitals lie at somewhat higher energy levels, and orbitals that hold unshared pairs, the **nonbonding (n) orbitals**, lie at even higher energies. The unoccupied, or **antibonding orbitals** ($\pi *$ and $\sigma *$), are the orbitals of highest energy. Figure 10.2a shows a typical progression of electronic energy levels.

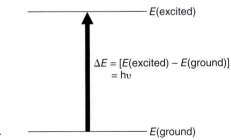

$$\Delta E = [E(\text{excited}) - E(\text{ground})]$$
$$= h\upsilon$$

FIGURE 10.1 The excitation process.

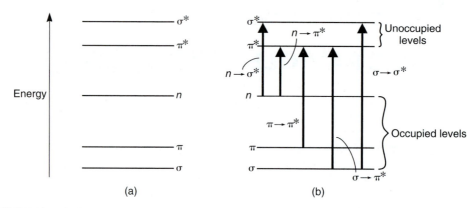

FIGURE 10.2 Electronic energy levels and transitions.

In all compounds other than alkanes, the electrons may undergo several possible transitions of different energies. Some of the most important transitions are

Figure 10.2b illustrates these transitions. Electronic energy levels in aromatic molecules are more complicated than the ones depicted here. Section 10.14 will describe the electronic transitions of aromatic compounds.

Clearly, the energy required to bring about transitions from the highest occupied energy level (HOMO) in the ground state to the lowest unoccupied energy level (LUMO) is less than the energy required to bring about a transition from a lower occupied energy level. Thus, in Figure 10.2b an $n \rightarrow \pi^*$ transition would have a lower energy than a $\pi \rightarrow \pi^*$ transition. For many purposes, the transition of lowest energy is the most important.

Not all of the transitions that at first sight appear possible are observed. Certain restrictions, called **selection rules,** must be considered. One important selection rule states that transitions that involve a change in the spin quantum number of an electron during the transition are not allowed to take place; they are called **"forbidden" transitions.** Other selection rules deal with the numbers of electrons that may be excited at one time, with symmetry properties of the molecule and of the electronic states, and with other factors that need not be discussed here. Transitions that are formally forbidden by the selection rules are often not observed. However, theoretical treatments are rather approximate, and in certain cases forbidden transitions *are* observed, although the intensity of the absorption tends to be much lower than for transitions that are **allowed** by the selection rules. The $n \rightarrow \pi^*$ transition is the most common type of forbidden transition.

10.2 THE ORIGIN OF UV BAND STRUCTURE

For an atom that absorbs in the ultraviolet, the absorption spectrum sometimes consists of very sharp lines, as would be expected for a quantized process occurring between two discrete energy levels. For molecules, however, the UV absorption usually occurs over a wide range of wavelengths because molecules (as opposed to atoms) normally have many excited modes of vibration and rotation at room temperature. In fact, the vibration of molecules cannot be completely "frozen out" even at very low temperatures. Consequently, a collection of molecules generally has its members in many states of vibrational and rotational excitation. The energy levels for these states are quite closely spaced, corresponding to energy differences considerably smaller than those of electronic levels. The rotational and vibrational levels are thus "superimposed" on the electronic levels. A molecule may therefore undergo electronic and vibrational–rotational excitation simultaneously, as shown in Figure 10.3.

Because there are so many possible transitions, each differing from the others by only a slight amount, each electronic transition consists of a vast number of lines spaced so closely that the spectrophotometer cannot resolve them. Rather, the instrument traces an "envelope" over the entire pattern. What is observed from these types of combined transitions is that the UV spectrum of a molecule usually consists of a broad **band** of absorption centered near the wavelength of the major transition.

10.3 PRINCIPLES OF ABSORPTION SPECTROSCOPY

The greater the number of molecules capable of absorbing light of a given wavelength, the greater the extent of light absorption. Furthermore, the more effectively a molecule absorbs light of a given wavelength, the greater the extent of light absorption. From these guiding ideas, the following empirical expression, known as the **Beer–Lambert Law,** may be formulated.

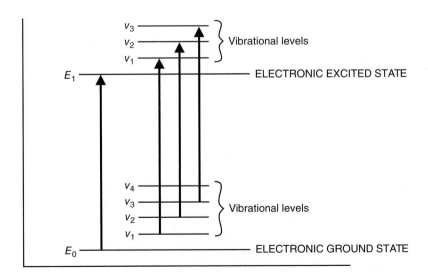

FIGURE 10.3 Electronic transitions with vibrational transitions superimposed. (Rotational levels, which are very closely spaced within the vibrational levels, are omitted for clarity.)

$$A = \log(I_0/I) = \varepsilon c l \text{ for a given wavelength}$$

Equation 10.1

A = absorbance

I_0 = intensity of light incident upon sample cell

I = intensity of light leaving sample cell

c = molar concentration of solute

l = length of sample cell (cm)

ε = molar absorptivity

The term $\log(I_0/I)$ is also known as the **absorbance** (or the **optical density** in older literature) and may be represented by A. The **molar absorptivity** (formerly known as the **molar extinction coefficient**) is a property of the molecule undergoing an electronic transition and is not a function of the variable parameters involved in preparing a solution. The size of the absorbing system and the probability that the electronic transition will take place control the absorptivity, which ranges from 0 to 10^6. Values above 10^4 are termed **high-intensity absorptions,** while values below 10^3 are **low-intensity absorptions.** Forbidden transitions (see Section 10.1) have absorptivities in the range from 0 to 1000.

The Beer–Lambert Law is rigorously obeyed when a *single species* gives rise to the observed absorption. The law may not be obeyed, however, when different forms of the absorbing molecule are in equilibrium, when solute and solvent form complexes through some sort of association, when *thermal* equilibrium exists between the ground electronic state and a low-lying excited state, or when fluorescent compounds or compounds changed by irradiation are present.

☐ 10.4 INSTRUMENTATION

The typical ultraviolet–visible spectrophotometer consists of a **light source,** a **monochromator,** and a **detector.** The light source is usually a deuterium lamp, which emits electromagnetic radiation in the ultraviolet region of the spectrum. A second light source, a tungsten lamp, is used for wavelengths in the visible region of the spectrum. The monochromator is a diffraction grating; its role is to spread the beam of light into its component wavelengths. A system of slits focuses the desired wavelength on the sample cell. The light that passes through the sample cell reaches the detector, which records the intensity of the transmitted light I. The detector is generally a photomultiplier tube, although in modern instruments photodiodes are also used. In a typical double-beam instrument, the light emanating from the light source is split into two beams, the **sample beam** and the **reference beam.** When there is no sample cell in the reference beam, the detected light is taken to be equal to the intensity of light entering the sample I_0.

The sample cell must be constructed of a material that is transparent to the electromagnetic radiation being used in the experiment. For spectra in the visible range of the spectrum, cells composed of glass or plastic are generally suitable. For measurements in the ultraviolet region of the spectrum, however, glass and plastic cannot be used because they absorb ultraviolet radiation. Instead, cells made of quartz must be used since quartz does not absorb radiation in this region.

The instrument design just described is quite suitable for measurement at only one wavelength. If a complete spectrum is desired, this type of instrument has some deficiencies. A mechanical system is required to rotate the monochromator and provide a scan of all desired wavelengths. This type of system operates slowly, and therefore considerable time is required to record a spectrum.

A modern improvement on the traditional spectrophotometer is the **diode-array spectrophotometer.** A diode array consists of a series of photodiode detectors positioned side by side on a silicon crystal. Each diode is designed to record a narrow band of the spectrum. The diodes are connected so that the entire spectrum is recorded at once. This type of detector has no moving parts and can record spectra very quickly. Furthermore, its output can be passed to a computer, which can process the information and provide a variety of useful output formats. Since the number of photodiodes is limited, the speed and convenience described here are obtained at some small cost in resolution. For many applications, however, the advantages of this type of instrument outweigh the loss of resolution.

10.5 PRESENTATION OF SPECTRA

The ultraviolet–visible spectrum is generally recorded as a plot of absorbance versus wavelength. It is customary to then replot the data with either ε or log ε plotted on the y-axis and wavelength plotted on the x-axis. Figure 10.4, the spectrum of benzoic acid, is typical of the manner in which spectra are displayed. However, very few electronic spectra are reproduced in the scientific literature; most are described by indications of the wavelength maxima and absorptivities of the principal absorption peaks. For benzoic acid, a typical description might be

$$\lambda_{max} = 230 \text{ nm} \qquad \log \varepsilon = 4.2$$
$$272 \qquad\qquad 3.1$$
$$282 \qquad\qquad 2.9$$

Figure 10.4 is the actual spectrum that corresponds to these data.

FIGURE 10.4 Ultraviolet spectrum of benzoic acid in cyclohexane. (From Friedel, R. A., and M. Orchin, *Ultraviolet Spectra of Aromatic Compounds*, John Wiley and Sons, New York, 1951. Reprinted by permission.)

10.6 SOLVENTS

The choice of the solvent to be used in ultraviolet spectroscopy is quite important. The first criterion for a good solvent is that it should not absorb ultraviolet radiation in the same region as the substance whose spectrum is being determined. Usually, solvents that do not contain conjugated systems are most suitable for this purpose, although they vary regarding the shortest wavelength at which they remain transparent to ultraviolet radiation. Table 10.1 lists some common ultraviolet spectroscopy solvents and their cutoff points or minimum regions of transparency.

Of the solvents listed in Table 10.1, water, 95% ethanol, and hexane are most commonly used. Each is transparent in the regions of the ultraviolet spectrum in which interesting absorption peaks from sample molecules are likely to occur.

A second criterion for a good solvent is its effect on the fine structure of an absorption band. Figure 10.5 illustrates the effects of polar and nonpolar solvents on an absorption band. A nonpolar solvent does not hydrogen bond with the solute, and the spectrum of the solute closely approximates the spectrum that would be produced in the gaseous state, in which fine structure is often observed. In a polar solvent, the hydrogen bonding forms a solute–solvent complex, and the fine structure may disappear.

TABLE 10.1
SOLVENT CUTOFFS

Acetonitrile	190 nm	*n*-Hexane	201 nm
Chloroform	240	Methanol	205
Cyclohexane	195	Isooctane	195
1,4-Dioxane	215	Water	190
95% Ethanol	205	Trimethyl phosphate	210

FIGURE 10.5 Ultraviolet spectra of phenol in ethanol and in isooctane. (From Coggeshall, N. D., and E. M. Lang, *Journal of the American Chemical Society, 70,* 1948: 3288. Reprinted by permission.)

TABLE 10.2
SOLVENT SHIFTS ON THE $n \rightarrow \pi^*$ TRANSITION OF ACETONE

Solvent	H_2O	CH_3OH	C_2H_5OH	$CHCl_3$	C_6H_{14}
λ_{max} (nm)	264.5	270	272	277	279

A third criterion for a good solvent is its ability to influence the wavelength of ultraviolet light that will be absorbed via stabilization of either the ground or the excited state. Polar solvents do not form hydrogen bonds as readily with the excited states of polar molecules as with their ground states, and these polar solvents increase the energies of electronic transitions in the molecules. Polar solvents shift transitions of the $n \rightarrow \pi^*$ type to shorter wavelengths. On the other hand, in some cases the excited states may form stronger hydrogen bonds than the corresponding ground states. In such a case, a polar solvent shifts an absorption to longer wavelength since the energy of the electronic transition is decreased. Polar solvents shift transitions of the $\pi \rightarrow \pi^*$ type to longer wavelengths. Table 10.2 illustrates typical effects of a series of solvents on an electronic transition.

10.7 WHAT IS A CHROMOPHORE?

Although the absorption of ultraviolet radiation results from the excitation of electrons from ground to excited states, the nuclei that the electrons hold together in bonds play an important role in determining which wavelengths of radiation are absorbed. The nuclei determine the strength with which the electrons are bound and thus influence the energy spacing between ground and excited states. Hence, the characteristic energy of a transition and the wavelength of radiation absorbed are properties of a group of atoms rather than of electrons themselves. The group of atoms producing such an absorption is called a **chromophore.** As structural changes occur in a chromophore, the exact energy and intensity of the absorption are expected to change accordingly. Very often, it is extremely difficult to predict from theory how the absorption will change as the structure of the chromophore is modified, and it is necessary to apply empirical working guides to predict such relationships.

Alkanes. For molecules, such as alkanes, that contain nothing but single bonds and lack atoms with unshared electron pairs, the only electronic transitions possible are of the $\sigma \rightarrow \sigma^*$ type. These transitions are of such a high energy that they absorb ultraviolet energy at very short wavelengths—shorter than the wavelengths that are experimentally accessible using typical spectrophotometers. Figure 10.6 illustrates this type of transition. The excitation of the σ-bonding electron to the σ^*-antibonding orbital is depicted at the right.

Alcohols, Ethers, Amines, and Sulfur Compounds. In saturated molecules that contain atoms bearing nonbonding pairs of electrons, transitions of the $n \rightarrow \sigma^*$ type become important. They are also

FIGURE 10.6 $\sigma \rightarrow \sigma^*$ transition.

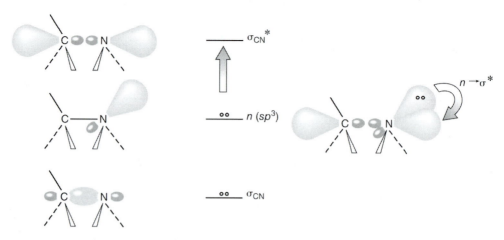

F I G U R E 1 0 . 7 $n \rightarrow \sigma^*$ transition.

rather high-energy transitions, but they do absorb radiation that lies within an experimentally accessible range. Alcohols and amines absorb in the range from 175 to 200 nm, while organic thiols and sulfides absorb between 200 and 220 nm. Most of the absorptions are below the cutoff points for the common solvents, so they are not observed in solution spectra. Figure 10.7 illustrates an $n \rightarrow \sigma^*$ transition for an amine. The excitation of the nonbonding electron to the antibonding orbital is shown at the right.

Alkenes and Alkynes. With unsaturated molecules, $\pi \rightarrow \pi^*$ transitions become possible. These transitions are of rather high energy as well, but their positions are sensitive to the presence of substitution, as will be clear later. Alkenes absorb around 175 nm, and alkynes absorb around 170 nm. Figure 10.8 shows this type of transition.

Carbonyl Compounds. Unsaturated molecules that contain atoms such as oxygen or nitrogen may also undergo $n \rightarrow \pi^*$ transitions. These are perhaps the most interesting and most studied transitions, particularly among carbonyl compounds. These transitions are also rather sensitive to substitution on the chromophore. The typical carbonyl compound undergoes an $n \rightarrow \pi^*$ transition around 280 to 290 nm ($\varepsilon = 15$). Most $n \rightarrow \pi^*$ transitions are forbidden and hence are of low intensity. Carbonyl compounds also have a $\pi \rightarrow \pi^*$ transition at about 188 nm ($\varepsilon = 900$). Figure 10.9 shows the $n \rightarrow \pi^*$ and $\pi \rightarrow \pi^*$ transitions of the carbonyl group.[1]

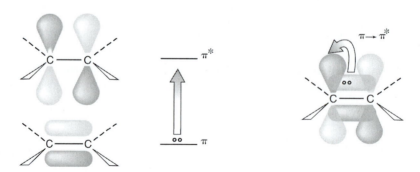

F I G U R E 1 0 . 8 $\pi \rightarrow \pi^*$ transition.

---

[1]Contrary to what you might expect from simple theory, the oxygen atom of the carbonyl group is *not* sp^2 hybridized. Spectroscopists have shown that although the carbon atom is sp^2 hybridized, the hybridization of the oxygen atom more closely approximates *sp*.

FIGURE 10.9 Electronic transitions of the carbonyl group.

Table 10.3 lists typical absorptions of simple isolated chromophores. You may notice that these *simple* chromophores nearly all absorb at approximately the same wavelength (160 to 210 nm).

The attachment of substituent groups in place of hydrogen on a basic chromophore structure changes the position and intensity of an absorption band of the chromophore. The substituent groups may not give rise to the absorption of the ultraviolet radiation themselves, but their presence modifies the absorption of the principal chromophore. Substituents that increase the intensity of the absorption, and possibly the wavelength, are called **auxochromes.** Typical auxochromes include methyl, hydroxyl, alkoxy, halogen, and amino groups.

Other substituents may have any of four kinds of effects on the absorption:

1. **Bathochromic shift** (red shift)—a shift to lower energy or longer wavelength.

2. **Hypsochromic shift** (blue shift)—a shift to higher energy or shorter wavelength.

3. **Hyperchromic effect**—an increase in intensity.

4. **Hypochromic effect**—a decrease in intensity.

TABLE 10.3
TYPICAL ABSORPTIONS OF SIMPLE ISOLATED CHROMOPHORES

Class	Transition	λ_{max} (nm)	$\log \varepsilon$	Class	Transition	λ_{max} (nm)	$\log \varepsilon$
R—OH	$n \rightarrow \sigma^*$	180	2.5	R—NO$_2$	$n \rightarrow \pi^*$	271	<1.0
R—O—R	$n \rightarrow \sigma^*$	180	3.5	R—CHO	$\pi \rightarrow \pi^*$	190	2.0
R—NH$_2$	$n \rightarrow \sigma^*$	190	3.5		$n \rightarrow \pi^*$	290	1.0
R—SH	$n \rightarrow \sigma^*$	210	3.0	R$_2$CO	$\pi \rightarrow \pi^*$	180	3.0
R$_2$C=CR$_2$	$\pi \rightarrow \pi^*$	175	3.0		$n \rightarrow \pi^*$	280	1.5
R—C≡C—R	$\pi \rightarrow \pi^*$	170	3.0	RCOOH	$n \rightarrow \pi^*$	205	1.5
R—C≡N	$n \rightarrow \pi^*$	160	<1.0	RCOOR′	$n \rightarrow \pi^*$	205	1.5
R—N=N—R	$n \rightarrow \pi^*$	340	<1.0	RCONH$_2$	$n \rightarrow \pi^*$	210	1.5

☐ 10.8 THE EFFECT OF CONJUGATION

One of the best ways to bring about a bathochromic shift is to increase the extent of conjugation in a double-bonded system. In the presence of conjugated double bonds, the electronic energy levels of a chromophore move closer together. As a result, the energy required to produce a transition from an occupied electronic energy level to an unoccupied level decreases, and the wavelength of the light absorbed becomes longer. Figure 10.10 illustrates the bathochromic shift that is observed in a series of conjugated polyenes as the length of the conjugated chain is increased.

Conjugation of two chromophores not only results in a bathochromic shift but increases the intensity of the absorption. These two effects are of prime importance in the use and interpretation of electronic spectra of organic molecules because conjugation shifts the selective light absorption of isolated chromophores from a region of the spectrum that is not readily accessible to a region that

FIGURE 10.10 CH$_3$—(CH=CH)$_n$—CH$_3$ ultraviolet spectra of dimethylpolyenes. (a) $n = 3$; (b) $n = 4$; (c) $n = 5$. (From Nayler, P., and M. C. Whiting, *Journal of the Chemical Society*, 1955: 3042.)

TABLE 10.4
EFFECT OF CONJUGATION ON ELECTRONIC TRANSITIONS

	λ_{max} (nm)	ε
Alkenes		
Ethylene	175	15,000
1,3-Butadiene	217	21,000
1,3,5-Hexatriene	258	35,000
β-Carotene (11 double bonds)	465	125,000
Ketones		
Acetone		
$\pi \rightarrow \pi^*$	189	900
$n \rightarrow \pi^*$	280	12
3-Buten-2-one		
$\pi \rightarrow \pi^*$	213	7,100
$n \rightarrow \pi^*$	320	27

is easily studied with commercially available spectrophotometers. The exact position and intensity of the absorption band of the conjugated system can be correlated with the extent of conjugation in the system. Table 10.4 illustrates the effect of conjugation on some typical electronic transitions.

10.9 THE EFFECT OF CONJUGATION ON ALKENES

The bathochromic shift that results from an increase in the length of a conjugated system implies that an increase in conjugation decreases the energy required for electronic excitation. This is true and can be explained most easily by the use of molecular orbital theory. According to molecular orbital (MO) theory, the atomic p orbitals on each of the carbon atoms combine to make π molecular orbitals. For instance, in the case of ethylene (ethene), we have two atomic p orbitals, ϕ_1 and ϕ_2. From these two p orbitals we form two π molecular orbitals, ψ_1 and ψ_2^*, by taking linear combinations. The bonding orbital ψ_1 results from the addition of the wave functions of the two p orbitals, and the antibonding orbital ψ_2^* results from the subtraction of these two wave functions. The new bonding orbital, a **molecular orbital,** has an energy lower than that of either of the original p orbitals; likewise, the antibonding orbital has an elevated energy. Figure 10.11 illustrates this diagrammatically.

Notice that *two* atomic orbitals were combined to build the molecular orbitals, and as a result, *two* molecular orbitals were formed. There were also two electrons, one in each of the atomic p orbitals. As a result of combination, the new π system contains *two* electrons. Since we fill the lower-energy orbitals first, these electrons end up in ψ_1, the bonding orbital, and they constitute a new π bond. Electronic transition in this system is a $\pi \rightarrow \pi^*$ transition from ψ_1 to ψ_2^*.

Now, moving from this simple two-orbital case, consider 1,3-butadiene, which has *four* atomic p orbitals that form its π system of two conjugated double bonds. Since we had four atomic orbitals with which to build, *four* molecular orbitals result. Figure 10.12 represents the orbitals of ethylene on the same energy scale as the new orbitals for the sake of comparison.

Notice that the transition of lowest energy in 1,3-butadiene, $\psi_2 \rightarrow \psi_3^*$, is a $\pi \rightarrow \pi^*$ transition and that it has a *lower energy* than the corresponding transition in ethylene, $\psi_1 \rightarrow \psi_2^*$. This result is general. As we increase the number of p orbitals making up the conjugated system, the transition from the highest occupied molecular orbital (HOMO) to the lowest unoccupied molecular orbital (LUMO) has

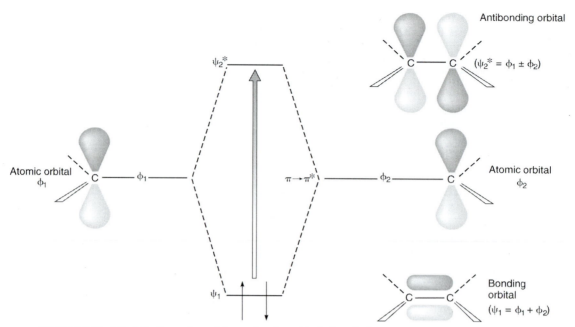

F I G U R E 1 0 . 1 1 Formation of the molecular orbitals for ethylene.

F I G U R E 1 0 . 1 2 A comparison of the molecular orbital energy levels and the energy of the $\pi \rightarrow \pi^*$ transitions in ethylene and 1,3-butadiene.

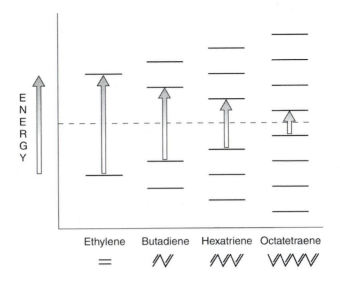

FIGURE 10.13 A comparison of the $\pi \rightarrow \pi^*$ energy gap in a series of polyenes of increasing chain length.

progressively lower energy. The energy gap dividing the bonding and antibonding orbitals becomes progressively smaller with increasing conjugation. Figure 10.13 plots the molecular orbital energy levels of several conjugated polyenes of increasing chain length on a common energy scale. Arrows indicate the HOMO–LUMO transitions. The increased conjugation shifts the observed wavelength of the absorption to higher values.

In a qualitatively similar fashion, many auxochromes exert their bathochromic shifts by means of an extension of the length of the conjugated system. The strongest auxochromes invariably possess a pair of unshared electrons on the atom attached to the double-bond system. Resonance interaction of this lone pair with the double bond(s) increases the length of the conjugated system.

As a result of this interaction, as just shown, the nonbonded electrons become part of the π system of molecular orbitals, increasing its length by one extra orbital. Figure 10.14 depicts this interaction for ethylene and an unspecified atom, B, with an unshared electron pair. However, any of the typical auxochromic groups, $-OH$, $-OR$, $-X$, or $-NH_2$, could have been illustrated specifically.

In the new system, the transition from the highest occupied orbital ψ_2 to the antibonding orbital ψ_3^* always has lower energy than the $\pi \rightarrow \pi^*$ transition would have in the system without the interaction. Although MO theory can explain this general result, it is beyond the scope of this book.

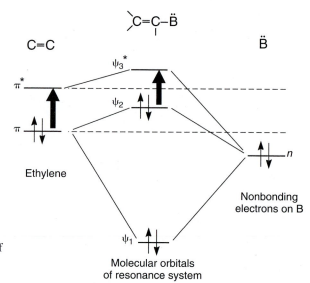

FIGURE 10.14 Energy relationships of the new molecular orbitals and the interacting π system and its auxochrome.

In similar fashion, methyl groups also produce a bathochromic shift. However, as methyl groups do not have unshared electrons, the interaction is thought to result from overlap of the C—H bonding orbitals with the π system as follows:

This type of interaction is often called **hyperconjugation.** Its net effect is an extension of the π system.

10.10 THE WOODWARD–FIESER RULES FOR DIENES

In butadiene, two possible $\pi \rightarrow \pi^*$ transitions can occur: $\psi_2 \rightarrow \psi_3^*$ and $\psi_2 \rightarrow \psi_4^*$. We have already discussed the easily observable $\psi_2 \rightarrow \psi_3^*$ transition (see Fig. 10.12). The $\psi_2 \rightarrow \psi_4^*$ transition is not often observed, for two reasons. First, it lies near 175 nm for butadiene; second, it is a forbidden transition for the *s-trans* conformation of double bonds in butadiene.

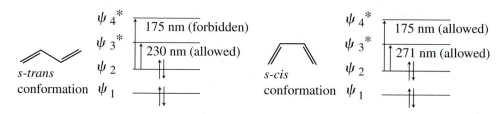

A transition at 175 nm lies below the cutoff points of the common solvents used to determine UV spectra (Table 10.1) and therefore is not easily detectable. Furthermore, the *s-trans* conformation is more favorable for butadiene than is the *s-cis* conformation. Therefore, the 175-nm band is not usually detected.

In general, conjugated dienes exhibit an intense band (ε = 20,000 to 26,000) in the region from 217 to 245 nm, owing to a $\pi \rightarrow \pi^*$ transition. The position of this band appears to be quite insensitive to the nature of the solvent.

Butadiene and many simple conjugated dienes exist in a planar *s-trans* conformation, as noted. Generally, alkyl substitution produces bathochromic shifts and hyperchromic effects. However, with certain patterns of alkyl substitution, the wavelength increases but the intensity decreases. The 1,3-dialkylbutadienes possess too much crowding between alkyl groups to permit them to exist in the *s-trans* conformation. They convert, by rotation around the single bond, to an *s-cis* conformation, which absorbs at longer wavelengths but with lower intensity than the corresponding *s-trans* conformation.

s-trans   *s-cis*

In cyclic dienes, where the central bond is a part of the ring system, the diene chromophore is usually held rigidly in either the *s-trans* (transoid) or the *s-cis* (cisoid) orientation. Typical absorption spectra follow the expected pattern:

Homoannular diene (cisoid or *s-cis*)
Less intense, ε = 5,000–15,000
λ longer (273 nm)

Heteroannular diene (transoid or *s-trans*)
More intense, ε = 12,000–28,000
λ shorter (234 nm)

By studying a vast number of dienes of each type, Woodward and Fieser devised an empirical correlation of structural variations that enables us to predict the wavelength at which a conjugated diene will absorb. Table 10.5 summarizes the rules. Following are a few sample applications of these rules. Notice that the pertinent parts of the structures are shown in bold face.

Transoid: 214 nm
Observed: 217 nm

Transoid: 214 nm
Alkyl groups: $3 \times 5 =$ 15
————
229 nm

Observed: 228 nm

Exocyclic double bond

Transoid: 214 nm
Ring residues: $3 \times 5 =$ 15
Exocyclic double bond: 5
————
234 nm

Observed: 235 nm

CH_3CH_2O — Exocyclic double bond

Transoid: 214 nm
Ring residues: $3 \times 5 =$ 15
Exocyclic double bond: 5
—OR: 6
————
240 nm

Observed: 241 nm

TABLE 10.5
EMPIRICAL RULES FOR DIENES

	Homoannular (cisoid)	Heteroannular (transoid)
Parent	$\lambda = 253$ nm	$\lambda = 214$ nm
Increments for:		
Double-bond-extending conjugation	30	30
Alkyl substituent or ring residue	5	5
Exocyclic double bond	5	5
Polar groupings:		
—$OCOCH_3$	0	0
—OR	6	6
—Cl, —Br	5	5
—NR_2	60	60

In this context, an *exocyclic double bond* is a double bond that lies outside a given ring. Notice that the exocyclic bond may lie within one ring even though it is outside another ring. Often, an exocyclic double bond will be found at a junction point on rings. Here is an example of a compound with the exocyclic double bonds labeled with asterisks:

Three exocyclic double bonds = 3 × 5 = 15 nm

Cisoid:	253 nm
Alkyl substituent:	5
Ring residues: 3 × 5 =	15
Exocyclic double bond:	5
	278 nm
Observed:	275 nm

Cisoid:	253 nm
Ring residues: 5 × 5 =	25
Double-bond-extending conjugation: 2 × 30 =	60
Exocyclic double bond: 3 × 5 =	15
CH₃COO—:	0
	353 nm
Observed:	355 nm

10.11 CARBONYL COMPOUNDS; ENONES

As discussed in Section 10.7, carbonyl compounds have two principal UV transitions, the allowed $\pi \rightarrow \pi^*$ transition and the forbidden $n \rightarrow \pi^*$ transition.

Of these, only the $n \rightarrow \pi^*$ transition, although it is weak (forbidden), is commonly observed above the usual cutoff points of solvents. Substitution on the carbonyl group by an auxochrome with a lone pair of electrons, such as $-NR_2$, $-OH$, $-OR$, $-NH_2$, or $-X$, as in amides, acids, esters, or acid chlorides, gives a pronounced hypsochromic effect on the $n \rightarrow \pi^*$ transition and a lesser, bathochromic effect on the $\pi \rightarrow \pi^*$ transition. Such bathochromic shifts are caused by resonance interaction similar to that discussed in Section 10.9. Seldom, however, are these effects large enough to bring the $\pi \rightarrow \pi^*$ band into the region above the solvent cutoff point. Table 10.6 lists the hypsochromic effects of an acetyl group on the $n \rightarrow \pi^*$ transition.

The hypsochromic shift of the $n \rightarrow \pi^*$ is due primarily to the inductive effect of the oxygen, nitrogen, or halogen atoms. They withdraw electrons from the carbonyl carbon, causing the lone pair of electrons on oxygen to be held more firmly than they would be in the absence of the inductive effect.

If the carbonyl group is part of a conjugated system of double bonds, both the $n \rightarrow \pi^*$ and the $\pi \rightarrow \pi^*$ bands are shifted to longer wavelengths. However, the energy of the $n \rightarrow \pi^*$ transition does not decrease as rapidly as that of the $\pi \rightarrow \pi^*$ band, which is more intense. If the conjugated chain becomes long enough, the $n \rightarrow \pi^*$ band is "buried" under the more intense $\pi \rightarrow \pi^*$ band. Figure 10.15 illustrates this effect for a series of polyene aldehydes.

Figure 10.16 shows the molecular orbitals of a simple enone system, along with those of the noninteracting double bond and the carbonyl group.

TABLE 10.6
HYPSOCHROMIC EFFECTS OF LONE-PAIR AUXOCHROMES ON THE $n \rightarrow \pi^*$ TRANSITION OF A CARBONYL GROUP

	λ_{max}	ε_{max}	Solvent
CH_3-CO-H	293 nm	12	Hexane
$CH_3-CO-CH_3$	279	15	Hexane
$CH_3-CO-Cl$	235	53	Hexane
$CH_3-CO-NH_2$	214	—	Water
$CH_3-CO-OCH_2CH_3$	204	60	Water
$CH_3-CO-OH$	204	41	Ethanol

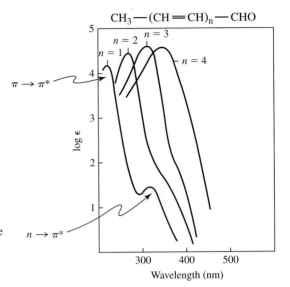

FIGURE 10.15 The spectra of a series of polyene aldehydes. (From Murrell, J. N., *The Theory of the Electronic Spectra of Organic Molecules*, Methuen and Co., Ltd., London, 1963. Reprinted by permission.)

FIGURE 10.16 The orbitals of an enone system compared to those of the noninteracting chromophores.

10.12 WOODWARD'S RULES FOR ENONES

The conjugation of a double bond with a carbonyl group leads to intense absorption ($\varepsilon = 8,000$ to 20,000) corresponding to a $\pi \rightarrow \pi^*$ transition of the carbonyl group. The absorption is found between 220 and 250 nm in simple enones. The $n \rightarrow \pi^*$ transition is much less intense ($\varepsilon = 50$ to 100) and appears at 310 to 330 nm. Although the $\pi \rightarrow \pi^*$ transition is affected in predictable fashion by structural modifications of the chromophore, the $n \rightarrow \pi^*$ transition does not exhibit such predictable behavior.

Woodward examined the ultraviolet spectra of numerous enones and devised a set of empirical rules that enable us to predict the wavelength at which the $\pi \rightarrow \pi^*$ transition occurs in an unknown enone. Table 10.7 summarizes these rules.

TABLE 10.7
EMPIRICAL RULES FOR ENONES

$$\overset{\beta}{\underset{|}{\beta}}-\overset{\alpha}{\underset{|}{C}}=\overset{}{\underset{|}{C}}-C=O \qquad\qquad \overset{\delta}{\underset{|}{\delta}}-\overset{\gamma}{\underset{|}{C}}=\overset{}{\underset{|}{C}}-\overset{\beta}{\underset{|}{C}}=\overset{\alpha}{\underset{|}{C}}-C=O$$

Base values:		
Six-membered ring or acyclic parent enone		= 215 nm
Five-membered ring parent enone		= 202 nm
Acyclic dienone		= 245 nm
Increments for:		
Double-bond-extending conjugation		30
Alkyl group or ring residue	α	10
	β	12
	γ and higher	18
Polar groupings:		
—OH	α	35
	β	30
	δ	50
—OCOCH$_3$	α,β,δ	6
—OCH$_3$	α	35
	β	30
	γ	17
	δ	31
—Cl	α	15
	β	12
—Br	α	25
	β	30
—NR$_2$	β	95
Exocyclic double bond		5
Homocyclic diene component		39
Solvent correction		Variable
	λ_{max}^{EtOH}(calc) = Total	

The following are a few sample applications of these rules. The pertinent parts of the structures are shown in bold face.

Acyclic enone:	215 nm
α-CH$_3$:	10
β-CH$_3$: $2 \times 12 =$	24
	249 nm
Observed:	249 nm

Six-membered enone:	215 nm
Double-bond-extending conjugation:	30
Homocyclic diene:	39
δ-Ring residue:	18
	302 nm
Observed:	300 nm

Five-membered enone:	202 nm
β-Ring residue: $2 \times 12 =$	24
Exocyclic double bond:	5
	231 nm
Observed:	226 nm

Five-membered enone:	202 nm
α-Br:	25
β-Ring residue: $2 \times 12 =$	24
Exocyclic double bond:	5
	256 nm
Observed:	251 nm

Six-membered enone:	215 nm
Double-bond-extending conjugation:	30
β-Ring residue:	12
δ-Ring residue:	18
Exocyclic double bond:	5
	280 nm
Observed:	280 nm

10.13 α,β-UNSATURATED ALDEHYDES, ACIDS, AND ESTERS

α,β-Unsaturated aldehydes generally follow the same rules as enones (see the preceding section) except that their absorptions are displaced by about 5 to 8 nm toward shorter wavelength than those of the corresponding ketones. Table 10.8 lists the empirical rules for unsaturated aldehydes.

Nielsen developed a set of rules for α,β-unsaturated acids and esters that are similar to those for enones (Table 10.9).

Consider 2-cyclohexenoic and 2-cycloheptenoic acids as examples:

COOH	α,β-dialkyl	217 nm calc.
	Double bond is in a six-membered ring, adds nothing	217 nm obs.
COOH	α,β-dialkyl	217 nm
	Double bond is in a seven-membered ring	+ 5
		222 nm calc.
		222 nm obs.

TABLE 10.8
EMPIRICAL RULES FOR UNSATURATED ALDEHYDES

Parent	208 nm
With α or β alkyl groups	220
With α,β or β,β alkyl groups	230
With α,β,β alkyl groups	242

10.14 AROMATIC COMPOUNDS

The absorptions that result from transitions within the benzene chromophore can be quite complex. The ultraviolet spectrum contains three absorption bands, which sometimes contain a great deal of fine structure. The electronic transitions are basically of the $\pi \rightarrow \pi^*$ type, but their details are not as simple as in the cases of the classes of chromophores described in earlier sections of this chapter.

Figure 10.17a shows the molecular orbitals of benzene. If you were to attempt a simple explanation for the electronic transitions in benzene, you would conclude that there are four possible transitions, but each transition has the same energy. You would predict that the ultraviolet spectrum of benzene consists of one absorption peak. However, owing to electron–electron repulsions and symmetry considerations, the actual energy states from which electronic transitions occur are somewhat modified. Figure 10.17b shows the energy-state levels of benzene. Three electronic transitions take

TABLE 10.9
EMPIRICAL RULES FOR UNSATURATED ACIDS AND ESTERS

Base values for:

With α or β alkyl group	208 nm
With α,β or β,β alkyl groups	217
With α,β,β alkyl groups	225
For an exocyclic α,β double bond	Add 5 nm
For an endocyclic α,β double bond in a five- or seven-membered ring	Add 5 nm

place to these excited states. Those transitions, which are indicated in Figure 10.17b, are the so-called **primary bands** at 184 and 202 nm and the **secondary** (fine-structure) **band** at 255 nm. Figure 10.18 is the spectrum of benzene. Of the primary bands, the 184-nm band (the **second primary band**) has a molar absorptivity of 47,000. It is an allowed transition. Nevertheless, this transition is not observed under usual experimental conditions because absorptions at this wavelength are in the vacuum ultraviolet region of the spectrum, beyond the range of most commercial instruments. In polycyclic aromatic compounds, the second primary band is often shifted to longer wavelengths,

(a) Molecular orbitals

(b) Energy states

FIGURE 10.17 Molecular orbitals and energy states for benzene.

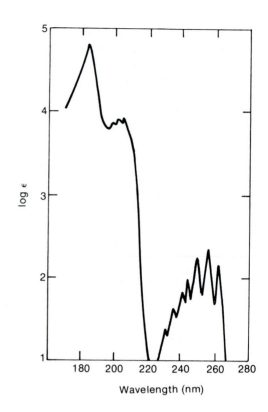

FIGURE 10.18 Ultraviolet spectrum of benzene. (From Petruska, J., *Journal of Chemical Physics, 34*, 1961: 1121. Reprinted by permission.)

in which case it can be observed under ordinary conditions. The 202-nm band is much less intense ($\varepsilon = 7400$), and it corresponds to a forbidden transition. The secondary band is the least intense of the benzene bands ($\varepsilon = 230$). It also corresponds to a symmetry-forbidden electronic transition. The secondary band, caused by interaction of the electronic energy levels with vibrational modes, appears with a great deal of fine structure. This fine structure is lost if the spectrum of benzene is determined in a polar solvent or if a single functional group is substituted onto the benzene ring. In such cases, the secondary band appears as a broad peak, lacking in any interesting detail.

Substitution on the benzene ring can cause bathochromic and hyperchromic shifts. Unfortunately, these shifts are difficult to predict. Consequently, it is impossible to formulate empirical rules to predict the spectra of aromatic substances as was done for dienes, enones, and the other classes of compounds discussed earlier in this chapter. You may gain a qualitative understanding of the effects of substitution by classifying substituents into groups.

A. Substituents with Unshared Electrons

Substituents that carry nonbonding electrons (*n* electrons) can cause shifts in the primary and secondary absorption bands. The nonbonding electrons can increase the length of the π system through resonance.

The more available these *n* electrons are for interaction with the π system of the aromatic ring, the greater the shifts will be. Examples of groups with *n* electrons are the amino, hydroxyl, and methoxy groups, as well as the halogens.

Interactions of this type between the *n* and π electrons usually cause shifts in the primary and secondary benzene absorption bands to longer wavelength (extended conjugation). In addition, the presence of *n* electrons in these compounds gives the possibility of $n \rightarrow \pi^*$ transitions. If an *n* electron is excited into the extended π^* chromophore, the atom from which it was removed becomes electron deficient, while the π system of the aromatic ring (which also includes atom Y) acquires an extra electron. This causes a separation of charge in the molecule and is generally represented as regular resonance, as was shown earlier. However, the extra electron in the ring is, in fact, in a π^* orbital and would be better represented by structures of the following type, with the asterisk representing the excited electron:

Such an excited state is often called a **charge-transfer** or an **electron-transfer** excited state.

In compounds that are acids or bases, pH changes can have very significant effects on the positions of the primary and secondary bands. Table 10.10 illustrates the effects of changing the pH of the solution on the absorption bands of various substituted benzenes. In going from benzene to phenol, notice the shift from 203.5 to 210.5 nm—a 7-nm shift—in the primary band. The secondary band shifts from 254 to 270 nm—a 16-nm shift. However, in phenoxide ion, the conjugate base of phenol, the primary band shifts from 203.5 to 235 nm (a 31.5-nm shift), and the secondary band shifts from 254 to 287 nm (a 33-nm shift). The intensity of the secondary band also increases. In phenoxide ion, there are more *n* electrons, and they are more available for interaction with the aromatic π system than in phenol.

TABLE 10.10
pH EFFECTS ON ABSORPTION BANDS

Substituent	Primary		Secondary	
	λ (nm)	ε	λ (nm)	ε
—H	203.5	7,400	254	204
—OH	210.5	6,200	270	1,450
—O⁻	235	9,400	287	2,600
—NH$_2$	230	8,600	280	1,430
—NH$_3^+$	203	7,500	254	169
—COOH	230	11,600	273	970
—COO⁻	224	8,700	268	560

The comparison of aniline and anilinium ion illustrates a reverse case. Aniline exhibits shifts similar to those of phenol. From benzene to aniline, the primary band shifts from 203.5 to 230 nm (a 26.5-nm shift), and the secondary band shifts from 254 to 280 nm (a 26-nm shift). However, these large shifts are not observed in the case of anilinium ion, the conjugate acid of aniline. For anilinium ion, the primary and secondary bands do not shift at all. The quaternary nitrogen of anilinium ion has no unshared pairs of electrons to interact with the benzene π system. Consequently, the spectrum of anilinium ion is almost identical to that of benzene.

B.   Substituents Capable of π-Conjugation

Substituents that are themselves chromophores usually contain π electrons. Just as in the case of n electrons, interaction of the benzene-ring electrons and the π electrons of the substituent can produce a new electron transfer band. At times, this new band may be so intense as to obscure the secondary band of the benzene system. Notice that this interaction induces the opposite polarity; the ring becomes electron deficient.

Table 10.10 demonstrates the effect of acidity or basicity of the solution on such a chromophoric substituent group. In the case of benzoic acid, the primary and secondary bands are shifted substantially from those noted for benzene. However, the magnitudes of the shifts are somewhat smaller in the case of benzoate ion, the conjugate base of benzoic acid. The intensities of the peaks are lower than for benzoic acid as well. We expect electron transfer of the sort just shown to be less likely when the functional group already bears a negative charge.

C.   Electron-Releasing and Electron-Withdrawing Effects

Substituents may have differing effects on the positions of absorption maxima, depending on whether they are electron releasing or electron withdrawing. Any substituent, regardless of its influence on the electron distribution elsewhere in the aromatic molecule, shifts the primary absorption band to longer wavelength. Electron-withdrawing groups have essentially no effect on the position of the secondary absorption band unless, of course, the electron-withdrawing group is also capable of acting as a chromophore. However, electron-releasing groups increase both the wavelength and the intensity of the secondary absorption band. Table 10.11 summarizes these effects, with electron-releasing and electron-withdrawing substituents grouped together.

D.   Disubstituted Benzene Derivatives

With disubstituted benzene derivatives, it is necessary to consider the effect of each of the two substituents. For *para*-disubstituted benzenes, two possibilities exist. If both groups are electron releasing or if they are both electron withdrawing, they exert effects similar to those observed with monosubstituted benzenes. The group with the stronger effect determines the extent of shifting of

TABLE 10.11
ULTRAVIOLET MAXIMA FOR VARIOUS AROMATIC COMPOUNDS

Substituent		Primary		Secondary	
		λ (nm)	ε	λ (nm)	ε
—H		203.5	7,400	254	204
Electron-releasing substituents	—CH$_3$	206.5	7,000	261	225
	—Cl	209.5	7,400	263.5	190
	—Br	210	7,900	261	192
	—OH	210.5	6,200	270	1,450
	—OCH$_3$	217	6,400	269	1,480
	—NH$_2$	230	8,600	280	1,430
Electron-withdrawing substituents	—CN	224	13,000	271	1,000
	—COOH	230	11,600	273	970
	—COCH$_3$	245.5	9,800		
	—CHO	249.5	11,400		
	—NO$_2$	268.5	7,800		

the primary absorption band. If one of the groups is electron releasing while the other is electron withdrawing, the magnitude of the shift of the primary band is greater than the sum of the shifts due to the individual groups. The enhanced shifting is due to resonance interactions of the following type:

If the two groups of a disubstituted benzene derivative are either *ortho* or *meta* to each other, the magnitude of the observed shift is approximately equal to the sum of the shifts caused by the individual groups. With substitution of these types, there is no opportunity for the kind of direct resonance interaction between substituent groups that is observed with *para* substituents. In the case of *ortho* substituents, the steric inability of both groups to achieve coplanarity inhibits resonance.

For the special case of substituted benzoyl derivatives, an empirical correlation of structure with the observed position of the primary absorption band has been developed (Table 10.12). It provides a means of estimating the position of the primary band for benzoyl derivatives within about 5 nm.

TABLE 10.12
EMPIRICAL RULES FOR BENZOYL DERIVATIVES

Parent chromophore:

R = alkyl or ring residue		246
R = H		250
R = OH or OAlkyl		230

Increment for each substituent:

—Alkyl or ring residue	*o, m*	3
	p	10
—OH, —OCH$_3$, or —OAlkyl	*o, m*	7
	p	25
—O$^-$	*o*	11
	m	20
	p	78
—Cl	*o, m*	0
	p	10
—Br	*o, m*	2
	p	15
—NH$_2$	*o, m*	13
	p	58
—NHCOCH$_3$	*o, m*	20
	p	45
—NHCH$_3$	*p*	73
—N(CH$_3$)$_2$	*o, m*	20
	p	85

Following are two sample applications of these rules:

Parent chromophore:	246 nm
o-Ring residue:	3
m-Br:	2
	251 nm
Observed:	253 nm

Parent chromophore:	230 nm
m-OH: 2 × 7 =	14
p-OH:	25
	269 nm
Observed:	270 nm

E. Polynuclear Aromatic Hydrocarbons and Heterocyclic Compounds

Researchers have observed that the primary and secondary bands in the spectra of polynuclear aromatic hydrocarbons shift to longer wavelength. In fact, even the second primary band, which appears at 184 nm for benzene, is shifted to a wavelength within the range of most UV spectrophotometers. This band lies at 220 nm in the spectrum of naphthalene. As the extent of conjugation increases, the magnitude of the bathochromic shift also increases.

The ultraviolet spectra of the polynuclear aromatic hydrocarbons possess characteristic shapes and fine structure. In the study of spectra of substituted polynuclear aromatic derivatives, it is common practice to compare them with the spectrum of the unsubstituted hydrocarbon. The nature of the chromophore can be identified on the basis of similarity of peak shapes and fine structure. This technique involves the use of model compounds. Section 10.15 will discuss it further.

Figure 10.19 shows the ultraviolet spectra of naphthalene and anthracene. Notice the characteristic shape and fine structure of each spectrum, as well as the effect of increased conjugation on the positions of the absorption maxima.

FIGURE 10.19 Ultraviolet spectra of naphthalene and anthracene. (From Friedel, R. A., and M. Orchin, *Ultraviolet Spectra of Aromatic Compounds*, John Wiley and Sons, New York, 1951. Reprinted by permission.)

Heterocyclic molecules have electronic transitions that include combinations of $\pi \rightarrow \pi^*$ and $n \rightarrow \pi^*$ transitions. The spectra can be rather complex, and analysis of the transitions involved will be left to more advanced treatments. The common method of studying derivatives of heterocyclic molecules is to compare them to the spectra of the parent heterocyclic systems. Section 10.15 will further describe the use of model compounds in this fashion.

Figure 10.20 includes the ultraviolet spectra of pyridine, quinoline, and isoquinoline. You may wish to compare the spectrum of pyridine with that of benzene (Fig. 10.18) and the spectra of quinoline and isoquinoline with the spectrum of naphthalene (Fig. 10.19).

FIGURE 10.20 The ultraviolet spectra of pyridine, quinoline, and isoquinoline. (From Friedel, R. A., and M. Orchin, *Ultraviolet Spectra of Aromatic Compounds*, John Wiley and Sons, New York, 1951. Reprinted by permission.)

10.15 MODEL COMPOUND STUDIES

Very often, the ultraviolet spectra of several members of a particular class of compounds are very similar. Unless you are thoroughly familiar with the spectroscopic properties of each member of the class of compounds, it is very difficult to distinguish the substitution patterns of individual molecules by their ultraviolet spectra. You can, however, determine the gross nature of the chromophore of an unknown substance by this method. Then, based on knowledge of the chromophore, you can employ the other spectroscopic techniques described in this book to elucidate the precise structure and substitution of the molecule.

This approach—the use of model compounds—is one of the best ways to put the technique of ultraviolet spectroscopy to work. By comparing the UV spectrum of an unknown substance with that of a similar but less highly substituted compound, you can determine whether or not they contain the same chromophore. Many of the books listed in the references at the end of this chapter contain large collections of spectra of suitable model compounds, and with their help you can establish the general structure of the part of the molecule that contains the π electrons. You can then utilize infrared or NMR spectroscopy to determine the detailed structure.

As an example, consider an unknown substance that has the molecular formula $C_{15}H_{12}$. A comparison of its spectrum (Fig. 10.21) with that of anthracene (Fig. 10.19) shows that the two spectra are nearly identical. Disregarding minor bathochromic shifts, the same general peak shape and fine structure appear in the spectra of both the unknown and anthracene, the model compound. You may then conclude that the unknown is a substituted anthracene derivative. Further structure determination reveals that the unknown is 9-methylanthracene. The spectra of model compounds can be obtained from published catalogues of ultraviolet spectra. In cases in which a suitable model compound is not available, a model compound can be synthesized and its spectrum determined.

FIGURE 10.21 The ultraviolet spectrum of 9-methylanthracene. (From Friedel, R. A., and M. Orchin, *Ultraviolet Spectra of Aromatic Compounds*, John Wiley and Sons, New York, 1951. Reprinted by permission.)

10.16 VISIBLE SPECTRA: COLOR IN COMPOUNDS

The portion of the electromagnetic spectrum lying between about 400 and 750 nm is the **visible** region. Light waves with wavelengths between these limits appear colored to the human eye. As anyone who has seen light diffracted by a prism or the diffraction effect of a rainbow knows, one end of the visible spectrum is violet, and the other is red. Light with wavelengths near 400 nm is violet, while that with wavelengths near 750 nm is red.

The phenomenon of color in compounds, however, is not as straightforward as the preceding discussion would suggest. If a substance absorbs visible light, it appears to have a color; if not, it appears white. However, compounds that absorb light in the visible region of the spectrum do not possess the color corresponding to the wavelength of the absorbed light. Rather, there is an inverse relationship between the observed color and the color absorbed.

When we observe light **emitted** from a source, as from a lamp or an emission spectrum, we observe the color corresponding to the wavelength of the light being emitted. A light source emitting violet light emits light at the high-energy end of the visible spectrum. A light source emitting red light emits light at the low-energy end of the spectrum.

However, when we observe the color of a particular object or substance, we do not observe that object or substance emitting light. (Certainly, the substance does not glow in the dark.) Rather, we observe the light that is being **reflected**. The color that our eye perceives is not the color corresponding to the wavelength of the light absorbed but its **complement**. When white light falls on an object, light of a particular wavelength is absorbed. The remainder of the light is reflected. The eye and brain register all of the reflected light as the color complementary to the color that was absorbed.

In the case of transparent objects or solutions, the eye receives the light that is **transmitted**. Again, light of a particular wavelength is absorbed, and the remaining light passes through to reach the eye. As before, the eye registers this transmitted light as the color complementary to the color that was absorbed. Table 10.13 illustrates the relationship between the wavelength of light absorbed by a substance and the color perceived by an observer.

Some familiar compounds may serve to underscore these relationships between the absorption spectrum and the observed color. The structural formulas of these examples are shown. Notice that each of these substances has a highly extended conjugated system of electrons. Such extensive conjugation shifts their electronic spectra to such long wavelengths that they absorb visible light and appear colored.

TABLE 10.13

RELATIONSHIP BETWEEN THE COLOR OF LIGHT ABSORBED BY A COMPOUND AND THE OBSERVED COLOR OF THE COMPOUND

Color of Light Absorbed	Wavelength of Light Absorbed (nm)	Observed Color
Violet	400	Yellow
Blue	450	Orange
Blue-green	500	Red
Yellow-green	530	Red-violet
Yellow	550	Violet
Orange-red	600	Blue-green
Red	700	Green

β-Carotene (pigment from carrots): $\lambda_{max} = 452$ nm, **orange**
Cyanidin (blue pigment of cornflower): $\lambda_{max} = 545$ nm, **blue**
Malachite green (a triphenylmethane dye): $\lambda_{max} = 617$ nm, **green**

β-Carotene (a carotenoid, which is a class of plant pigments)
$\lambda_{max} = 452$ nm

Cyanidin chloride (an anthocyanin, another class of plant pigments)
$\lambda_{max} = 545$ nm

Malachite green (a triphenylmethane dye)
$\lambda_{max} = 617$ nm

10.17 WHAT TO LOOK FOR IN AN ULTRAVIOLET SPECTRUM: A PRACTICAL GUIDE

It is often difficult to extract a great deal of information from a UV spectrum used by itself. It should be clear by now that a UV spectrum is most useful when at least a general idea of the structure is already known; in this way, the various empirical rules can be applied. Nevertheless, several generalizations can serve to guide our use of UV data. These generalizations are a good deal more meaningful when combined with infrared and NMR data—which can, for instance, definitely identify carbonyl groups, double bonds, aromatic systems, nitro groups, nitriles, enones, and other important chromophores. In the absence of infrared or NMR data, the following observations should be taken only as guidelines:

1. *A single band of low-to-medium intensity (ε = 100 to 10,000) at wavelengths less than 220 nm* usually indicates an $n \rightarrow \sigma^*$ transition. Amines, alcohols, ethers, and thiols are possibilities, provided the nonbonded electrons are not included in a conjugated system. An exception to this generalization is that the $n \rightarrow \pi^*$ transition of cyano groups ($-C\equiv N :$) appears in this region. However, this is a weak transition ($\varepsilon < 100$), and the cyano group is easily identified in the infrared. Do not neglect to look for N—H, O—H, C—O, and S—H bands in the infrared spectrum.

2. *A single band of low intensity (ε = 10 to 100) in the region 250 to 360 nm, with no major absorption at shorter wavelengths (200 to 250 nm),* usually indicates an $n \rightarrow \pi^*$ transition. Since the absorption does not occur at long wavelength, a simple, or unconjugated, chromophore is indicated, generally one that contains an O, N, or S atom. Examples of this may include C=O, C=N, N=N, $-NO_2$, $-COOR$, $-COOH$, or $-CONH_2$. Once again, infrared and NMR spectra should help a great deal.

3. *Two bands of medium intensity (ε = 1,000 to 10,000), both with λ_{max} above 200 nm,* generally indicate the presence of an aromatic system. If an aromatic system is present, there may be a good deal of fine structure in the longer-wavelength band (in nonpolar solvents only). Substitution on the aromatic rings increases the molar absorptivity above 10,000, particularly if the substituent increases the length of the conjugated system.

   In polynuclear aromatic substances, a third band appears near 200 nm, a band that in simpler aromatics occurs below 200 nm, where it cannot be observed. Most polynuclear aromatics (and heterocyclic compounds) have very characteristic intensity and band-shape (fine-structure) patterns, and they may often be identified via comparison to spectra that are available in the literature. The textbooks by Jaffé and Orchin and by Scott, which are listed in the references at the end of this chapter, are good sources of spectra.

4. *Bands of high intensity (ε = 10,000 to 20,000) that appear above 210 nm* generally represent either an α,β-unsaturated ketone (check the infrared spectrum), a diene, or a polyene. The greater the length of the conjugated system, the longer the observed wavelength. For dienes, the λ_{max} may be calculated using the Woodward–Fieser Rules (Section 10.10).

5. *Simple ketones, acids, esters, amides, and other compounds containing both π systems and unshared electron pairs show two absorptions:* an $n \rightarrow \pi^*$ transition at longer wavelengths (>300 nm, low intensity) and a $\pi \rightarrow \pi^*$ transition at shorter wavelengths (<250 nm, high intensity). With conjugation (enones), the λ_{max} of the $\pi \rightarrow \pi^*$ band moves to longer wavelengths and can be predicted by Woodward's Rules (Section 10.12). The ε value usually rises above 10,000 with conjugation, and as it is very intense, it may obscure or bury the weaker $n \rightarrow \pi^*$ transition.

   For α,β-unsaturated esters and acids, Nielsen's Rules (Section 10.13) may be used to predict the position of λ_{max} with increasing conjugation and substitution.

6. *Compounds that are highly colored* (have absorption in the visible region) are likely to contain a long-chain conjugated system or a polycyclic aromatic chromophore. Benzenoid compounds may be colored if they have enough conjugating substituents. For nonaromatic systems, usually a minimum of four to five conjugated chromophores are required to produce absorption in the visible region. However, some simple nitro, azo, nitroso, α-diketo, polybromo, and polyiodo compounds may also exhibit color, as may many compounds with quinoid structures.

P R O B L E M S

*1. The ultraviolet spectrum of benzonitrile shows a primary absorption band at 224 nm and a secondary band at 271 nm.

  (a) If a solution of benzonitrile in water, with a concentration of 1×10^{-4} molar, is examined at a wavelength of 224 nm, the absorbance is determined to be 1.30. The cell length is 1 cm. What is the molar absorptivity of this absorption band?

  (b) If the same solution is examined at 271 nm, what will be the absorbance reading ($\varepsilon = 1000$)? What will be the intensity ratio, I_0/I?

*2. Draw structural formulas that are consistent with the following observations:

  (a) An acid $C_7H_4O_2Cl_2$ shows a UV maximum at 242 nm.

  (b) A ketone $C_8H_{14}O$ shows a UV maximum at 248 nm.

  (c) An aldehyde $C_8H_{12}O$ absorbs in the UV with $\lambda_{max} = 244$ nm.

*3. Predict the UV maximum for each of the following substances:

(a)

$$CH_2{=}CH{-}\overset{\overset{\displaystyle O}{\|}}{C}{-}CH_3$$

(b)

(c)

(d)

(e)

(f)

(g)

(h)

(i)

(j)

*4. The UV spectrum of acetone shows absorption maxima at 166, 189, and 279 nm. What type of transition is responsible for each of these bands?

*5. Chloromethane has an absorption maximum at 172 nm, bromomethane shows an absorption at 204 nm, and iodomethane shows a band at 258 nm. What type of transition is responsible for each band? How can the trend of absorptions be explained?

*6. What types of electronic transitions are possible for each of the following compounds?

(a) Cyclopentene

(b) Acetaldehyde

(c) Dimethyl ether

(d) Methyl vinyl ether

(e) Triethylamine

(f) Cyclohexane

7. Predict and explain whether UV/visible spectroscopy can be used to distinguish between the following pairs of compounds. If possible, support your answers with calculations.

8. (a) Predict the UV maximum for the reactant and product of the following photochemical reaction:

(b) Is UV spectroscopy a good way to distinguish the reactant from the product?

(c) How would you use infrared spectroscopy to distinguish between the reactant and the product?

(d) How would you use proton NMR to distinguish between the reactant and the product (two ways)?

(e) How could you distinguish between the reactant and the product by using DEPT NMR (see Chapter 9)?

REFERENCES

American Petroleum Institute Research Project 44, *Selected Ultraviolet Spectral Data*, Vols. I–IV, Thermodynamics Research Center, Texas A&M University, College Station, Texas, 1945–1977.

Friedel, R. A., and M. Orchin, *Ultraviolet Spectra of Aromatic Compounds*, John Wiley and Sons, New York, 1951.

Graselli, J. G., and W. M. Ritchey, eds., *Atlas of Spectral Data and Physical Constants*, CRC Press, Cleveland, OH, 1975.

Grinter, H. C., T. L. Threlfall (translator), *UV-VIS Spectroscopy and Its Applications*, Springer Laboratory, 1992.

Hershenson, H. M., *Ultraviolet Absorption Spectra: Index for 1954–1957*, Academic Press, New York, 1959.

Jaffé, H. H., and M. Orchin, *Theory and Applications of Ultraviolet Spectroscopy*, John Wiley and Sons, New York, 1964.

Parikh, V. M., *Absorption Spectroscopy of Organic Molecules*, Addison–Wesley Publishing Co., Reading, MA, 1974, Chapter 2.

Scott, A. I., *Interpretation of the Ultraviolet Spectra of Natural Products*, Pergamon Press, New York, 1964.

Silverstein, R. M., F. X. Webster, and D. J. Kiemle *Spectrometric Identification of Organic Compounds*, 7th ed., John Wiley and Sons, New York, 2005.

Stern, E. S., and T. C. J. Timmons, *Electronic Absorption Spectroscopy in Organic Chemistry*, St. Martin's Press, New York, 1971.

Selected Websites

http://webbook.nist.gov/chemistry/

The National Institute of Standards and Technology (NIST) has developed the WebBook. This site includes UV/visible spectra, gas phase infrared spectra, and mass spectral data for compounds.

COMBINED STRUCTURE PROBLEMS

In this chapter, you will employ jointly all of the spectroscopic methods we have discussed so far to solve structural problems in organic chemistry. Forty-five problems are provided to give you practice in applying the principles learned in earlier chapters. The problems involve analysis of the mass spectrum (MS), the infrared (IR) spectrum, and proton and carbon (^{1}H and ^{13}C) nuclear magnetic resonance (NMR). Ultraviolet (UV) spectral data, if provided in the problem, appear in a tabular form rather than as a spectrum. You will notice as you proceed through this chapter that the problems use different "mixes" of spectral information. Thus, you may be provided with a mass spectrum, an infrared spectrum, and a proton NMR spectrum in one problem, and in another you may have available the infrared spectrum and both proton and carbon NMR.

All ^{1}H (proton) NMR spectra were determined at 300 MHz, while the ^{13}C NMR spectra were obtained at 75 MHz. The ^{1}H and ^{13}C spectra were determined in CDCl$_3$ unless otherwise indicated. In some cases, the ^{13}C spectral data have been tabulated, along with the DEPT-135 and DEPT-90 data. Some of the proton NMR spectra have been expanded to show the details. Finally, all infrared spectra on *liquid* samples were obtained neat (with no solvent) on KBr salt plates. The infrared spectra of *solids* have either been melted (cast) onto the salt plate or else determined as a mull (suspension) in Nujol (mineral oil).

The compounds in these problems may contain the following elements: C, H, O, N, S, Cl, Br, and I. In most cases if halogens are present, the mass spectrum should provide you with information regarding which halogen atom is present and the number of halogen atoms (Chapter 3, Section 3.7).

There are a number of possible approaches that you may take in solving the problems in this chapter. There are no "right" ways of solving them. In general, however, you should first try to gain an overall impression by looking at the gross features of the spectra provided in the problem. As you do so, you will observe evidence for pieces of the structure. Once you have identified pieces, you can assemble them and test against each of the spectra the validity of the structure you have assembled.

1. **Mass Spectrum.** You should be able to use the mass spectrum to obtain a molecular formula by performing the Rule of Thirteen calculation (Chapter 1, Section 1.5) on the molecular ion peak (*M*) labeled on the spectrum. In most cases, you will need to convert the hydrocarbon formula to one containing a functional group. For example, you may observe a carbonyl group in the infrared spectrum or ^{13}C spectrum. Make appropriate adjustments to the hydrocarbon formula so that it fits the spectroscopic evidence. When the mass spectrum is not provided in the problem, you will be given the molecular formula. Some of the labeled fragment peaks may provide excellent evidence for the presence of a particular feature in the compound being analyzed. Fragmentation patterns for common functional groups are discussed in Chapter 4, Sections 4.4 (alcohols), 4.6 (carbonyl-containing compounds), and 4.9 (halides).

2. **Infrared Spectrum.** The infrared spectrum provides some idea of the functional group or groups that are present or absent. Look first at the left-hand side of the spectrum to identify functional groups such as O—H, N—H, C≡N, C≡C, C=C, C=O, NO$_2$, and aromatic rings. See Chapter 2, Sections 2.8 and 2.9 for tips on what to look for in the spectrum. Ignore C—H stretching bands during this first "glance" at the spectrum as well as the right-hand side of

the spectrum. Determine the type of C=O group you have and also check to see if there is conjugation with a double bond or aromatic ring. Remember that you can often determine the substitution patterns on alkenes (Chapter 2, Figure 2.22) and aromatic rings (Chapter 2, Figure 2.28) by using the out-of-plane bending bands (oops). A complete analysis of the infrared spectrum is seldom necessary.

3. **Proton NMR Spectrum.** The proton (^{1}H) NMR spectrum gives information on the numbers and types of hydrogen atoms attached to the carbon skeleton. Chapter 5, Section 5.19 provides information on proton NMR spectra of various functional groups, especially expected chemical shift values. You will need to determine the integral ratios for the protons by using the integral traces shown. See Chapter 5, Section 5.9 to see how you can obtain the numbers of protons attached to the carbon chain. In most cases, it is not easy to see the splitting patterns of multiplets in the full 300-MHz spectrum. We have therefore indicated the multiplicities of peaks as doublet, triplet, quartet, quintet, and sextet on the full spectrum. Singlets are usually easy to see, and they have not been labeled. Many problems have been provided with proton expansions. When expansions are provided, Hertz values have been shown so that you can calculate the coupling constants. Often, the magnitude of the proton coupling constants will help you assign structural features to the compound such as the relative position of hydrogen atoms in alkenes (*cis/trans* isomers).

4. **Carbon NMR Spectra.** The carbon (^{13}C) NMR spectrum indicates the total number of nonequivalent carbon atoms in the molecule. In some cases, because of symmetry, carbon atoms may have identical chemical shifts. In this case, the total number of carbons is less than that found in the molecular formula. Chapter 6 contains important correlation charts that you should review. Figure 6.1 and Table 6.1 show the chemical shift ranges that you should expect for various structural features. Expected ranges for carbonyl groups are shown in Figure 6.2. In addition, you may find it useful to calculate approximate ^{13}C chemical shift values as shown in Appendix 8. Commonly, sp^3 carbon atoms appear to the upfield (right) side of the CDCl$_3$ solvent peak, while the sp^2 carbon atoms in an alkene or in an aromatic ring appear to the left of the solvent peak. Carbon atoms in a C=O group appear furthest to the left in a carbon spectrum. You should first look on the left-hand side of the carbon spectrum to see if you can identify potential carbonyl groups.

5. **Advanced NMR Techniques.** In some cases, the problems list information that can provide valuable information on the types of carbon atoms present in the unknown compound. Review Chapter 6, Section 6.10, A Quick Dip into DEPT, for information on how to determine the presence of CH$_3$, CH$_2$, CH, and C atoms in a carbon spectrum. You may also need to consult Chapter 9, Section 9.4 (DEPT), Section 9.7 (COSY) and Section 9.8 (HETCOR) for more information on these advanced techniques.

6. **Ultraviolet/Visible Spectrum.** The ultraviolet spectrum becomes useful when unsaturation is present in a molecule. See Chapter 10, Section 10.17 for information on how to interpret a UV spectrum.

7. **Determining a Final Structure.** A complete analysis of the information provided in the problems should lead to a unique structure for the unknown compound. Four solved examples are presented first. Note that more than one approach may be taken to the solution of these example problems. Since the problems near the beginning of this chapter are easier, you should attempt them before you move on. Have fun (no kidding)! You may even find that you have as much fun as the authors of this book.

■ EXAMPLE 1

Problem

The UV spectrum of this compound shows only end absorption. Determine the structure of the compound.

¹H

quartet
d

quartet
c

triplets

b a

¹³C

CDCl₃

Solution

Notice that this problem does not provide a molecular formula. We need to obtain it from the spectral evidence. The molecular ion peak appears at $m/z = 102$. Using the Rule of Thirteen (Chapter 1, Section 1.5), we can calculate a formula of C_7H_{18} for the peak at 102. The infrared spectrum shows a strong absorption at 1740 cm⁻¹, suggesting that a simple unconjugated ester is present in the compound. The presence of a C—O (strong and broad) at 1200 cm⁻¹ confirms the ester. We now know that there are two oxygen atoms in the formula. Returning to the mass spectral evidence, the formula calculated via the Rule of Thirteen was C_7H_{18}. We can modify this formula by converting carbons and hydrogens (one carbon and four hydrogens per oxygen atom) to the two oxygen atoms, yielding the formula $C_5H_{10}O_2$. This is the molecular formula for the compound. We can now calculate the index of hydrogen deficiency for this compound, which equals one, and that corresponds to the un-saturation in the C=O group. The infrared spectrum also shows sp^3 (aliphatic) C—H absorption at less than 3000 cm⁻¹. We conclude that the compound is an aliphatic ester with formula $C_5H_{10}O_2$.

Notice that the ¹³C NMR spectrum shows a total of five peaks, corresponding exactly to the number of carbons in the molecular formula! This is a nice check on our calculation of the formula via the Rule of Thirteen (five carbon atoms). The peak at 174 ppm corresponds to the ester C=O carbon. The peak at 60 ppm is a deshielded carbon atom caused by a neighboring single-bonded oxygen atom. The rest of the carbon atoms are relatively shielded. These three peaks correspond to the remaining part of the carbon chain in the ester.

We could probably derive a couple of possible structures at this point. The ¹H NMR spectrum should provide confirmation. Using the integral traces on the spectrum, we should conclude that the peaks shown have the ratio 2:2:3:3 (downfield to upfield). These numbers add up to the 10 total hydrogen atoms in the formula. Now, using the splitting patterns on the peaks, we can determine the structure of the compound. It is ethyl propanoate.

$$\underset{a}{CH_3}-\underset{c}{CH_2}-\overset{\displaystyle\overset{O}{\|}}{C}-O-\underset{d}{CH_2}-\underset{b}{CH_3}$$

The downfield quartet at 4.1 ppm (**d** protons) results from splitting with the neighboring protons on carbon **b**, while the other quartet at 2.4 ppm (**c** protons) results from spin-spin splitting with the protons on carbon **a**. Thus, the proton NMR is consistent with the final structure.

The UV spectrum is uninteresting but supports the identification of structure. Simple esters have weak $n \rightarrow \pi^*$ transitions (205 nm) near the solvent cutoff point. Returning to the mass spectrum, the strong peak at 57 mass units results from an α-cleavage of an alkoxy group to yield the acylium ion $(CH_3-CH_2-\overset{+}{C}=O)$, which has a mass of 57.

■ EXAMPLE 2

Problem

Determine the structure of a compound with the formula $C_{10}H_{12}O_2$. In addition to the infrared spectrum and ^1H NMR, the problem includes tabulated data for the normal ^{13}C NMR, DEPT-135, and DEPT-90 spectral data.

Normal Carbon	DEPT-135	DEPT-90
29 ppm	Positive	No peak
50	Negative	No peak
55	Positive	No peak
114	Positive	Positive
126	No peak	No peak
130	Positive	Positive
159	No peak	No peak
207	No peak	No peak

Solution

We calculate an index of hydrogen deficiency (Chapter 1, Section 1.4) of five. The ^{1}H and ^{13}C NMR spectra, as well as the infrared spectrum, suggest an aromatic ring (unsaturation index = four). The remaining index of one is attributed to a C=O group found in the infrared spectrum at 1711 cm^{-1}. This value for the C=O is close to what you might expect for an unconjugated carbonyl group in a ketone and is too low for an ester. The ^{13}C NMR confirms the ketone C=O; the peak at 207 ppm is typical for a ketone. The ^{13}C NMR spectrum shows only 8 peaks, while 10 are present in the molecular formula. This suggests some symmetry that makes some of the carbon atoms equivalent.

When inspecting the ^{1}H NMR spectrum, notice the nice *para* substitution pattern between 6.8 and 7.2 ppm, which appears as a nominal "pair of doublets", integrating for two protons in each pair. The electron-donating nature of the methoxy (or ^{1}H chemical shift calculations) allow us to assign the more upfield resonance at 6.8 ppm to the protons (d) adjacent to the $-OCH_3$ group on the aromatic ring. Notice in the ^{1}H NMR that the upfield portion of the spectrum has protons that integrate for 3:2:3 for a CH_3, a CH_2, and a CH_3, respectively. Also, notice that these peaks are unsplit, indicating that there are no neighboring protons. The downfield methyl at 3.8 ppm is next to an oxygen atom, suggesting a methoxy group. The ^{13}C DEPT NMR spectra results confirm the presence of two methyl groups and one methylene group. The methyl group at 55 ppm is deshielded by the presence of an oxygen atom ($O-CH_3$). Keeping in mind the *para*-disubstituted pattern and the singlet peaks in the ^{1}H NMR, we derive the following structure for 4-methoxyphenylacetone:

Further confirmation of the *para*-disubstituted ring is obtained from the carbon spectral results. Notice the presence of four peaks in the aromatic region of the ^{13}C NMR spectrum. Two of these peaks (126 and 159 ppm) are *ipso* carbon atoms (no attached protons) that do not appear in the DEPT-135 or DEPT-90 spectra. The remaining two peaks at 114 and 130 ppm are assigned to the remaining four carbons (two each equivalent by symmetry). The two carbon atoms **d** show peaks in both of the DEPT experiments, which confirms that they have attached protons (C—H). Likewise, the two carbon atoms **e** have peaks in both DEPT experiments confirming the presence of C—H. The infrared spectrum has a *para* substitution pattern in the out-of-plane region (835 cm^{-1}), which helps confirm the 1,4-disubstitution on the aromatic ring.

■ **EXAMPLE 3**

Problem

This compound has the molecular formula $C_9H_{11}NO_2$. Included in this problem are the infrared spectrum, ^1H NMR with expansions, and ^{13}C NMR spectra data.

1708 cm^{-1}

Normal Carbon	DEPT-135	DEPT-90
14 ppm	Positive	No peak
61	Negative	No peak
116	Positive	Positive
119	Positive	Positive
120	Positive	Positive
129	Positive	Positive
131	No peak	No peak
147	No peak	No peak
167	No peak	No peak

Solution

We calculate an index of hydrogen deficiency of five. All of the spectra shown in this problem suggest an aromatic ring (unsaturation index = four). The remaining index of one is assigned to the C=O group found at 1708 cm^{-1}. This value for the carbonyl group is too high for an amide. It is in a reasonable place for a conjugated ester. While the NO_2 present in the formula suggests a possible nitro group, this cannot be the case because we need the two oxygens for the ester functional group. The doublet at about 3400 cm^{-1} in the infrared spectrum is perfect for a primary amine.

The ^{13}C NMR spectrum has nine peaks, which correspond to the nine carbon atoms in the molecular formula. The ester C=O carbon atom appears at 167 ppm. The remaining downfield carbons are attributed to the six unique aromatic ring carbons. From this, we know that the ring is not symmetrically substituted. The DEPT results confirm the presence of two carbon atoms with no attached protons (131 and 147 ppm) and four carbon atoms with one attached proton (116, 199, 120, and 129 ppm). From this information, we now know that the ring is disubstituted.

We must look carefully at the aromatic region between 6.8 and 7.5 ppm in the ^1H spectrum. Notice that there are four protons on the aromatic ring with each integrating for one proton (see integral lines drawn on the ^1H spectrum). Since it is difficult to determine the splitting pattern for the protons shown in the ^1H spectrum, an expansion of the 6.8 to 7.5 ppm region is shown above. The ring must be disubstituted because four protons appear on the aromatic ring. The pattern suggests a 1,3 disubstituted pattern rather than 1,4- or 1,2-disubstitution (see Chapter 7, Section 7.10). The key observation is that proton **f** is a narrowly spaced triplet (or dd), suggesting ^4J couplings, but with no ^3J couplings. In other words, that proton must not have any adjacent protons! It is "sandwiched" between two non-proton groups: amino (-NH_2) and carbonyl (C=O). Protons **g** and **f** appear down field relative to protons **e** and **d** because of the deshielding effect of the anisotropy of the C=O group (see Figure 7.61). Although not as reliable as the protonNMR evidence, the aromatic out-of-plane bending bands in the infrared spectrum suggests *meta*-disubstitution: 680, 760, and 880 cm^{-1}.

The ^{1}H NMR spectrum shows an ethyl group because of the quartet and triplet found upfield in the spectrum (4.3 and 1.4 ppm, respectively, for the CH$_2$ and CH$_3$ groups). Finally, a broad NH$_2$ peak, integrating for two protons, appears in the Proton NMR spectrum at 3.8 ppm. The compound is ethyl 3-aminobenzoate.

We need to look at the proton expansions provided in the problem to confirm the assignments made for the aromatic protons. The Hertz values shown on the expansions allow us the opportunity to obtain coupling constants that confirms the 1,3-disubstitution pattern. The splittings observed in the expansions can be explained by looking at the coupling constants ^3J and ^4J present in the compound. ^5J couplings are either zero or too small to be observed in the expansions.

7.42 ppm (H$_g$)	Doublet of triplets (dt) or doublet of doublets of doublets (ddd); $^3J_{eg} = 7.8$ Hz, $^4J_{fg}$ and $^4J_{dg} \approx 1.5$ Hz.
7.35 ppm (H$_f$)	This proton is located between the two attached groups. The only proton couplings that are observed are small ^4J couplings that result in a closely spaced triplet or, more precisely, a doublet of doublets; $^4J_{fg}$ and $^4J_{df} \approx 1.5$ to 2 Hz.
7.19 ppm (H$_e$)	This proton appears as a widely spaced "triplet." One of the coupling constants, $^3J_{eg} = 7.8$ Hz, was obtained from the pattern at 7.42 ppm. The other coupling constant, $^3J_{de} = 8.1$ Hz, was obtained from the pattern at 6.84 ppm. The pattern appears as a triplet because the coupling constants are nearly equal, resulting in an accidental overlap of the center peak in the "triplet." More precisely, we should describe this "triplet" as a doublet of doublets (dd).
6.84 ppm (H$_d$)	Doublet of doublets of doublets (ddd); $^3J_{de} = 8.1$ Hz, $^4J_{dg} \neq {}^4J_{df}$.

■ **EXAMPLE 4**

Problem

This compound has the molecular formula $C_5H_7NO_2$. Following are the infrared, ^1H NMR, and ^{13}C NMR spectra.

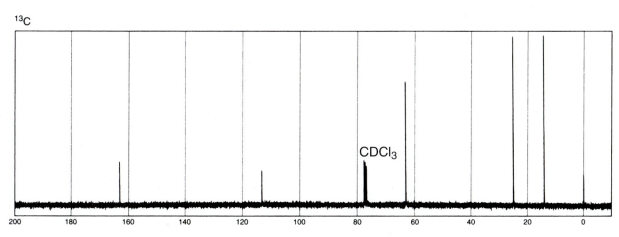

Solution

We calculate an index of hydrogen deficiency of three. A quick glance at the infrared spectrum reveals the source of unsaturation implied by an index of three: a nitrile group at 2260 cm^{-1} (unsaturation index = two) and a carbonyl group at 1747 cm^{-1} (unsaturation index = one). The frequency of the carbonyl absorption indicates an unconjugated ester. The appearance of several strong C—O bands near 1200 cm^{-1} confirms the presence of an ester functional group. We can rule out a C≡C bond because they usually absorb at a lower value (2150 cm^{-1}) and have a weaker intensity than compounds that contain C≡N.

The ^{13}C NMR spectrum shows five peaks and thus is consistent with the molecular formula, which contains five carbon atoms. Notice that the carbon atom in the C≡N group has a characteristic value of 113 ppm. In addition, the carbon atom in the ester C=O appears at 163 ppm. One of the remaining carbon atoms (63 ppm) probably lies next to an electronegative oxygen atom. The remaining two carbon atoms, which absorb at 25 and 14 ppm, are attributed to the remaining methylene and methyl carbons. The structure is

$$N\equiv C-\underset{\textbf{b}}{CH_2}-\overset{\overset{\textstyle O}{\|}}{C}-O-\underset{\textbf{c}}{CH_2}-\underset{\textbf{a}}{CH_3}$$

The ^{1}H NMR spectrum shows a classic ethyl pattern: a quartet (2 H) at 4.3 ppm and a triplet (3 H) at 1.3 ppm. The quartet is strongly influenced by the electronegative oxygen atom, which shifts it downfield. There is also a two-proton singlet at 3.5 ppm.

PROBLEMS

*1. The structure for the naphthalene ring derivative with the formula $C_{13}H_{12}O_2$ is provided in this question. The IR spectrum shows a strong band at 1680 cm^{-1} for the C=O group. The normal ^{13}C NMR spectrum is shown in a stacked plot along with the DEPT-135 and DEPT-90 spectra. You should consult Chapter 7, Section 7.10 to determine the effect of a carbonyl group *versus* a methoxy group on the relative expected chemical shifts for protons attached to a benzene ring (also consult Table 6.3 in Appendix 6). In your analysis of the structure, consider three individual steps. Complete assignments have been provided in the Answers to Selected Problems.

(A) **DEPT analysis**: The 13 carbon atoms are numbered on the structure. As part of this exercise, you will need to assign a ppm value to as many of the carbon atoms as possible. Using the DEPT spectra, you should be able to assign some of the ^{13}C atoms in groups, for example all the C—H in one group and all the quaternary carbons (no attached hydrogen atoms) in another group. The methyl groups can each be assigned based upon their chemical shift values (see Table 5.4 and Figure 5.20 in Chapter 5).

(B) **COSY analysis**: The ^{1}H NMR spectrum and expansions are provided in the problem along with the COSY spectrum. The top axis of the COSY spectrum is labeled with letters that correspond to the position of individual hydrogen atoms in the ^{1}H spectrum. Since this spectrum only covers the range of 7.0 to 8.6 ppm, the two methyl groups are not included in the COSY spectrum. Using the COSY spectrum assign all of the C—H hydrogen atoms in the compound. Assign the two methyl groups based upon their chemical shift values. On the structure shown below, label the hydrogen atoms with the appropriate letter, a, b, c, d, e, or f consistent with the COSY spectrum.

(C) **Coupling constants**: Calculate the ^3J and ^4J coupling constants for the each of the C—H hydrogen atoms on the naphthalene ring. The values obtained should confirm the assignments made from the correlations obtained from the COSY spectrum.

2. Determine the structure for a compound with formula C_3H_5ClO. The IR spectrum, ^1H NMR, ^{13}C NMR, DEPT, COSY, and HETCOR (HSQC) spectra are included in this problem. The infrared spectrum has a trace of water that should be ignored (region from 3700 to 3400 cm^{-1}). The HETCOR spectrum should be carefully examined, for it provides very important information. You will find it helpful to consult Appendix 5 (alkanes and cyclic alkanes) for values of coupling constants. Determine the coupling constants from the ^1H NMR spectrum, except for proton c, and compare the calculated values to those shown in Appendix 5. Draw the structure of the compound, and label the protons on the structure.

Infrared spectrum

DEPT

COSY

F2
(ppm)

F1 (ppm)

HETCOR

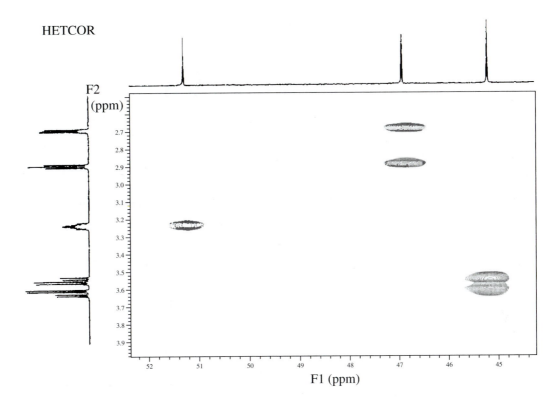

F2
(ppm)

F1 (ppm)

*3. The UV spectrum of this compound is determined in 95% ethanol: λ_{max} 290 nm (log ε =1.3).

(a)

(b)

1718 cm^{-1}

(c)

quartet

triplet

*4. The UV spectrum of this compound shows no maximum above 205 nm. When a drop of aqueous acid is added to the sample, the pattern at 3.6 ppm in the ^{1}H NMR spectrum simplifies to a triplet, and the pattern at 3.2 ppm simplifies to a singlet.

(a)

(b)

(c)

*5. UV spectrum of this compound is determined in 95% ethanol: λ_{max} 280 nm (log $\varepsilon = 1.3$).

(a)

(b)

(c)

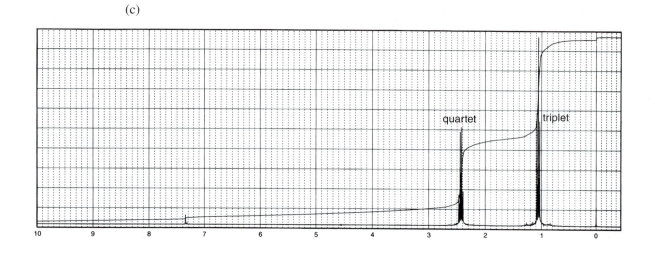

*6. The formula for this compound is $C_6H_{12}O_2$.

(a)

1737 cm^{-1}

(b)

(c)

CDCl$_3$

*7. The UV spectrum of this compound is determined in 95% ethanol: strong end absorption and a band with fine structure appearing at λ_{max} 257 nm (log ε = 2.4). The IR spectrum was obtained as a Nujol mull. The strong bands at about 2920 and 2860 cm^{-1} from the C—H stretch in Nujol overlap the broad band that extends from 3300 to 2500 cm^{-1}.

(a)

(b)

(c)

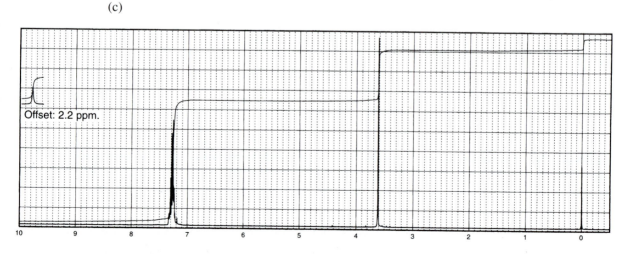

*8. The mass spectrum of this compound shows an intense molecular ion at 172 mass units and an
  $M + 2$ peak of approximately the same size. The IR spectrum of this solid unknown was obtained
  in Nujol. The prominent C—H stretching bands centering on about 2900 cm^{-1} are derived from
  the Nujol and are not part of the unknown. The peak appearing at about 5.3 ppm in the ^{1}H NMR
  spectrum is solvent dependent. It shifts readily when the concentration is changed.

(a)

(b)

(c)

*9. This compound has the molecular formula $C_{11}H_{14}O$.

Normal Carbon	DEPT-135	DEPT-90
14 ppm	Positive	No peak
22	Negative	No peak
26	Negative	No peak
38	Negative	No peak
128	Positive	Positive
129	Positive	Positive
133	Positive	Positive
137	No peak	No peak
200	No peak	No peak

(ppm)

*10. Determine the structures of the isomeric compounds that show strong infrared bands at 1725 cm^{-1} and several strong bands in the range 1300–1200 cm^{-1}. Each isomer has the formula $C_9H_9BrO_2$. Following are the ^{1}H NMR spectra for both compounds, **A** and **B**. Expansions have been included for the region from 8.2 to 7.2 ppm for compound **A**.

A.

B.

*11. This compound has the molecular formula $C_4H_{11}N$.

(a)

(b)

(c)

CDCl$_3$

*12. The UV spectrum of this compound is determined in 95% ethanol: λ_{max} 280 nm (log $\varepsilon = 1.3$). This compound has the formula C$_5$H$_{10}$O.

(a)

(b)

(c)

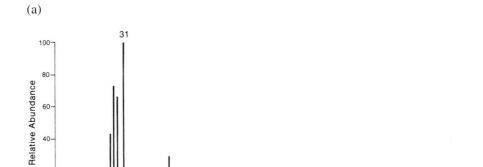

*13. This compound has the formula $C_3H_6O_2$. The UV spectrum of this compound shows no maximum above 205 nm. The ^{13}C NMR spectrum shows peaks at 14, 60, and 161 ppm. The peak at 161 ppm appears as a positive peak in the DEPT-90 spectrum.

(a)

(b)

(c)

*14. Determine the structures of the isomeric compounds *A* and *B*, each of which has the formula C_8H_7BrO. The infrared spectrum for compound *A* has a strong absorption band at 1698 cm^{-1}, while compound *B* has a strong band at 1688 cm^{-1}. The ^{1}H NMR spectrum for compound *A* is shown, along with expansions for the region from 7.7 to 7.2 ppm. The ^{1}H NMR spectrum of compound *B* is also shown.

A.

2285.26
2283.88
2277.40
2276.12
2244.41
2242.49
2236.83
2234.91
2217.82
2216.54
2210.42
2209.14
2202.93
2201.56
2195.07
2193.15
2187.22
2185.39
2179.81
2177.99

(ppm)

B.

doublets

*15. This compound has the formula C_4H_8O. When expanded, the singlet peak at 9.8 ppm in the ^1H NMR spectrum is, in fact, a triplet. The triplet pattern at 2.4 ppm turns out to be a triplet of doublets when expanded.

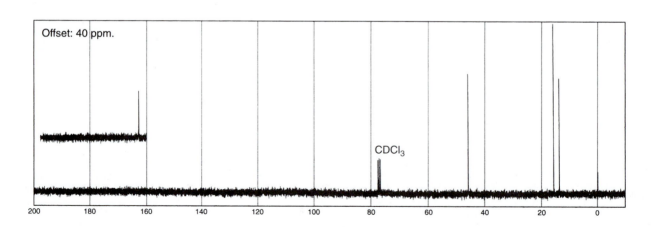

*16. This compound has the formula $C_5H_{12}O$. When a trace of aqueous acid is added to the sample, the ^1H NMR spectrum resolves into a clean triplet at 3.6 ppm, and the broad peak at 2.2 ppm moves to 4.5 ppm.

*17. Determine the structures of the isomeric compounds with the formula $C_5H_9BrO_2$. The ^1H NMR spectra for both compounds follow. The IR spectrum corresponding to the first ^1H NMR spectrum has strong absorption bands at 1739, 1225, and 1158 cm^{-1}, and that corresponding to the second one has strong bands at 1735, 1237, and 1182 cm^{-1}.

*18. This compound has the molecular formula $C_{10}H_9NO_2$.

*19. This compound has the formula C_9H_9ClO. The full ^1H NMR spectrum is shown along with expansions of individual patterns.

20. The anesthetic procaine (Novocaine) has the formula $C_{13}H_{20}N_2O_2$. In the ^{1}H NMR spectrum, each pair of triplets at 2.8 and 4.3 ppm has a coupling constant of 6 Hz. The triplet at 1.1 and the quartet at 2.6 ppm have coupling constants of 7 Hz. The IR spectrum was determined in Nujol. The C–H absorption bands of Nujol at about 2920 cm^{-1} in the IR spectrum obscure the entire C–H stretch region. The carbonyl group appearing at 1669 cm^{-1} in the IR spectrum has an unusually low frequency. Why?

Normal Carbon	DEPT-135	DEPT-90
12 ppm	Positive	No peak
48	Negative	No peak
51	Negative	No peak
63	Negative	No peak
114	Positive	Positive
120	No peak	No peak
132	Positive	Positive
151	No peak	No peak
167	No peak	No peak

21. The UV spectrum of this compound shows no maximum above 250 nm. In the mass spectrum, notice that the patterns for the M, $M + 2$, and $M + 4$ peaks have a ratio of 1:2:1 (214, 216, and 218 m/z). Draw the structure of the compound and comment on the structures of the mass 135 and 137 fragments.

(a)

(b)

(c)

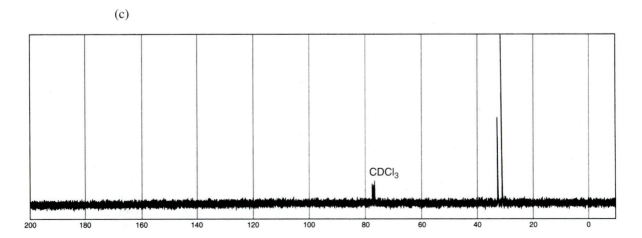

22. The UV spectrum of this compound is determined in 95% ethanol: λ_{max} 225 nm (log $\varepsilon = 4.0$) and 270 nm (log $\varepsilon = 2.8$). This compound has the formula $C_9H_{12}O_3S$.

(a)

(b)

(c)

23. This compound has the molecular formula $C_9H_{10}O$. We have supplied you with the IR and ^1H NMR spectra. The expansions of the interesting sets of peaks centering near 4.3, 6.35, and 6.6 ppm in the ^1H NMR are provided as well. Do not attempt to interpret the messy pattern near 7.4 ppm for the aromatic protons. The broad peak at 2.3 ppm (one proton) is solvent and concentration dependent.

24. This compound has the formula C_3H_4O. We have supplied you with the IR and ^{1}H NMR spectra. Notice that a single peak at 3300 cm^{-1} overlaps the broad peak there. The expansions of the interesting sets of peaks centering near 2.5 and 4.3 ppm in the ^{1}H NMR are provided as well. The peak at 3.25 ppm (one proton) is solvent and concentration dependent.

2.4-Hz
spacings

25. This compound has the molecular formula $C_7H_8N_2O_3$. We have supplied you with the IR and ^1H NMR spectra (run in DMSO-d_6). The expansions of the interesting sets of peaks centering near 7.75, 7.6, and 6.7 ppm in the ^1H NMR are provided as well. The peak at 6.45 ppm (two protons) is solvent and concentration dependent. The UV spectrum shows peaks at 204 nm ($\varepsilon = 1.68 \times 10^4$), 260 nm ($\varepsilon = 6.16 \times 10^3$), and 392 nm ($\varepsilon = 1.43 \times 10^4$). The presence of the intense band at 392 nm is an important clue regarding the positions of groups on the ring. This band moves to a lower wavelength when acidified. The IR spectrum was determined in Nujol. The C—H bands for Nujol at about 2920 cm^{-1} obscure the C—H bands in the unknown compound.

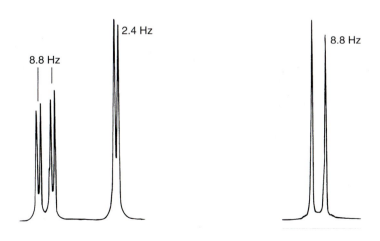

26. This compound has the formula $C_6H_{12}N_2$.

Normal Carbon	DEPT-135	DEPT-90
13 ppm	Positive	No peak
41	Negative	No peak
48	Negative	No peak
113	No peak	No peak

27. This compound has the formula $C_6H_{11}BrO_2$. Determine the structure of this compound. Draw the structures of the fragments observed in the mass spectrum at 121/123 and 149/151. The ^{13}C NMR spectrum shows peaks at 14, 31, 56, 62, and 172 ppm.

28. This compound has the formula $C_9H_{12}O$. The ^{13}C NMR spectrum shows peaks at 28, 31, 57, 122, 124, 125, and 139 ppm.

29. This compound has the formula $C_6H_{10}O$. The ^{13}C NMR spectrum shows peaks at 21, 27, 31, 124, 155, and 198 ppm.

1690 cm^{-1}

5.96 5.94 5.92
(ppm)

2.00 1.98 1.96
(ppm)

1.74 1.72 1.70
(ppm)

7.0   6.5   6.0   5.5   5.0   4.5   4.0   3.5   3.0   2.5   2.0   1.5   1.0   0.0
(ppm)

30. This compound has the formula $C_{10}H_{10}O_2$. The ^{13}C NMR spectrum shows peaks at 52, 118, 128, 129, 130, 134, 145, and 167 ppm.

(ppm)

31. This compound has the formula $C_5H_8O_2$. The ^{13}C NMR spectrum shows peaks at 14, 60, 129, 130, and 166 ppm.

(ppm)

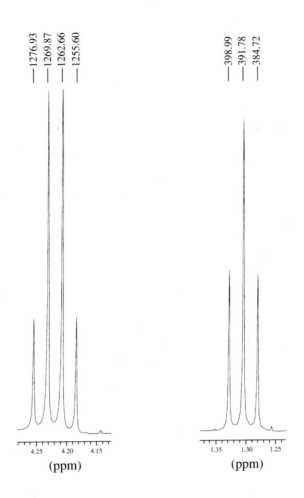

1276.93
1269.87
1262.66
1255.60

398.99
391.78
384.72

4.25   4.20   4.15
(ppm)

1.35   1.30   1.25
(ppm)

32. This compound has the formula $C_6H_{12}O$. Interpret the patterns centering on 1.3 and 1.58 ppm in the ^1H NMR spectrum.

NEAT

MICRONS

NICOLET 20SX FT-IR

1715 cm^{-1}

WAVENUMBERS

—719.63
—712.78
—705.90
—699.06
—692.18
—685.36

494.05
486.52
479.77
479.06
472.77
466.04
465.36
458.60
451.25

410.29
402.73
395.92
395.33
388.98
382.19
381.54
374.78
367.58

(ppm)

—293.65	—239.61	
—286.68	—232.20	
	—224.73	

Normal Carbon	DEPT-135	DEPT-90
12 ppm	Positive	No peak
16	Positive	No peak
26	Negative	No peak
28	Positive	No peak
49	Positive	Positive
213	No peak	No peak

33. This compound has the formula $C_9H_{10}O_2$.

1690 cm^{-1}

34. This compound has the formula $C_8H_{14}O$.

Normal Carbon	DEPT-135	DEPT-90
18 ppm	Positive	No peak
23	Negative	No peak
26	Positive	No peak
30	Positive	No peak
44	Negative	No peak
123	Positive	Positive
133	No peak	No peak
208	No peak	No peak

35. This compound has the formula $C_6H_6O_3$. The ^{13}C NMR spectrum shows peaks at 52, 112, 118, 145, 146, and 159 ppm.

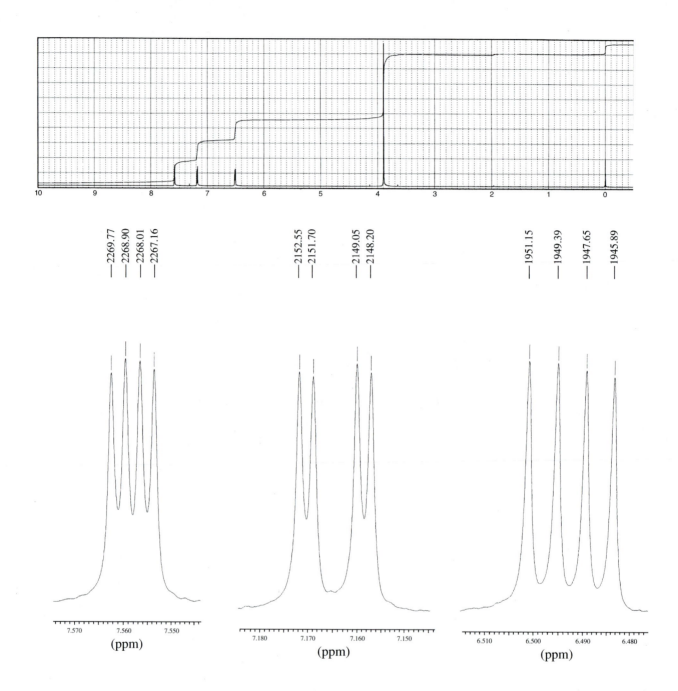

2269.77
2268.90
2268.01
2267.16

2152.55
2151.70
2149.05
2148.20

1951.15
1949.39
1947.65
1945.89

7.570   7.560   7.550
(ppm)

7.180   7.170   7.160   7.150
(ppm)

6.510   6.500   6.490   6.480
(ppm)

36. A compound with the formula $C_9H_8O_3$ shows a strong band at 1661 cm^{-1} in the infrared spectrum. The ^{1}H NMR spectrum is shown, but there is a small impurity peak at 3.35 ppm that should be ignored. Expansions are shown for the downfield protons. In addition, the normal ^{13}C NMR, DEPT-135, and DEPT-90 spectral results are tabulated.

Normal Carbon	DEPT-135	DEPT-90
26 ppm	Positive	No peak
102	Negative	No peak
107	Positive	Positive
108	Positive	Positive
125	Positive	Positive
132	No peak	No peak
148	No peak	No peak
151	No peak	No peak
195	No peak	No peak

37. A compound with the formula $C_5H_{10}O_2$ shows a very broad band that extends from about 3500 to 2500 cm^{-1} in the infrared spectrum. Another prominent band appears at 1710 cm^{-1}. The ^1H and ^{13}C NMR spectra are shown. Draw the structure for this compound.

CDCl$_3$

CDCl$_3$

38. A compound with the formula $C_8H_{14}O_2$ shows several bands in the infrared spectrum in the region from 3106 to 2876 cm^{-1}. In addition there are strong peaks that appear at 1720 and 1170 cm^{-1}. A medium-sized peak appears at 1640 cm^{-1}. The ^1H and ^{13}C NMR spectra are shown along with the DEPT data. Draw the structure for this compound.

Normal Carbon	DEPT-135	DEPT-90
13.73 ppm	Positive	No peak
18.33	Positive	No peak
19.28	Negative	No peak
30.76	Negative	No peak
64.54	Negative	No peak
125.00	Negative	No peak
136.63	No peak	No peak
167.51	No peak	No peak

1.03   1.03   2.08   3.07 2.03 1.98   2.91

6.0   5.5   5.0   4.5   4.0   3.5   3.0   2.5   2.0   1.5   1.0

39. A compound with the formula $C_8H_{10}O$ shows a broad peak centering on about 3300 cm^{-1} in the infrared spectrum. In addition, there are several bands appearing in the region from 3035 to 2855 cm^{-1}. There are also medium-sized peaks appearing in the range of 1595 to 1445 cm^{-1}. The ^{1}H and ^{13}C NMR spectra are shown. Draw the structure for this compound.

40. A compound with the formula $C_8H_6O_3$ shows weak peaks between 3100 and 2716 cm^{-1} in the infrared spectrum. Very strong peaks appear at 1697 and 1260 cm^{-1}. There are also several medium-sized peaks appearing in the range of 1605 to 1449 cm^{-1}. The ^1H and ^{13}C NMR spectra are shown. The DEPT results are tabulated. Draw the structure for this compound.

Normal Carbon	DEPT-135	DEPT-90
102.10 ppm	Negative	No peak
106.80	Positive	Positive
108.31	Positive	Positive
128.62	Positive	Positive
131.83	No peak	No peak
148.65	No peak	No peak
153.05	No peak	No peak
190.20	Positive	Positive (C=O)

41. The ^1H and ^{13}C NMR spectra of a compound with formula $C_{11}H_8O_2$ are shown. The DEPT experimental results are tabulated. The infrared spectrum shows a broad peak centering on about 3300 cm^{-1} and a strong peak at 1670 cm^{-1}. Draw the structure of this compond. (*Hint*: There are two substituents on the same ring of a naphthalene system.)

Normal Carbon	DEPT-135	DEPT-90
111.88 ppm	No peak	No peak
118.69	Positive	Positive
120.68	Positive	Positive
124.13	Positive	Positive
127.52	No peak	No peak
128.85	Positive	Positive
128.95	Positive	Positive
132.18	No peak	No peak
138.41	Positive	Positive
164.08	No peak	No peak
193.28	Positive	Positive (C=O)

42. The ^1H and ^{13}C NMR spectra of a compound with formula $C_3H_8O_3$ are shown. The infrared spectrum shows a broad peak centering on about 3350 cm^{-1} and strong peaks at 1110 and 1040 cm^{-1}. Draw the structure of this compound and determine the coupling constants for pattern at 3.55 and 3.64 ppm to support the structure that you have drawn.

43. The ^{1}H and ^{13}C NMR spectra of a compound with formula $C_5H_3ClN_2O_2$ are shown. The infrared spectrum shows medium-sized peaks at 3095, 3050, 1590, 1564, and 1445 cm^{-1} and strong peaks at 1519 and 1355 cm^{-1}. Determine the coupling constants from the Hertz values printed on the ^{1}H NMR spectrum. The coupling constant data listed in Appendix 5 should allow you to determine the structure(s) of compounds that fit the data.

44. The ^1H NMR spectrum of a compound with formula $C_6H_{12}O_2$ is shown. The DEPT experimental results are tabulated. The infrared spectrum is rather uninteresting. There are four strong bands that appear in the range of 1200 to 1020 cm^{-1}. The compound is prepared from the reaction of 1,2-ethanediol and 2-butanone. Draw the structure of this compound.

Normal Carbon	DEPT-135	DEPT-90
8.35 ppm	Positive	No peak
23.31	Positive	No peak
31.98	Negative	No peak
64.70	Negative	No peak
110.44	No peak	No peak

45. The ^{1}H NMR spectrum of a compound with formula $C_7H_{14}O$ is shown. The DEPT experimental results are tabulated. The infrared spectrum shows bands at 3080, 2960, 2865, and 1106 cm^{-1} and a medium-intensity band at 1647 cm^{-1}. Draw the structure of this compound.

Normal Carbon	DEPT-135	DEPT-90
13.93 ppm	Positive	No peak
19.41	Negative	No peak
31.91	Negative	No peak
70.20	Negative	No peak
71.80	Negative	No peak
116.53	Negative	No peak
135.16	Positive	Positive

SOURCES OF ADDITIONAL PROBLEMS

Books that Contain Combined Spectral Problems

Field, L. D., S. Sternhell, and J. R. Kalman, *Organic Structures from Spectra*, 5th ed., John Wiley and Sons, New York, 2013.

Fuchs, P. L., and C. A. Bunnell, *Carbon-13 NMR-Based Organic Spectral Problems*, John Wiley and Sons, New York, 1979.

Huggins, M., J. E. Gurst, and D. A Lightner, *2D NMR-Based Organic Spectroscopy Problems*, Prentice Hall, 2010.

Silverstein, R. M., F. X. Webster, and D. J. Kiemle, *Spectrometric Identification of Organic Compounds*, 7th ed., John Wiley and Sons, New York, 2005.

Simpson, J. H., *Organic Structure Determination Using 2-D NMR Spectroscopy: A Problem-Based Approach*, 2nd ed., Academic Press, San Diego, 2012.

Taber, D. F. *Organic Spectroscopic Structure Determination: A Problem-Based Learning Approach*, Oxford University Press, 2007.

Tomasi, R. A., *A Spectrum of Spectra*, Sunbelt R&T, Inc., Sunbelt R&T, Inc., 1946 S. 74 E. Ave., Tulsa, OK 741112-7716, phone 918-627-9655. A CD-ROM version is available.

Williams, D. H., and I. Fleming, *Spectroscopic Methods in Organic Chemistry*, 6th ed., McGraw-Hill Book Co. Ltd., London, 2008.

Compilations of Spectra and Tables of Spectral Data

Pouchert, C. J., and J. Behnke, *The Aldrich Library of ^{13}C and ^{1}H FT-NMR Spectra, 300 MHz*, Aldrich Chemical Company, Milwaukee, 1993.

Pretsch, E., J. P. Buhlmann, and C. Affolter, *Structure Determination of Organic Compounds. Tables of Spectral Data*, 3rd ed., Springer-Verlag, Berlin, 2000.

Websites That Have Combined Spectral Problems

http://orgchem.colorado.edu/Spectroscopy/Spectroscopy.html
This website at the University of Colorado at Boulder, Department of Chemistry and Biochemistry includes infrared and nmr tutorials along with 20 problems that include molecular formula, infrared, and low field ^{1}H spectra. The problems are graded from easy to more difficult and include problem answers. This site is highly recommended.

http://www.chem.ucla.edu/~webspectra/
The UCLA Department of Chemistry and Biochemistry in connection with Cambridge University Isotopes Laboratories presents WebSpectra, an outstanding collection of 75 combined problems for students to interpret. All problems include ^{1}H NMR and ^{13}C NMR spectra, but some problems also include infrared spectra and DEPT and COSY spectra, as well. Formulas for each

of the unknowns are provided. No mass spectral data is included. This website is highly recommended. Solutions to the problems are provided. They provide links to other sites with problems.

http://www3.nd.edu/~smithgrp/structure/workbook.html

The Smith group at the University of Notre Dame provide 64 combined problems that include molecular formulas, infrared spectra, 500 MHz ^{1}H and 125 MHz ^{13}C nmr spectra and mass spectral data. The problems are graded according to difficulty. Answers are not provided.

http://www2.chemistry.msu.edu/faculty/reusch/VirtTxtJml/S pectrpy/spectro.htm

The excellent website presented by Professor William Reusch from the Department of Chemistry at Michigan State University includes background material for infrared spectroscopy, NMR spectroscopy, UV spectroscopy, and mass spectrometry. Some problems are included. The site includes links to other available sites for spectral problems.

Websites That Have Spectral Data

http://sdbs.riodb.aist.go.jp/sdbs/cgi-bin/cre_index.cgi? lang=eng

This excellent free website presents an Integrated Spectral Data Base System for Organic Compounds (SDBS). It is organized and maintained by the National Institute of Advanced Industrial Science and Technology (AIST), Tsukuba, Ibaraki 305-8565, Japan. This database includes infrared, mass spectra, and NMR data (^{1}H and ^{13}C) for a number of compounds. This site provides the actual spectrum when the formula or name is provided.

http://webbook.nist.gov/chemistry/

National Institute of Standards and Technology (NIST). The site includes gas phase infrared spectra and mass spectral data.

ANSWERS TO SELECTED PROBLEMS

CHAPTER 1

1. (a) 90.50% carbon; 9.50% hydrogen   (b) C_4H_5

2. 32.0% carbon; 5.4% hydrogen; 62.8% chlorine; $C_3H_6Cl_2$

3. $C_2H_5NO_2$

4. 180.2 = molecular mass. Molecular formula is $C_9H_8O_4$.

5. Equivalent weight = 52.3

6. (a) 6   (b) 1   (c) 3   (d) 6   (e) 12

7. The index of hydrogen deficiency = 1. There cannot be a triple bond because the presence of a triple bond would require an index of hydrogen deficiency of at least 2.

8. (a) 59.96% carbon; 5.75% hydrogen; 34.29% oxygen   (b) $C_7H_8O_3$ (c) $C_{21}H_{24}O_9$
   (d) A maximum of two aromatic (benzenoid) rings

9. (a) $C_8H_8O_2$ (b) $C_8H_{12}N_2$ (c) $C_7H_8N_2O$ (d) $C_5H_{12}O_4$

10. Molecular formula = $C_8H_{10}N_4O_2$
    Index of hydrogen deficiency = 6

11. Molecular formula = $C_{21}H_{30}O_2$
    Index of hydrogen deficiency = 7

CHAPTER 2

1. (a) Propargyl chloride (3-chloropropyne)   (b) *p*-Cymene (4-isopropyltoluene)
   (c) *o*-Cresol (2-methylphenol)   (d) *N*-Ethylaniline
   (e) 2-Chlorotoluene   (f) 2-Chloropropanoic acid
   (g) 1,2,3,4-Tetrahydronaphthalene   (h) 1,2-Epoxybutane

2. Citronellal

3. *trans*-Cinnamaldehyde (*trans*-3-phenyl-2-propenal)

4. Upper spectrum, *trans*-3-hexen-1-ol; Lower spectrum, *cis*-3-hexen-1-ol

5. (a) Structure B (ethyl cinnamate)   (b) Structure C (cyclobutanone)
   (c) Structure D (2-ethylaniline)   (d) Structure A (propiophenone)
   (e) Structure D (butanoic anhydride)

6. Poly(acrylonitrile-styrene); poly(methyl methacrylate); polyamide (nylon)

C H A P T E R   3

1. $C_{43}H_{50}N_4O_6$

2. $C_{34}H_{44}O_{13}$

3. $C_{12}H_{10}O$

4. C_6H_{12}

5. C_7H_9N

6. C_3H_7Cl

C H A P T E R   4

1. (a) Methylcyclohexane   (b) 2-Methyl-1-pentene   (c) 2-Methyl-2-hexanol
   (d) Ethyl isobutyl ether   (e) 2-Methylpropanal   (f) 3-Methyl-2-heptanone
   (g) Ethyl octanoate   (h) 2-Methylpropanoic acid   (i) 4-Methylbenzoic acid
   (j) Butylamine   (k) 2-Propanethiol   (l) Nitroethane
   (m) Propanenitrile   (n) Iodoethane   (o) Chlorobenzene
   (p) 1-Bromobutane   (q) Bromobenzene   (r) 1,1-Dichloroethane
   (s) 1,2,3-Trichloro-1-propene

C H A P T E R   5

1. (a) $-1, 0, +1$ (b) $-\frac{1}{2}, +\frac{1}{2}$ (c) $-\frac{5}{2}, -\frac{3}{2}, -\frac{1}{2}, +\frac{1}{2}, +\frac{3}{2}, +\frac{5}{2}$ (d) $-\frac{1}{2}, +\frac{1}{2}$

2. 128 Hz/60 MHz = 2.13 ppm

3. (a) 180 Hz   (b) 1.50 ppm

4. See Figures 5.22 and 5.23. The methyl protons are in a shielding region. Acetonitrile shows similar anisotropic behavior to acetylene.

5. *o*-Hydroxyacetophenone is intramolecularly hydrogen bonded. The proton is deshielded (12.05 ppm). Changing concentration does not alter the extent of hydrogen bonding. Phenol is intermolecularly hydrogen bonded. The extent of hydrogen bonding depends upon concentration.

6. The methyl groups are in a shielding region of the double bonds. See Figure 5.23.

7. The carbonyl group deshields the *ortho* protons owing to anisotropy.

8. The methyl groups are in the shielding region of the double-bonded system. See Figure 5.24.

9. The spectrum will be similar to that in Figure 5.25, with some differences in chemical shifts. Spin arrangements: H_A will be identical to the pattern in Figure 5.32 (triplet); H_B will see one adjacent proton and will appear as a doublet ($+\frac{1}{2}$ and $-\frac{1}{2}$).

10. The isopropyl group will appear as a septet for the α-H (methine). From Pascal's triangle, the intensities are 1:6:15:20:15:6:1. The CH_3 groups will be a doublet.

11. Downfield doublet, area = 2, for the protons on carbon 1 and carbon 3; upfield triplet, area = 1, for the proton on carbon 2.

12. $X-CH_2-CH_2-Y$, where $X \neq Y$.

13. Upfield triplet for the C-3 protons, area = 3; intermediate sextet for the C-2 protons, area = 2; and downfield triplet for the C-1 protons, area = 2

14. Ethyl acetate (ethyl ethanoate)

15. Isopropylbenzene

16. 2-Bromobutanoic acid

17. 1,3-Dibromopropane

18. 2,2-Dimethoxypropane

19. (a) Isobutyl propanoate   (b) *t*-Butyl propanoate   (c) Butyl propanoate

20. (a) 2-Phenylbutane (*sec*-butylbenzene)   (b) 1-Phenylbutane (butylbenzene)

21. 2-Phenylethylamine

CHAPTER 6

1. Methyl acetate

2. (e) 7 peaks   (f) 10 peaks
   (g) 4 peaks   (h) 4 peaks
   (i) 5 peaks   (j) 6 peaks
   (k) 8 peaks

3. (a) 2-Methyl-2-propanol   (b) 2-Butanol   (c) 2-Methyl-1-propanol

4. Methyl methacrylate (methyl 2-methyl-2-propenoate)

5. (a) 2-Bromo-2-methylpropane   (b) 2-Bromobutane   (c) 1-Bromobutane
   (d) 1-Bromo-2-methylpropane

6. (a) 4-Heptanone   (b) 2,4-Dimethyl-3-pentanone   (c) 4,4-Dimethyl-2-pentanone

18. 2,3-Dimethyl-2-butene. A primary cation rearranges to a tertiary cation via a hydride shift. E1 elimination forms the tetrasubstituted alkene.

19. (a) Three equal-sized peaks for ^{13}C coupling to a single D atom; quintet for ^{13}C coupling to two D atoms.

    (b) Fluoromethane: doublet for ^{13}C coupling to a single F atom ($^1J > 180$ Hz).
    Trifluoromethane: quartet for ^{13}C coupling to three F atoms ($^1J > 180$ Hz).

1,1-Difluoro-2-chloroethane: triplet for carbon-1 coupling to two F atoms ($^1J > 180$ Hz); triplet for carbon-2 coupling to two F atoms ($^2J \approx 40$ Hz).
1,1,1-trifluoro-2-chloroethane: quartet for carbon-1 coupling to three F atoms ($^1J > 180$ Hz); quartet for carbon-2 coupling to three F atoms ($^2J \approx 40$ Hz).

23. C1 = 128.5 + 9.3 = 137.8 ppm; C2 = 128.5 + 0.7 = 129.2 ppm; C3 = 128.5 − 0.1 = 128.4 ppm; C4 = 128.5 − 2.9 = 125.6 ppm.

24. All carbons are numbered according to IUPAC rules. The following information is given: the name of the compound, the number of the table used (A8.2–A8.7, Appendix 8), and, where needed, the name of the reference compound used (from A8.1, Appendix 8). If the actual values are known, they are given in parentheses.

(a) Methyl vinyl ether, A8.2 (actual: 153.2, 84.2 ppm)
C1 = 123.3 + 29.4 = 152.7   C2 = 123.3 − 38.9 = 84.4

(b) Cyclopentanol, A8.3-cyclopentane (actual: 73.3, 35.0, 23.4 ppm)
C1 = 25.6 + 41 = 66.6   C2 = 25.6 + 8 = 33.6   C3 = 25.6 − 5 = 20.6

(c) 2-Pentene, A8.5 (actual: 123.2, 132.7 ppm)
C2 = 123.3 + 10.6 − 7.9 − 1.8 = 124.2   C3 = 123.3 + 10.6 + 7.2 − 7.9 = 133.2
Using Table A8.4:
C2 = 123.3 + 12.9 − 9.7 = 126.5   C3 = 123.3 + 17.2 − 7.4 = 133.1

(d) *ortho*-Xylene, A8.7
C1,C2 = 128.5 + 9.3 + 0.7 = 138.5
C3,C6 = 128.5 + 0.7 − 0.1 = 129.1
C4,C5 = 128.5 − 0.1 − 2.9 = 125.5
meta-Xylene, A8.7 (actual: 137.6, 130.0, 126.2, 128.2 ppm)
C1,C3 = 128.5 + 9.3 − 0.1 = 137.7
C2 = 128.5 + 0.7 + 0.7 = 129.9
C4,C6 = 128.5 + 0.7 − 2.9 = 126.3
C5 = 128.5 − 0.1 − 0.1 = 128.3
para-Xylene, T7
C1,C4 = 128.5 + 9.3 − 2.9 = 134.9
C2,C3,C5,C6 = 128.5 + 0.7 − 0.1 = 129.1

(e) 3-Pentanol, A8.3-pentane (actual: 9.8, 29.7, 73.8 ppm)
C1,C5 = 13.9 − 5 = 8.9   C2,C4 = 22.8 + 8 = 30.8   C3 = 34.7 + 41 = 75.7

(f) 2-Methylbutanoic acid, A8.3-butane
C1 = 13.4 + 2 = 15.4   C2 = 25.2 + 16 = 41.2   C3 = 25.2 + 2 = 27.2   C4 = 13.4 − 2 + 11.4

(g) 1-Phenyl-1-propene, A8.4
C1 = 123.3 + 12.5 − 7.4 = 128.4   C2 = 123.3 + 12.9 − 11 = 125.2

(h) 2,2-Dimethylbutane, A8.3 or A8.2 (actual: 29.1, 30.6, 36.9, 8.9 ppm)
Using Table A8.3: C1 = 13.4 + 8 + 8 = 29.4   C2 = 25.2 + 6 + 6 = 37.2
C3 = 25.2 + 8 + 8 = 41.2   C4 = 13.4 − 2 − 2 = 9.4
Using Table A8.2:
C1 = −2.3 + [9.1(1) + 9.4(3) − 2.5(1)] + [(−3.4)] = 29.1
C2 = −2.3 + [9.1(4) + 9.4(1)] + [3(−1.5) + (−8.4)] = 30.6
C3 = −2.3 + [9.1(2) + 9.4(3)] + [(0) + (−7.5)] = 36.6
C4 = −2.3 + [9.1(1) + 9.4(1) − 2.5(3)] + [(0)] = 8.7

(i) 2,3-Dimethyl-2-pentenoic acid, A8.6
C2 = 123.3 + 4 + 10.6 − 7.9 − 7.9 − 1.8 = 120.3
C3 = 123.3 + 10.6 + 10.6 + 7.2 + 9 − 7.9 = 152.8

(j) 4-Octene, A8.5, and assume *trans* geometry
   C4,C5 = 123.3 + [10.6 + 7.2 − 1.5] − [7.9 + 1.8 − 1.5] = 131.4
   To estimate *cis*, correct as follows: 131.4 − 1.1 = 130.3

(k) 4-Aminobenzoic acid, A8.7
   C1 = 128.5 + 2.1 − 10.0 = 120.6   C2 = 128.5 + 1.6 + 0.8 = 130.9
   C3 = 128.5 − 13.4 + 0.1 = 115.2   C4 = 128.5 + 18.2 + 5.2 = 151.9

(l) 1-Pentyne, A8.3-propane
   C3 = 15.8 + 4.5 = 20.3   C4 = 16.3 + 5.4 = 21.7   C5 = 15.8 − 3.5 = 12.3

(m) Methyl 2-methylpropanoate, A8.3-propane
   C2 = 16.3 + 17 = 33.3   C3 = 15.8 + 2 = 17.8

(n) 2-Pentanone, A8.3-propane
   C3 = 15.8 + 30 = 45.8   C4 = 16.3 + 1 = 17.3   C5 = 15.8 − 2 = 13.8

(o) Bromocyclohexane, A8.3-cyclohexane
   C1 = 26.9 + 25 = 51.9   C2 = 26.9 + 10 = 36.9   C3 = 26.9 − 3 = 23.9
   C4 = 26.9 (no correction)

(p) 2-Methylpropanoic acid, A8.3-propane
   C1 = 15.8 + 2 = 17.8   C2 = 16.3 + 16 = 32.3

(q) 4-Nitroaniline, A8.7 (actual: 155.1, 112.8, 126.3, 136.9 ppm)
   C1 = 128.5 + 18.2 + 6.0 = 152.7   C2 = 128.5 − 13.4 + 0.9 = 116.0
   C3 = 128.5 + 0.8 − 4.9 = 124.4   C4 = 128.5 + 19.6 − 10.0 = 138.1
   2-Nitroaniline, A8.7
   C1 = 128.5 + 18.2 − 4.9 = 141.8   C2 = 128.5 − 13.4 + 19.6 = 134.7
   C3 = 128.5 − 4.9 + 0.8 = 124.4   C4 = 128.5 + 0.9 − 10.0 = 119.4
   C5 = 128.5 + 0.8 + 6 = 135.3   C6 = 128.5 − 13.4 + 0.9 = 114.2

(r) 1,3-Pentadiene, A8.4
   C3 = 123.3 + 13.6 − 13.6 = 129.5   C4 = 123.3 + 12.9 − 7 = 129.2

(s) Cyclohexene, A8.5 (actual: 127.3 ppm)
   C1,C2 = 123.3 + [10.6 + 7.2 − 1.5] − [7.9 + 1.8 − 1.5] + [−1.1] = 130.3

(t) 4-Methyl-2-pentene, A8.5, and assume *trans*
   C2 = 123.3 + [10.6(1)] − [7.9(1) + 1.8(2)] = 122.4
   C3 = 123.3 + [10.6(1) + 7.2(2)] − [7.9(1)] + 2.3 = 142.7

C H A P T E R   7

1. Refer to Sections 7.6 and 7.9 for instructions on measuring coupling constants using the Hertz values that are printed above the expansions of the proton spectra.

(a) Vinyl acetate (Fig. 7.50): all vinyl protons are doublets of doublets.

   H_a = 4.57 ppm, $^3J_{ac}$ = 6.25 Hz and $^2J_{ab}$ = 1.47 Hz.

   H_b = 4.88 ppm. The coupling constants are not consistent; $^3J_{bc}$ =13.98 or 14.34 Hz from the spacing of the peaks. $^2J_{ab}$ = 1.48 or 1.84 Hz. It is often the case that the coupling constants are not consistent (see Section 7.9). More consistent coupling constants can be obtained from analysis of proton H_c.

   H_c = 7.27 ppm, $^3J_{bc}$ = 13.97 Hz and $^3J_{ac}$ = 6.25 Hz from the spacing of the peaks.

Summary of coupling constants from the analysis of the spectrum: $^3J_{ac} = 6.25$ Hz, $^3J_{bc} = 13.97$ Hz and $^2J_{ab} = 1.47$ Hz. They can be rounded off to: 6.3, 14.0 and 1.5 Hz, respectively.

(b) *trans*-Crotonic acid (Fig. 7.53).

$H_a = 1.92$ ppm (methyl group at C-4). It appears as a doublet of doublets (dd) because it shows both ^3J and ^4J couplings; $^3J_{ac} = 6.9$ Hz and $^4J_{ab}$ allylic $= 1.6$ Hz.

$H_b = 5.86$ ppm (vinyl proton at C-2). It appears as a doublet of quartets (dq); $^3J_{bc}$ trans $= 15.6$ Hz and $^4J_{ab}$ allylic $= 1.6$ Hz.

$H_c = 7.10$ ppm (vinyl proton at C-3). It appears as a doublet of quartets (dq), with some partial overlap of the quartets; $^3J_{bc}$ trans $= 15.6$ Hz and $^3J_{ac} = 6.9$ Hz. Notice that H_c is shifted further downfield than H_b because of the resonance effect of the carboxyl group and also a through-space deshielding by the oxygen atom in the carbonyl group.

$H_d = 12.2$ ppm (singlet, acid proton on carboxyl group).

(c) 2-Nitrophenol (Fig. 7.69). H_a and H_b are shielded by the electron releasing effect of the hydroxyl group caused by the non-bonded electrons on the oxygen atom being involved in resonance. They can be differentiated by their appearance: H_a is a triplet with some fine structure and H_b is a doublet with fine structure. H_d is deshielded by the electron withdrawing effect and by the anisotropy of the nitro group. Notice that the pattern is a doublet with some fine structure. H_c is assigned by a process of elimination. It lacks any of the above effects that shields or deshields that proton. It appears as a triplet with some fine structure.

$H_a = 7.00$ ppm (ddd); $^3J_{ac} \cong {}^3J_{ad} = 8.5$ Hz and $^4J_{ab} = 1.5$ Hz. H_a could also be described as a triplet of doublets (td) since $^3J_{ac}$ and $^3J_{ad}$ are nearly equal.

$H_b = 7.16$ ppm (dd); $^3J_{bc} = 8.5$ Hz and $^4J_{ab} = 1.5$ Hz.

$H_c = 7.60$ ppm (ddd or td); $^3J_{ac} \cong {}^3J_{bc} = 8.5$ Hz and $^4J_{cd} = 1.5$ Hz.

$H_d = 8.12$ ppm (dd); $^3J_{ad} = 8.5$ Hz and $^4J_{cd} = 1.5$ Hz; $^5J_{bd} = 0$.

The OH group is not shown in the spectrum.

(d) 3-Nitrobenzoic acid (Fig. 7.70). H_d is significantly deshielded by the anisotropy of both the nitro and carboxyl groups and appears furthest downfield. It appears as a narrowly spaced triplet. This proton only shows ^4J couplings. H_b is *ortho* to a carboxyl group while H_c is *ortho* to a nitro group. Both protons are deshielded, but the nitro group shifts a proton further downfield than for a proton next to a carboxyl group (see Appendix 6). Both H_b and H_c are doublets with fine structure consistent with their positions on the aromatic ring. H_a is relatively shielded and appears upfield as a widely spaced triplet. This proton does not experience any anisotropy effect because of its distance away from the attached groups. H_a has only ^3J couplings ($^5J_{ad} = 0$).

$H_a = 7.72$ ppm (dd); $^3J_{ac} = 8.1$ Hz and $^3J_{ab} = 7.7$ Hz (these values come from analysis of H_b and H_c, below). Since the coupling constants are similar, the pattern appears as an accidental triplet.

$H_b = 8.45$ ppm (ddd or dt); $^3J_{ab} = 7.7$ Hz; $^4J_{bd} \cong {}^4J_{bc} = 1.5$ Hz. The pattern is an accidental doublet of triplets.

$H_c = 8.50$ ppm (ddd); $^3J_{ac} = 8.1$ Hz and $^4J_{cd} \neq {}^4J_{bc}$.

$H_d = 8.96$ ppm (dd). The pattern appears to be a narrowly spaced triplet, but is actually an accidental triplet because $^4J_{bd} \neq {}^4J_{cd}$.

The carboxyl proton is not shown in the spectrum.

(e) Furfuryl alcohol (Fig. 7.71). The chemical shift values and coupling constants for a furanoid ring are given in Appendix 4 and 5.

$H_a = 6.24$ ppm (doublet of quartets); $^3J_{ab} = 3.2$ Hz and $^4J_{ac} = 0.9$ Hz. The quartet pattern results from a nearly equal ^4J coupling of H_a to the two methylene protons in the CH_2OH group and the ^4J coupling of H_a to H_c (n + 1 rule, three protons plus one equals four, a quartet).

$H_b = 6.31$ ppm (dd); $^3J_{ab} = 3.2$ Hz and $^3J_{bc} = 1.9$ Hz.

$H_c = 7.36$ ppm (dd); $^3J_{bc} = 1.9$ Hz and $^4J_{ac} = 0.9$ Hz.

The CH_2 and OH groups are not shown in the spectrum.

(f) 2-Methylpyridine (Fig. 7.72). Typical chemical shift values and coupling constants for a pyridine ring are given in Appendix 4 and 5.

$H_a = 7.08$ ppm (dd); $^3J_{ac} = 7.4$ Hz and $^3J_{ad} = 4.8$ Hz.

$H_b = 7.14$ ppm (d); $^3J_{bc} = 7.7$ Hz and $^4J_{ab} \cong 0$ Hz.

$H_c = 7.56$ ppm (ddd or td). This pattern is a likely accidental triplet of doublets because $^3J_{ac} \cong {}^3J_{bc}$ and $^4J_{cd} = 1.8$ Hz.

$H_d = 8.49$ ppm ("doublet"). Because of the broadened peaks in this pattern, it is impossible to extract the coupling constants. We expect a doublet of doublets, but $^4J_{cd}$ is not resolved from $^3J_{ad}$. The adjacent nitrogen atom may be responsible for the broadened peaks.

2. (a) $J_{ab} = 0$ Hz   (b) $J_{ab} \sim 10$ Hz   (c) $J_{ab} = 0$ Hz   (d) $J_{ab} \sim 1$ Hz
   (e) $J_{ab} = 0$ Hz   (f) $J_{ab} \sim 10$ Hz   (g) $J_{ab} = 0$ Hz   (h) $J_{ab} = 0$ Hz
   (i) $J_{ab} \sim 10$ Hz; $J_{ac} \sim 16$ Hz; $J_{bc} \sim 1$ Hz

3.

$H_a = 2.80$ ppm (singlet, CH_3).

$H_b = 5.98$ ppm (doublet); $^3J_{bd} = 9.9$ Hz and $^2J_{bc} = 0$ Hz.

$H_c = 6.23$ ppm (doublet); $^3J_{cd} = 16.6$ Hz and $^2J_{bc} = 0$ Hz.

$H_d = 6.61$ ppm (doublet of doublets); $^3J_{cd} = 16.6$ Hz and $^3J_{bd} = 9.9$ Hz.

4.

$H_a = 0.88$ ppm (triplet, CH_3); $^3J_{ac} = 7.4$ Hz.

$H_c = 2.36$ ppm (quartet, CH_2; $^3J_{ac} = 7.4$ Hz.

$H_b = 1.70$ ppm (doublet of doublets, CH_3); $^3J_{be} = 6.8$ Hz and $^4J_{bd} = 1.6$ Hz.

$H_d = 5.92$ ppm (doublet of quartets, vinyl proton). The quartets are narrowly spaced, suggesting a four bond coupling, ^4J; $^3J_{de} = 15.7$ Hz and $^4J_{bd} = 1.6$ Hz.

$H_e = 6.66$ ppm (doublet of quartets, vinyl proton). The quartets are widely spaced, suggesting a three bond coupling, ^3J; $^3J_{de} = 15.7$ Hz and $^3J_{be} = 6.8$ Hz. H_e appears further downfield than H_d (see the answer to problem 1b for an explanation).

5.

$H_a = 0.96$ ppm (triplet, CH_3); $^3J_{ab} = 7.4$ Hz.

$H_d = 6.78$ ppm (doublet of triplets, vinyl proton). The triplets are widely spaced suggesting a three bond coupling, ^3J; $^3J_{cd} = 15.4$ Hz and $^3J_{bd} = 6.3$ Hz. H_d appears further downfield than H_c (see the answer to problem 1b for an explanation).

$H_b = 2.21$ ppm (quartet of doublets of doublets, CH_2) resembles a quintet with fine structure. $^3J_{ab} = 7.4$ Hz and $^3J_{bd} = 6.3$ Hz are derived from the H_a and H_d patterns while $^4J_{bc} = 1.5$ Hz is obtained from the H_b pattern (left hand doublet at 2.26 ppm) or from the H_c pattern.

$H_c = 5.95$ ppm (doublet of doublets of triplets, vinyl proton). The triplets are narrowly spaced, suggesting a four bond coupling, ^4J; $^3J_{cd} = 15.4$ Hz, $^3J_{ce} = 7.7$ Hz and $^4J_{bc} = 1.5$ Hz.

$H_e = 9.35$ ppm (doublet, aldehyde proton); $^3J_{ce} = 7.7$ Hz.

6. Structure **A** would show allylic coupling. The C—H bond orbital is parallel to the π system of the double bond leading to more overlap. A stronger coupling of the two protons results.

13. 3-Bromoacetophenone. The aromatic region of the proton spectrum shows one singlet, two doublets, and one triplet consistent with a 1,3-disubstituted (*meta*) pattern. Each carbon atom in the aromatic ring is unique leading to the observed six peaks in the carbon spectrum. The downfield peak at near 197 ppm is consistent with a ketone C=O. The integral value (3H) in the proton spectrum and the chemical shift value (2.6 ppm) indicates that a methyl group is present. The most likely possibility is that there is an acetyl group attached to the aromatic ring. A bromine atom is the other substituent on the ring.

14. Valeraldehyde (pentanal). The aldehyde peak on carbon 1 appears at 9.8 ppm. It is split into a triplet by the two methylene protons on carbon 2 ($^3J = 1.9$ Hz). Aldehyde protons often have smaller three-bond (vicinal) coupling constants than typically found. The pattern at 2.4 ppm

(triplet of doublets) is formed from coupling with the two protons on carbon 3 ($^3J = 7.4$ Hz) and with the single aldehyde proton on carbon 1 ($^3J = 1.9$ Hz).

15. The DEPT spectral results indicate that the peak at 15 ppm is a CH_3 group; 40 and 63 ppm peaks are CH_2 groups; 115 and 130 ppm peaks are CH groups; 125 and 158 ppm peaks are quaternary (*ipsi* carbons). The 179 ppm peak in the carbon spectrum is a C=O group at a value typical for esters and carboxylic acids. A carboxylic acid is indicated because a broad peak appears at 12.5 ppm in the proton spectrum. The value for the chemical shift of the methylene carbon peak at 63 ppm indicates an attached oxygen atom. Confirmation of this is seen in the proton spectrum (4 ppm, a quartet), leading to the conclusion that the compound has an ethoxy group (triplet at 1.4 ppm for the CH_3 group). A *para* disubstituted aromatic ring is indicated with the carbon spectrum (two C—H and two C with no protons). This substitution pattern is also indicated in the proton spectrum (two doublets at 6.8 and 7.2 ppm). The remaining methylene group at 40 ppm in the carbon spectrum is a singlet in the proton spectrum indicating no adjacent protons. The compound is 4-ethoxyphenylacetic acid.

24. (a) In the *proton* NMR, one fluorine atom splits the CH_2 ($^2J_{HF}$) into a doublet. This doublet is shifted downfield because of the influence of the electronegative fluorine atom. The CH_3 group is too far away from the fluorine atom and thus appears upfield as a singlet.

    (b) Now the operating frequency of the NMR is changed so that only *fluorine* atoms are observed. The fluorine NMR would show a triplet for the single fluorine atom because of the two adjacent protons ($n + 1$ Rule). This would be the only pattern observed in the spectrum. Thus, we do not see protons directly in a fluorine spectrum because the spectrometer is operating at a different frequency. We do see, however, the *influence* of the protons on the fluorine spectrum. The J values would be the same as those obtained from the *proton* NMR.

25. The aromatic proton spectral data indicates a 1,3-disubstituted (*meta* substituted) ring. One attached substituent is a methyl group (2.35 ppm, integrating for 3H). Since the ring is disubstituted, the remaining substituent would be an oxygen atom attached to the remaining two carbon atoms with one proton and four fluorine atoms in the "ethoxy" group. This substituent would most likely be a 1,1,2,2-tetrafluoroethoxy group. The most interesting pattern is the widely spaced triplet of triplets centering on 5.85 ppm; $^2J_{HF} = 53.1$ Hz for the proton on carbon 2 of the ethoxy group coupled to two adjacent fluorine atoms (two bond, ^2J) and $^3J_{HF} = 2.9$ Hz for this same proton on carbon 2 coupled to the remaining two fluorine atoms on carbon 1 (three bond, ^3J) from this proton. The compound is 1-methyl-3-(1,1,2,2-tetrafluoroethoxy) benzene.

27. In the *proton* NMR, the attached deuterium, which has a spin = 1, splits the methylene protons into a triplet (equal intensity for each peak, a 1 : 1 : 1 pattern). The methyl group is too far removed from deuterium to have any influence, and it will be a singlet. Now change the frequency of the NMR to a value where only *deuterium* undergoes resonance. Deuterium will see two adjacent protons on the methylene group, splitting it into a triplet (1 : 2 : 1 pattern). No other peaks will be observed because at this NMR frequency, the only atom observed is deuterium. Compare the results to the answers in Problem 24.

28. Two singlets will appear in the proton NMR spectrum: a downfield CH_2 and an upfield CH_3 group. Compare this result to the answer in problem 24a.

31. (a) δ_H ppm = 0.23 + 1.70 = 1.93 ppm

    (b) δ_H ppm (α to two C=O groups) = 0.23 + 1.70 + 1.55 = 3.48 ppm
        δ_H ppm (α to one C=O group) = 0.23 + 1.70 + 0.47 = 2.40 ppm

(c) δ_H ppm = 0.23 + 2.53 + 1.55 = 4.31 ppm

(d) δ_H ppm = 0.23 + 1.44 + 0.47 = 2.14 ppm

(e) δ_H ppm = 0.23 + 2.53 + 2.53 + 0.47 = 5.76 ppm

(f) δ_H ppm = 0.23 + 2.56 + 1.32 = 4.11 ppm

32. (a) δ_H ppm (*cis* to $COOCH_3$) = 5.25 + 1.15 − 0.29 = 6.11 ppm
     δ_H ppm (*trans* to $COOCH_3$) = 5.25 + 0.56 − 0.26 = 5.55 ppm

   (b) δ_H ppm (*cis* to CH_3) = 5.25 + 0.84 − 0.26 = 5.83 ppm
     δ_H ppm (*cis* to $COOCH_3$) = 5.25 + 1.15 + 0.44 = 6.84 ppm

   (c) δ_H ppm (*cis* to C_6H_5) = 5.25 + 0.37 = 5.62 ppm
     δ_H ppm (*gem* to C_6H_5) = 5.25 + 1.35 = 6.60 ppm
     δ_H ppm (*trans* to C_6H_5) = 5.25 − 0.10 = 5.15 ppm

   (d) δ_H ppm (*cis* to C_6H_5) = 5.25 + 0.37 + 1.10 = 6.72 ppm
     δ_H ppm (*cis* to $COCH_3$) = 5.25 + 1.13 + 1.35 = 7.73 ppm

   (e) δ_H ppm (*cis* to CH_3) = 5.25 + 0.67 − 0.26 = 5.66 ppm
     δ_H ppm (*cis* to CH_2OH) = 5.25 − 0.02 + 0.44 = 5.67 ppm

   (f) δ_H ppm = 5.25 + 1.10 − 0.26 − 0.29 = 5.80 ppm

CHAPTER 8

1. The methylene group is a quartet of doublets. Draw a tree diagram where the quartet has spacings of 7 Hz. This represents the ^3J (three bond coupling) to the CH_3 group from the methylene protons. Now split each leg of the quartet into doublets (5 Hz). This represents the ^3J (three bond coupling) of the methylene protons to the O—H group. The pattern can also be interpreted as a doublet of quartets, where the doublet (5 Hz) is constructed first, followed by splitting each leg of the doublet into quartets (7 Hz spacings).

2. 2-Methyl-3-buten-2-ol. H_a = 1.3 ppm; H_b = 1.9 ppm; H_c = 5.0 ppm (doublet of doublets, $^3J_{ce}$ = 10.7 Hz (*cis*) and $^2J_{cd}$ = 0.9 Hz (*geminal*)); H_d = 5.2 ppm (doublet of doublets, $^3J_{de}$ = 17.4 Hz (*trans*) and $^2J_{cd}$ = 0.9 Hz (*geminal*)); H_e = 6.0 ppm (doublet of doublets, $^3J_{de}$ = 17.4 Hz and $^3J_{ce}$ = 10.7 Hz.

3. 2-Bromophenol. The unexpanded spectrum shows two doublets and two triplets, consistent with a 1,2-disubstituted (*ortho*) pattern. Each shows fine structure in the expansions (^4J). Assignments can be made by assuming that the two upfield protons (shielded) are *ortho* and *para* with respect to the electron-releasing OH group. The other two patterns can be assigned by a process of elimination.

4. The two structures shown here are the ones that can be derived from 2-methylphenol. The infrared spectrum shows a significantly shifted conjugated carbonyl group which suggests that the OH group is releasing electrons and providing single bond character to the C=O group,

consistent with 4-hydroxy-3-methylacetophenone (the other compound would not have as significant a shift in the C=O). The 3136 cm^{-1} peak is an OH group, also seen in the NMR spectrum as a solvent-dependent peak. Both structures shown would be expected to show a singlet and two doublets in the aromatic region of the NMR spectrum. The positions of the downfield singlet and doublet in the spectrum fit the calculated values from Appendix 6 for 4-hydroxy-3-methylacetophenone more closely than for 3-hydroxy-4-methylacetophenone (calculated values are shown on each structure). The other doublet appearing at 6.9 ppm is a reasonable fit to the calculated value of 6.79 ppm. It is interesting to note that the two *ortho* protons in 3-hydroxy-4-methylacetophenone are deshielded by the C=O group and shielded by the OH group leading to little shift from the base value of 7.27 (Appendix 6). In conclusion, the NMR spectrum and calculated values best fit 4-hydroxy-3-methylacetophenone.

4-hydroxy-3-methylacetophenone   3-hydroxy-4-methylacetophenone

5. All of the compounds would have a singlet and two doublets in the aromatic portion of the NMR spectrum. When comparing the calculated values to the observed chemical shifts, it is important to compare the *relative* positions of each proton (positions of doublet, singlet, and doublet). Don't be concerned with slight differences (about ± 0.10 Hz) in the calculated *vs* observed values. The calculated values for third compound fits the observed spectral data better than the first two.

			observed
6.51 s	6.59 d	6.48 d	6.57 d
6.62 d	6.84 d	6.51 s	6.64 s
6.95 d	6.87 s	6.95 d	6.97 d

6. 3-Methyl-3-buten-1-ol. The DEPT spectral results show a CH$_3$ group at 22 ppm, two CH$_2$ groups at 41 and 60 ppm. The peaks at 112 ppm (CH$_2$) and 142 ppm (C with no attached H) are part of a vinyl group. The peaks at 4.78 and 4.86 ppm in the proton spectrum are the protons on the terminal double bond. The 4.78 ppm pattern (fine structure) shows long range coupling (^4J) to the methyl and methylene groups. The methylene group at 2.29 ppm is broadened because of non-resolved ^4J coupling.

9. 4-Butylaniline

10. 2,6-Dibromoaniline

12. 2,4-Dichloroaniline. The broad peak at about 4 ppm is assigned to the −NH$_2$ group. The doublet at 7.23 ppm is assigned to the proton on carbon 3 (it appears as a near singlet in the

upper trace). Proton 3 is coupled, long range, to the proton on carbon 5 ($^4J = 2.3$ Hz). The doublet of doublets centering on 7.02 ppm is assigned to the proton on carbon 5. It is coupled to the proton on carbon 6 ($^3J = 8.6$ Hz) and also to proton 3 ($^4J = 2.3$ Hz). Finally, the doublet at 6.65 ppm is assigned to the proton on carbon 6 ($^3J = 8.6$ Hz), which arises from coupling to the proton on carbon 5. There is no sign of ^5J coupling in this compound.

13. Alanine

21. Rapid equilibration at room temperature between chair conformations leads to one peak. As one lowers the temperature, the interconversion is slowed down until, at temperatures below −66.7°C, peaks due to the axial and equatorial hydrogens are observed. Axial and equatorial hydrogens have different chemical shifts under these conditions.

22. The *t*-butyl-substituted rings are conformationally locked. The hydrogen at C4 has different chemical shifts, depending upon whether it is axial or equatorial. 4-Bromocyclohexanes are conformationally mobile. No difference between axial and equatorial hydrogens is observed until the rate of chair–chair interconversion is decreased by lowering the temperature.

CHAPTER 9

1.

$$\overset{1}{CH_3}-\overset{2}{CH}-\overset{3}{CH_2}-\overset{4}{CH_3}$$
$$\overset{|}{Cl}$$

Proton #1: 1.5 ppm   Carbon #1: 24 ppm
Proton #2: 4.0 ppm   Carbon #2: 60 ppm
Proton #3: 1.7 ppm   Carbon #3: 33 ppm (inverted peak indicates CH_2)
Proton #4: 1.0 ppm   Carbon #4: 11 ppm

2.

$$\overset{6}{}CH_3$$
$$CH_3-CH_2-CH_2-\overset{|}{CH}-CH_2-OH$$
$$54321$$

Carbon #1: 68 ppm
Carbon #2: 35.2 ppm
Carbon #3: 35.3 ppm
Carbon #4: 20 ppm
Carbon #5: 14 ppm
Carbon #6: 16 ppm

3-Methyl-1-pentanol and 4-methyl-1-pentanol would be expected to give similar DEPT spectra. They are also acceptable answers based on the information provided.

3.

Citronellol

The DEPT spectrum shown in Figure 9.12 yields the number of protons on each of the ^{13}C peaks. There are three methyl groups, four methylene groups, and two methine groups. One of methine peaks appears downfield at about 124 ppm and is assigned to C6. The other methine peak is assigned to C3. The remaining carbon atom is in the double bond, and must be C7. It does not appear in the DEPT spectrum because there is no attached proton.

The COSY spectra are shown in Figures 9.15a and 9.15b. The assignments were determined as described on pages 531 to 534, and are shown on the top axes in these figures. Notice that C2 overlaps with C3 and one of the methyl groups, either C8 or C9. C2 and C4 also overlap. There are three sets of diastereotopic protons on three methylene groups: C2, C4 and C5.

The peaks are identified on the top axes in the COSY spectra. The HETCOR (HSQC) spectrum shown in this problem allows us to easily determine all of the ^1H and ^{13}C chemical shift values. The doublet for the methyl group on C10 appears at 9.1 ppm. The protons on C2 are diastereotopic and appear at 1.37 and 1.60 ppm. They correlate to only one ^{13}C peak at 40 ppm. The protons on C4 are also diastereotopic and appear at 1.17 and 1.35 ppm, and they correlate to the single ^{13}C peak at 37 ppm. Finally, the remaining set of diastereotopic protons are best observed in the COSY spectrum (Figure 9.15b) at 1.96 and 2.02 ppm for C5. In the expanded HETCOR spectrum in Problem 3, the oblong peak centering on about 2.0 ppm shows that the two protons on C5 nearly overlap each other at 25.1 ppm on the ^{13}C scale. The assignments are tabulated below.

Proton C1	3.7 ppm	Carbon C1	61 ppm
Proton C2	1.37 and 1.60 ppm	Carbon C2	40 ppm
Proton C3	1.57 ppm	Carbon C3	28 ppm
Proton C4	1.17 and 1.35 ppm	Carbon C4	37 ppm
Proton C5	1.96 and 2.02 ppm	Carbon C5	25.1 ppm
Proton C6	5.10 ppm	Carbon C6	125 ppm
Proton C7	—	Carbon C7	131 ppm
Proton C8 or C9	1.60 ppm	Carbon 8 or 9	18.0 ppm
(unassigned)	1.68 ppm	(unassigned)	25.2 ppm
Proton 10	0.91 ppm	Carbon 10	19.5 ppm

Some of the chemical shifts for the ^{13}C atoms differ slightly from those determined on page 520–522. They agree within experimental error.

4.

Geraniol

You should inspect the values assigned to citronellol in the answers to Problem 3, along with the spectral data in Problem 4 for geraniol. The stacked full-edited DEPT on page 555 shows three methyl groups, three methylene groups, and two methine groups. C3 and C7 do not show in this spectrum.

Begin by drawing a diagonal line from the upper right to the lower left on the COSY spectrum shown on page 556. Start the analysis at the lower left end of the diagonal line where the two double bonded carbon atoms each with one attached proton are located (5.40 and 5.20 ppm). Draw a horizontal line from the lower spot at 5.40 ppm (C2) showing that this proton correlates with the two methylene protons on C1 at 4.20 ppm (a doublet on the upper scale). The proton at 5.40 also correlates by long range coupling (^4J) with the protons on the methyl group at 1.67 ppm on C10.

Now draw another horizontal line from the other proton at 5.20 ppm (C6) on the diagonal line that connects with spots at 2.10 ppm (C5) and the two methyl groups at 1.60 and 1.70 ppm (C8 and C9) by long range coupling (^4J). The HETCOR spectrum shown on page 557 provides the ^{13}C chemical shifts for these two methyl groups C8 and C9 (18.0 and 26.0 ppm). These two methyl groups have nearly values to those found in citronellol (18.0 and 25.2). The third methyl group, C10 appearing at 1.67 ppm shows a ^{13}C chemical shift value of 16.0 ppm.

Another horizontal line can be drawn from the spot at 4.20 ppm on the diagonal line for the methylene group on C1 showing that it correlates with the C–H on the double bond at 5.40 ppm (C2). They also correlate with the methyl group on C10 by long range coupling (^4J).

Unfortunately, it is more difficult to use the COSY spectrum for the protons on the two methylene groups located at 2.10 and 2.05 ppm for the protons on C5 and C4. These protons have nearly the same chemical shifts values. The HETCOR spectrum shown on page 557 helps make the assignments. C5 has a ^{13}C chemical shift of 27 ppm, which is close to the value of 25.1 ppm in citronellol for C5 (see the answer to Problem 3). C4 has a much larger value of 39 ppm which is close to the value in citronellol for C4 (37 ppm).

Interestingly, the OH group appears on the diagonal line at about 2.4 ppm. Notice that the group does not correlate with anything! The assignments for geraniol are tabulated below.

Proton C1	4.20 ppm	Carbon C5	59 ppm
Proton C2	5.40 ppm	Carbon C2	124 ppm
Proton C3	—	Carbon C3	132 or 139 ppm

Proton C4	2.05 ppm	Carbon C4	39 ppm
Proton C5	2.10 ppm	Carbon C5	27 ppm
Proton C6	5.20 ppm	Carbon C6	125 ppm
Proton C7	—	Carbon C7	132 or 139 ppm
Proton C8 or C9	1.60 ppm	Carbon 8 or 9	18.0 ppm
(unassigned)	1.70 ppm		26.0 ppm
Proton 10	1.67 ppm	Carbon 10	16.0 ppm

5.

Proton #3: 6.95 ppm   Carbon #3: 117 ppm
Proton #4: 7.40 ppm   Carbon #4: 136 ppm
Proton #5: 6.82 ppm   Carbon #5: 119 ppm
Proton #6: 7.75 ppm   Carbon #6: 130 ppm
$J_{3,4} = 8$ Hz $J_{3,5} = 1$ Hz $J_{3,6} \sim 0$ Hz
$J_{4,5} = 7$ Hz $J_{4,6} = 2$ Hz $J_{5,6} = 8$ Hz

C H A P T E R   1 0

1. (a) $\varepsilon = 13,000$ (b) $I_0/I = 1.26$

2. (a) 2,4-Dichlorobenzoic acid or 3,4-dichlorobenzoic acid   (b) 4,5-Dimethyl-4-hexen-3-one
   (c) 2-Methyl-1-cyclohexenecarboxaldehyde

3. (a) Calculated: 215 nm   observed: 213 nm
   (b) Calculated: 249 nm   observed: 249 nm
   (c) Calculated: 214 nm   observed: 218 nm
   (d) Calculated: 356 nm   observed: 348 nm
   (e) Calculated: 244 nm   observed: 245 nm
   (f) Calculated: 303 nm   observed: 306 nm
   (g) Calculated: 249 nm   observed: 245 nm
   (h) Calculated: 281 nm   observed: 278 nm
   (i) Calculated: 275 nm   observed: 274 nm
   (j) Calculated: 349 nm   observed: 348 nm

4. 166 nm: $n \rightarrow \sigma^*$
   189 nm: $\pi \rightarrow \pi^*$
   279 nm: $n \rightarrow \pi^*$

5. Each absorption is due to $n \rightarrow \sigma^*$ transitions. As one goes from the *chloro* to the *bromo* to the *iodo* group, the electronegativity of the halogens decreases. The orbitals interact to different degrees, and the energies of the n and the σ^* states differ.

6. (a) $\sigma \rightarrow \sigma^*$, $\sigma \rightarrow \pi^*$, $\pi \rightarrow \pi^*$, and $\pi \rightarrow \sigma^*$

   (b) $\sigma \rightarrow \sigma^*$, $\sigma \rightarrow \pi^*$, $\pi \rightarrow \pi^*$, $\pi \rightarrow \sigma^*$, $n \rightarrow \sigma^*$, and $n \rightarrow \pi^*$

   (c) $\sigma \rightarrow \sigma^*$ and $n \rightarrow \sigma^*$

   (d) $\sigma \rightarrow \sigma^*$, $\sigma \rightarrow \pi^*$, $\pi \rightarrow \pi^*$, $\pi \rightarrow \sigma^*$, $n \rightarrow \sigma^*$, and $n \rightarrow \pi^*$

   (e) $\sigma \rightarrow \sigma^*$ and $n \rightarrow \sigma^*$

   (f) $\sigma \rightarrow \sigma^*$

C H A P T E R   1 1

*1. The normal ^1H spectrum of the compound on page 625 clearly show two CH_3 groups integrating for 3H each at 2.7 ppm (OCH_3) and 3.9 ppm ($CH_3-C=O$). All of the remaining protons between 7.0 and 8.5 ppm are assigned to six aromatic proton peaks, integrating for 1H each. The methoxy group is an electron donating group (Chapter 7, Section 7.10). The non-bonded electrons on the oxygen atom in the methoxy group release electrons into the left hand ring to shield the protons H_a and H_b. The $C=O$ in the acetyl group in the right hand ring withdraws electrons and deshields H_e and H_f (Chapter 7, Section 7.10, and Figure 7.61). We can conclude that protons H_a and H_b will appear upfield relative to H_e and H_f.

The COSY spectrum on page 628 provide valuable information. Starting at the upper right side of the diagonal line, we see the following correlations:

H_a and H_b correlate with H_d.
H_c correlates with H_e.
H_d correlates with H_a and H_b
H_e correlates with H_c and H_f
H_f correlates with H_e

From these correlations, we now know that protons H_a, H_b and H_d are attached to the left hand ring, while H_c, H_e and H_f are attached to the right hand ring.

The expansions of H_f and H_e on page 626 yield the following values:

H_f 8.40 ppm, doublet, $^4J_{ef} = 1.83$ Hz, shows only long range coupling.
H_e 8.01 ppm, doublet of doublets, $^3J_{ce} = 8.77$ Hz and $^4H_{ef} = 1.83$ Hz.

The expansions on page 627 yield the following values:

H_c 7.77 ppm, doublet, $^3J_{ce}$ = 8.77 Hz.
H_d 7.86 ppm, doublet, $^3J_{bd}$ = 8.77 Hz.
H_b 7.21 ppm, doublet of doublets, $^3J_{bd}$ = 8.77 Hz and $^4J_{ab}$ ≈ 2.5 Hz.
H_a 7.16 ppm, doublet, $^4J_{ab}$ ≈ 2.5 Hz, shows only long range coupling.

The DEPT 135 spectrum on page 628 shows the six aromatic C—H and the aliphatic CH_3 groups. There are no CH_2 groups. Since the HETCOR spectrum was not included in the problem, the best we can do is to make as many assignments as possible. The normal carbon spectrum provides some chemical shifts for the unprotonated carbon atoms.

C11, CH_3, 26.4 ppm
C12, C=O, 197.8 ppm
C13, CH_3, 55.3 ppm
C1, C3, C4, C5, C7 and C8 are the unassigned C—H carbon atoms: 105.6, 119.6, 124.5, 127.0, 129.9, and 131.0.
C2, C6, C9, and C10 are the unassigned unprotonated carbon atoms: 127.7, 132.5, 137.2, and 159.6.

3. 2-Butanone

4. 1-Propanol

5. 3-Pentanone

6. Methyl trimethylacetate (methyl 2,2-dimethylpropanoate)

7. Phenylacetic acid

8. 4-Bromophenol

9. Valerophenone (1-phenyl-1-pentanone)

10. Ethyl 3-bromobenzoate; ethyl 4-bromobenzoate

11. *N,N*-dimethylethylamine

12. 2-Pentanone

13. Ethyl formate

14. 2-Bromoacetophenone; 4-bromoacetophenone

15. Butyraldehyde (butanal)

16. 3-Methyl-1-butanol

17. Ethyl 2-bromopropionate (ethyl 2-bromopropanoate)
    ethyl 3-bromopropionate (ethyl 3-bromopropanoate)

18. Ethyl 4-cyanobenzoate

19. 3-Chloropropiophenone (3-chloro-1-phenyl-1-propanone)

APPENDICES

A P P E N D I X   1

Infrared Absorption Frequencies of Functional Groups

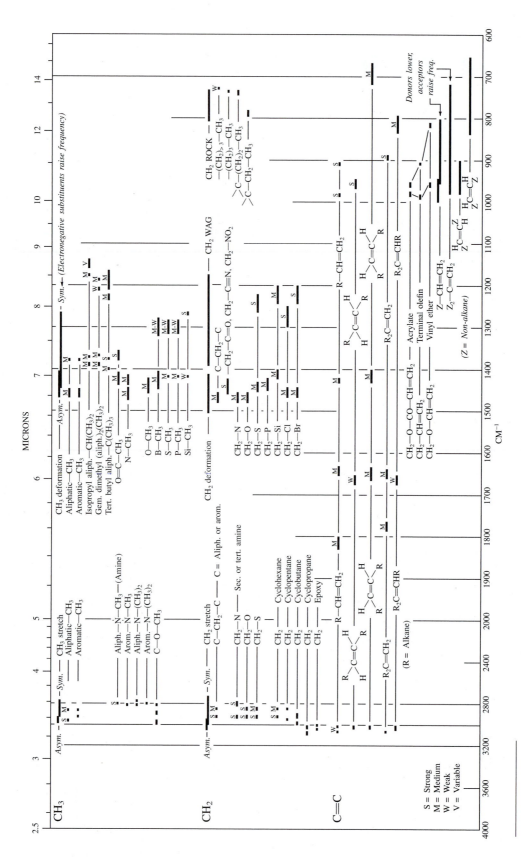

Colthup spectra–structure correlation charts for infrared frequencies in the 4000–600 cm⁻¹ region from Lin-Vien, D., N. B. Colthup, W. G. Fateley, and J. G. Grasselli, *The Handbook of Infrared and Raman Characteristic Frequencies of Organic Molecules*, Academic Press, New York, 1991.

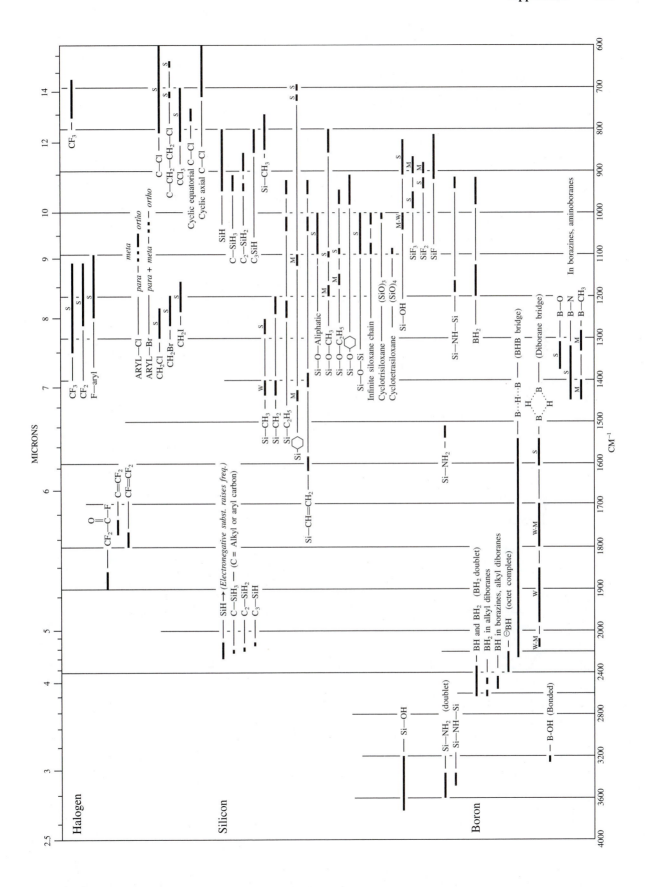

APPENDIX 2

Approximate ^{1}H Chemical Shift Ranges (ppm) for Selected Types of Protons[a]

R−CH$_3$		0.7 – 1.3	R−N−C−H	2.2 – 2.9
R−CH$_2$−R		1.2 – 1.4		
R$_3$CH		1.4 – 1.7	R−S−C−H	2.0 – 3.0
R−C=C−C−H		1.6 – 2.6	I−C−H	2.0 – 4.0
R−C(=O)−C−H, H−C(=O)−C−H		2.1 – 2.4	Br−C−H	2.7 – 4.1
RO−C(=O)−C−H, HO−C(=O)−C−H		2.1 – 2.5	Cl−C−H	3.1 – 4.1
N≡C−C−H		2.1 – 3.0	R−S(=O)(=O)−O−C−H	ca. 3.0
⬡−C−H		2.3 – 2.7	RO−C−H, HO−C−H	3.2 – 3.8
R−C≡C−H		1.7 – 2.7	R−C(=O)−O−C−H	3.5 – 4.8
R−S−H	var	1.0 – 4.0[b]	O$_2$N−C−H	4.1 – 4.3
R−N−H	var	0.5 – 4.0[b]	F−C−H	4.2 – 4.8
R−O−H	var	0.5 – 5.0[b]		
⬡−O−H	var	4.0 – 7.0[b]	R−C=C−H	4.5 – 6.5
⬡−N−H	var	3.0 – 5.0[b]	⬡−H	6.5 – 8.0
R−C(=O)−N−H	var	5.0 – 9.0[b]	R−C(=O)−H	9.0 – 10.0
			R−C(=O)−OH	11.0 – 12.0

[a]*For those hydrogens shown as* −C−H, *if that hydrogen is part of a methyl group (CH$_3$) the shift is generally at the low end of the range given, if the hydrogen is in a methylene group (−CH$_2$−) the shift is intermediate, and if the hydrogen is in a methine group (−CH−) the shift is typically at the high end of the range given.*

[b]*The chemical shift of these groups is variable, depending not only on the chemical environment in the molecule, but also on concentration, temperature, and solvent.*

APPENDIX 3

Some Representative ¹H Chemical Shift Values[a] for Various Types of Protons[b]

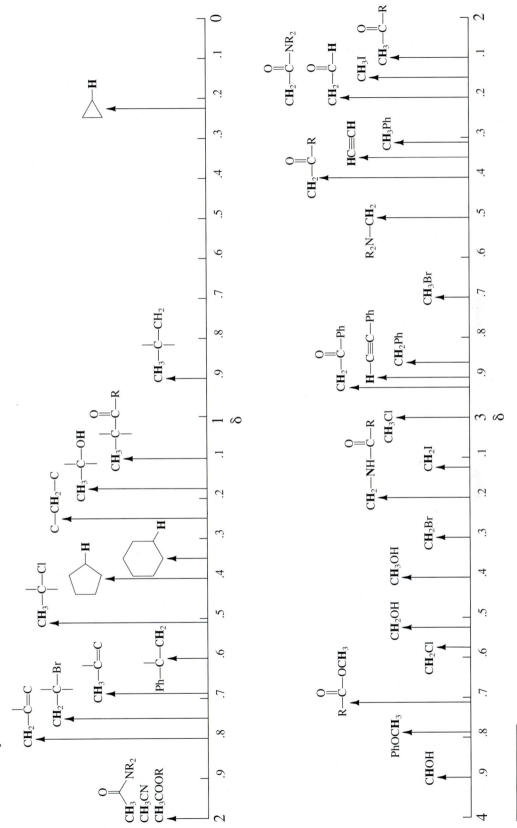

---

[a]Chemical shift values refer to the boldface protons **H**, not to regular H.

[b]Adapted with permission from Landgrebe, J. A., *Theory and Practice in the Organic Laboratory*, 4th ed., Brooks/Cole Publishing, Pacific Grove, CA, 1993.

APPENDIX 4

¹H Chemical Shifts of Selected Heterocyclic and Polycyclic Aromatic Compounds

A P P E N D I X   5

Typical Proton Coupling Constants

ALKANES AND CYCLIC COMPOUNDS

Type		Typical Value (Hz)	Range (Hz)	
^2J geminal		12	12–15	(For a 109° H—C—H angle)
^3J vicinal		7	6–8	(Depends on HCCH dihedral angle)
^3J a,a		10	8–14	In conformationally rigid systems
^3J a,e		5	0–7	(in systems undergoing inversion,
^3J e,e		3	0–5	all $J \approx$ 7–8 Hz)
^3J cis (H_bH_c)		9	6–12	
^3J trans (H_aH_c)		6	4–8	
^2J gem (H_aH_b)		6	3–9	
^3J cis (H_bH_c)		4	2–5	
^3J trans (H_aH_c)		2.5	1–3	
^2J gem (H_aH_b)		5	4–6	
^4J		0	0–7	(W-configuration obligatory—strained systems have the larger values)

ALKENES AND CYCLOALKENES (^2J AND ^3J)

Type		Typical Value (Hz)	Range (Hz)	Type		Typical Value (Hz)	Range (Hz)
	^2J gem	<1	0–5		^3J	2	0–2
	^3J cis	10	6–15		^3J	4	2–4
	^3J trans	16	11–18				
	^3J	5	4–10		^3J	6	5–7
	^3J	10	9–13		^3J	10	8–11

ALKENES AND ALKYNES (^4J AND ^5J)

Type			Typical Value (Hz)	Range (Hz)	Type		Typical Value (Hz)	Range (Hz)
H–C=C–C–H Allylic	^4J	(cis or trans)	1	0–3	H–C≡C–C–H Allylic	^4J	2	2–3
H–C–C=C–C–H Homoallylic	^5J		0	0–1.5	H–C–C≡C–C–H Homoallylic	^5J	2	2–3

AROMATICS AND HETEROCYCLES

Type		Typical Value (Hz)	Range (Hz)	Type		Range (Hz)
	^3J ortho	8	6–10		^3J $\alpha\beta$	4.6–5.8
	^4J meta	3	1–4		^4J $\alpha\beta'$	1.0–1.5
	^5J para	<1	0–2		^4J $\alpha\alpha'$	2.1–3.3
					^3J $\beta\beta'$	3.0–4.2
	^3J $\alpha\beta$		1.6–2.0		^3J $\alpha\beta$	4.9–5.7
	^4J $\alpha\beta'$		0.3–0.8		^4J $\alpha\gamma$	1.6–2.0
	^4J $\alpha\alpha'$		1.3–1.8		^5J $\alpha\beta'$	0.7–1.1
	^3J $\beta\beta'$		3.2–3.8		^4J $\alpha\alpha'$	0.2–0.5
					^3J $\beta\gamma$	7.2–8.5
					^4J $\beta\beta'$	1.4–1.9
	^3J $\alpha\beta$		2.0–2.6			
	^4J $\alpha\beta'$		1.0–1.5			
	^4J $\alpha\alpha'$		1.8–2.3			
	^3J $\beta\beta'$		2.8–4.0			

ALCOHOLS

Type		Typical Value (Hz)	Range (Hz)
	^3J	5	4–10
(No exchange occurring)			

ALDEHYDES

Type		Typical Value (Hz)	Range (Hz)
	^3J	2	1–3
	^3J	6	5–8

PROTON–OTHER NUCLEUS COUPLING CONSTANTS

Type	Typical Value (Hz)	Type	Typical Value (Hz)	Type	Typical Value (Hz)
$C\!\!<^{H}_{F}$ ^2J	44–81	$R\!-\!P\!\!<^{H}_{H}$ ^1J	~190	N–H	~52
				$\overset{\displaystyle H}{\underset{\displaystyle C}{\mid}}\!-\!\overset{\displaystyle H}{\underset{\displaystyle N}{\mid}}$	0
$\overset{H}{\underset{\mid}{C}}\!-\!\overset{F}{\underset{\mid}{C}}$ ^3J	3–25	$-\overset{O}{\overset{\parallel}{P}}-$ ^1J	~650		
$\overset{H}{\underset{\mid}{C}}\!-\!C\!-\!\overset{F}{\underset{\mid}{C}}$ ^4J	~0				
$\overset{H\;O}{-\overset{\mid\;\parallel}{C}\!-\!\overset{}{P}-}$ ^2J	~17				
$C\!\!<^{D}_{H}$ ^2J	~2				
$-\overset{H}{\underset{\mid}{C}}\!-\!\overset{}{\underset{\mid}{C}}\!-\!O\!-\!\overset{O}{\overset{\parallel}{P}}-$ ^4J	~0				
$\overset{H}{\underset{\mid}{C}}\!-\!\overset{D}{\underset{\mid}{C}}$ ^3J	<1 (Leads only to peak broadening)	$-\overset{H}{\underset{\mid}{C}}\!-\!O\!-\!\overset{O}{\overset{\parallel}{P}}-$ ^3J	~8		

Proton to Fluorine Coupling Constants (^1J, ^2J and longer distances)

C_a 4.45 ppm, doublet of triplets 2H (^2J H_aF = 47.4 Hz, H_aH_b = 6.3 Hz)
C_b 1.70 ppm, doublet of quintets 2H (^3J H_bF = 25.0 Hz, H_bH_a = H_bH_c = 6 Hz)
C_c and C_d, 1.40 ppm, multiplet 4H (^4J HF = 0)
C_e 0.95 ppm, triplet (^5J HF = 0, H_eH_d = 7.0 Hz)

7.03 ppm, doublet of doublets 2H (H_aH_b = 8.8 Hz, ^3J H_aF = 8.9 Hz). Looks like a triplet with fine structure

7.30 ppm, triplet of doublets, 2H (H_bH_a and H_bH_c = 7.8, ^4J H_bF = 5.8. Looks like a quartet, with fine structure

7.10 ppm, triplet of doublets 1H (H_cH_b = 7.4, $^5J H_cF$ = 0.8. Looks like a triplet

Proton to Phosphorus Coupling Constants (^1J, ^2J, ^3J, and ^4J)

These coupling constants depend on the oxidation state of the phosphorus atom.

Phosphonium Salts

Tetramethyl phosphonium chloride

$^2J = 14.7$ Hz

Trivalent Phosphorus

Trimethyl phosphine

$^2J = 1.8$ Hz

Trimethyl phosphite

$^3J = 10.7$ Hz

Phosphate, Phosphonate, and Phosphite Esters

Trimethoxy phosphate

$^3J = 11.0$ Hz

Ethyl methylphosphonate

$^2J = 17.6$ Hz
$^3J = 8$ Hz
$^4J = 0$ Hz

Dimethyl phosphite

$^1J = 697.5$ Hz
$^3J = 11.8$ Hz

APPENDIX 6

Calculation of Proton (^{1}H) Chemical Shifts

TABLE A6.1
^{1}H CHEMICAL-SHIFT CALCULATIONS FOR DISUBSTITUTED METHYLENE COMPOUNDS

$$X-CH_2-X \qquad or \qquad X-CH_2-Y \qquad \boxed{\delta_H \text{ ppm} = 0.23 + \Sigma \text{ constants}}$$

Substituents	Constants	Substituents	Constants
Alkanes, alkenes, alkynes, aromatics		Bonded to oxygen	
—R	0.47	—OH	2.56
C=C	1.32	—OR	2.36
—C≡C—	1.44	—OCOR	3.13
—C$_6$H$_5$	1.85	—OC$_6$H$_5$	3.23
Bonded to nitrogen and sulfur		Bonded to halogen	
—NR$_2$	1.57	—F	4.00
—NHCOR	2.27	—Cl	2.53
—NO$_2$	3.80	—Br	2.33
—SR	1.64	—I	1.82
Ketones		Derivatives of carboxylic acids	
—COR	1.70	—COOR	1.55
—COC$_6$H$_5$	1.84	—CONR$_2$	1.59
		—C≡N	1.70

Example Calculations

The formula allows you to calculate the *approximate* chemical-shift values for protons (^{1}H) based on methane (0.23 ppm). Although it is possible to calculate chemical shifts for any proton (methyl, methylene, or methine), agreement with actual experimental values is best with *disubstituted* compounds of the type $X-CH_2-Y$ or $X-CH_2-X$.

Cl—**CH$_2$**—Cl $\delta_H = 0.23 + 2.53 + 2.53 = 5.29$ ppm; actual $= 5.30$ ppm

C$_6$H$_5$—**CH$_2$**—O—C—CH$_3$ (‖ O) $\delta_H = 0.23 + 1.85 + 3.13 = 5.21$ ppm; actual $= 5.10$ ppm

C$_6$H$_5$—**CH$_2$**—C—O—CH$_3$ (‖ O) $\delta_H = 0.23 + 1.85 + 1.55 = 3.63$ ppm; actual $= 3.60$ ppm

CH$_3$—CH$_2$—**CH$_2$**—NO$_2$ $\delta_H = 0.23 + 3.80 + 0.47 = 4.50$ ppm; actual $= 4.38$ ppm

TABLE A6.2
^{1}H CHEMICAL-SHIFT CALCULATIONS FOR SUBSTITUTED ALKENES

$$\delta_H \text{ ppm} = 5.25 + \delta_{gem} + \delta_{cis} + \delta_{trans}$$

Substituents (—R)	δ_{gem}	δ_{cis}	δ_{trans}
Saturated carbon groups			
Alkyl	0.44	−0.26	−0.29
—CH$_2$—O—	0.67	−0.02	−0.07
Aromatic groups			
—C$_6$H$_5$	1.35	0.37	−0.10
Carbonyl, acid derivatives, and nitrile			
COR	1.10	1.13	0.81
—COOH	1.00	1.35	0.74
—COOR	0.84	1.15	0.56
—C≡N	0.23	0.78	0.58
Oxygen groups			
—OR	1.18	−1.06	−1.28
—OCOR	2.09	−0.40	−0.67
Nitrogen groups			
—NR$_2$	0.80	−1.26	−1.21
—NO$_2$	1.87	1.30	0.62
Halogen groups			
—F	1.54	−0.40	−1.02
—Cl	1.08	0.19	0.13
—Br	1.04	0.40	0.55
—I	1.14	0.81	0.88

Example Calculations

H_{gem} = 5.25 + 2.09 = 7.34 ppm; actual = 7.25 ppm

H_{cis} = 5.25 − 0.40 = 4.85 ppm; actual = 4.85 ppm

H_{trans} = 5.25 − 0.67 = 4.58 ppm; actual = 4.55 ppm

H_{gem} = 5.25 + 0.84 = 6.09 ppm; actual = 6.14 ppm

H_{cis} = 5.25 + 1.15 = 6.40 ppm; actual = 6.42 ppm

H_{trans} = 5.25 + 0.56 = 5.81 ppm; actual = 5.82 ppm

$$\left\{ \begin{array}{l} \delta_{gem} \text{ for } -COOR = 0.84 \\ \delta_{cis} \text{ for } -C_6H_5 = 0.37 \\ H_a = 5.25 + 0.84 + 0.37 = 6.46 \text{ ppm;} \\ \qquad\qquad\qquad\qquad \text{actual} = 6.43 \text{ ppm} \end{array} \right.$$

$$H_b \left\{ \begin{array}{l} \delta_{gem} \text{ for } -C_6H_5 = 1.35 \\ \delta_{cis} \text{ for } -COOR = 1.15 \\ H_b = 5.25 + 1.35 + 1.15 = 7.75 \text{ ppm;} \\ \qquad\qquad\qquad\qquad \text{actual} = 7.69 \text{ ppm} \end{array} \right.$$

TABLE A6.3
^{1}H CHEMICAL-SHIFT CALCULATIONS FOR SUBSTITUTED BENZENE RINGS

$$\delta_H \text{ ppm} = 7.27 + \Sigma\delta$$

Substituents (−R)	δ_{ortho}	δ_{meta}	δ_{para}
Saturated carbon groups			
Alkyl	−0.14	−0.06	−0.17
−CH$_2$OH	−0.07	−0.07	−0.07
Aldehydes and ketones			
−CHO	0.61	0.25	0.35
−COR	0.62	0.14	0.21
Carboxylic acids and derivatives			
−COOH	0.85	0.18	0.34
−COOR	0.71	0.10	0.21
−C≡N	0.25	0.18	0.30
Oxygen groups			
−OH	−0.53	−0.17	−0.45
−OCH$_3$	−0.48	−0.09	−0.44
−OCOCH$_3$	−0.19	−0.03	−0.19
Nitrogen groups			
−NH$_2$	−0.80	−0.25	−0.65
−NO$_2$	0.95	0.26	0.38
Halogen groups			
−F	−0.29	−0.02	−0.23
−Cl	0.03	−0.02	−0.09
−Br	0.18	−0.08	−0.04
−I	0.38	−0.23	−0.01

Example Calculations

The formula allows you to calculate the *approximate* chemical-shift values for protons (^1H) on a benzene ring. Although the values given in the table are for *monosubstituted benzenes*, it is possible to estimate chemical shifts for disubstituted and trisubstituted compounds by adding values from the table. The calculations for *meta-* and *para*-disubstituted benzenes often agree closely with actual values. More significant deviations from the experimental values are expected with *ortho*-disubstituted and trisubstituted benzenes. With these types of compounds, steric interactions cause groups such as carbonyl and nitro to turn out of the plane of the ring and thereby lose conjugation. Calculated values are often lower than the actual chemical shifts for *ortho*-disubstituted and trisubstituted benzenes.

$H_{ortho} = 7.27 + 0.71 = 7.98$ ppm; actual $= 8.03$ ppm

$H_{meta} = 7.27 + 0.10 = 7.37$ ppm; actual $= 7.42$ ppm

$H_{para} = 7.27 + 0.21 = 7.48$ ppm; actual $= 7.53$ ppm

$H_a \begin{cases} \delta_{ortho} \text{ for } -Cl = 0.03 \\ \delta_{meta} \text{ for } -NO_2 = 0.26 \\ H_a = 7.27 + 0.03 + 0.26 = 7.56 \text{ ppm; actual} = 7.50 \text{ ppm} \end{cases}$

$H_b \begin{cases} \delta_{meta} \text{ for } -Cl = -0.02 \\ \delta_{ortho} \text{ for } -NO_2 = 0.95 \\ H_b = 7.27 - 0.02 + 0.95 = 8.20 \text{ ppm; actual} = 8.20 \text{ ppm} \end{cases}$

$H_a \begin{cases} \delta_{meta} \text{ for } -Cl = -0.02 \\ \delta_{meta} \text{ for } -NO_2 = 0.26 \\ H_a = 7.27 - 0.02 + 0.26 = 7.51 \text{ ppm; actual} = 7.51 \text{ ppm} \end{cases}$

$H_b \begin{cases} \delta_{ortho} \text{ for } -Cl = 0.03 \\ \delta_{para} \text{ for } -NO_2 = 0.38 \\ H_b = 7.27 + 0.03 + 0.38 = 7.68 \text{ ppm; actual} = 7.69 \text{ ppm} \end{cases}$

$H_c \begin{cases} \delta_{para} \text{ for } -Cl = -0.09 \\ \delta_{ortho} \text{ for } -NO_2 = 0.95 \\ H_c = 7.27 - 0.09 + 0.95 = 8.13 \text{ ppm; actual} = 8.12 \text{ ppm} \end{cases}$

$H_d \begin{cases} \delta_{ortho} \text{ for } -Cl = 0.03 \\ \delta_{ortho} \text{ for } -NO_2 = 0.95 \\ H_d = 7.27 + 0.03 + 0.95 = 8.25 \text{ ppm; actual} = 8.21 \text{ ppm} \end{cases}$

A P P E N D I X   7

Approximate ^{13}C Chemical-Shift Values (ppm) for Selected Types of Carbon

Types of Carbon	Range (ppm)	Types of Carbon	Range (ppm)
R–CH$_3$	8–30	C≡C	65–90
R$_2$CH$_2$	15–55	C=C	100–150
R$_3$CH	20–60	C≡N	110–140
C–I	0–40	(benzene ring)	110–175
C–Br	25–65	R–C(=O)–OR, R–C(=O)–OH	155–185
C–N	30–65	R–C(=O)–NH$_2$	155–185
C–Cl	35–80	R–C(=O)–Cl	160–170
C–O	40–80	R–C(=O)–R, R–C(=O)–H	185–220

APPENDIX 8

Calculation of ^{13}C Chemical Shifts

TABLE A8.1
^{13}C CHEMICAL SHIFTS OF SELECTED HYDROCARBONS (PPM)

Compound	Formula	C1	C2	C3	C4	C5
Methane	CH_4	−2.3				
Ethane	CH_3CH_3	5.7				
Propane	$CH_3CH_2CH_3$	15.8	16.3			
Butane	$CH_3CH_2CH_2CH_3$	13.4	25.2			
Pentane	$CH_3CH_2CH_2CH_2CH_3$	13.9	22.8	34.7		
Hexane	$CH_3(CH_2)_4CH_3$	14.1	23.1	32.2		
Heptane	$CH_3(CH_2)_5CH_3$	14.1	23.2	32.6	29.7	
Octane	$CH_3(CH_2)_6CH_3$	14.2	23.2	32.6	29.9	
Nonane	$CH_3(CH_2)_7CH_3$	14.2	23.3	32.6	30.0	30.3
Decane	$CH_3(CH_2)_8CH_3$	14.2	23.2	32.6	31.1	30.5
2-Methylpropane		24.5	25.4			
2-Methylbutane		22.2	31.1	32.0	11.7	
2-Methylpentane		22.7	28.0	42.0	20.9	14.3
2,2-Dimethylpropane		31.7	28.1			
2,2-Dimethylbutane		29.1	30.6	36.9	8.9	
2,3-Dimethylbutane		19.5	34.4			
Ethylene	$CH_2=CH_2$	123.3				
Cyclopropane		−3.0				
Cyclobutane		22.4				
Cyclopentane		25.6				
Cyclohexane		26.9				
Cycloheptane		28.4				
Cyclooctane		26.9				
Cyclononane		26.1				
Cyclodecane		25.3				
Benzene		128.5				

TABLE A8.2
¹³C CHEMICAL-SHIFT CALCULATIONS FOR LINEAR AND BRANCHED ALKANES

$$\delta_C = -2.3 + 9.1\alpha + 9.4\beta - 2.5\gamma + 0.3\delta + 0.1\varepsilon + \Sigma \text{ (steric corrections) ppm}$$

α, β, γ, δ, and ε are the numbers of carbon atoms in the α, β, γ, δ, and ε positions relative to the carbon atom being observed.

$$\cdots\cdots C_\varepsilon - C_\delta - C_\gamma - C_\beta - C_\alpha - \underset{\uparrow}{C} - C_\alpha - C_\beta - C_\gamma - C_\delta - C_\varepsilon \cdots\cdots$$

Steric corrections are derived from the following table (use all that apply, even if they apply more than once).

Steric Corrections (ppm)				
	Type of Carbons Attached			
Carbon Atom Observed	Primary	Secondary	Tertiary	Quaternary
Primary	0	0	−1.1	−3.4
Secondary	0	0	−2.5	−7.5
Tertiary	0	−3.7	−8.5	−10.0
Quaternary	−1.5	−8.4	−10.0	−12.5

Example

$$\overset{\displaystyle CH_3}{\underset{\displaystyle CH_3}{\overset{1}{CH_3}-\overset{2}{C}-\overset{3}{CH_2}-\overset{4}{CH_3}}}$$

2,2-Dimethylbutane

Actual values: C1   29.1 ppm
                   C2   30.6 ppm
                   C3   36.9 ppm
                   C4   8.9 ppm

C1 $= -2.3 + 9.1(1) + 9.4(3) - 2.5(1) + 0.3(0) + 0.1(0)$ **$+ [1(-3.4)]$** $= 29.1$ ppm
    Steric correction (boldface) = primary with 1 adjacent quaternary

C2 $= -2.3 + 9.1(4) + 9.4(1) - 2.5(0) + 0.3(0) + 0.1(0)$ **$+ [3(-1.5)] + [1(-8.4)]$** $= 30.6$ ppm
    Steric corrections = quaternary/3 adj. primary, and quaternary/1 adj. secondary

C3 $= -2.3 + 9.1(2) + 9.4(3) - 2.5(0) + 0.3(0) + 0.1(0)$ **$+ [1(0)] + [1(-7.5)]$** $= 36.6$ ppm
    Steric corrections = secondary/1 adj. primary, and secondary/1 adj. quaternary

C4 $= -2.3 + 9.1(1) + 9.4(1) - 2.5(3) + 0.3(0) + 0.1(0)$ **$+ [1(0)]$** $= 8.7$ ppm
    Steric correction = primary/1 adj. secondary

TABLE A8.3
^{13}C SUBSTITUENT INCREMENTS FOR ALKANES AND CYCLOALKANES (PPM)[a]

Substituent Y	Terminal: $Y-C_\alpha-C_\beta-C_\gamma$			Internal: $C_\gamma-C_\beta-\overset{\overset{Y}{\shortmid}}{C_\alpha}-C_\beta-C_\gamma$		
	α	β	γ	α	β	γ
$-D$	-0.4	-0.1	0			
$-CH_3$	9	10	-2	6	8	-2
$-CH=CH_2$	19.5	6.9	-2.1			-0.5
$-C\equiv CH$	4.5	5.4	-3.5			-3.5
$-C_6H_5$	22.1	9.3	-2.6	17	7	-2
$-CHO$	29.9	-0.6	-2.7			
$-COCH_3$	30	1	-2	24	1	-2
$-COOH$	20.1	2	-2.8	16	2	-2
$-COOR$	22.6	2	-2.8	17	2	-2
$-CONH_2$	22	2.5	-3.2			-0.5
$-CN$	3.1	2.4	-3.3	1	3	-3
$-NH_2$	29	11	-5	24	10	-5
$-NHR$	37	8	-4	31	6	-4
$-NR_2$	42	6	-3			-3
$-NO_2$	61.6	3.1	-4.6	57	4	
$-OH$	48	10	-6.2	41	8	-5
$-OR$	58	8	-4	51	5	-4
$-OCOCH_3$	56.5	6.5	-6.0	45	5	-3
$-F$	70.1	7.8	-6.8	63	6	-4
$-Cl$	31	10	-5.1	32	10	-4
$-Br$	20	11	-3	25	10	-3
$-I$	-7.2	10.9	-1.5	4	12	-1

[a]*Add these increments to the values given in Table A8.1.*

Example 1

$$\overset{1}{CH_3}-\overset{2}{CH}-\overset{3}{CH_2}-\overset{4}{CH_3}\quad \textbf{2-Butanol}$$
$$|$$
$$OH$$

Using the values for butane listed in Table A8.1 and the internal substituent corrections from Table A8.3, we calculate:

	Actual value
C1 = 13.4 + 8 = 21.4 ppm	22.6 ppm
C2 = 25.2 + 41 = 66.2 ppm	68.7 ppm
C3 = 25.2 + 8 = 33.2 ppm	32.0 ppm
C4 = 13.4 + (−5) = 8.4 ppm	9.9 ppm

Example 2

$$HO-\overset{1}{CH_2}-\overset{2}{CH_2}-\overset{3}{CH_2}-\overset{4}{CH_3}\quad \textbf{1-Butanol}$$

Using the values for butane listed in Table A8.1 and the terminal substituent corrections from Table A8.3, we calculate:

	Actual value
C1 = 13.4 + 48 = 61.4 ppm	61.4 ppm
C2 = 25.2 + 10 = 35.2 ppm	35.0 ppm
C3 = 25.2 + (−6.2) = 19.0 ppm	19.1 ppm
C4 = 13.4 = 13.4 ppm	13.6 ppm

Example 3

$$Br-\overset{1}{CH_2}-\overset{2}{CH_2}-\overset{3}{CH_3}\quad \textbf{1-Bromopropane}$$

Using the values for propane listed in Table A8.1 and the terminal substituent corrections from Table A8.3, we calculate:

	Actual value
C1 = 15.8 + 20 = 35.8 ppm	35.7 ppm
C2 = 16.3 + 11 = 27.3 ppm	26.8 ppm
C3 = 15.8 + (−3) = 12.8 ppm	13.2 ppm

TABLE A8.4
^{13}C SUBSTITUENT INCREMENTS FOR ALKENES (PPM)[a,b]

$$Y - \overset{1}{C} = \overset{2}{C} - X$$

Substituent	Y	X
—H	0	0
—CH$_3$	12.9	−7.4
—CH$_2$CH$_3$	19.2	−9.7
—CH$_2$CH$_2$CH$_3$	15.7	−8.8
—CH(CH$_3$)$_2$	22.7	−12.0
—C(CH$_3$)$_3$	26.0	−14.8
—CH=CH$_2$	13.6	−7
—C$_6$H$_5$	12.5	−11
—CH$_2$Cl	10.2	−6.0
—CH$_2$Br	10.9	−4.5
—CH$_2$I	14.2	−4.0
—CH$_2$OH	14.2	−8.4
—COOH	5.0	9.8
—NO$_2$	22.3	−0.9
—OCH$_3$	29.4	−38.9
—OCOCH$_3$	18.4	−26.7
—CN	−15.1	14.2
—CHO	15.3	14.5
—COCH$_3$	13.8	4.7
—COCl	8.1	14.0
—Si(CH$_2$)$_3$	16.9	6.7
—F	24.9	−34.3
—Cl	2.6	−6.1
—Br	−8.6	−0.9
—I	−38.1	7.0

[a]Corrections for C1; add these increments to the base value of ethylene (123.3 ppm).

[b]Calculate C1 as shown in the diagram. Redefine C2 as C1 when estimating values for C2.

Example 1

$$Br-\overset{1}{C}H=\overset{2}{C}H-\overset{3}{C}H_3 \quad \textbf{1-Bromopropene}$$

	Actual values	
	cis	*trans*
C1 = 123.3 + (−8.6) + (−7.4) = 107.3 ppm	108.9	104.7 ppm
C2 = 123.3 + 12.9 + (−0.9) = 135.3 ppm	129.4	132.7 ppm

Example 2

$$HOOC-\overset{1}{C}H=\overset{2}{C}H-\overset{3}{C}H-\overset{4}{C}H_3 \quad \textbf{Crotonic acid}$$

	Actual value (trans)
C2 = 123.3 + 5 + (−7.4) = 120.9 ppm	122.0 ppm
C3 = 123.3 + 12.9 + 9.8 = 146.0 ppm	147.0 ppm

TABLE A8.5
¹³C CHEMICAL-SHIFT CALCULATIONS FOR LINEAR AND BRANCHED ALKENES[a]

$$\delta_{C1} = 123.3 + [10.6\alpha + 7.2\beta - 1.5\gamma] - [7.9\alpha' + 1.8\beta' - 1.5\gamma'] + \Sigma \text{ (steric corrections)}$$

α, β, γ and α', β', γ' are the numbers of carbon atoms in those same positions relative to C1:

$$C\gamma-C\beta-C\alpha-\overset{1}{C}=\overset{2}{C}-C\alpha'-C\beta'-C\gamma'$$

Steric corrections are applied as follows (use all that apply):

Cα and Cα' are *trans* (*E*-configuration)	0
Cα and Cα' are *cis* (*Z*-configuration)	−1.1
Two alkyl substituents at C1 (two Cα)	−4.8
Two alkyl substituents at C2 (two Cα')	+2.5
Two or three alkyl substituents at Cβ	+2.3

[a] *Calculate C1 as shown in the diagram. Redefine C2 as C1 when calculating values for C2.*

Example 1

$$\overset{1}{C}H_3-\overset{2}{C}=\overset{3}{C}H-\overset{4}{C}H_3 \quad \textbf{2-Methyl-2-butene}$$
$$| \\ CH_3$$

Actual value

C2 = 123.3 + [10.6(2)] − [7.9(1)] + [(−4.8) + (−1.1)] = 130.7 ppm   131.4 ppm
C3 = 123.3 + [10.6(1)] − [7.9(2)] + [(+2.5) + (−1.1)] = 119.5 ppm   118.7 ppm

Example 2

$$\overset{1}{C}H_2=\overset{2}{C}H-\overset{3}{C}H-\overset{4}{C}H_2-\overset{5}{C}H_3 \quad \textbf{3-Methyl-1-pentene}$$
$$| \\ CH_3$$

Actual value

C1 = 123.3 + [0] − [7.9(1) + 1.8(2) − 1.5(1)]   = 113.3 ppm   112.9 ppm
C2 = 123.3 + [10.6(1) + 7.2(2) − 1.5(1)] − [0] + [(+2.3)] = 149.1 ppm   144.9 ppm

Example 3

$$\overset{1}{C}H_3-\overset{2}{C}H=\overset{3}{C}H-\overset{4}{C}H_3 \quad \textbf{2-Butene}$$

Actual value

C2 (*cis* isomer)   = C3 = 123.3 + [10.6(1)] − [7.9(1)] + [(−1.1)] = 124.9 ppm   124.6 ppm
C2 (*trans* isomer) = C3 = 123.3 + [10.6(1)] − [7.9(1)] + [0]   = 126.0 ppm   126.0 ppm

TABLE A8.6
^{13}C SUBSTITUENT INCREMENTS FOR ALKENE (VINYL) CARBONS[a,b]

$$\gamma-\beta-\alpha \diagdown \quad \diagup \alpha'-\beta'-\gamma'$$
$$C=C$$
$$\alpha \diagup \quad \diagdown \alpha'$$

Substituent	α	β	γ	α'	β'	γ'
Carbon	10.6	7.2	−1.5	−7.9	−1.8	−1.5
−C$_6$H$_5$	12			−11		
−OR	29	2		−39	−1	
−OCOR	18			−27		
−COR	15			6		
−COOH	4			9		
−CN	−16			15		
−Cl	3	−1		−6	2	
−Br	−8	0		−1	2	
−I	−38			7		

[a]In the upper chains, if a group is in the β or γ position, the preceding atoms (α and/or β) are assumed to be carbon atoms. Add these increments to the base value of ethylene (123.3 ppm).

[b]Calculate C1 as shown in the diagram. Redefine C2 as C1 when estimating values for C2.

Example 1

$$\overset{1}{\text{Br}-\text{CH}}=\overset{2}{\text{CH}}-\overset{3}{\text{CH}_3} \quad \textbf{1-Bromopropene}$$

	Actual values	
	cis	*trans*
C1 = 123.3 − 8 − 7.9 = 107.4 ppm	108.9	104.7 ppm
C2 = 123.3 + 10.6 − 1 = 132.9 ppm	129.4	132.7 ppm

Example 2

$$\begin{array}{ccc} \text{CH}_3 & & \text{O} \\ | & & \| \\ \text{CH}_3-\underset{5}{\text{C}}=\underset{4}{\underset{3}{\text{CH}}}-\underset{2}{\text{C}}-\underset{1}{\text{CH}_3} \end{array} \quad \textbf{Mesityl oxide}$$

	Actual value
C3 = 123.3 + 15 − 7.9 − 7.9 = 122.5 ppm	124.3 ppm
C4 = 123.3 + 10.6 + 10.6 + 6 = 150.5 ppm	154.6 ppm

TABLE A8.7
^{13}C SUBSTITUENT INCREMENTS FOR BENZENE RINGS (PPM)[a]

Substituent Y	α (*ipso*)	o (*ortho*)	m (*meta*)	p (*para*)
−CH$_3$	9.3	0.7	−0.1	−2.9
−CH$_2$CH$_3$	11.7	−0.5	0	−2.6
−CH(CH$_2$)$_2$	20.1	−2.0	−0.3	−2.5
−C(CH$_3$)$_3$	18.6	−3.4	−0.4	−3.1
−CH=CH$_2$	9.1	−2.4	0.2	−0.5
−C≡CH	−6.2	3.6	−0.4	−0.3
−C$_6$H$_5$	8.1	−1.1	−0.5	−1.1
−CHO	8.2	1.2	0.6	5.8
−COCH$_3$	8.9	−0.1	−0.1	4.4
−COC$_6$H$_5$	9.1	1.5	−0.2	3.8
−COOH	2.1	1.6	−0.1	5.2
−COOCH$_3$	2.0	1.2	−0.1	4.3
−CN	−16.0	3.6	0.6	4.3
−NH$_2$	18.2	−13.4	0.8	−10.0
−N(CH$_3$)$_2$	16.0	−15.7	0.8	−10.5
−NHCOCH$_3$	9.7	−8.1	0.2	−4.4
−NO$_2$	19.6	−4.9	0.9	6.0
−OH	28.8	−12.7	1.6	−7.3
−OCH$_3$	33.5	−14.4	1.0	−7.7
−OCOCH$_3$	22.4	−7.1	−0.4	−3.2
−F	33.6	−13.0	1.6	−4.5
−Cl	5.3	0.4	1.4	−1.9
−Br	−5.4	3.4	2.2	−1.0
−I	−31.2	8.9	1.6	−1.1

[a]*Add these increments to the base value for benzene-ring carbons (128.5 ppm).*

Example 1

Mesitylene

Observed

C1,C3,C5 = 128.5 + 9.3 − 0.1 − 0.1 = 137.6 ppm   137.4 ppm
C2,C4,C6 = 128.5 + 0.7 + 0.7 − 2.9 = 127.0 ppm   127.1 ppm

Example 2

Salicylaldehyde

Observed

C1 = 128.5 +   8.2 − 12.7 = 124.0 ppm   121.0 ppm
C2 = 128.5 + 28.8 +   1.2 = 158.5 ppm   161.4 ppm
C3 = 128.5 − 12.7 +   0.6 = 116.4 ppm   117.4 ppm
C4 = 128.5 +   1.6 +   5.8 = 135.9 ppm   136.6 ppm
C5 = 128.5 −   7.3 +   0.6 = 121.8 ppm   119.6 ppm
C6 = 128.5 +   1.2 +   1.6 = 131.3 ppm   133.6 ppm

Example 3

4-Nitrophenol

Observed

C1 = 128.5 + 28.8 + 6.0 = 163.3 ppm   161.5 ppm
C2 = 128.5 − 12.7 + 0.9 = 116.7 ppm   115.9 ppm
C3 = 128.5 +   1.6 − 4.9 = 125.2 ppm   126.4 ppm
C4 = 128.5 + 19.6 + 7.3 = 140.8 ppm   141.7 ppm

APPENDIX 9

^{13}C Coupling Constants to Proton, Deuterium, Fluorine and Phosphorus

^{13}C-Proton Coupling Constants (^1J and ^2J)

^{13}C-proton coupling constants (^1J) are very dependent on the hybridization of the carbon atom. Strained ring compounds have larger coupling constants because of increased s-character.
Reich, Hans J., University of Wisconsin
http://www.chem.wise.edu/areas/reich/nmr/06-cmr-05-1jch.htm
$^1J_{CH}$, $^2J_{CH}$, $^3J_{CH}$ coupling constants

^{13}C–proton coupling constants (^1J)		^{13}C–proton coupling constants (^2J)	
sp^3 ^{13}C−H	115–125 Hz	^{13}C−C−H	0–60 Hz
sp^2 ^{13}C−H	150–170 Hz		
sp ^{13}C−H	250–270 Hz		

sp^3 C−H = 125.0 Hz   sp^2 C−H = 156 Hz   sp^2 C−H = 159 Hz   sp C−H = 248 Hz

^1J = 127 Hz

Unstrained rings have normal sp^3 H−C−H bond angles of about 109°

^1J = 161 Hz

Strained rings compounds have larger H−C−H bond angles leading to values close to sp^2 values

^{13}C-Deuterium Coupling Constants (^1J)

^{13}C−D   20–30 Hz

^{13}C-Fluorine Coupling Constants (^1J, ^2J, and longer distances)

^{13}C–fluorine coupling constants (^1J)
^{13}C–F   165–370 Hz

^{13}C–fluorine coupling constants (^2J)
^{13}C–C–F   18–45 Hz

$$F-\underset{\underset{CH_3}{|}}{\overset{\overset{CH_3}{|}}{C}}-CH_3$$

$$^1J_{CF} = 162 \text{ Hz}$$
$$^2J_{CF} = 24 \text{ Hz}$$

C1 162.9 ppm, doublet, ^1J = 245 Hz
C2 115.3 ppm, doublet, ^2J = 20.7 Hz
C3 129.9 ppm, doublet, ^3J = 8.5 Hz
C4 124.0 ppm, doublet, ^4J = 2.5 Hz

C1 = 84.2 ppm, doublet, ^1J = 165 Hz
C2 = 30.2 ppm, doublet, ^2J = 19.5 Hz
C3 = 27.4 ppm, doublet, ^3J = 6.1 Hz
C4 = 22.4 ppm, singlet, ^4J = 0 Hz
C5 = 13.9 ppm, singlet, ^5J = 0 Hz

^{13}C-Phosphorus Coupling Constants (^1J and ^2J)

These coupling constants depend on the oxidation state of the phosphorus atom.

Phosphonium Salts

^{13}C–phosphorus coupling constants (^1J)
^{13}C–P   48–56 Hz

^{13}C–phosphorus coupling constants (^2J)
^{13}C–C–P   4–6 Hz

$$CH_3-CH_2-\overset{+}{\underset{\underset{CH_3}{|}\underset{CH_2}{|}}{\overset{\overset{CH_3}{|}\overset{CH_2}{|}}{P}}}-CH_2-CH_3 \quad Cl^-$$

$$^1J = 48.9 \text{ Hz}$$
$$^2J = 4.9 \text{ Hz}$$

Trivalent Phosphorus

$$^1J = 12.2 \text{ Hz}$$

$$^2J = 9.8 \text{ Hz}$$

$$^1J = 15.9 \text{ Hz}$$
$$^2J = 12.2 \text{ Hz}$$

Phosphonate and Phosphite Esters

$$^1J = 145.3 \text{ Hz}$$
$$^2J = 4.8 \text{ Hz}$$
$$^3J = 6.1 \text{ Hz}$$

$$^1J = 146.5 \text{ Hz}$$
$$^2J = 6.1 \text{ Hz}$$
$$^3J = 7.3 \text{ Hz}$$

$$^2J = 6.1 \text{ Hz}$$

APPENDIX 10

^{1}H and ^{13}C Chemical Shifts for Common NMR Solvents

TABLE A10.1
^{1}H CHEMICAL-SHIFT VALUES (PPM) FOR SOME COMMON NMR SOLVENTS

Solvent	Deuterated Form	Chemical Shift (Multiplicity)[a]
Acetone	Acetone-d_6	2.05 (5)
Acetonitrile	Acetonitrile-d_3	1.93 (5)
Benzene	Benzene-d_6	7.15 (broad)
Carbon tetrachloride	—	—
Chloroform	Chloroform-d	7.25 (1)
Dimethylsulfoxide	Dimethylsulfoxide-d_6	2.49 (5)
Water	Deuterium oxide	4.82 (1)
Methanol	Methanol-d_4	4.84 (1) hydroxyl
		3.30 (5) methyl
Methylene chloride	Methylene chloride-d_2	5.32 (3)

[a]Where multiplets apply, the center peak is given and the number of lines is indicated in parentheses. No proton peak should be observed in the completely deuterated solvents listed. However, multiplets will arise from coupling of a proton with deuterium because the solvents are not 100% isotopically pure. For example, acetone-d_6 has a trace of acetone-d_5 in it, while $CDCl_3$ has some $CHCl_3$ present.

TABLE A10.2
^{13}C CHEMICAL-SHIFT VALUES FOR SOME COMMON NMR SOLVENTS (PPM)

Solvent	Deuterated Form	Chemical Shift (Multiplicity)[a]
Acetone	Acetone-d_6	206.0 (1) carbonyl
		29.8 (7) methyl
Acetonitrile	Acetonitrile-d_3	118.3(1) CN
		1.3(7) methyl
Benzene	Benzene-d_6	128.0 (3)
Chloroform	Chloroform-d	77.0 (3)
Dimethylsulfoxide	Dimethylsulfoxide-d_6	39.5 (7)
Dioxane	Dioxane-d_8	66.5 (5)
Methanol	Methanol-d_4	49.0 (7)
Methylene chloride	Methylene chloride-d_2	54.0 (5)

[a]Where multiplets apply, the center peak is given and the number of lines is indicated in parentheses. These multiplets arise from the coupling of carbon with the deuterium.

APPENDIX 11

Common Fragment Ions under Mass 105[a]

m/z	Ions	m/z	Ions
14	CH_2	44	$CH_2CH{=}O + H$
15	CH_3		CH_3CHNH_2
16	O		CO_2
17	OH		$NH_2C{=}O$
18	H_2O		$(CH_3)_2N$
	NH_4	45	CH_3CHOH
19	F		CH_2CH_2OH
	H_3O		CH_2OCH_3
26	$C{\equiv}N$		$\overset{\displaystyle O}{\overset{\displaystyle \|}{C}}{-}OH$
27	C_2H_3		
28	C_2H_4		$CH_3CH{-}O + H$
	CO	46	NO_2
	N_2 (air)	47	CH_2SH
	$CH{=}NH$		CH_3S
29	C_2H_5	48	$CH_3S + H$
	CHO	49	CH_2Cl
30	CH_2NH_2	51	CHF_2
	NO		C_4H_3
31	CH_2OH	53	C_4H_5
	OCH_3	54	$CH_2CH_2C{\equiv}N$
32	O_2 (air)	55	C_4H_7
33	SH		$CH_2{=}CHC{=}O$
	CH_2F	56	C_4H_8
34	H_2S	57	C_4H_9
35	Cl		$C_2H_5C{=}O$
36	HCl	58	$CH_3{-}\underset{\underset{\displaystyle CH_2}{\displaystyle \|}}{C}{=}O \;\; +H$
39	C_3H_3		
40	$C{\equiv}N$		
41	C_3H_5		
	$CH_2C{=}H + H$		$C_2H_5CHNH_2$
	C_2H_2NH		$(CH_3)_2NHCH_2$
42	C_3H_6		$C_2H_5NHCH_2$
43	C_3H_7		C_2H_2S
	$CH_3C{=}O$		
	C_2H_5N		

[a]Adapted with permission from Silverstein, R. M. and F. X. Webster, Spectrometric Identification of Organic Compounds, 6th ed., John Wiley & Sons, New York, 1998.

m/z	Ions	m/z	Ions
59	$(CH_3)_2COH$	74	$CH_2-\overset{O}{\overset{\|}{C}}-OCH_3 + H$
	$CH_2OC_2H_5$	75	$\overset{O}{\overset{\|}{C}}-OC_2H_5 + 2H$
	$\overset{O}{\overset{\|}{C}}-OCH_3$		$CH_2SC_2H_5$
	$NH_2C=O\ \underset{CH_2}{\|} + H$		$(CH_3)_2CSH$
	CH_3OCHCH_3		$(CH_3O)_2CH$
	CH_3CHCH_2OH	77	C_6H_5
60	$CH_2C=O\ \underset{OH}{\|} + H$	78	$C_6H_5 + H$
	CH_2ONO	79	$C_6H_5 + 2H$
61	$\overset{O}{\overset{\|}{C}}-OCH_3 + 2H$		Br
	CH_2CH_2SH	80	$CH_3SS + H$
	CH_2SCH_3	81	C_6H_9
65	(or C_5H_5)		
66	(or C_5H_6)	82	$CH_2CH_2CH_2CH_2C\equiv N$
			CCl_2
67	C_5H_7		C_6H_{10}
68	$CH_2CH_2CH_2C\equiv N$	83	C_6H_{11}
69	C_5H_9		$CHCl_2$
	CF_3	85	C_6H_{13}
	$CH_3CH=CHC=O$		$C_4H_9C=O$
	$CH_2=C(CH_3)C=O$		$CClF_2$
70	C_5H_{10}	86	$C_3H_7\overset{O}{\overset{\|}{C}}-CH_2 + H$
71	C_5H_{11}		$C_4H_9CHNH_2$ and isomers
	$C_3H_7C=O$	87	$C_3H_7\overset{O}{\overset{\|}{C}}O$
72	$C_2H_5\overset{O}{\overset{\|}{C}}-CH_2$		Homologs of 73
	$C_3H_7CHNH_2$		$CH_2CH_2\overset{}{C}OCH_3$
	$(CH_3)N=C=O$		
	$C_2H_5NHCHCH_3$ and isomers	88	$CH_2-\overset{O}{\overset{\|}{C}}-OC_2H_5 + H$
73	Homologs of 59		

m/z	Ions	m/z	Ions
89	$\overset{O}{\underset{\|\|}{C}}-OC_3H_7 + 2H$; benzene ring—C	94	benzene ring—O + H
90	CH_3CHONO_2 ; benzene ring—CH	96	$CH_2CH_2CH_2CH_2CH_2C\equiv N$
		97	C_7H_{13}
91	benzene ring—CH_2 or (cycloheptatrienyl, with H) ; benzene ring—CH + H ; benzene ring—C + 2H ; $(CH_2)_4Cl$	99	C_7H_{15} ; $C_6H_{11}O$
		100	$C_4H_9\overset{O}{\underset{\|\|}{C}}-CH_2 + H$; $C_5H_{11}CHNH_2$
92	benzene ring—N ; pyridine ring—CH_2 ; benzene ring—CH_2 + H ; CH_2Br	101	$\overset{O}{\underset{\|\|}{C}}-OC_4H_9$
		102	$CH_2\overset{O}{\underset{\|\|}{C}}-OC_3H_7 + H$
93	benzene ring—OH ; C_7H_9 ; benzene ring—O	103	$\overset{O}{\underset{\|\|}{C}}-OC_4H_9 + 2H$; $C_5H_{11}S$; $CH(OCH_2CH_3)_2$
		104	$C_2H_5CHONO_2$
		105	benzene ring—C=O ; benzene ring—CH_2CH_2 ; benzene ring—$CHCH_3$

APPENDIX 12

A Handy-Dandy Guide to Mass Spectral Fragmentation Patterns

Alkanes

Good M^+

14-amu fragments

Alkenes

Distinct M^+

Loss of 15, 29, 43, and so on

Cycloalkanes

Strong M^+

Loss of $CH_2=CH_2$ $M - 28$

Loss of alkyl

Aromatics

Strong M^+

$C_7H_7^+$ $m/z = 91$, weak $m/z = 65$ $(C_5H_5^+)$

$m/z = 92$ Transfer of *gamma* hydrogens

Halides

Cl and Br doublets (M^+ and $M + 2$)

$m/z = 49$ or 51 $CH_2=Cl^+$

$m/z = 93$ or 95 $CH_2=Br^+$

$M - 36$ Loss of HCl

$m/z = 91$ or 93

$m/z = 135$ or 137

$M - 79$ ($M - 81$)   Loss of Br·

$M - 127$ Loss of I·

Alcohols

M^+ weak or absent

Loss of alkyl

$CH_2=OH^+$ $m/z = 31$

$RCH=OH^+$ $m/z = 45, 59, 73, \ldots$

$R_2C=OH^+$ $m/z = 59, 73, 87, \ldots$

$M - 18$ Loss of H_2O

$M - 46$ Loss of $H_2O + CH_2=CH_2$

Phenols

Strong M^+
Strong $M - 1$ Loss of H·
$M - 28$ Loss of CO

Ethers

M^+ stronger than alcohols
Loss of alkyl
Loss of OR' $M - 31, M - 45, M - 59,$ and so on
$CH_2=OR'^+$ $m/z = 45, 59, 73, \ldots$

Amines

M^+ weak or absent
Nitrogen Rule
$m/z = 30$ $CH_2=NH_2^+$ (base peak)
Loss of alkyl

Aldehydes

Weak M^+
$M - 29$ Loss of HCO
$M - 43$ Loss of $CH_2=CHO$
$m/z = 44$

Transfer of *gamma* hydrogens

or 58, 72, 86, . . .

Aromatic Aldehydes

Strong M^+
$M - 1$ Loss of H·
$M - 29$ Loss of H· and CO

Ketones

M^+ intense
$M - 15, M - 29, M - 43, \ldots$ Loss of alkyl group
$m/z = 43$ CH_3CO^+
$m/z = 58, 72, 86, \ldots$ Transfer of *gamma* hydrogens
$m/z = 55$ $^+CH_2-CH=C=O$ Base peak for cyclic ketones

$m/z = 83$ $C\equiv O^+$ in cyclohexanone

$m/z = 42$ in cyclohexanone

$m/z = 105$ $-C\equiv O^+$ in aryl ketones

$m/z = 120$ Transfer of *gamma* hydrogens

Carboxylic Acids

M^+ weak but observable

$M - 17$ Loss of OH

$M - 45$ Loss of COOH

$m/z = 45$ $^+$COOH

$m/z = 60$

$$HO-\overset{\overset{\displaystyle +OH}{\|}}{C}-CH_2 \cdot$$

Transfer of *gamma* hydrogens

Aromatic Acids

M^+ large

$M - 17$ Loss of OH

$M - 45$ Loss of COOH

$M - 18$ *Ortho* effect

Methyl Esters

M^+ weak but observable

$M - 31$ Loss of OCH_3

$m/z = 59$ $^+COOCH_3$

$m/z = 74$

$$CH_3O-\overset{\overset{\displaystyle +OH}{\|}}{C}-CH_2$$

Transfer of *gamma* hydrogens

Higher Esters

M^+ weaker than for $RCOOCH_3$

Same pattern as in methyl esters

$M - 45, M - 59, M - 73$ Loss of OR

$m/z = 73, 87, 101$ ^+COOR

$m/z = 88, 102, 116$

$$RO-\overset{\overset{\displaystyle +OH}{\|}}{C}-CH_2 \cdot$$

Transfer of *gamma* hydrogens

$m/z = 28, 42, 56, 70$ *Beta* hydrogens on alkyl group

$m/z = 61, 75, 89$

$$R-\overset{\overset{\displaystyle +OH}{\|}}{C}-OH$$

Long alkyl chain

$m/z = 108$ Loss of $CH_2{=}C{=}O$ Benzyl or acetate ester

$m/z = 105$

$-C{\equiv}O^+$

$m/z = 77$

$+$ weak

$M - 32, M - 46, M - 60$ *ortho* effect—loss of ROH

A P P E N D I X   1 3

Index of Spectra

^{1}H NMR Spectra

^{13}C NMR Spectra

COSY Spectra

DEPT Spectra

HETCOR Spectra

NOE Difference Spectra

Ultraviolet-Visible Spectra

INDEX